Basic
Nuclear
Engineering

Fourth Edition

Arthur R. Foster

Northeastern University

Robert L. Wright, Jr.

Allyn and Bacon, Inc.
Boston, London, Sydney, Toronto

This book is part of the
ALLYN AND BACON SERIES IN
ENGINEERING
Consulting Editor: Frank Kreith
University of Colorado

Copyright © 1983, 1977, 1973, 1968 by Allyn and Bacon, Inc.,
7 Wells Avenue, Newton, Massachusetts 02159. All rights reserved.
No part of the material protected by this copyright notice may be
reproduced or utilized in any form or by any means, electronic or
mechanical, including photocopying, recording, or by any information storage and retrieval system, without written permission from
the copyright owner.

LIBRARY OF CONGRESS CATALOGING IN PUBLICATION DATA

Foster, Arthur R.
  Basic nuclear engineering.

  (Allyn and Bacon series in engineering)
  Bibliography: p.
  Includes index.
  1. Nuclear engineering.   I.   Wright, Robert L.
  II.   Title.   III.   Series.
TK9145.F6 1983   621.48   82-11577
ISBN 0-205-07886-9

Printed in the United States of America

10 9 8 7 6 5 4 3 2 1   87 86 85 84 83 82

Contents

9

The Steady-State Reactor Core   243

10

Transient Reactor Behavior and Control   299

11

Radiation Damage and Reactor Materials Problems   340

12

Nuclear Heat Transfer   400

13

Nuclear Reactors   441

14

Selected Topics in Reactor and Fuel Cycle Technology   536

Appendix A   576

Appendix B   583

Appendix C   584

Appendix D   586

Appendix E   589

Appendix F   590

Appendix G   592

Answers to Selected Problems   593

Index   598

Preface

Basic Nuclear Engineering is designed to be used as a text at the advanced under-graduate and graduate levels for a first pair of courses covering the spectrum of nuclear activities, but with particular emphasis on nuclear power, the major nuclear activity from the point of view of both manpower and capital investment. For students who have a suitable nuclear physics background, a single-term course might omit Chapters 2, 3, and 4. Other special topics might be omitted to allow a single course to conform to the time available.

The past several years have been a period of trial and tribulation for the proponents of nuclear power. The ravages of inflation on capital costs, the less than enthusiastic stance of the Carter administration on nuclear power, and regulatory delays have largely offset much of the advantage of lower nuclear fuel cycle costs in comparison to other forms of energy. Three Mile Island did not particularly enhance the image of the nuclear power industry, despite the fact that there was no significant exposure as the result of the release of radioactivity. In a positive sense, it focused on the need for better operator training and the desirability of a central bank of information on all plant malfunctions with appropriate analysis of both their frequency and their consequences. The result has been the formation of two new groups by the nuclear industry: the Institute for Nuclear Power Operations (INPO) to improve and maintain the proficiency of nuclear plant operators and plant operations and the Nuclear Safety Analysis Center (NSAC) to collect information on plant malfunctions and to analyze their causes and develop effective means for their prevention.

Since the inception of the energy crisis, the rate of growth in the demand for energy has slowed to approximately 2 percent per year from a value of approxi-mately 5 percent in the preceding decade, much of this due to improved means of conservation. This has revised downward the estimates for necessary new plant capacity. The result has been the cancellation of several planned new nuclear generating stations and a lack of new orders for plants in the United States. Abroad, nuclear activity continues with the 1300 MW(e) French Super Phénix due to go critical in 1983 and the 600 MW(e) Canadian CANDU reactors in Wolsung-Kun, Korea, and in Rio Tercero, Argentina, which should go critical in December 1982 and August 1983, respectively.

Completion of reactors under construction and operation of plants already in operation will require a steady supply of new engineers to design, construct, and operate these facilities. Hopefully, new sales here and abroad will require additional personnel. It is for these people for whom this book is written, as

well as those who wish to have an understanding of the nuclear field as an integral part of an engineering education.

The authors are grateful to the many companies, technical publications, other authors, and government agencies that have most generously allowed us to use various illustrations and tables. We are also most appreciative of the many constructive suggestions from users of the previous editions, both professors and our students.

1

Introduction

Energy Sources

The utilization of energy for the performance of tasks has always been integral with our standard of living. This utilization can be traced all the way from the first use of the wheel, to the great works of the ancient Egyptians, through the power-producing devices of Newcomen and Watt, up to the present advanced state of our technology and standard of living. The great wars of history provide examples of vast amounts of energy expended.

Historically, except for the last few hundred years, our primary energy source has been wood. In fact, in many parts of the world today wood is still one of the chief sources of power. In the mid-1800s fossil fuels replaced wood; oil replaced coal in the mid-1900s. Because of the increased cost of oil and the U.S. dependence on imported oil in the 1970s, engineers and scientists are looking toward other sources to meet our energy requirements.

Most forecasts of energy requirements for the immediate future are based on an analysis of population growth and per capita energy consumption. Two of the more interesting forecasts have been made by P. Putnam and H. Brown. Their predictions are shown in Fig. 1.1. Meeting these tremendous demands will require diverse sources of energy. The present practical sources of energy are (a) fossil fuels, (b) flowing water, (c) winds, (d) ocean tides, (e) solar energy, (f) geothermal energy, (g) fissionable nuclei, and (h) fusionable nuclei (questionable as to its practicability). At some time all fossil fuels will be used up and energy will be derived solely from the remaining sources and the nuclear fuels. Various estimates have been made of the time when the supply of fossil fuels will be exhausted. They range from 25 to 200 years. The important point, however, is that the world's energy requirements will outstrip the economically recoverable fossil fuels. Another source of energy must then be developed to meet the world's energy demands.

The contributions of wind, ocean, tides, solar and geothermal energy, and flowing water, although interesting, are too limited and costly to become a large percentage of the total energy produced in the near future. This leaves nuclear fuels and coal as the economical sources for most of the energy requirements shown in Fig. 1.1. Forecasts made concerning future nuclear capacity are summarized in Fig. 1.2.

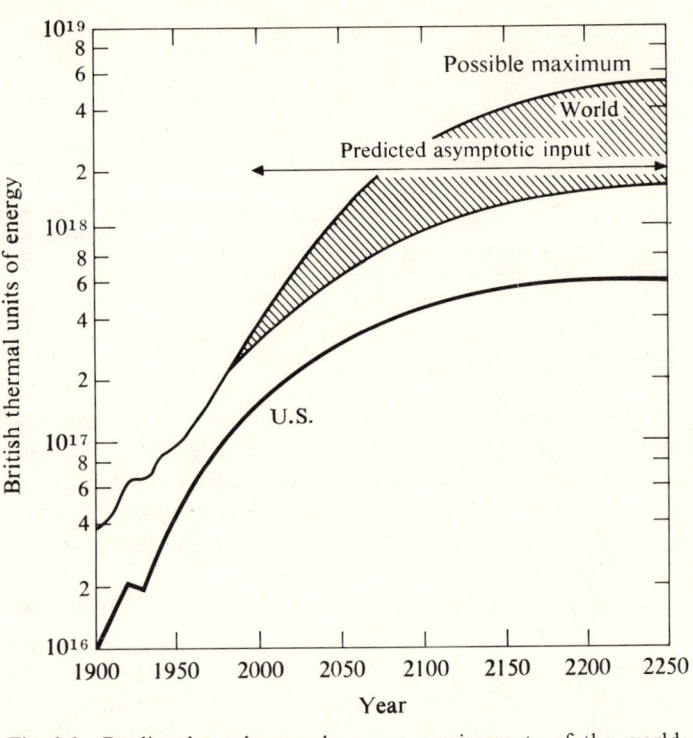

Fig. 1.1   Predicted total annual energy requirements of the world, based on population growth predictions of References 1 and 2 and per capita energy requirement predictions of the U.S. Atomic Energy Commission.

Because of environmental considerations, concerns over the safety of nuclear power plants, fears about nuclear weapons proliferation, and inflationary costs of energy sources, great emphasis has been placed on energy conservation measures. Consequently, it does not appear that the rosy predictions for nuclear energy growth shown in Fig. 1.2 will be met. The more recent predictions are shown as dashed bar graphs in the figure.

This book concerns the applications of nuclear fission to power production and isotope technology. Nuclear fusion, although theoretically a possible source of energy, has presented formidable problems. Until these problems (described in some detail in Chapters 4 and 13) are completely solved, fusion reactions will not be a useful source of energy.

There is a vast amount of energy available from fission (and fusion). Various estimates place the amount of fissionable ore reserves at about 15 million metric tons (excluding Communist countries). If 1 lb of uranium underwent complete fission, it would produce about 30 000 000 000 Btu of energy. It is believed that the earth's crust contains about 36×10^7 million

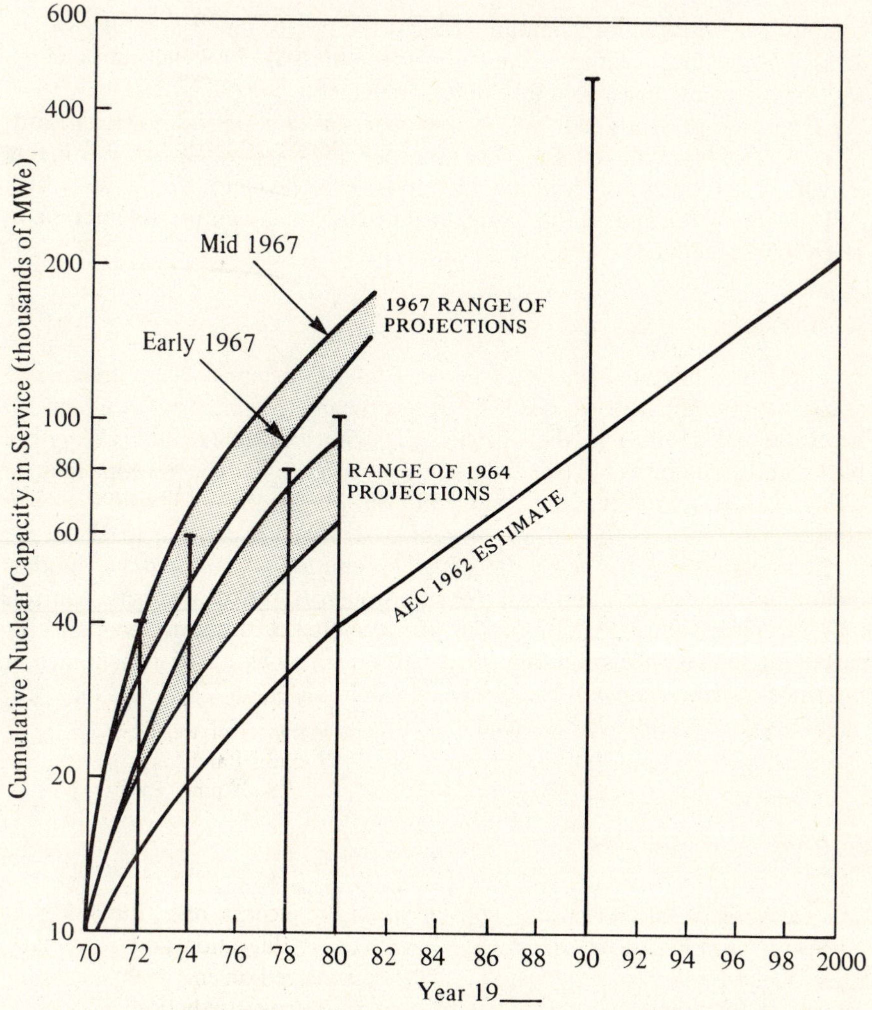

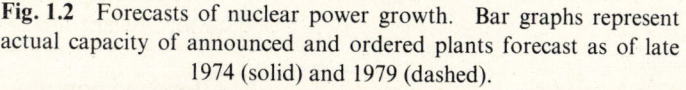

Fig. 1.2   Forecasts of nuclear power growth.   Bar graphs represent actual capacity of announced and ordered plants forecast as of late 1974 (solid) and 1979 (dashed).

tons of uranium and thorium compounds.   The challenge is to recover this fissionable material economically, convert it safely into energy, and properly dispose of the radioactive wastes.

Nuclear Fuels

In a chemical combustion reaction there is a rearrangement in the atoms of fuel and oxygen to form molecules of combustion products.   During this

combustion an insignificant amount of mass is converted into energy. However, in fission reactions a larger proportion of the mass is converted into energy, and the fissioning nuclei split into different elements.

There are presently four radioactive materials that are suitable for fission: ^{233}U, ^{235}U, ^{239}Pu, and ^{241}Pu. The isotopes ^{238}U and ^{232}Th are fissionable by what are termed *fast neutrons*, as opposed to *thermal neutrons* used with ^{233}U, ^{235}U, ^{239}Pu, and ^{241}Pu. A detailed discussion of neutron interactions is given in Chapter 8.

Fission

Fission (illustrated schematically in Fig. 1.3) occurs when a fissionable nucleus captures a neutron. Capture upsets the internal force balance between neutrons and protons in the nucleus. The nucleus splits into two lighter nuclei, and an average of two or three neutrons is emitted. The resulting mass of products is less than that of the original nucleus plus neutron. The difference in masses appears as energy in an amount determined according to Einstein's formula, $E = mc^2$. If one of the neutrons emitted is captured by another fissionable nucleus, a second fission occurs similar to the first, another neutron may produce a third fission, and so on. When the reaction becomes self-sustaining so that one fission triggers at least one more fission, the phenomenon is termed a *chain reaction*. The device in which this chain reaction is initiated, maintained, and controlled is called a *nuclear reactor*.

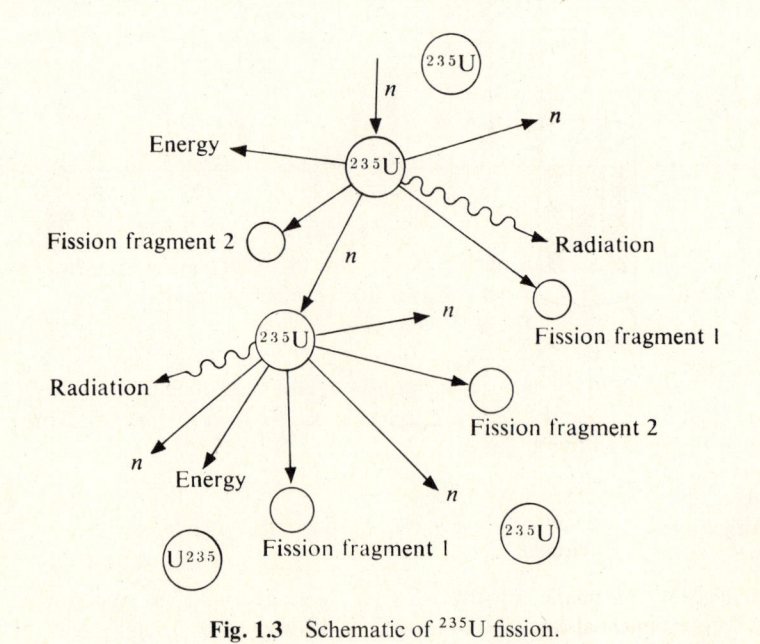

Fig. 1.3   Schematic of ^{235}U fission.

Three results of a fission chain reaction are important. The first is the energy released. About 80 percent of this energy appears as kinetic energy of the product nuclei (fission fragments) and neutrons. Through numerous collisions among fission fragments, neutrons, structural material, and fuel, their kinetic energy is removed and released in the form of heat.

The second important result is the radiation emitted during the reaction. Succeeding chapters treat radiation in more detail. It suffices to say here that the radiation appears in the form of electrons, helium nuclei, or pure electromagnetic radiation. The radiation from a nuclear reactor comes directly from fission fragments. Radioactive isotopes, that is, isotopes that are unstable and emit radiation, can be formed by placing stable nuclei in a reactor. In the reactor they are bombarded by neutrons, and their nuclear structure is altered. The radiation is potentially harmful to living organisms and to materials. With proper application, however, the radiation can also be extremely useful to materials and living organisms. The branch of nuclear engineering that treats radiation and its uses is termed *radioisotope technology*.

Finally, the neutrons released during fission are important in the overall fission process. In order for a fission reaction to be self-sustaining (a chain reaction) at least one neutron released per fission must be captured and cause fission in another nucleus. Depending on the fate of the neutrons released, the reaction becomes subcritical, critical, or supercritical. *Critical* describes a chain reaction in which fission is maintained at a constant rate per unit time.

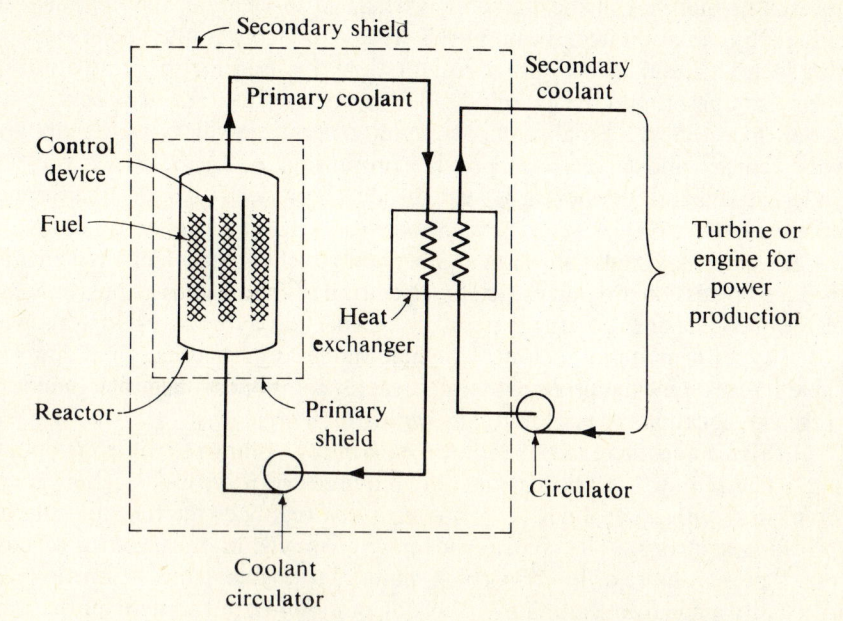

Fig. 1.4 Schematic of a nuclear reactor power plant. The primary coolant might also be used as a working substance in the power plant.

A reactor in which the rate of fissioning is decreasing is *subcritical*, and one in which the rate of fissioning is increasing is *supercritical*. An atomic bomb, obviously, is *super-super-critical*.

The problem in nuclear engineering appears to be how to harness and control the nuclear chain reaction. After this is done, one must find a way to convert the released energy to a useful form or utilize the radiation emitted and safely dispose of the waste. Figure 1.4 illustrates how a nuclear reactor might be designed. It is readily observed that the reactor takes the place of the combustion device in a conventional power cycle.

Reactor Use and Classification

The uses for nuclear reactors are varied. The major application, however, is for the production of electric power. Here the objective is to have a high rate of energy release at a high enough temperature for the attainment of a high thermal efficiency in the associated power cycle. Research reactors supply a large quantity of neutrons for irradiation of various materials in the core or to be allowed to stream from the reactor through a beam port. The temperatures in most research reactors are relatively low when compared with power reactors and disposal of the heat generated may be a nuisance. University research reactors are often used as training reactors in addition to performing their research mission. A medical reactor is designed to facilitate the treatment of patients by radiation from beam ports. Hospital facilities must be associated with such an installation. Other reactors may be used for the irradiation of stable isotopes to produce useful radioisotopes. Reactors also contribute to our space effort. Small compact units generate auxiliary power aboard space vehicles and larger units provide propulsion, as in the Rover project. Under study and development are in-pile thermoelectric and thermionic reactors.

Reactors may be classified in a number of ways: the use to which the neutrons produced by fission are put, the energy spectrum of the neutron population, the degree of conversion of fertile material, the dispersion of the core materials, and by the types of materials selected for fuel, moderator, cladding, and control. Table 1.1 lists some materials of thermal reactor components and their function. These classifications are not mutually exclusive.

In all reactors there must be some means of controlling the neutron population. The removal of all neutrons not required for a critical reaction is accomplished with control devices. A *control rod* regulates the reaction rate by absorbing neutrons. The control rods may be moved in or out of the reactor core to present more or less absorbing material to the neutrons. Control rods are usually made from cadmium or boron compounds, but sometimes they are made from fertile material (such as ^{238}U) to utilize the neutrons released by fissioning. A *fertile material* is a substance that can be converted into fission-

Table 1.1 Reactor Components and Materials

Component	Material	Function
Fuel	^{233}U, ^{235}U, ^{239}Pu, ^{241}Pu	Fission reaction
Moderator	Light water, heavy water, carbon, beryllium	To reduce energy of fast neutrons to thermal neutrons
Coolant	Light water, heavy water, air, CO_2, He, sodium, bismuth, sodium potassium, organic	To remove heat
Reflector	Same as moderator	To minimize neutron leakage
Shielding	Concrete, water, steel, lead, polyethylene	To provide protection from radiation
Control rods	Cadmium, boron, hafnium	To control neutron production rate
Structure	Aluminum, steel, zirconium, stainless steel	To provide physical support of reactor structure and components, containment of fuel elements

able material by neutron absorption and subsequent decay to a fissionable species.

In addition to control rods, a *burnable poison* is introduced into reactors to offset long-term changes in fuel concentration. The poison (or chemical shim) absorbs neutrons and is gradually reduced in concentration.

The *energy of the neutrons inducing fission* constitutes one of the major classes of nuclear reactors. Thus, a *thermal reactor* is one in which fission is induced by neutrons in thermal equilibrium with the reactor core material. Most of today's reactors are thermal. The neutron produced in the ^{235}U fission of Fig. 1.3 has much energy and a very small probability of interacting with another ^{235}U atom. The probability of an interaction of a neutron with a bombarded nucleus is referred to as the *neutron cross section*. A moderator is put into the reactor core to slow the neutron to thermal energies by collision (called *scattering*). A good moderator reduces the energy (speed) of neutrons in a small number of collisions (possesses a high scattering cross section).

Fast reactors make no attempt to slow neutrons and thus contain no moderator. The average neutron energy is on the order of 0.5 to 1.0 MeV. At these energies it is easier to convert a larger fraction of any fertile ^{238}U and ^{232}Th to fissionable ^{239}Pu and ^{233}U. Today's thermal reactors convert much less fertile material to fissionable material than they consume and would be called *burner reactors*. Perhaps a conversion ratio of 50 percent is reasonable for today's *power reactors*. If this conversion ratio exceeds 100 percent, the reactor is a *breeder*. A fast-neutron spectrum favors breeding and opens the possibility of converting our vast supplies of ^{232}Th and ^{238}U to nuclear fuel. Fast breeder reactors would be designed to produce large blocks of power, as well as a surplus of fuel. EBR-II, shown in Fig. 1.5, was an experimental

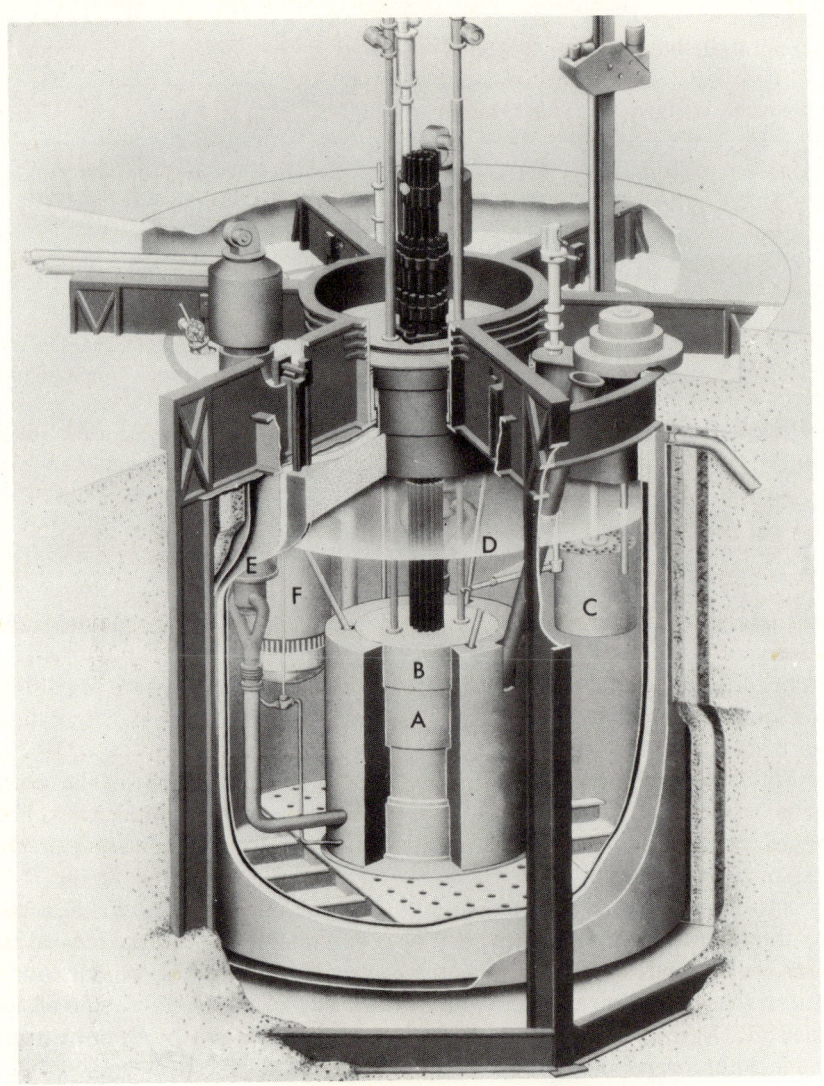

Fig. 1.5   Cutaway drawing showing the major reactor components of Argonne National Laboratory's Experimental Breeder Reactor II. [Courtesy of Argonne National Laboratory.]

prototype of a breeder that produced 20 MW(e), as well as plutonium. It continues to be useful in the testing of fast reactor materials. A demonstration commercial breeder reactor has been proposed for development in Tennessee. The French utility EdeF placed a 250-MW(e) prototype breeder reactor (Phénix) into commercial operation in 1974 and is constructing a 1200-MW(e) fast breeder reactor scheduled for operation in 1983.

An intermediate reactor would have the average energy of its neutrons in the epithermal (above thermal) region. Reactors with this type of energy spectrum have been investigated, but the results have not justified active development.

Materials consisting of low-mass-number atoms usually make the best moderators. In this respect hydrogen makes an ideal moderator (except that hydrogen absorbs some neutrons). Light water is desirable because it is cheap and plentiful. Furthermore, the light water can be used as both moderator and coolant; however, the water must be completely free of impurities in order to avoid neutron absorption and possible radioactivity. Water also has a relatively low boiling point; thus, pressures must be high if high temperatures are desired. Heavy water, D_2O, is also an excellent moderator and coolant. Heavy water has a smaller probability for neutron absorption than light water, but is not quite as effective as light water in slowing down the neutrons. Its chief disadvantage is its high cost. Carbon is another good moderator because it does not absorb many neutrons and does scatter neutrons well. Carbon is readily available in the form of graphite. One disadvantage is that graphite may oxidize at high temperatures. Beryllium is one of the best solid moderators and is used either as metallic beryllium or as beryllium oxide. Beryllium has a low absorption cross section, a high scattering cross section, and a high melting point, 1158 K.

The reactor coolant removes the heat released by fission from the reactor. To do this efficiently the coolant should have a high specific heat, high conductivity, good stability, good pumping characteristics, and a low neutron absorption cross section. Coolants can be either liquid or gaseous. The popularity of light and heavy water as coolants was mentioned in the preceding paragraph.

Liquid metals such as sodium, sodium–potassium (NaK), and bismuth make desirable coolants. They have a high boiling point and therefore can be used at low pressures. The heat transfer and nuclear properties are good, but the metals must be preheated before reactor startup. Some organic substances (terphenyl, $C_{12}H_{10}$) have been tried experimentally as reactor coolants, but they have not yet had commercial application. Their chief disadvantage has been a tendency to decompose.

The motion of neutrons in a reactor is completely random. Unless the reactor core is surrounded by a reflector, neutrons may leak out. The amount of fissionable material in a reactor may be reduced by placing this reflector around the core to "reflect" the neutrons back into the core. A good reflector normally has the same characteristics as a good moderator.

A shield is necessary to prevent or reduce passage of radiation to the outside of the reactor. All types of radiation are dangerous to personnel and must be reduced to tolerable levels. Neutrons and gamma radiation are the most penetrating radiations; if these two are stopped the other radiations and fission products will be stopped as well. Shielding is dependent on the purpose

of the reactor. For example, a zero-power reactor requires very little or no shielding. In general the best shield for neutrons is a low-atomic-weight material and for gamma rays a high-atomic-weight material. Frequently, a shield is made of layers of heavy and light material such as concrete and polyethylene or concrete and water.

As mentioned previously, control materials are used to regulate the neutron density in the reactor. The reactor power is directly proportional to the neutron density; removing neutrons from the reactor core, therefore, will decrease the power and reaction rate. Conversely, not removing as many neutrons will increase the power. The control materials are usually in the form of rods. The rod material must have a huge neutron absorption cross section. Control rods are made from cadmium, boron, hafnium, and rare earths.

Three types of control rods are used:

(1) *Shim rods* are used for making occasional coarse adjustments in neutron density.
(2) *Regulating rods* are used for fine adjustments and to maintain the desired power output.
(3) *Safety rods* are capable of shutting down the entire reactor in case of failure of the normal control system. The control rods are normally suspended above or below the reactor core. In the case of shutdown, they move into the reactor core, absorb neutrons, and stop the chain reaction.

The reactor structural materials must have high structural impact and tensile strength, high corrosion resistance, high rupture strength, low neutron absorption, little interaction with radiation, and ease of fabrication. Aluminum, stainless steel, zirconium, nickel, magnesium, concrete, and many others are used as reactor structural materials. All the components are assembled into the complete nuclear reactor, whether it be for power, irradiation, production of fissile material, research, or other purposes.

It is enlightening to examine reactors according to the major class of *nuclear configuration.* Based on nuclear configuration a reactor is either homogeneous or heterogeneous. In a *homogeneous reactor* the core materials are distributed in such a manner that the neutron characteristics can be accurately described by the assumption of homogeneous distribution of materials throughout the core. In a *heterogeneous reactor* the core materials are segregated to such an extent that the neutron characteristics cannot be accurately described by the assumption of homogeneous distributions of materials throughout the core.

From the standpoint of neutron economy a spherical core is best. A spherical core, however, is economical only in a homogeneous reactor. Since the coolant–fuel mixture becomes heated in the reactor core, it must be circulated through a heat exchanger to remove this heat.

Analytically, it is simplest to treat a homogeneous reactor and, for the most part, the theoretical and design considerations of this book are for homog-

eneous reactors. (Heterogeneous reactors with small fuel elements may often be treated as being homogeneous without introducing serious errors.) A homogeneous reactor uses a uranium salt mixed with coolant (with coolant sometimes doubling as the moderator).

A homogeneous reactor is not practical for power production purposes because the radioactive fission products in the fuel–coolant mixture must be circulated through the heat exchanger. The whole plant must then be shielded; the mass of circulating mixture will be enormous, and maintenance will be extremely difficult and expensive.

A heterogeneous reactor is much more practical from the standpoints of both power and economy. The solid-fuel heterogeneous reactor is usually cubical or cylindrical. The fuel is uniformly spaced throughout the moderator while the coolant circulates through passages between fuel elements. As mentioned previously, the coolant and moderator are frequently one and the same. For economy, effective heat transfer, and ease of refueling, fuel elements have been fabricated as thin rods, plates, and small-diameter pins. These are assembled into bundles of parallel elements.

In the case of a pressurized water- or organic-cooled reactor, the heated coolant, after passing from the fuel element passages, gives up its heat in a heat exchanger. Figure 1.6 is a cutaway view of the Connecticut Yankee Atomic Electric Company's pressurized water power reactor core. If the coolant is gas (GCR) or steam as in a boiling water reactor (BWR), the coolant may pass directly to a turbine for the production of useful work. In this type of reactor, heat exchangers and pumps may be eliminated, reducing the total plant equipment necessary. Figure 1.7 is a cutaway view of a boiling water reactor.

In any of the heterogeneous reactor configurations, heavy shielding is put around the reactor. Lighter shielding is used around the coolant piping, heat exchangers, and pumps. By judicious selection of coolant almost 100 percent of the radioactivity present is confined inside the heavy, or primary shield. As indicated in Table 1.1, heavy materials make the best radiation shields. It is this factor which has severely handicapped the development of a nuclear aircraft engine.

Finally, research and irradiation reactors are another important major class of reactors. Irradiation reactors take several forms, but essentially they all use either neutrons or gamma radiation from the fission reaction for food irradiation, biomedical irradiation, material processing and testing, or isotope production for industrial uses.

Practically a separate industry has grown from the field of radioisotope technology. Neutron bombardment of certain stable isotopes produces radioactive isotopes. New uses of these radioisotopes are constantly being developed. Some commercially available radioactive isotopes produced in reactors are ^{60}Co, ^{131}I, ^{3}H, ^{59}Fe, ^{90}Sr, ^{63}Zn, and ^{206}Tl. ^{60}Co is used in cancer irradiation. Other isotopes are used as tracers in medical research. Industrially, radioisotopes have a wide variety of uses. They serve as thickness gages,

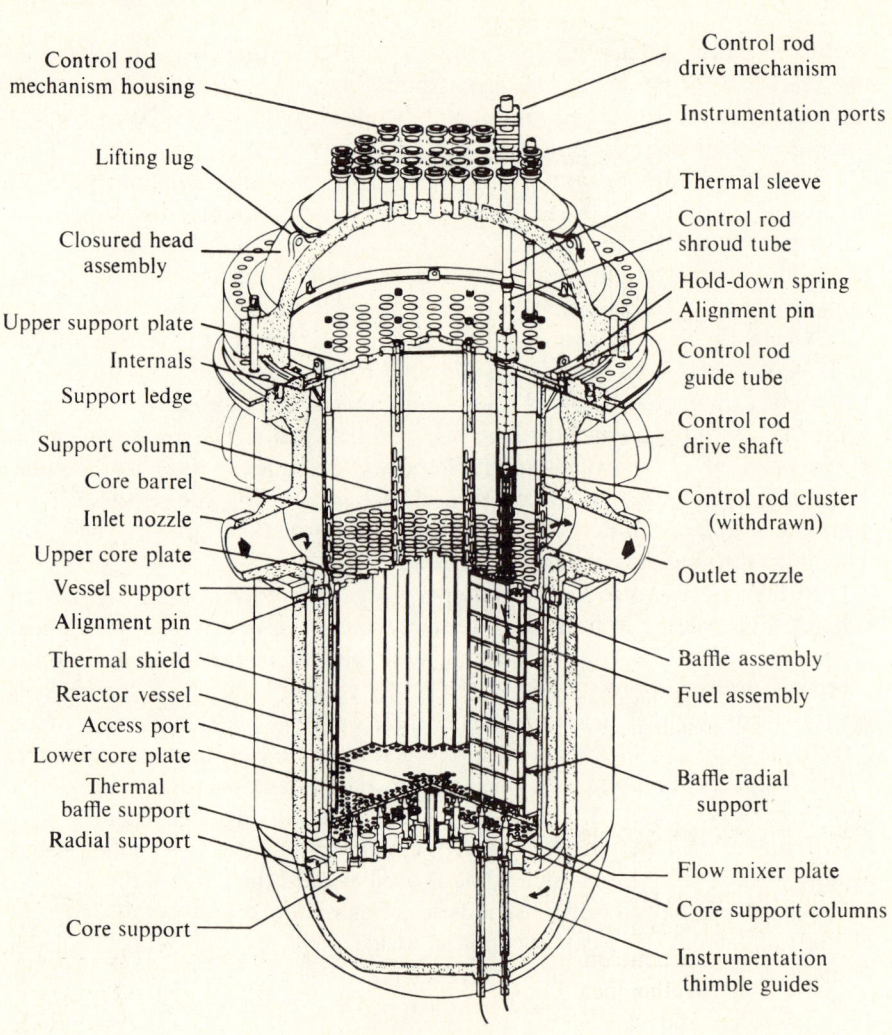

Control rod
mechanism housing

Lifting lug

Closured head
assembly

Upper support plate

Internals

Support ledge

Support column

Core barrel

Inlet nozzle

Upper core plate

Vessel support

Alignment pin

Thermal shield

Reactor vessel

Access port

Lower core plate

Thermal
baffle support

Radial support

Core support

Control rod
drive mechanism

Instrumentation ports

Thermal sleeve

Control rod
shroud tube

Hold-down spring

Alignment pin

Control rod
guide tube

Control rod
drive shaft

Control rod cluster
(withdrawn)

Outlet nozzle

Baffle assembly

Fuel assembly

Baffle radial
support

Flow mixer plate

Core support columns

Instrumentation
thimble guides

Fig. 1.6 Pressurized water reactor core. [Courtesy of Westinghouse Hanford Company.]

tracers, level gages, and sources of energy. They are used for promotion of chemical reactions and for radiographic inspections. The field of radio-isotope technology requires not only a knowledge of interactions of radiation with matter, but knowledge of detection and production of isotopes as well as an understanding of the design and operation of reactors.

Only a few general remarks concerning general types of reactors have been made. A more detailed description of some reactor types and design considerations is presented in Chapter 13.

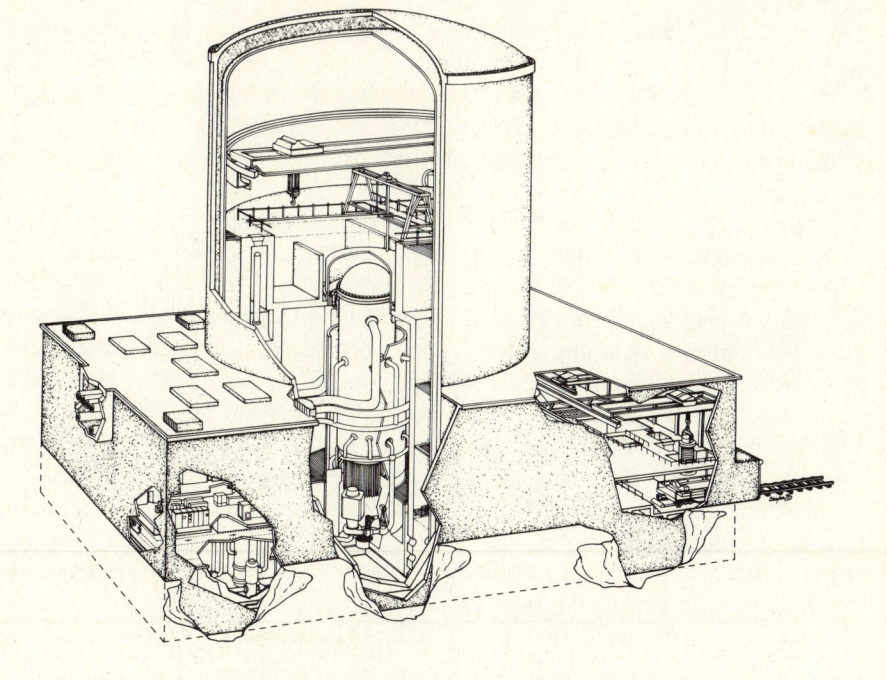

Fig. 1.7 Boiling water reactor building showing pressure suppression containment. [Courtesy of General Electric Company.]

Reactor Design Problems

In a heterogeneous reactor the fuel and moderator are lumped throughout the reactor. Since the heat is produced in the fuel elements, it is desirable to have a uniform heat and neutron flux. To do this the fuel elements are sometimes spaced closer together near the edge of the reactor. This levels off the heat flux somewhat and causes most of the fuel elements to operate near their maximum temperature limits. Theoretically, a reactor is a source of energy at unlimited temperature, and the higher the temperature at which heat is supplied to the working fluid of a cycle the higher the thermal efficiency of the cycle will be. However, considering the hydraulic, nuclear, and heat transfer characteristics of the core, the temperature must be kept below melting-point temperatures of metals in the core and below boiling-point temperatures for the coolant, in the case of a liquid-cooled reactor.

When a reactor is designed, care must be exercised to select materials that will not become excessively radioactive. Manganese, for example, a component of many steels, becomes radioactive on exposure to neutrons. Some electrical insulation loses its strength on exposure to radiation. The strength characteristics of metals change and some lubricating oils turns heavy and gummy after

exposure. These and many other problems must be accounted for and solved before a reactor is successfully operated.

Care must be taken to ensure that pipes will not make a "hole" in the shield. The piping and the material in it are not good shields; therefore, shielding must be built over the pipes, or the pipes must be bent inside the shield, or both.

Eventually, the fuel in a reactor becomes used to a point when a critical chain reaction can no longer be sustained. When this point is reached, the reactor must be refueled. The reactor now contains a valuable amount of usable fuel, radioactive fission products, and stable fission products. The fuel elements must be removed from the reactor and reprocessed or stored indefinitely. Since they are radioactive they must be handled remotely after they have "cooled" somewhat, usually by being stored under water for about 90 days. After the cladding material is removed the uranium can be reclaimed by a solvent extraction process. The U.S. Department of Energy built a large plant on the Savannah River in South Carolina to do this reprocessing. President Carter ordered a halt to development of commercial reprocessing facilities in 1977. His action came as a result of fears over possible nuclear proliferation. Commercial reprocessing has been carried out successfully in Europe.

The fission products that remain are radioactive wastes and must be disposed of so as to never endanger persons. Contaminated reactor components and materials also come under the category of radioactive wastes.

One last design problem will be posed here. A reactor that has once been critical always produces some heat from the decay of fission products and spontaneous fission. This heat must be removed or the reactor will overheat. A thermal loop is built into the reactor so that this heat, produced after shutdown, will be removed. This is particularly important if the safety devices cause a sudden or emergency shutdown (called a *scram*).

From the beginning it was recognized that nuclear reactors present inherent hazards to the general public. Furthermore, the nuclear explosions of World War II left an indelible mark on the minds of the public regarding the dangerous capability of nuclear effects. Unfortunately, there was no such milestone to imprint the beneficial aspects of nuclear energy on the public.

The development of nuclear power has always been characterized by an overriding concern for health and safety. More attention in the forms of research, debate, legislation, and development has been given to environmental and health effects of nuclear power than any other technology.

To make the safety problem more difficult, nuclear safety depends on a theoretical potential for conceivable damage. Other areas of technology have been able to rely on practical experience to demonstrate hazards or lack of safety. Since nuclear reactor safety requirements must be based on theoretical potentials, an ultra-ultra-conservative approach has been taken. In many instances the regulatory agencies have insisted on safety precautions that have nearly precluded design or construction of the reactor.

The two basic premises that have been applied in reactor siting are isolation by distance from inhabited areas and prevention of accidental release of radioactivity by safety devices and design features. In the past, reactor siting has been made on a combination of these two. Newer reactors have been located much closer to centers of population than were their predecessors. However, the accident at Three Mile Island has caused regulators to reconsider the advantages of isolation.

Despite the outstanding safety record and ultraconservativeness of safety requirements, there is little likelihood or desire to change safety requirements to more practical levels.

On March 28, 1979, a serious accident occurred at the Three Mile Island Nuclear Power Station. The site of the accident was Unit 2 of the Three Mile Island Nuclear Power Station about 10 miles from Harrisburg, Pennsylvania. A loss of feedwater to the steam generators resulted in a steam turbine trip and reactor trip. Through a succession of operator and equipment malfunctions a portion of the water in the reactor core boiled away. The result was fuel overheating and damage to the fuel cladding. Some fission products mixed with the water then spilled into the containment and auxiliary building, contaminating these areas.

A presidential commission (Reference 25) reporting on the accident said: "The accident at Three Mile Island occurred as a result of a series of human, institutional, and mechanical failures." The commission also found that "most of the radiation was contained and the actual release will have a negligible effect on the physical health of individuals."

The average radiation dose to a person living within 5 mi of the plant was calculated to be 10 percent or less of annual background. Outside the 5-mi radius it was even less. Unfortunately, a difficulty in communications on March 30 resulted in an erroneous report of a high radiation level. As a result of this error the governor of Pennsylvania recommended that pregnant women and preschool children living within a 5-mi radius of the plant leave the area. Schools within the 5-mi radius of the area were also closed.

Although the immediate accident situation ended within a few days, risks still remain. The cleanup will take several years and probably will cost 1×10^9 for plant cleanup and repair alone. Estimates put replacement power costs at about 14×10^6 per month and total refurbishment costs from $1 to $2 billion. However, other estimates for replacement with a coal-fired plant put the costs from $2 to $2.5 billion. A comprehensive discussion of the accident is presented in Chapter 14.

Multiple systems for protection from the escape of radioactivity are always provided for in the design of nuclear power plants. This has been termed "defense-in-depth." The first system is the design itself. Each system and component of the plant is designed so that it will operate as intended under all normal and anticipated accident conditions. To accomplish this requires that the plant be very carefully designed, constructed, tested, and maintained

according to high-quality engineering standards and effective quality assurance standards.   The second protection system is safety systems to protect and prevent damage if an abnormal or accident condition should occur.   These systems are designed so that no single failure of a component would result in an accident condition.   Examples of these systems include a scram system to shut the reactor down quickly, an emergency core coolant system to provide cooling in the event of a major accident, and limits on the rate at which power can be increased.   Finally, additional safety systems supplement the normal redundant safety and operational systems.   That is, features are added, based on hypothetical accidents, which preclude release of radioactivity should some protection systems fail simultaneously during the accident they are designed to control.   Included in this type of accident are earthquakes, steam generator failure, missiles, and floods.

Environmental Considerations

Concern about the discharges of nuclear power plants to the environment continues to be of great interest to a wide spectrum of people, including government, citizen groups, conservationists, scientists and engineers, and the general public.   Concern for the environment is brought to focus for the public almost daily by television and other media.   It is not surprising to find opposition to nuclear power plants from people whose attention has been focused on routine releases of radioactive materials to the atmosphere and the polluting aspects of thermal discharges.   Many people do not have a thorough understanding of built-in safeguards, of the range of modifications possible, or the need for power plants to meet demands for power.

There is no doubt that nuclear power plants are a potential hazard due to radiation discharges and thermal pollution.   Utilities have the responsibility of proving to regulatory agencies, environmentalists, and other groups that the power plant will be built and operated safely and without significant effects on the environment.   Designers of plants where inadequate cooling water is available are being forced to eliminate the direct release of waste heat by the use of cooling towers.

Every thermal system of converting energy will produce some wasted energy and thermal loss. After all, the Second Law of Thermodynamics guarantees that some energy will be discharged from the system.   Because of lower thermal efficiencies, current nuclear power plants discharge about 25 to 30 percent greater volumes of heated water than a fossil fuel plant of the same capacity. Also, the size of the plants being installed has increased.   The resulting larger discharges are what is leading to the thermal effects problem.

Thermal effects problems are undergoing continued research and are manageable.   Better efficiencies for advanced reactors, such as breeder and gas-cooled reactors, should reduce the thermal problem as well as conserve

natural supplies of fissionable fuels. Solutions to the thermal problems are chiefly economics and siting. The least costly means of discharging waste heat is to dissipate it in large rivers, lakes, or the ocean. The resultant thermal effects may be insignificant, beneficial, or detrimental depending on the use of the water, amount of water available, ecology, and so on.

If economics permit or ecology demands, methods other than discharge to bodies of water may be used. Examples are artificial ponds, cooling towers, and use of heat as a by-product.

Other methods will add to the cost of generating power. Unquestionably, these costs will be worth it when there will be significant detrimental effects by direct discharge of heat. Unfortunately, not very much is known about environmental effects of the discharge of large amounts of condenser cooling water. Methods for predicting temperature variations must be better developed and checked so that meaningful knowledge on the effects of waste heat on aquatic life can be determined. Studies of both short-term and long-term effects should be made. The short-term information can help determine how individual power plants should be designed while long-term information will affect regional power growth and plant siting. The chief concern should be on the water quality criteria, not on numbers of fish killed. Established water temperature standards are well below the lethal temperatures for fish. The fish were apparently killed as a result of mechanical or chemical problems, not thermal or radiation problems. Therefore, improvements in water intake and chemical treatments are being made.

It has been suggested that heated water be used for irrigating, thereby extending the growing season for agriculture, to delay freezing of waterways in northern latitudes, and to warm lower depths of the ocean to promote more biological growth. However, any of these alternatives may produce other undesirable environmental effects. Whatever methods of heat removal are used and whatever the power plant—fossil or nuclear—the Second Law of Thermodynamics must be satisfied.

It seems clear, however, that the total environmental effects of nuclear power plants are no worse than those of fossil fuel plants.

Every energy conversion device leaves other waste residue also. Even the good old horse of the horse and buggy left much waste. The internal combustion engine replaced the horse or else the streets would undoubtedly be uninhabitable! Fossil plants produce ash; nuclear plants produce fission products.

The problem of disposing of radioactive waste material produced by nuclear plants is also very important. There are two general categories of management of radioactive waste: (a) treatment and disposal of low-level radioactive material, and (b) treatment and storage of material with high levels of radioactivity.

The low-level radioactive solid wastes are generally separated from the nonradioactive liquids and gases, concentrated, and then encased in concrete or other material.

High-level wastes are produced during fuel reprocessing at a few specially selected reprocessing sites. The several ways in which radioactive wastes are disposed of are as follows:

(1) Short-lived isotopes can be diluted and stored in remote locations until their activity is reduced to very low limits. The Nuclear Regulation Commission (NRC) sets limits of radioactivity that can be released to the environment.

(2) Wastes may be concentrated in tanks and stored on tank farms or buried underground. These methods are dangerous because of leakage and contamination problems. Also, aboveground storage requires a very large restricted area.

(3) Wastes may be concentrated and encased in concrete, then dumped at sea. This is dangerous because leakage may be ingested by fish or plants and spread throughout the world.

(4) The radioactive materials may be stored in large caves or deep craters formed by underground explosions. The danger here is that an earthquake might release contamination.

At present all the methods described above are being used for waste disposal. One other proposal has been made: that radioactive wastes be built into rockets and placed in outer space or into orbit. This has not yet been tried because of the obvious problems it presents.

The National Environmental Policy Act of 1969 (NEPA) became effective on January 1, 1970. NEPA requires that the NRC regulate and assess the impact of nuclear power on the total environment in terms of alternatives and the need for increased electric power. In an interpretation of NEPA, the U.S. Court of Appeals for the District of Columbia in 1971 made a far-reaching ruling (commonly referred to as the Calvert Cliffs decision). This ruling resulted in the NRC revising and strengthening its regulations to make a more rigorous implementation of NEPA. Some of the more significant aspects of this ruling were as follows:

(1) Environmental aspects must be considered at each stage in the licensing process. The environmental reports are circulated to government agencies and other interested persons.

(2) At each licensing stage a cost–benefit analysis is made of environmental costs vs. economic and technical benefits. Consideration must be given to alternatives that will minimize environmental impacts.

(3) Even if federal or state agencies certify that their environmental standards are satisfied, the NRC independently evaluates the total environmental impact. The NRC can, if it chooses, require controls more strict than those of a local agency.

(4) Each nuclear power plant that was under construction but not granted an operating license was required to submit an environmental impact statement showing why its construction permit should not be suspended until a complete environmental and impact review was made. This in effect opened the possibility that plants already under construction might have to be redesigned or even abandoned.

Even if power plant operation and construction complies with environmental quality standards and requirements that have been imposed by other agencies, the NRC independently assesses the environmental impact. In addition to the normal consideration of radiological effects, the cost–benefit analysis required by NEPA assesses the environmental costs of the radiological effects of the construction and operation of the power plant.

The increased review procedures and care which will be taken to protect the environment means that more time will be required to construct and begin operation of nuclear power plants.

One of the most complex problems challenging the ingenuity of the engineering profession is that of air pollution. Any and all forms of combustion yield products which are undesirable in our environment. Among the pollutants produced by conventional power plants are oxides of sulfur and nitrogen, hydrocarbons, and particulate matter. Air pollution varies from season to season, place to place, and daylight to dark. In addition to the irritating haze, it produces nitrogen oxides which may contribute to respiratory disease. Sunlight combines the hydrocarbons discharged by automobiles and nitrogen oxides to form the infamous photochemical smog. Smog damages crops, trees, materials, causes eyes to smart, and reduces resistance to respiratory disease. Sulfur oxides, on the other hand, corrode stone and metal, and injure plants and animals. Particulate matter produced by combustion dirties the surrounding terrain and reduces visibility. More important, however, it acts as a catalyst in formation of other pollutants.

While not all pollutants are caused by discharge from power plants, a significant amount of air pollution can come from these power plants. About 10 percent of the heat discharged from a fossil fuel plant is dissipated directly into the atmosphere through the stack. Latest estimates indicate these fossil fuel plants emit approximately 20 million tons of pollutants to the atmosphere each year, but over 90 percent of pollutants are gases. Until the problem was completely recognized, little effort was made to reduce air pollution. Now the Environmental Protection Agency has extensive programs under way to sample, study, and reduce air pollution. Short of stopping the discharge of pollutants to the atmosphere air pollution can be reduced somewhat, but not eliminated. A nuclear power plant has some inherent environmental advantages over conventional power plants. Since the nuclear reactor replaces the conventional combustion device, there will be no burning of hydrocarbon and oxygen resources and there will be no sulfur or nitrogen oxides and no particulate matter released to the atmosphere. Through careful design and operation there are essentially no radioactive particles released to the atmosphere, either. (In fact, studies have shown that a conventional power plant—and operation of a television set—produce significant radiation.) If nuclear power production replaces conventional power production there should be a reduction in the total number of pollutants introduced into the atmosphere by power plants.

At present, regulations express the maximum allowable concentration of radioisotopes in water and air at the boundary of the plant. The values are taken from *National Bureau of Standards Handbook 69* and reduced by a factor of 10. If a group of people live near the boundary, the values are further reduced by a factor of 3. The resultant average dose to the population will be no more than 170 mrem per year, approximately equal to the average national background radiation.

In January 1971, the National Council on Radiation Protection issued a report stating that the limits cited above provide adequate safety factors. Nevertheless, the (then) Atomic Energy Commission (AEC) in 1971 proposed an amendment to its regulations (10 CFR, Part 50) to "as low as practicable" to give guidance for lowering radioactive releases. Regulations limit the total quantity of radioactive material in liquid and gaseous discharges from nuclear reactors.

The regulations specify a calculated design maximum annual dose from liquid effluents from each reactor to the total body of 3 mrem (except tritium and dissolved gases) and 10 mrem to any organ.

In gaseous discharges the regulations limit annual releases to 10 mrads for gamma radiation and 20 mrads for beta radiation. Restrictions also are imposed concerning the maximum amount of radioactive iodine and other particulates that can be released to the atmosphere.

It is expected for the foreseeable future that total body doses to individuals living near a nuclear power plant will be less than 5 percent of the average natural background. It is expected also that the annual average total body dose to the U.S. population will be less than 1 percent of the dose from natural background radiation.

That there is no danger to the population from radioactivity in a nuclear power plant is a testimonial to the AEC and nuclear power equipment manufacturers. Over 400 reactors have been built in the United States since the CP-1 reactor of 1941. In this time period there have been remarkably few significant accidents. Only one accident affecting the general public has occurred in any civilian nuclear power plant in the United Sates. In fact, through 1980 there had been only five accidents, all minor (except Three Mile Island) in power, production, or propulsion reactors.

History

The history of the neutron (and, consequently, of atomic energy) dates from 1930. In that year German physicists W. Bothe and H. Becker discovered that when beryllium or boron is bombarded with high-energy helium nuclei, a highly penetrating radiation is produced. J. Chadwick, in 1932, proved that this radiation is not gamma radiation but an unchanged particle with a mass about the same as a proton. It is called a *neutron*.

Enrico Fermi used neutrons to bombard several elements. In 1934, he reported that about 40 different target nuclei would become radioactive when bombarded by neutrons. In experiments with uranium he found that different elements were formed with the radioactive isotopes. During the course of his experiments he also suggested that neutron energies were reduced when the neutron passed through water or paraffin. In January 1939, the German physicists Lise Meitner and Otto Frisch deduced that bombardment of uranium by neutrons causes the uranium to split into two approximately equal parts, each with an enormous amount of energy. This was the discovery of fission. It had been preceded by the work of many experimenters, but none of them individually was able to correctly interpret his results.

On January 26, 1939, a conference on theoretical physics was convened in Washington, D.C. The theories of fission were discussed and many physicists began experiments to detect the fission products. Within approximately three months the fission process was confirmed by many people and most of the fission products had been identified. The emission of neutrons in the fission process was also discovered. It is, of course, the emission of neutrons that permits the fission process to be self-sustaining and makes possible a nuclear reactor. In October 1939, President Roosevelt established the "Advisory Committee on Uranium." Six thousand dollars was granted to this committee for the procurement of 50 tons of uranium and 4 tons of graphite.

By June 1940 it was known that uranium, thorium, and protactinium could be fissioned. The fragments produced were isotopes of elements with atomic numbers ranging from 34 to 57. These fission fragments have large kinetic energies and are unstable, emitting beta particles through successive elements to a stable isotope. It was known that ^{235}U has a larger probability of fission by low-energy neutrons than by high-energy neutrons. It was also established that ^{238}U, thorium, and protactinium can be fissioned only by fast neutrons, and that there is a resonance absorption of neutrons by ^{238}U with certain energies between thermal and fast. Finally, it was known that one, two, three, or more neutrons are emitted in each fission. About this time physicists were also investigating the possibility of using uranium in a thermal reactor to make plutonium. The plutonium could then be separated chemically from the uranium and used in a thermal neutron reaction.

The third report of the National Academy of Sciences Committee on Atomic Fission (dated November 6, 1941) stated that from 2 to 100 kg of ^{235}U would be required to make a bomb. The report also stated that two methods of separating ^{235}U isotopes from natural uranium were possible (natural uranium is 0.7115 weight percent ^{235}U). It was estimated that three to four years would be required to produce a significant number of bombs. In August 1942, President Roosevelt set up the Manhattan Engineer District of the U.S. Army Corps of Engineers to develop an atomic bomb.

By the end of 1941 there were only a few grams of pure metallic uranium available. Experiments indicated that the commercial grade of black uranium

Table 1.2 Early Nuclear Reactors

Name	Location	Operation	Fuel	Moderator
CP-1	Chicago, Illinois	Dec. 1942	50 ton natural U	Graphite
X-10	Oak Ridge, Tennessee	Nov. 1943	Natural U cylinders	Graphite
CP-3	Argonne, Illinois	May 1944	Natural U rods	D_2O
LOPO	Los Alamos, New Mexico	May 1944	14.6% enriched ^{235}U, UO_2SO_4 dissolved in H_2O	H_2O
	Hanford, Washington	Sept. 1944	Natural U slugs	Graphite
ZEEP	Chalk River, Canada	Apr. 1945	Natural U	D_2O
Clementine	Los Alamos, New Mexico	Nov. 1946	^{239}Pu rods	—
GLEEP	Harwell, England	Aug. 1947	Natural U bars	Graphite
NRX	Chalk River, Canada	Aug. 1947	Natural U cylinder rod	D_2O
BSR	Oak Ridge, Tennessee	Nov. 1950	Enriched sandwich plates	H_2O
JEEP	Kjeller, Norway	Aug. 1951	Natural U rods	D_2O
EBR-I	NRTS, Arco, Idaho	Dec. 1951	90% enriched U rods	—
MTR	NRTS, Arco, Idaho	Mar. 1952	Enriched ^{235}U sandwich plate	H_2O
APS-I	USSR	1954	5% enriched U rods	Graphite
PWR	USS *Nautilus*	Jan. 1955	Highly enriched ^{235}U	H_2O
Calder Hall	Cumberland, England	Oct. 1956	Natural U rods	Graphite
EBWR	Argonne, Illinois	Dec. 1956	1.5% ^{235}U plates	H_2O
PWR	Shippingport, Pennsylvania	Dec. 1957	Highly enriched ^{235}U and natural U	H_2O
Dresden I	Morris, Illinois	Oct. 1959	Enriched ^{235}U	H_2O
Phénix	France	1974	Mixed PUO_2–UO_2	—

Coolant	Reflector	Power	Remarks
Free-air convection	Graphite	200 W	First controlled chain reaction
Forced-air convection	Graphite	4000 kW	Pilot plant for Pu production; used for isotope production and research
D_2O	Graphite	300 kW	First D_2O reactor
H_2O	Beryllium	1 W	First enriched uranium and first homogeneous water boiler; spherical core
H_2O	—	100 000 kW (est.)	Plutonium production reactors; no power used
D_2O	—	—	First Canadian reactor
Hg	Natural U	25 kW	First fast reactor; first ^{239}Pu reactor
Air	Graphite	100 kW	First English reactor; plutonium production only
H_2O	Graphite	40 000 kW	High neutron flux
H_2O	H_2O	100 kW	Free convection cooling; swimming pool type; prototype for research
D_2O	Graphite	100 kW	No power used
NaK	Natural U	100 kW	First production of electric power; breeder reactor
H_2O	H_2O	40 000 kW	High neutron flux to study effects of radiation on materials
H_2O	Graphite	1000 kW(e)	First Soviet reactor
H_2O	H_2O	—	First propulsion reactor; small size, high power
CO_2	Graphite	4200 kW(e)	First commercial production of power; plutonium production
H_2O	H_2O	5000 kW(e)	First boiling water reactor
H_2O	H_2O	230 000 kW(th) 60 000 kW(e)	First central station nuclear plant in United States
H_2O	H_2O	200 000 kW(e)	First commercial boiling water reactor in United States
Na	—	250 MW(e)	First commercial fast breeder reactor

oxide available would not sustain a critical reaction. By May 1942, purer uranium oxide was available, but it was not pure enough to sustain a critical reaction. It was not until July 1942 that uranium oxide pure enough to sustain a critical reaction was made. It was planned to use graphite as moderator with the natural uranium oxide as fuel for the first critical reaction. By this same time (summer 1942) the problems of purifying graphite were also solved.

In autumn 1942, enough purified uranium oxide, pure uranium metal, and graphite moderator were assembled in Chicago to build a reactor. Since the purity of the materials was in doubt, there was concern as to whether or not a self-sustaining reaction could be achieved. There need have been no fear because the *pile*, as the reactor was called, became critical before it was finished. This first reactor, designed and assembled by Enrico Fermi and his staff, was known as CP-1 and was put into operation on December 2, 1942. It was constructed in a squash court under the west stands of the University of Chicago's Stagg Field and operated at a power of $\frac{1}{2}$ watt until December 12, when the power was increased to 200 W. The reactor has since been dismantled, and a bronze sculpture has been erected to commemorate the event. The reactor was made by gradually building up graphite blocks. Uranium metal and uranium oxide were inserted into holes in the graphite blocks. Natural air circulation provided the cooling. Because of its construction methods, the reactor was called a pile. When it was completed it contained about 3200 uranium metal cylinders and 14 500 uranium oxide lumps, for a total weight of about 50 tons of uranium and 470 tons of graphite. There were five cadmium-coated control rods.

Now that it was certain that a chain reaction could be effected, it was decided to use the chain reaction to produce plutonium for atomic bombs. It was necessary to separate the plutonium from the uranium by chemical means. To study reactor operation and isotope separation, a graphite-moderated natural uranium–fueled reactor was built in Oak Ridge, Tennessee. It had forced-air cooling and was designed with an initial power of 1000 kW. It was called the X-10 or Clinton pile and subsequently operated at a power of about 4000 kW. It went into operation November 4, 1943, and was used for extensive production of isotopes and research purposes and the early studies of plutonium production. It was shut down permanently in November 1963.

Hanford, Washington, was selected as the site for large-scale production reactors. Originally, the plan was to use helium as the coolant, but because of problems in supplying helium the coolant was changed to water. Construction of the first water-cooled, graphite-moderated reactor was started in June 1943, and criticality was achieved in September 1944. Several similar reactors were built at the same location. In the early 1950s more efficient heavy water-moderated plutonium production reactors were built at Savannah River, South Carolina.

There were so many uncertainties in the production of plutonium at Hanford and in the subsequent construction of a bomb that it was decided to

undertake the mechanical separation of ^{235}U from ^{238}U. A gaseous diffusion plant was constructed at Oak Ridge, Tennessee. The plant started operation in June 1945. Its operation is based on the principle that more of the lighter molecules of UF_6 will pass through a porous barrier. The molecules on the far side of the barrier become slightly enriched in the lighter gas ($^{235}UF_6$). It takes approximately 4000 stages of diffusion to produce 99 percent pure ^{235}U. A second plant utilizing a mass spectrograph method was also built at Oak Ridge. The uranium is ionized, collimated, and focused into a magnetic field. Since the ^{235}U ions are lighter, they are deflected more by the magnetic field. Thus, the lighter ^{235}U ions are separated from heavier ^{238}U ions. For more information concerning isotope separation, References 3 and 4 may be consulted.

Nuclear reactor technology and isotope technology are still in their beginning stages. Much more design and development will be done before the full potential of reactor technology is realized. Quantitatively, progress has been very rapid, with the construction of well over 100 reactors within the first 20 years of reactor development. Table 1.2 lists in chronological order some of the early reactors and their significant characteristics. There are a multitude of possible combinations of fuel, moderator, coolant, reflector, and so on, for reactor construction. So far no ideal combination has evolved, even though 50 or more combinations have been studied. There are both proponents and opponents of different styles; each country seems to have followed lines determined by its own experience. For example, the United States has favored pressurized water and boiling water reactors using slightly enriched uranium, whereas the United Kingdom has favored gas-cooled reactors. Canada has concentrated on heavy water–moderated natural uranium–fueled reactors.

The Nuclear Industry

The nuclear industry is an extremely complex overlapping of many diversified industries and services. It includes applications in every category of science and industry. It is convenient to break the nuclear industry down into three divisions: (a) output and uses of reactors, (b) fuels and materials; and (c) services and applications.

The useful output of reactors may be power, radiation, steam, or gas. These outputs are used for training, research, testing, production of isotopes, propulsion, public utilities, manufacturing, and in space vehicles.

Nuclear fuels and materials may be the most complex of the three divisions because they involve not only the fissionable and fertile materials, but all materials that are associated with the fission process. This division is composed of exploration, mining, milling, refining, and fabrication. In addition, there are enrichment, processing, and the development of special materials.

By services and applications are meant fuel reprocessing, reactor hazards, licensing, and waste disposal, plus the myriad applications of reactor output. Applications of reactor output can be identified in agriculture, medicine, radiation source preparation, explosives, fusion power, industry, and public relations.

It can easily be seen that nearly every industry and service organization in the world can make use of radiation advantageously.

References

1. Putnam, P. C., *Energy in the Future.*   Princeton, N. J.: D. Van Nostrand Company, 1953.
2. Brown, H., et al., *The Next Hundred Years.*   New York: The Viking Press, Inc., 1957.
3. Smyth, H. D., *Atomic Energy for Military Purposes.*   Princeton, N.J.: Princeton University Press, 1945.
4. Benedict, M., *Separation of Stable Isotopes*, TID-5031, U.S. Atomic Energy Commission, 1951.
5. "Regulation of Nuclear Power Reactors and Related Facilities," *Nuclear Safety* 15, No. 1 (January–February 1974), pp. 1–13.
6. U.S. Bureau of the Census, *Statistical Abstract of the United States*, Washington, D.C., 1969.
7. Glasstone, S., *Sourcebook on Atomic Energy*, 2nd ed.   Princeton, N.J.: D. Van Nostrand Company, 1958.
8. Kaufmann, A. R., ed., *Nuclear Reactor Fuel Elements: Metallurgy and Fabrication.* New York: John Wiley & Sons, Inc., 1962.
9. *Nuclear News*, published monthly by American Nuclear Society, Inc., Chicago.
10. *Reactor Handbook*, rev. ed., U.S. Atomic Energy Commission, 1962.
11. Kulcinski, G. L., et al., "Energy for the Long Run: Fission or Fusion?" *American Scientist* 67 (January–February 1979), pp. 78–89.
12. Benedict, M., and A. Pigford, *Nuclear Chemical Engineering.*   New York: McGraw-Hill Book Company, 1957.
13. Bethe, H. A., "The Necessity of Fission Power," *Scientific American* 234, No. 1 (January 1976), pp. 21–31.
14. *TMI-2 Lessons Learned: Task Force Final Report*, NUREG-0585, U.S. Nuclear Regulatory Commission, October 1979.
15. *Proceedings of the Third United Nations International Conference on the Peaceful Uses of Atomic Energy*, Geneva, 1965.
16. *Thermal Effects and U.S. Nuclear Power Stations*, U.S. Aromic Energy Commission, 1971.
17. Jordan, W. H., "The Issues Concerning Nuclear Power," *Nuclear News* 14, No. 10 (October 1971), pp. 43–49.
18. Weinberg, A. M., "The Moral Imperatives of Nuclear Energy," *Nuclear News* 14, No. 12 (December 1971), pp. 33–37.
19. Jaffe, L., *Technical Staff Analysis Report Summary*, President's Commission on the Accident at Three Mile Island, October 1979.
20. *Plan for the Management of AEC—Generated Radioactive Wastes*, U.S. Atomic Energy Commission, 1972.

21. *The Nation's Energy Future*, U.S. Atomic Energy Commission, 1973.
22. *The Safety of Nuclear Power Reactors and Related Facilities*, U.S. Atomic Energy Commission, 1973.
23. *Nuclear Power Growth*, U.S. Atomic Energy Commission, 1974.
24. *Rules and Regulations*, U.S. Nuclear Regulatory Commission, Title 10, Chapter 1, Code of Federal Regulations—Energy.
25. *The Accident at Three Mile Island*, Report of the President's Commission on the Accident at Three Mile Island, October 1979.
26. *NRC Action Plan Developed as a Result of the TMI-2 Accident*, NUREG-0660, U.S. Nuclear Regulatory Commission, August 1980.
27. *Nuclear Accident and Recovery at Three Mile Island*, Subcommittee on Nuclear Regulation, U.S. Senate, June 1980.

2

Atomic Structure

The Atom

The atom is an assemblage of neutrons and protons tightly clustered in a nucleus and surrounded by electrons whirling in a variety of orbits. The protons are positively charged particles, each having a unit charge exactly opposite that of a negatively charged electron. The mass of the proton is 1836 times that of an electron. Neutrons have no electrical charge and a mass just slightly larger than that of a proton. Outside the nucleus a neutron cannot exist alone; it is unstable and will decay into a proton and an electron. Figure 2.1 shows a schematic arrangement of a helium atom with two protons and two neutrons in the nucleus, plus two orbiting electrons.

A particular atom may be designated by its atomic number, Z, which represents the number of protons present, and by its mass number, A, which is equal to the number of neutrons, N, plus the number of protons.

$$A = N + Z \tag{2.1}$$

It is possible for an element to have various values of the mass number, A, for a given number of protons, Z. These differing atoms of the same element are known as *isotopes*. Although they may behave similarly in chemical reactions, their nuclear characteristics may differ markedly. An example of this is found in natural uranium, whose three isotopes are listed below with their abundances (*abundance* is the atom percentage of an isotope present in a naturally occurring mixture).

$$^{234}_{92}\text{U} \qquad 0.006\%$$
$$^{235}_{92}\text{U} \qquad 0.714\%$$
$$^{238}_{92}\text{U} \qquad 99.28\%$$

Note that an isotope may be designated by its chemical symbol with the Z value as a presubscript and the mass number, A, as a presuperscript, $^{A}_{Z}X$.

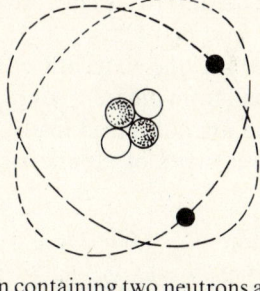

Fig. 2.1 Helium atom containing two neutrons and two protons in the nucleus. Two electrons are in orbit outside the nucleus.

Enrichment is an artificially increased abundance. In light-water reactors, enrichments of 2 to 5 percent are common, and in fast reactors the enrichment is greater, 15 to 25 percent.

The difference in nuclear characteristics of two isotopes is illustrated by ^{238}U, which will not fission by absorption of a slow (thermal) neutron, whereas ^{235}U fissions readily when a thermal neutron is absorbed.

Some elements have only a single stable isotope, such as ^{23}Na, ^{27}Al, and ^{9}Be. Other elements have many stable isotopes. For example, molybdenum has 7, cadmium has 8, and tin has 10. It is interesting to observe that nuclei with even numbers of protons and neutrons have a better probability of stability, as indicated by Table 2.1.

Hydrogen and its isotopes are of particular interest since $^{1}_{1}$H, light hydrogen, is a common reactor moderating material either in water or in an organic compound. Deuterium is heavy hydrogen formed when a neutron joins the proton in the nucleus to form $^{2}_{1}$H. It is also designated as $^{2}_{1}$D. A second neutron addition produces tritium, $^{3}_{1}$H (or $^{3}_{1}$T). Deuterium has much less probability of neutron absorption than light hydrogen, thus making heavy water an effective moderating medium. Tritium becomes important when fusion reactions are considered.

Two other terms are used to describe related nuclear species (nuclides): (a) *Isobars* are nuclides having the same mass number (A), but different numbers of protons (Z). In radioactive decay the ejection of an electron from the nucleus forms an isobar of the parent. (b) *Isotones* are nuclides with like numbers of neutrons.

Table 2.1 Stable Isotopes

		Number of Neutrons	
		Odd	Even
Number of Protons	Odd	8	52
	Even	56	167

The physical atomic mass unit (u) is equal to one-twelfth of the mass of a $^{12}_{6}C$ atom. This is equal to $1.660\,438 \times 10^{-27}$ kg. Formerly, the physical atomic mass unit was based on one-sixteenth of the mass of an atom of the $^{16}_{8}O$ isotope. In selecting data from different sources, caution must be exercised that ^{12}C and ^{16}O based data are not mixed inadvertently.

The masses for the three types of atomic particles based on the carbon system are listed below.

Proton	1.007 276 u
Neutron	1.008 665 u
Electron	$5.485\,80 \times 10^{-4}$ u

Avogadro's hypothesis states that equal volumes of gases at the same pressure and temperature contain equal numbers of molecules. This permits the determination of the relative masses of various molecules. Thus, when the molecular mass of a particular type of molecule is expressed in grams, it represents a definite number of molecules (or atoms of a monatomic substance). This is known as *Avogadro's number*, which is $6.022\,52 \times 10^{23}$ molecules per gram mole (or atoms per gram atom).

The number density, N, of atoms (or molecules) may be found by using the density of the material, ρ, its atomic (or molecular) mass, and Avogadro's number:

$$N = \frac{\rho \times 6.023 \times 10^{23}}{\text{atomic mass}} \qquad (2.2)$$

Example 1a If natural uranium has a density of 19.0 g/cm³, find the number density of ^{235}U atoms present.

$$N_{235} = \frac{19.0 \text{ (g U/cm}^3) \times 6.023 \times 10^{23} \text{ (atoms U/g atom U)}}{238 \text{ (g U/g atom U)}}$$

$$\times 0.007\,14 \text{ (atoms } ^{235}U/\text{atom U)}$$

$$= 3.43 \times 10^{20} \text{ atoms } ^{235}U/\text{cm}^3$$

Example 1b If a 3 percent enriched uranium–aluminum alloy contains 10 mass percent U and 90 mass percent Al and has a density of 3.0 g/cm³, compute the number density of ^{235}U atoms present.

$$A_U = 0.03(235.0439) + 0.97(238.0508) = 238.0 \frac{\text{g U}}{\text{g atom U}}$$

$$N_{235} = 3.0 \text{ (g alloy/cm}^3) \times 0.1 \text{ (g U/g alloy)}$$

$$\times \frac{6.023 \times 10^{23} \text{ (atoms U/g atom U)}}{238 \text{ (g U/g atom U)}} \times 0.03 \text{ atom } ^{235}U/\text{atom U}$$

$$= 2.278 \times 10^{19} \text{ atoms 235 U/cm}^3$$

Note that in Example 1a, where the number density of the uranium as a whole has been multiplied by the abundance of ^{235}U, the mass number used is for the composite material. When a mass percent is given, the density is multiplied by the mass percent of the particular element and the atomic mass used is for that element, as illustrated by Example 1b.

Binding Energies

If the masses of the neutrons and the protons making up the nucleus of an atom are added, the total will exceed the experimentally determined mass for that nucleus. This loss in mass is due to its conversion to binding energy in accordance with Einstein's famous equation

$$\Delta E = \Delta mc^2 \tag{2.3}$$

where ΔE represents the binding energy, Δm is the loss in mass or *mass defect*, and c is the speed of light. This same amount of energy would need to be supplied to the nucleus to separate all the nucleons.

In evaluating the mass defect, note that the isotopic masses given in Appendix A are given for the complete atom. To obtain the mass of the nucleus, the electron masses must be subtracted from the isotopic mass, M_x.

$$\Delta m = Z(m_p) + N(m_n) - (M_x - Zm_e) \tag{2.4a}$$

If the electron and proton masses are combined, the mass of the hydrogen atom is approximately correct in the following alternative relation. It is in error only by the amount of the electron binding energy of a few electron volts.

$$\Delta m = Z(m_H) + N(m_n) - M_x \tag{2.4b}$$

Example 2. Compute the mass defect and the binding energy per nucleon for $^7_3$Li. From Appendix A the isotopic mass of ^{7}Li is 7.016 01 u. Using Eq. (2.4a), we obtain

$$\Delta m = 3 \times 1.007\,277 + 4 \times 1.008\,665 - (7.016\,01 - 3 \times 0.000\,549)$$
$$= 0.042\,13 \text{ u}$$

From this mass defect the total binding energy is determined using Eq. (2.3).

$$\Delta E = \Delta mc^2 = 0.042\,13 \text{ u} \times 1.660\,438 \times 10^{-27} \text{ kg/u}$$
$$\times (3.0 \times 10^8)^2 \text{ m}^2/\text{s}^2 \times (1/1.602\,10 \times 10^{-13})$$
$$\text{MeV/J} \times 1\frac{\text{kg m}^2}{\text{s}^2 \text{ J}} = 39.3 \text{ MeV}$$

The binding energy per nucleon is then

$$BE/\text{nucleon} = \frac{39.3}{7} = 5.61 \text{ MeV/nucleon}$$

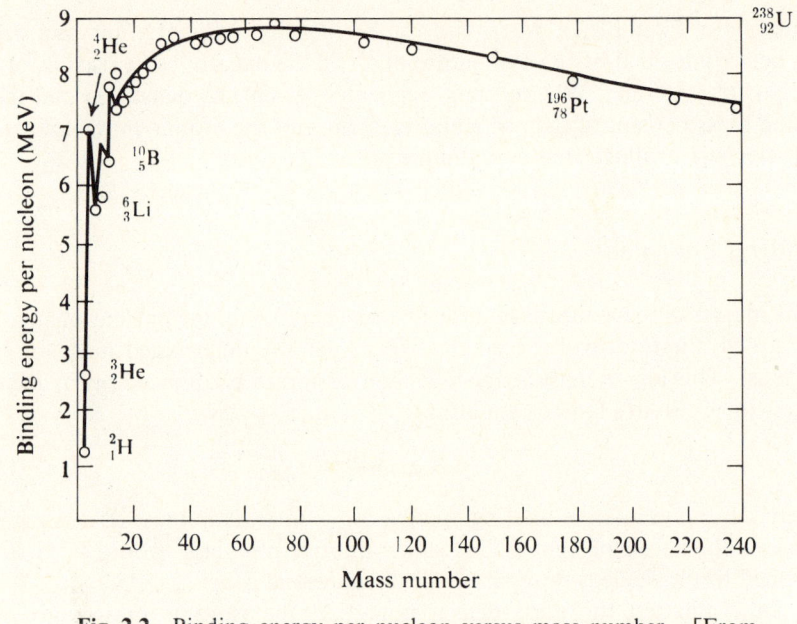

Fig. 2.2 Binding energy per nucleon versus mass number. [From G. Cahen and P. Treille, *Nuclear Engineering*, trans. G. B. Melese. Boston: Allyn and Bacon, Inc., 1961.]

Figure 2.2 shows the variation in binding energy per nucleon with mass number. The magnitude of the binding energy per nucleon increases somewhat erratically from zero for ^1_1H to a rather flat peak in the vicinity of $A = 56$ (iron), where the value is approximately 8.7 MeV/nucleon. It then falls slowly to a value of about 7.5 MeV/nucleon for the uranium isotopes.

Both the *fission* process, where a heavy nucleus splits into two fragments, and the *fusion* process, where two light nuclei are joined to form a heavier nucleus, tend to move the elements formed toward the region of greater stability. The resultant energy release of these processes accounts for the development of nuclear power as a reality and the harnessing of fusion energy as a hope for the future.

As mass numbers become larger, the ratio of neutrons to protons in the nucleus becomes larger. For ^4He and ^{16}O this ratio is unity. For ^{115}In it has increased to a value of 1.35 and for ^{238}U it is 1.59. This variation is shown by Fig. 2.3. When the heavy uranium nucleus splits into two fission fragments, each will have an excess of neutrons, accounting for the intense radioactivity of these species in irradiated nuclear fuels. Radioactive decay is discussed in detail in Chapter 3.

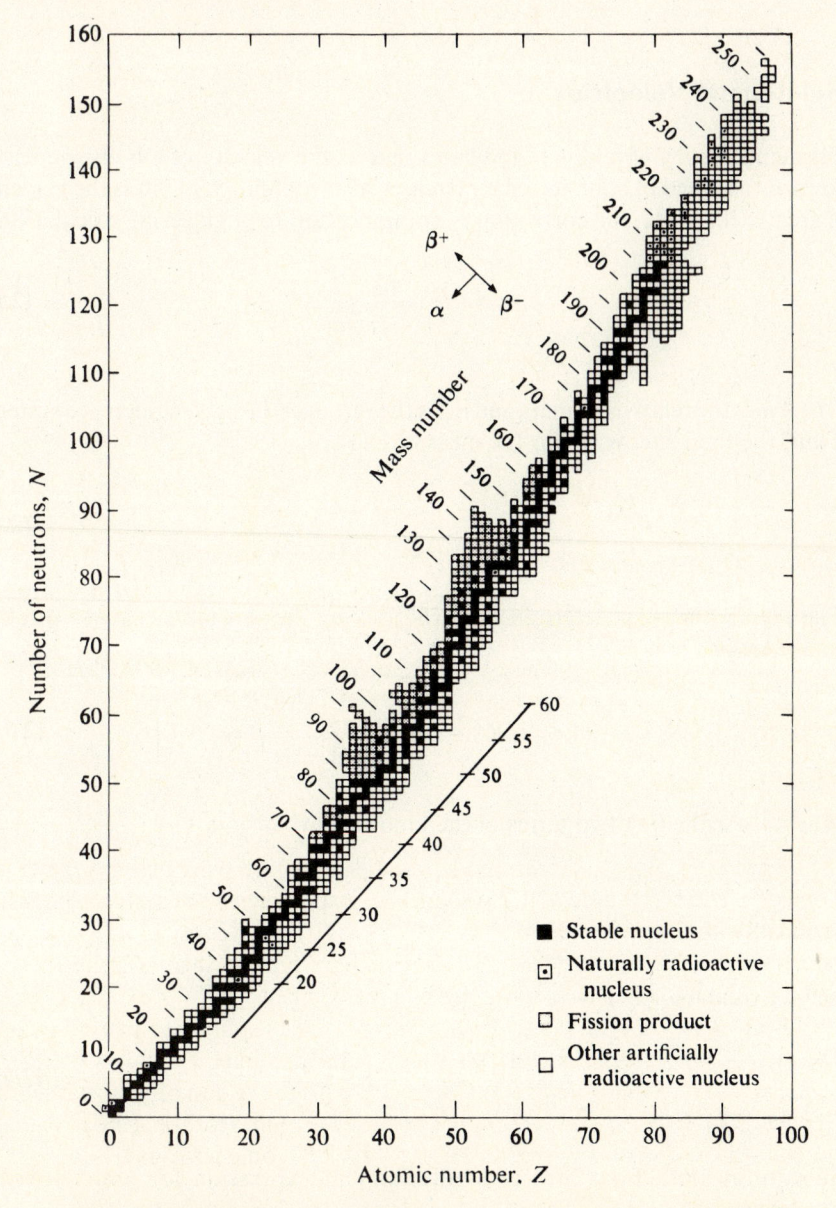

Fig. 2.3 Distribution of stable and radioactive nuclides. [From G. Cahen and P. Treille, *Nuclear Engineering*, trans. G. B. Melese. Boston: Allyn and Bacon, Inc., 1961.]

Relativistic Velocities

Einstein's theory of relativity indicates that as the velocity of a body increases toward the speed of light, its mass increases toward infinity. Unless the velocity is greater than $0.1c$, the correction is not important to engineering calculations.

$$m = \frac{m_0}{\sqrt{1 - (v^2/c^2)}} \tag{2.5}$$

where m is the relativistic mass and m_0 is the rest mass in a given reference system. Thus, the total energy, E, for the mass is

$$E = mc^2 = \frac{m_0 c^2}{\sqrt{1 - (v^2/c^2)}} \tag{2.6}$$

The kinetic energy is determined by subtracting the rest mass energy from the total energy.

$$KE = (m - m_0)c^2 = m_0 c^2 \left[\frac{1}{\sqrt{1 - (v^2/c^2)}} - 1 \right] \tag{2.7a}$$

When $v \ll c$ the first two terms of the binomial expansion

$$\left(1 + \frac{1}{2}\frac{v^2}{c^2} + \cdots \right)$$

may be used for $1/\sqrt{1 - (v^2/c^2)}$.

$$\frac{1}{\sqrt{1 - (v^2/c^2)}} = 1 + \frac{1}{2}\frac{v^2}{c^2} + \cdots \tag{2.7b}$$

Substituting into Eq. (2.7a) gives the familiar expression for nonrelativistic kinetic energy:

$$KE_{nonrel} = m_0 c^2 \left(1 + \frac{1}{2}\frac{v^2}{c^2} - 1 \right)$$

$$= \frac{m_0 v^2}{2} \tag{2.7c}$$

$$v_{rel} = c\sqrt{1 - \left(\frac{KE}{m_0 c^2} + 1 \right)^{-2}} \qquad v_{nonrel} = \sqrt{\frac{2KE}{m_0}}$$

Example 3 If an electron travels at 90 percent of the speed of light, compute its mass and its kinetic energy.

$$m = \frac{m_0}{\sqrt{1 - (v^2/c^2)}} = \frac{0.000\ 549}{\sqrt{1 - (0.9)^2}} = 0.001\ 263\ u$$

$$KE = (m - m_0)c^2 = (0.001\ 263 - 0.000\ 549)\ u \times 1.660\ 438$$

$$\times 10^{-27}\ kg/u \times (2.997\ 925 \times 10^8)^2\ m^2/s^2$$

$$\times \frac{1}{1.602\ 10 \times 10^{-13}} \frac{MeV}{J} \times \frac{1\ J\ s^2}{kg\ m^2}$$

$$= 0.660\ MeV$$

Note that it is convenient to combine the square of the constant speed of light and the conversion constants just used into a single constant, 931.482 MeV/u. This is the energy equivalent of the atomic mass unit.

The large particle accelerators, such as that at the Fermi National Accelerator Laboratory, produce particle velocities approaching the speed of light. The Fermi Accelerator is designed to produce 400-GeV protons with a velocity 99.999 726 percent of the speed of light.

Energy Levels in an Atom

Electromagnetic radiation is given off as an electron drops from a higher energy level to a lower atomic energy level. The first theory which successfully accounted for the certain discrete energy levels of this radiation was proposed by Nils Bohr. Although the *Bohr theory* of the atom has been supplanted by solutions to the Schrödinger wave equation, it is still useful in understanding and approximating the energy emitted (or absorbed) as an electron jumps from one energy level to another.

The energy emitted as an electron changes its state is equal to the product of Planck's constant, h, and the frequency of the radiation, v.

$$E_2 - E_1 = hv \tag{2.8}$$

Here E_2 represents the final electron energy and E_1 the initial electron energy. Such an amount of discrete energy emission (or absorption) is called a *photon*.

The electromagnetic radiation travels at the speed of light and this velocity is equal to the product of the frequency of the radiation times its wavelength, λ.

$$c = v\lambda \tag{2.9}$$

The wave number is the reciprocal of the wavelength and is denoted by $\bar{v}$.

$$\bar{v} = \frac{1}{\lambda} = \frac{v}{c} \tag{2.10}$$

The force of attraction between the electron and the positively charged nucleus is balanced by the centrifugal force of the electron as it whirls in an orbit about the nucleus.

$$\frac{Ze^2}{r^2} = \frac{m_e V^2}{r} \qquad (2.11)$$

Z represents the number of protons in the nucleus, e the unit electronic charge, and r the orbital radius; m_e is the mass of the electron and V its velocity.

The angular momentum ($m_e Vr$) of an electron must be an integral multiple of $h/2\pi$. This in effect fixes the radii at which the electron may travel.

$$m_e Vr_n = \frac{nh}{2\pi} \qquad (2.12)$$

where r_n is the radius of an allowable orbit and $n = 1, 2, 3, \dots$. Eliminating first the radius and then the velocity by combining Eqs. (2.11) and (2.12) allows one to solve for the orbital velocity and radius of an electron's orbit.

$$V = \frac{2\pi Ze^2}{nh} \qquad (2.13)$$

$$r_n = \frac{n^2 h^2}{4\pi^2 m_e Ze^2} \qquad (2.14)$$

Figure 2.4 shows the allowable electron orbits for the hydrogen atom. The electron in the hydrogen atom is shown as having just dropped from the third shell to the first with the emission of a photon. Since it drops from the third, rather than from the second orbit, to the first, the frequency is higher.

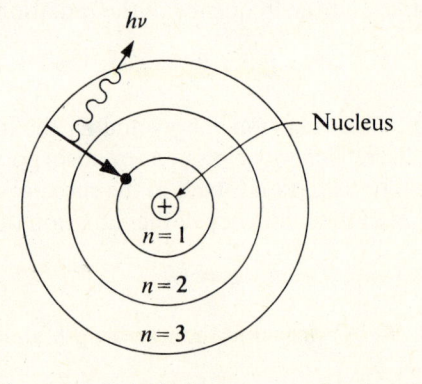

Fig. 2.4 Bohr's model of the hydrogen atom showing allowable electron orbits.

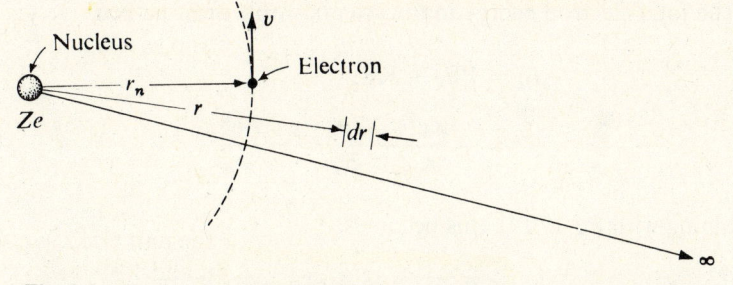

Fig. 2.5 Sketch showing relation of an electron in the *n*th allowable orbit to the nucleus of an atom.

To determine the energy of an electron in a given orbit, its kinetic and potential energies must be added.

Consider first that an electron completely divorced from a nucleus has zero potential energy. As a free electron at an infinite radius falls into an allowable orbit of radius r_n it will lose potential energy (Fig. 2.5). At radius r the force of attraction between the nucleus and the electron is

$$F = \frac{Ze^2}{r^2} \qquad (2.15a)$$

It will acquire a differential amount of potential energy $d\text{PE}$ as it moves a differential distance dr. Integrating between ∞ and r_n gives the potential energy for the electron.

$$\int_0^{PE_n} d\text{PE} = \int_\infty^{r_n} \frac{Ze^2}{r^2}\, dr$$

$$\text{PE}_n = -\frac{Ze^2}{r_n} \qquad (2.15b)$$

From Eq. (2.11)

$$V_n^2 = \frac{Ze^2}{r_n m_e}$$

and thus

$$\text{KE}_n = \frac{m_e V_n^2}{2} = \frac{Ze^2}{2r_n} \qquad (2.15c)$$

The total electron energy in the nth allowable orbit is then

$$E_n = PE_n + KE_n$$

$$= -\frac{Ze^2}{r_n} + \frac{Ze^2}{2r_n} = -\frac{Ze^2}{2r_n} \tag{2.15d}$$

Combining with Eq. (2.14), this becomes

$$E_n = -\frac{2\pi^2 Z^2 m_e e^4}{n^2 h^2} \tag{2.15e}$$

The change in energy from one orbit, n_1, to another, n_2, is

$$\Delta E = \left(\frac{1}{n_1^2} - \frac{1}{n_2^2}\right)\frac{2\pi^2 Z^2 m_e e^4}{h^2} \tag{2.16}$$

When an electron falls from an outer orbit to an inner one, ΔE is negative and energy is released as a photon. If it jumps to a higher orbit, energy is absorbed.

Example 4   Compute the radius of the innermost electron orbit in a carbon atom. What will be the energy of the photon given off if a free electron ($n = \infty$) falls into the innermost orbit?

$$r_1 = \frac{n^2 h^2}{4\pi^2 m_e Ze^2}$$

$$= \frac{1^2 \times (6.626 \times 10^{-34})^2 \text{ J}^2 \text{ s}^2 \times 1 \text{ kg}^2 \text{ m}^4 \text{ s}^{-4} \text{ J}^{-2}}{4\pi^2 \times 5.4858 \times 10^{-4}\, u \times 1.660\,438 \times 10^{-27} \text{ kg u}^{-1}}$$
$$\times (4.80 \times 10^{-10})^2 \text{ esu}^2$$

$$\times \left(\frac{1}{3.1623 \times 10^{-5}}\right)^2 \frac{\text{esu}^2 \text{ s}^2}{\text{kg m}^3}$$

(handwritten: divide by Z=6)

$$= 8.832 \times 10^{-12} \text{ m } (0.088\,32 \text{ Å})$$

$$\Delta E = \left(\frac{1}{n_1^2} - \frac{1}{n_2^2}\right)\frac{2\pi^2 Z^2 m_e e^4}{h^2}$$

$$= \left(\frac{1}{\infty^2} - \frac{1}{1^2}\right)\frac{2\pi^2 \times 6^2 \times 5.485 \times 10^{-4}\, u \times 1.660\,438 \times 10^{-27} \text{ kg u}^{-1}}{(6.6252 \times 10^{-27})^2 \text{ kg}^2 \text{ m}^4 \text{ s}^{-2}}$$

(handwritten: −34 over the exponent; −10)

$$\times \frac{(4.80 \times 10^{10})^4 \text{ esu}^4 \times (3.1623 \times 10^{-5}) \text{kg}^{1/2} \text{ m}^{3/2} \text{ s}^{-1}}{1.602\,10 \times 10^{-19} \text{ kg m}^2 \text{ s}^{-2} \text{ eV}^{-1}}$$

$$= -489 \text{ eV}$$

Wave mechanics shows that electrons are not always at a fixed distance from the nucleus, as predicted for the Bohr atom. Their location may be more properly described by a position probability curve. For an electron in a particular energy state, this plots the probability of being at a given radius versus distance from the center of the nucleus. Figure 2.6 shows such curves for the hydrogen atom.

To describe completely the energy state of an electron, four quantum numbers are required, rather than just the one used in Bohr theory. The principal quantum number, n, indicates the most probable radius of an electron orbit. It may have any integral value from 1 to ∞. The orbital quantum number, l, indicates the angular momentum of the electron and the eccentricity of the orbit. It may have values from 0 to $(n - 1)$. The magnetic orbital quantum number, m, indicates the "plane" of the electron's orbit about the nucleus. It may have values from $-l$ to $+l$. The spin number, s, has only two possible values for an electron, $\pm 1/2$.

Figure 2.6 shows for $_1^1$H the probability of finding an electron at a given radius for values of the principal quantum number $n = 1$, 2, or 3 and for the orbital quantum number $l = 0$, 1, or 2. From this figure it may be observed that the most probable radius is related to n. Electrons with a particular value of the principal quantum number are said to be in the same shell. The *Pauli exclusion principle* stipulates that no two electrons in a given atom may have the same set of quantum numbers. Therefore, the maximum number of electrons is said to be *filled*. The electrons in an unfilled outer shell are the valence electrons. Thus, magnesium with its twelve electrons has two in the first shell,

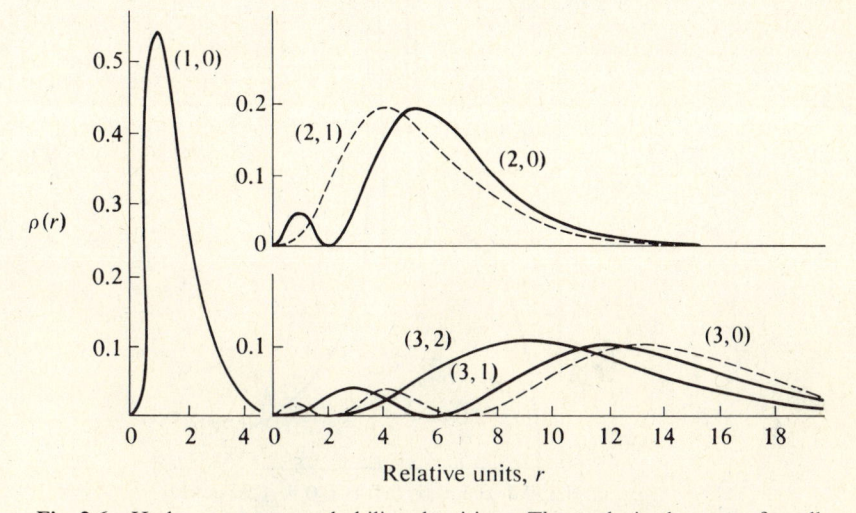

Fig. 2.6 Hydrogen atom-probability densities. The scale is the same for all curves. [From K. S. Pitzer, *Quantum Chemistry*. Englewood Cliffs, N.J.: Prentice-Hall, Inc., 1953. Reprinted by permission.]

eight in the second, and two valence electrons in the third shell. Elements with all shells filled, such as helium and neon, are inert. In heavier elements, if there are eight electrons in the outer shell, the configuration is very stable and valence may be figured as the difference between the electrons in the outer shell and eight.

In stable atoms, as protons are added to the nucleus, corresponding electrons are added in order of decreasing binding energies. Starting with the lightest elements, electrons fill first the energy states in the innermost shell, then those in the second shell, and so on. This takes the elements from an atomic number of 1 for hydrogen up to argon with a $Z = 18$ (3,1 state). After this point the 4,0 state will have a lower binding energy than the 3,2 state; thus, the transition elements will have the 4,0 state filled before electrons complete the third shell. Many important metals are included in these transition elements. Included are V, Cr, Mn, Fe, Co, and Ni. There is a similar premature filling of the outer shells in many of the heavier elements.

X-Rays and Bremsstrahlung

X-rays are electromagnetic radiation produced when energetic electrons interact with matter. In an X-ray tube, the electrons are emitted by a heated cathode whose potential may be the order of 30 000 to 50 000 V above the

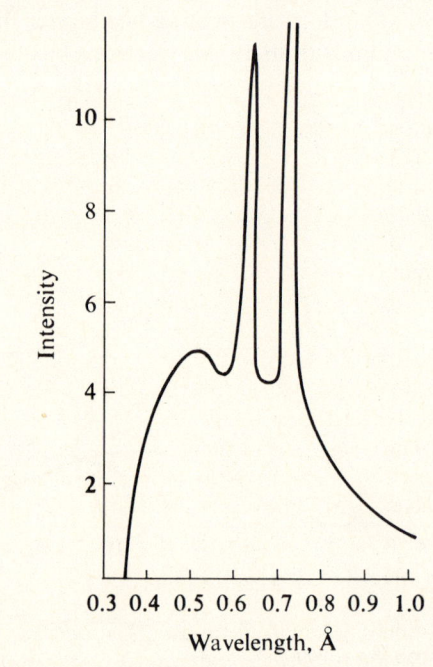

Fig. 2.7   X-ray spectrum for a molybdenum target operated at 35 000 V.

target, which is made of a material such as tungsten or molybdenum. X-rays are emitted by the target, which stops the electrons. They form a continuous spectrum with wavelengths longer than a minimum value dictated by the energy of the electrons as they strike the target. Superimposed on the continuous spectrum may be several sharp spikes. The continuous spectrum is due to *bremsstrahlung* or braking radiation. It is radiation emitted as electrons decelerate in the Coulomb fields of the target nuclei. The sharp peaks are caused by the electronic rearrangement which takes place when atoms are struck by energetic electrons. These characteristic X-rays are produced as electrons fall from excited energy states to lower energy levels. Figure 2.7 shows the X-ray spectrum for a molybdenum target operated with a 35 000-V potential. Two sharp peaks are superimposed on the continuous spectrum.

The X-ray spectrum for a tungsten target operated with a 35-kV potential would show no sharp peaks. It would have only a continuous spectrum since the electrons striking the target would lack the energy to raise the tungsten atoms to the lowest excited state. It would be necessary to double the operating voltage of the tube to produce the first two characteristic peaks for tungsten.

Atomic Bonding

The atomic bonding which holds molecules together and provides for the rigidity of solids may be one or a combination of the following:

(1) *Ionic bonding* occurs when there is an exchange of electrons where the donor becomes positively charged and the recipient is negatively charged. The bond is then electrostatic in nature. A simple example is found in magnesia (MgO). The magnesium ($Z = 12$) has two electrons in the third shell, which it gives up to an oxygen atom ($Z = 8$), which lacks two electrons to complete the second shell.

(2) *Covalent bonding* takes place when electrons are shared by the bound nuclei. The hydrogen molecule (H_2) illustrates this type of bond. The electrons are shared by the nuclei, in effect filling both inner shells.

(3) *Metallic bonding* leaves the valence electrons relatively free to wander through the lattice structure acting much like a gas. The negative charge of the free electrons and the net positive charge of the nucleus with its firmly bound electrons cause a mutual attraction binding the structure together.

(4) *Van der Waals bonds* are relatively weak bonds due to the nonsymmetrical distribution of electrons in atoms or molecules. This produces polarization and attraction of unlike charges. This accounts for the forces of attraction when an inert material such as helium or argon is solidified at cryogenic temperatures.

In many cases the bonding is not a simple matter of being merely one of the preceding types. A case in point is graphite, where in the basal planes of the hexagonal lattice the bonding is covalent but where van der Waals bonds exist between the planes.

Nuclear Structure and Binding Forces

In nature there are but four fundamental forces: electromagnetic forces; gravitational forces; the strong force, which accounts for the binding of neutrons and protons in the nucleus; and the weak force, which is involved with the decay of various particles. The first two are well known and understood, as their effects can both be observed and felt. However, the latter two operate over much shorter distances, as shown in Fig. 2.8. The range of gravitational and electromagnetic forces is unlimited, and their strength varies inversely as the square of the distance between two interacting bodies. The strong force operates at distances up to the nuclear radius and the weak force at distances no greater than one-hundredth of the nuclear radius.

The strong force not only accounts for the binding forces between nucleons, but also for the binding of subnucleon particles known as *quarks*. Each nucleon is a three-quark assembly. The strong force between nucleons drops pre-

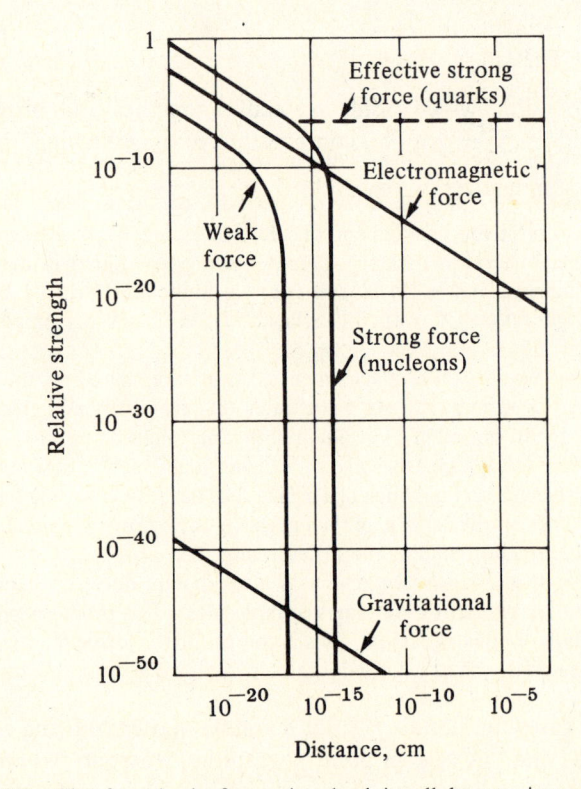

Fig. 2.8 The four basic forces involved in all known interactions between particles of matter.

cipitously at about 10^{-13} cm. However, between quarks it appears to remain constant at distances greater than 10^{-13} cm. It appears doubtful that a quark can ever be isolated as a separate entity outside a nucleon for observation. The nature of quarks, together with antimatter and various strange particles, is discussed further in Chapter 4.

To describe the state of the nucleus, a set of four quantum numbers is required for the protons and another set of four for the neutrons. One may infer from the previous discussion of the atom that when one of the nucleons drops to a lower energy state, electromagnetic radiation carries the energy away. This energy is transmitted by gamma rays, which are similar to X-rays except for their source. Gamma rays originate from the nucleus, whereas characteristic X-rays are due to a change in electronic configuration.

Figure 2.9 shows the energy levels for a ^{15}O nucleus. The lowest energy level is at 5.29 MeV. If the nucleus has a scattering collision with a neutron of less than 5.29 MeV, the process is elastic. However, if more than 5.29 MeV is imparted to the nucleus on the collision, a 5.29-MeV gamma will be given off. The remaining energy would be shared by the recoiling nucleus and the scattered neutron in such an inelastic scattering process.

For light elements the first excited state is normally several MeV above ground level. The interval between levels decreases to a few keV above 8 MeV and becomes nearly continuous above 15 MeV. Since the slowing down of neutrons is accomplished primarily by light atoms in the moderator of a thermal reactor, scattering may be considered elastic. This is so because the average energy of a fission neutron is only 2 MeV.

Fig. 2.9 Energy-level diagram for a ^{15}O nucleus.

For heavy elements inelastic scatter is more prevalent, since their energy levels are only about 0.1 MeV apart near ground level. The differences in energy levels decrease to a few eV at 8 MeV.

Problems

1. Compute the number density of atoms in the following cases:
   (a) Pure aluminum
   (b) ^{17}O in the atmosphere at 0.1 MPa and 20°C
   (c) Aluminum in an alloy of 90 mass percent Al and 10 mass percent Mg which has a density of 2.55 g/cm^3

2. Determine the binding energy per nucleon and the mass defect (u) for $^{232}_{90}Th$ and for $^{59}_{27}Co$.

3. The average KE of a fission neutron is 2 MeV. What is its velocity? Must it be considered relativistic?

4. Calculate the ratio of a particle's mass to its rest mass when the particle travels at the following fractions of the speed of light: 0.1, 0.5, 0.9, 0.999.

5. Compute the kinetic energy of the particle (MeV) in Prob. 4 for each of the velocities if it is (a) an electron; (b) a deuteron.*

6. Show that the relativistic momentum of a particle may be expressed as

$$p = (1/c)\sqrt{(KE)^2 + 2m_0 c^2 (KE)}$$

7. Show that the wave number for an X-ray may be expressed as

$$\bar{v} = RZ^2 \left(\frac{1}{n_1^2} - \frac{1}{n_2^2} \right)$$

where R represents the Rydberg constant.

$$R = \frac{2\pi^2 m_e e^4}{ch^3}$$

8. When a free electron ($n = \infty$) falls into the innermost shell of a chlorine atom, what will be the energy of the X-ray emitted and its wavelength?

9. Compute the velocity of an electron in the first shell (ground state) of a hydrogen atom. What would it be if it were in an excited state in the third shell? What would be the orbital radius for each of these shells and the energy of an X-ray emitted when the excited electron drops back to the ground state?

10. For a 100-kV X-ray machine using a tungsten target ($Z = 74$), compute the shortest wavelength due to bremsstrahlung and also the wavelength and energy for characteristic X-rays due to electrons dropping from the third and second shells to the first.

---

* A *deuteron* is the nucleus of a deuterium atom.

References

1. Kaplan, I., *Nuclear Physics*, 2nd ed. Reading, Mass.: Addison-Wesley Publishing Co., Inc., 1963.
2. Halliday, D., *Introductory Nuclear Physics*. New York: John Wiley & Sons, Inc., 1955.
3. Semat, H., *Introduction to Atomic and Nuclear Physics*. New York: Holt, Rinehart and Winston, 1962.
4. Frankel, J. P., *Principles of the Properties of Materials*. New York: McGraw-Hill Book Company, 1957.
5. Lapp, R. E., and H. L. Andrews, *Nuclear Radiation Physics*. Englewood Cliffs, N.J.: Prentice-Hall, Inc., 1963.
6. Liverhant, S. E., *Elementary Introduction to Nuclear Reactor Physics*. New York: John Wiley & Sons, Inc., 1960.
7. Cahen, G., and P. Treille, *Nuclear Engineering*, trans. by G. B. Melese. Boston: Allyn and Bacon, Inc., 1961.
8. Pitzer, K. S., *Quantum Chemistry*. Englewood Cliffs, N.J.: Prentice-Hall, Inc., 1953.
9. Kendall, H. W., and W. K. H. Panofsky, "The Structure of the Proton and the Neutron," *Scientific American* **224**, No. 6 (June 1971), pp. 60–77.
10. Arya, A. P., *Fundamentals of Nuclear Physics*. Boston: Allyn and Bacon, Inc., 1961.
11. Arya, A. P., *Fundamentals of Atomic Physics*. Boston: Allyn and Bacon, Inc., 1971.
12. Andrews, H. L., *Radiation Biophysics*. Englewood Cliffs, N.J.: Prentice-Hall, Inc., 1974.
13. Burcham, W. E., *Elements of Nuclear Physics*. New York: Longman, Inc., 1977.
14. 't Hooft, G., "Gauge Theories of the Forces Between Elementary Particles," *Scientific American* **242**, No. 6 (June 1980), pp. 104ff.
15. Marray, R. L., *Nuclear Energy*. Elmsford, N.Y.: Pergamon Press, Inc., 1980.
16. Choppin, G. R., and J. Rydberg, *Nuclear Chemistry*. Elmsford, N.Y.: Pergamon Press, Inc., 1980.

3

The Decay
of Radioactive
Nuclei

Radioactivity is due to the decay of unstable nuclei. These nuclei can be:

(1) Heavy elements such as uranium or thorium which have such a slow rate of decay that they have been present since their creation at the beginning of geological time.
(2) The daughter products of the initial heavy elements, which are, in turn, radioactive.
(3) Unstable fission products whose N/Z ratio is too high for stability.
(4) Unstable isotopes produced by particle bombardment. Neutron activation is particularly important in this regard. In a reactor the absorption of a neutron produces radiation effects that may be considered a nuisance. However, neutron activation can be used as a powerful analytical tool to identify unknown elements. Here the induced radiation spectrum can be examined to identify the decaying elements.

Natural Radioactivity

The heavy radioactive elements and their unstable daughters emit three types of radiation: alpha particles, beta particles, and gamma rays. The *alpha particle is a helium nucleus that has a double positive charge* due to its pair of protons, which combine with two uncharged neutrons to make the mass approximately 4 u. The naturally occurring *beta particle* is a negatively charged electron expelled by the nucleus, effectively converting a neutron to a proton. The *gamma ray* is electromagnetic radiation, as indicated in Chapter 2, which allows an excited nucleus to drop toward the ground state.

Figure 3.1 shows the series of elements formed as ^{232}Th and ^{235}U decay through their various daughter products to end up as stable ^{208}Pb and ^{207}Pb, respectively. Table 3.1 lists the half-lives and types of radiation emitted for each of the reactions in the actinium (^{235}U) decay chain.

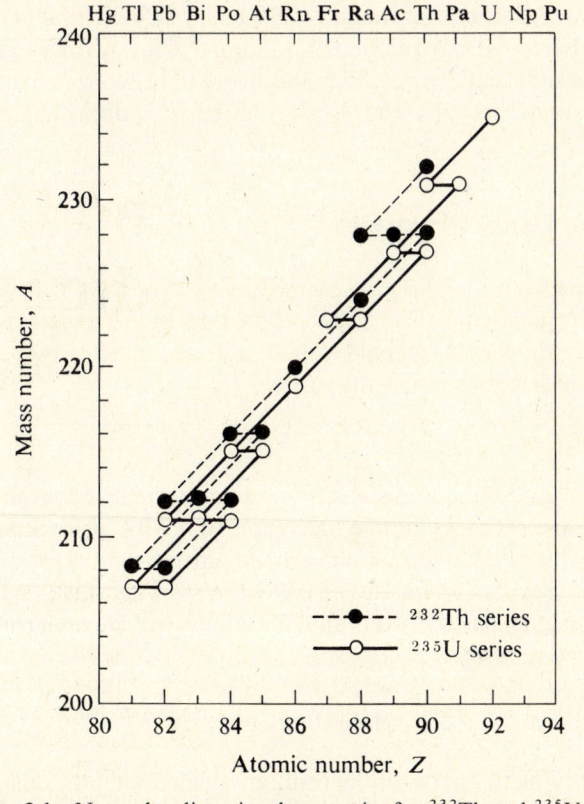

Fig. 3.1 Natural radioactive decay series for ^{232}Th and ^{235}U.

Table 3.1 Actinium Decay Series

Element	Half-Life	Type of Radiation (*Energy, MeV*)
$^{235}_{92}$U	7.07×10^8 yr	$\alpha(4.52)$, $\gamma(0.09)$
$^{231}_{90}$Th	25.6 h	β, $\gamma(0.03)$
$^{231}_{91}$Pa	3.4×10^4 yr	$\alpha(5.05)$, $\gamma(0.32)$
$^{227}_{89}$Ac	21.6 yr	$\alpha(5.0)$, $\beta(0.22)$
$^{227}_{90}$Th	18.1 days	$\alpha(6.05)$, γ
$^{223}_{88}$Ra	11.7 days	$\alpha(5.86)$, γ
$^{219}_{86}$Rn	3.29 s	$\alpha(6.82)$, γ
$^{215}_{84}$Po	1.83×10^{-3} s	$\alpha(7.36)$
$^{211}_{82}$Pb	36.1 min	$\beta(1.4)$, $\gamma(0.8)$
$^{211}_{83}$Bi	2.16 min	$\alpha(6.62)$, β, $\gamma(0.35)$
$^{211}_{84}$Po	0.5 s	$\alpha(7.43)$
$^{207}_{81}$Tl	4.76 min	$\beta(1.4)$, γ
$^{207}_{82}$Pb	Stable	

The fertile isotope of uranium, ^{238}U, has a half-life of 4.51 billion years. One of its important decay products is radium (^{226}Ra), with a 1620-yr half-life. It is present in sufficient quantity in uranium ore to be separated for medical and industrial usage.   The stable end product of the ^{238}U decay chain is ^{206}Pb.

Radioactive Decay Processes

Radioactive decay takes place in several ways.   In each of these processes the mass of the resultant particles is less than that of the parent nucleus.   The difference may show up as gamma radiation or as kinetic energy shared by the daughter nucleus and an emergent particle.

Alpha decay

When an unstable nucleus ejects an alpha particle, the atomic number is reduced by 2 and the mass number decreases by 4.   An example is ^{234}U, which may decay by the ejection of an alpha accompanied by the emission of a gamma, which in this case has an energy of 0.053 MeV.   This same isotope can also decay without gamma emission, and the alpha will have a correspondingly higher energy (see Appendix D).

$$^{234}_{92}\text{U} \rightarrow {}^{230}_{90}\text{Th} + {}^4_2\alpha + \gamma + \text{KE} \tag{3.1}$$

The combined kinetic energy of the resultant nucleus and the emergent particle (in this case the ^{230}Th and the α) is designated as KE.   The sum of the kinetic energy and the gamma energy is equal to the difference in mass between the original nucleus and the final particles.   This total mass–energy conversion is called Q.

$$Q = \text{KE} + \gamma = (m_\text{U} - m_\text{Th} - m_\text{He}) \times 931 \tag{3.2}$$

Substituting yields

$$Q = (234.0409 - 230.0331 - 4.002\,60) \times 931 = 4.84 \text{ MeV}$$

This represents the total mass–energy conversion for the process.   The kinetic energy shared by the particles is

$$\text{KE} = Q - \gamma = 4.84 - 0.053 = 4.79 \text{ MeV}$$

The disintegration of the parent ^{234}U nucleus is shown in Fig. 3.2.   The radioactive nucleus is considered to be stationary.   The momenta of the

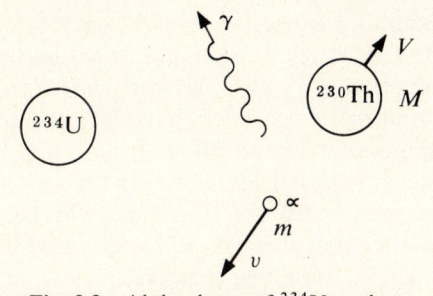

Fig. 3.2   Alpha decay of ^{234}U nucleus.

emergent particle and the recoiling daughter nucleus are therefore equal and opposite.

$$MV + mv = 0 \qquad\qquad (3.3a)$$

$$V = -(m/M)v \qquad\qquad (3.3b)$$

Here M and V represent the mass and velocity of the recoiling daughter and m and v represent the mass and velocity of the emergent alpha. The kinetic energy of the two particles is

$$KE = (M/2)V^2 + (m/2)v^2$$

$$= \left(\frac{m+M}{M}\right)\frac{mv^2}{2} = \left(\frac{m+M}{M}\right)KE_\alpha \qquad\qquad (3.4a)$$

The kinetic energy of the alpha is then

$$KE_\alpha = \left(\frac{M}{m+M}\right)KE \qquad\qquad (3.4b)$$

The alpha emitted by the ^{234}U will have a KE as follows:

$$KE_\alpha = (230/234)4.79 = 4.71 \text{ MeV}$$

Alpha radiation is not very penetrating, being effectively stopped by a piece of paper. Thus, external exposure to alpha radiation presents little hazard, but ingestion of an alpha emitter can be very serious. The radioactive material can collect preferentially in various organs of the body and cause great harm.

Beta emission

Beta particles are electrons that have been expelled by excited nuclei. They can have a charge of either sign. If both energy and momentum are to

be conserved, a third type of particle, the *neutrino*, v, is involved. The neutrino is associated with positive electron emission and its antiparticle, the *antineutrino*, $\bar{v}$, is emitted with a negative electron. They are similar in all respects, except that the spin vector is in the same direction as the direction of motion for the neutrino, while the spin vector for the antineutrino is opposite to the direction of motion. These uncharged particles have only the very weakest interactions with matter, their rest mass is zero, and they travel with the speed of light. For all practical purposes they pass through all materials with so few interactions that their energy is lost (i.e., cannot be recovered).

Negative electron emission effectively converts a neutron to a proton, thus increasing Z by 1 and leaving A unchanged. This is a common mode of decay for nuclei with an excess of neutrons, as with fission fragments on the high side of the N/Z stability line (refer again to Fig. 2.3). Velocities of the ejected electrons are often high enough to be a considerable fraction of the speed of light. This requires that they be treated in a relativistic fashion. There is a spectrum of beta energy values below some maximum. At the maximum beta energy the beta particle accounts for the total Q value. At lower energies the mass–energy conversion is shared with the antineutrino. The decay of ^{32}P illustrates the negative beta decay process.

$$^{32}_{15}P \rightarrow {}^{32}_{16}S + {}^{0}_{-1}e + \bar{v} \tag{3.5}$$

The average electron energy is about one-third of the maximum. The intensity of a stream of $^{32}_{15}P$ betas will be attenuated 50 percent by a thickness of 0.1 mm of aluminum.

Positively charged electrons are known as *positrons*. Except for sign, they are identical with their negatively charged cousins. When a positron is ejected from the nucleus, Z is decreased by 1 and A remains unchanged. A proton has been converted to a neutron. The decay of ^{13}N illustrates this process.

$$^{13}_{7}N \rightarrow {}^{13}_{6}C + {}^{0}_{+1}e + v \tag{3.6a}$$

In this case the neutrino shares with the positron the loss in mass which has been converted to energy. In the alpha decay and in the β^- decay discussed previously, atomic masses, rather than nuclear masses, could be used for computing the mass change since the electron masses canceled out. With positron emission this is not so, as is shown by the following example.

$$Q = [(m_N - 7m_e) - (m_C - 6m_e) - m_e] \times 931$$
$$= [(m_N - m_C) - 2m_e] \times 931 \tag{3.6b}$$

Note that an orbital electron is released in the conversion from nitrogen to carbon. This or an equivalent electron will meet later and annihilate the

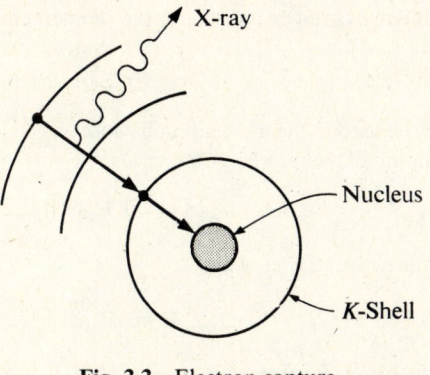

Fig. 3.3   Electron capture.

positron by producing two gammas with a combined energy equal to the rest mass of the $\pm$ electrons ($2 \times 0.000\ 549 \times 931 = 1.02$ MeV).

Electron capture

Nuclei having an excess of protons but lacking energy for positron emission can move an electron from one of the inner orbits (usually the K shell) into the nucleus, as shown in Fig. 3.3.   Positron emission and electron capture are competing processes.   However, if the parent's mass is not greater than that of the daughter by more than twice the electron mass, only electron capture is energetically possible [see Eq. (3.6b)]. $^{59}_{28}$Ni is such an isotope which can decay only by orbital electron capture.

$$^{59}_{28}\text{Ni} + {}^{0}_{-1}\text{e} \rightarrow {}^{59}_{27}\text{Co} + v + \text{X-rays} \tag{3.7}$$

A neutrino must be released by the nucleus to conserve the angular momentum of the nucleus.   There may sometimes be accompanying gamma radiation and there will always be characteristic X-rays given off when electrons fill the orbital vacancies.

Example 1   Consider the decay of $^{59}_{28}$Ni by electron capture.   Determine the mass of the parent $^{59}_{28}$Ni; and for the recoiling $^{59}_{27}$Co, determine its recoil velocity and kinetic energy.   Show that it is safe to assume that the neutrino gets all of the kinetic energy.

Q for this reaction is 1.07 MeV, and it is shared by the neutrino and the recoiling Co nucleus.

$$Q = [(m_{\text{Ni}} - 28m_e) + m_e - (m_{\text{Co}} - 27m_e)]c^2 = \text{KE}_v + \text{KE}_{\text{Co}}$$

$$1.07 = (m_{\text{Ni}} - 58.9332) \times 931.4$$

$$m_{\text{Ni}} = 58.9332 + 0.001\ 188 = 58.9344 \text{ u}$$

Since the neutrino rest mass is zero, the momentum equation of Prob. 2.6 reduces to

$$p_v = KE_v/c$$

which may be used in the momentum balance as being equal in magnitude to the momentum of the recoiling Co nucleus.

$$m_{Co} V_{Co} = KE_v/c$$

and thus the neutrino energy is

$$KE_v = m_{Co} V_{Co} c$$

The recoiling Co is nonrelativistic, so

$$KE_{Co} = m_{Co} V_{Co}^2/2$$

The sum of the kinetic energies is equal to Q.

$$1.07 = m_{Co} V_{Co} c + m_{Co} V_{Co}^2/2$$

Rearranging, we obtain

$$V_{Co}^2 + 2c V_{Co} - 2 \times 1.07/m_{Co} = 0$$

$$V_{Co}^2 + 2 \times 3 \times 10^8 V_{Co}$$

$$- \frac{2 \times 1.07 \text{ MeV} \times 1.602\,10 \times 10^{-13} \text{ kg m}^2 \text{ s}^{-2} \text{ MeV}^{-1}}{58.9332 \text{ u} \times 1.660\,438 \times 10^{-27} \text{ kg/u}} = 0$$

$$V_{Co}^2 + 6 \times 10^8 V_{Co} - 3.503\,54 \times 10^{16} = 0$$

$$V_{Co} = \frac{-6 \times 10^8 \pm \sqrt{(6 \times 10^8)^2 + 4 \times 3.503\,65 \times 10^{12}}}{2}$$

$$= 5.8393 \times 10^3 \text{ m/s}$$

The kinetic energy of the Co nucleus is then

$$KE_{Co} = m_{Co} V_{Co}^2/2 = \frac{58.9332 \times 1.660\,438 \times 10^{-27} \times (5.8393 \times 10^3)^2}{2 \times 1.6021 \times 10^{-19}}$$

$$= 10.41 \text{ eV}$$

Compared to the total of 1 070 000 eV, this is insignificant. Initially, it would have been safe to assume that the neutrino received the total energy ($Q \approx KE$), and the momentum balance will yield the recoil nucleus velocity directly.

$$V_{Co} = KE_v/m_{Co} c = \frac{1.07 \text{ MeV} \times 1.6021 \times 10^{-13} \text{ kg m}^2 \text{ s}^{-2} \text{ MeV}^{-1}}{3 \times 10^{10} \text{ cm/s} \times 58.9332 \text{ u} \times 1.660\,438 \times 10^{-27} \text{ kg/u}}$$

$$= 5.84 \times 10^3 \text{ m/s}$$

Proton decay

Previously, transmutation to a different chemical element was considered possible to occur only by alpha decay, $\pm$ beta decay, and K-capture. However, during 1970 direct proton decay was reported.

Bombardment of ^{40}Ca nuclei by ^{16}O nuclei has resulted in the formation of an unstable ^{53}Co nucleus which decays to ^{52}Fe by the emission of a proton. The same nucleus has been produced by proton bombardment of ^{54}Fe nuclei to yield the same ^{53}Co nuclei and again, proton decay. The half-life of the proton emitting ^{53}Co is very short (245 ms) and the decay process reduces both A and Z by one.

$$^{53}_{27}\text{Co} \rightarrow {}^{1}_{1}\text{p} + {}^{52}_{26}\text{Fe} \tag{3.8}$$

As more exotic nuclei are produced it is expected that others will be found that will decay by proton emission.

Neutron emission

Very energetic nuclei can expel a neutron (~ 8 MeV). Several such isotopes among the fission fragments provide delayed neutrons. This small but important group of neutrons simplifies the control of a reactor, as will be shown in Chapter 10. Iodine-137 is a negative beta emitter which decays to ^{137}Xe, which, in turn, immediately expels a neutron. The neutrons appear at a rate determined by the iodine decay.

$$^{137}_{53}\text{I} \xrightarrow[27\,\text{s}]{\beta^-} {}^{137}_{54}\text{Xe} \xrightarrow[\text{instantaneous}]{n} {}^{136}_{54}\text{Xe} \tag{3.9}$$

Isomeric transition

Often a daughter isotope is left in an excited state after a radioactive parent nucleus emits a particle. The nucleus will drop to the ground state by the emission of gamma radiation. This commonly occurs immediately on particle emission; however, the nucleus may remain in an excited state for a measurable period of time before dropping to the ground state at its own characteristic rate. A nucleus that remains in such an excited state before decay is known as an *isomer*.

Figure 3.4a shows the decay scheme for ^{107}Cd, which includes K-capture, gammas, and positron emission. Each of the three possible modes of Cd decay leaves the ^{107}Ag in an isomeric state from which it decays with a 44.3-s half-life by emitting a 0.0939-MeV gamma.

Internal conversion

Sometimes nuclear excitation energy is transferred to an orbital electron (usually in the innermost or K shell) rather than having a gamma emitted. This

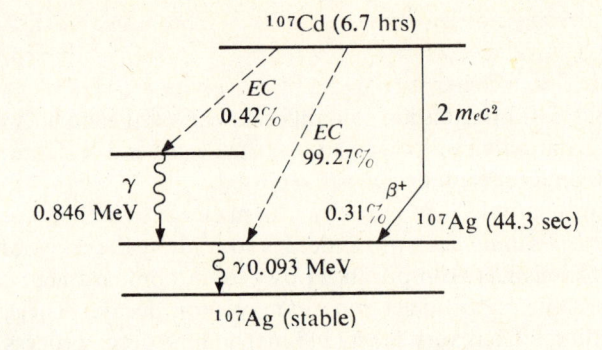

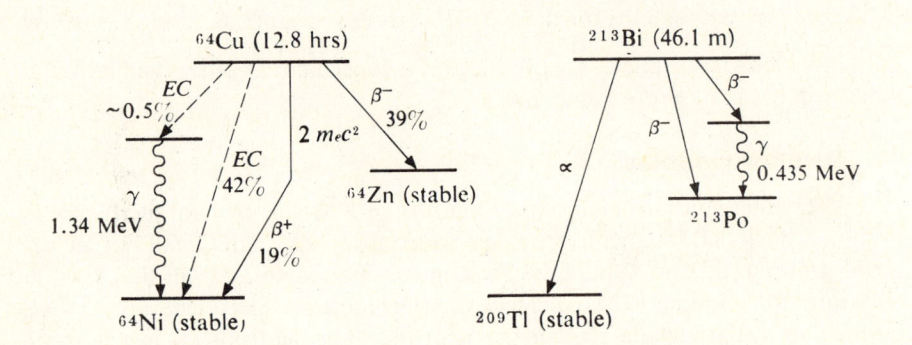

Fig. 3.4 Complex decay schemes: (a) ^{107}Cd, (b) ^{64}Cu, and (c) ^{213}Bi.

conversion electron is ejected with a discrete energy and without a neutrino. The orbital electrons drop to a lower energy state with the emission of X-rays. These in turn may eject other orbital electrons, which are known as *Auger electrons.*

Internal conversion sometimes occurs in conjunction with a beta–gamma decay, as with $^{127}_{52}$Te. In $^{145}_{61}$Pm, internal conversion accompanies electron capture and gamma emission. It also may precede an isomeric transition by another process, such as beta decay. The latter occurs with $^{79}_{34}$Se, where there is an internal conversion with a 3.89-min half-life followed by a long-lived beta decay having a 6.4×10^4-yr half-life.

Complex Decay Schemes

Several modes of decay may be available to a single nucleus. A case in point is ^{64}Cu, which can decay not only by $\pm \beta$ emission, but also by electron capture,

which may or may not leave the daughter ^{64}Ni in an excited state, so that a gamma can be produced as well (see Fig. 3.4b). A similar diagram is shown in Fig. 3.4c for ^{213}Bi, which emits alphas, betas, and gammas. The decay of ^{107}Cd (shown in Fig. 3.4a) was discussed previously as an example of decay to an isomeric state of the daughter silver isotope. Many other isotopes have even more complicated decay schemes.

Decay Rates and Half-Lives

The decay of radioactive isotopes occurs in a random manner. There is a certain probability that in a given time interval a certain fraction of the nuclei of a particular unstable isotope will decay. The rate of decay (dN/dt) is equal to minus the probability of decay, λ, times the number of unstable nuclei present, N.

$$dN/dt = -\lambda N \qquad (3.10)$$

The probability of decay is known as the *decay constant*. It is clear from Eq. (3.10) that the rate of decay of an isotope is proportional to the number of nuclei present. Separation of variables allows this simple first-order differential equation to be integrated.

$$\int_{N_0}^{N} \frac{dN}{N} = -\lambda \int_{0}^{t} dt \qquad (3.11)$$

Here N_0 represents the original number of unstable nuclei present at the initial time ($t = 0$) and N is the number of nuclei at some subsequent time, t.

$$\ln\left(\frac{N}{N_0}\right) = -\lambda t \qquad (3.12)$$

Taking the antilog of each side of Eq. (3.12), we obtain

$$N = N_0 e^{-\lambda t} \qquad (3.13)$$

It is well to note here that the activity of a sample, A, is the absolute magnitude of its decay rate.

$$A = \left|\frac{dN}{dt}\right| = \lambda N \qquad (3.14)$$

The activity of a sample is often expressed in *curies*, where 1 curie (Ci) is taken as 3.7×10^{10} disintegrations per second. A microcurie is taken as 3.7×10^{4} disintegrations per second. The curie was originally defined as the number of

disintegrations per second occurring in 1 gram of ^{226}Ra. This is slightly less than the value of 3.7×10^{10} disintegrations per second currently in use. The curie is being replaced by the *becquerel* (Bq) as the unit for activity, where 1 Bq is taken simply as 1 disintegration per second.

For any sample being monitored, only a fraction of the events are recorded. Thus, the count rate, CR, is the activity times the counter efficiency, e. For a given geometry and sample the efficiency is taken as constant.

$$CR = eA = e\lambda N \qquad (3.15)$$

so that CR $\propto A \propto N$. This proportionality of count rate, activity, and the number of unstable nuclei present permits a simple determination of the decay constant from the slope of a log count rate versus time plot for a decaying isotope.

$$\left(\frac{CR_1}{CR_2} = \frac{A_1}{A_2} = \frac{N_1}{N_2} \right) \qquad (3.16)$$

Figure 3.5 represents the decay of a particular isotope. Equation (3.11) can be integrated between times t_1 and t_2 when N_1 and N_2 unstable nuclei are present.

$$\int_{N_1}^{N_2} \frac{dN}{N} = -\lambda \int_{t_1}^{t_2} dt \qquad (3.17)$$

so that

$$\lambda = \frac{\ln N_1 - \ln N_2}{t_2 - t_1} = \frac{\ln (N_1/N_2)}{t_2 - t_1} \qquad (3.18)$$

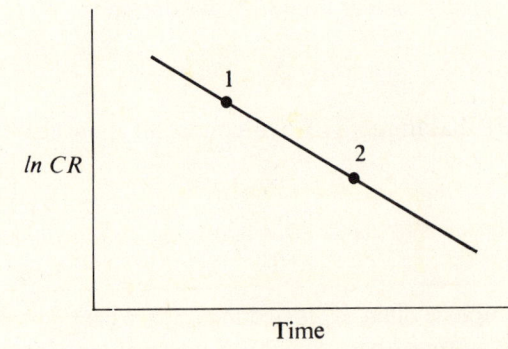

ln CR

Time

Fig. 3.5   Typical decay curve for a radioisotope.

Because of the proportionality between CR and N,

$$\lambda = \frac{\ln (CR_1/CR_2)}{t_2 - t_1} = \frac{\ln CR_1 - \ln CR_2}{t_2 - t_1} \qquad (3.19)$$

The half-life of a radioisotope, $t_{1/2}$, is defined as the time required to reduce the number of decaying nuclei to one-half their original value.

$$\frac{N_{1/2}}{N_0} = \frac{CR_{1/2}}{CR_0} = \frac{1}{2} = e^{-\lambda t_{1/2}} \qquad (3.20)$$

$$2 = e^{\lambda t_{1/2}} \qquad (3.21)$$

Taking the log of each side and then solving for the half-life yields

$$t_{1/2} = \frac{\ln 2}{\lambda} = \frac{0.693}{\lambda} \qquad (3.22)$$

A differential number of radioactive nuclei dN will decay during the time interval between t and $t + dt$ (see Fig. 3.6). The total lifetime for this small group of nuclei will be

$$T = -t \times dN \qquad (3.23a)$$

By differentiating Eq. (3.13), dN may be expressed as a function of time and then

$$T = N_0 \lambda t e^{-\lambda t} \, dt \qquad (3.23b)$$

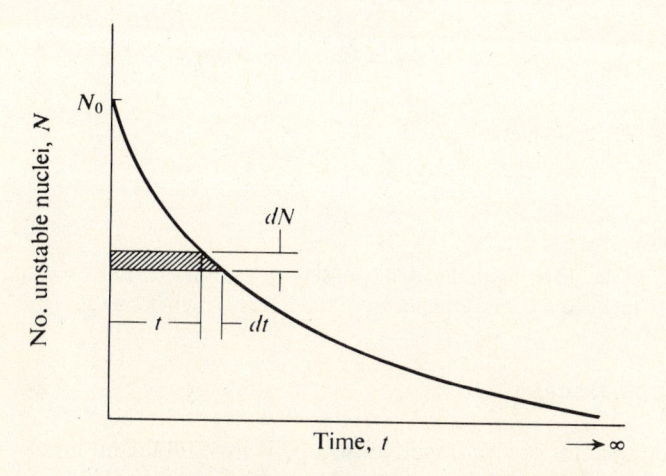

Fig. 3.6   Decay of a radioisotope: number of unstable nuclei versus time.

Thus, if we integrate this function between $t = 0$ and $t = \infty$, we will have the total lifetime of all the radioactive nuclei. Dividing by the original number of nuclei will give the **mean lifetime for decay**, t_m.

$$t_m = \frac{N_0 \lambda \int_0^\infty t e^{-\lambda t}\, dt}{N_0}$$

$$= \left[-t e^{-\lambda t} - \frac{e^{-\lambda t}}{\lambda} \right]_0^\infty$$

$$= \frac{1}{\lambda} \tag{3.23c}$$

The mean lifetime is thus seen to be the reciprocal of the probability of decay.

> **Example 2**   Cesium-132 is an isotope useful in medical research for determining bodily retention of cesium. How long will it take a sample of this isotope to decay to 1 percent of its original activity? What will be the mean lifetime of the ^{132}Cs atoms in a given sample?
>
> $$\text{For } ^{132}\text{Cs} \qquad t_{1/2} = 6.47 \text{ days}$$
>
> Since $A \propto CR \propto N$,
>
> $$\frac{CR}{CR_0} = \frac{N}{N_0} = e^{-\lambda t}$$
>
> $$0.01 = e^{-(0.693/6.47)t}$$
>
> Taking the log of both sides and solving for t, we obtain
>
> $$t = \frac{\ln 100 \times 6.47}{0.693} = 43.0 \text{ days}$$
>
> The mean lifetime is
>
> $$t_m = \frac{1}{\lambda} = \frac{6.47}{0.693} = 9.33 \text{ days}$$
>
> It should be noted that bodily rejection of any ingested ^{132}Cs would reduce the time required for attainment of the 1 percent activity level.

Compound Decay

Often the daughter of a radioactive isotope is not stable and decays to a third nuclide, which can also be unstable. Figure 3.1 illustrated the rather lengthy decay chains for ^{232}Th and ^{235}U.

For the case of an element decaying to an unstable daughter and thence to a third stable isotope, expressions can be written for the number of atoms of each species. If N_{10} represents the original number of parent atoms and N_1 is the number of parent atoms having a decay constant λ_1, at any subsequent time, t,

$$N_1 = N_{10}e^{-\lambda_1 t} \tag{3.24}$$

The rate of change of parent nuclei is

$$\frac{dN_1}{dt} = -N_{10}\lambda_1 e^{-\lambda_1 t} \tag{3.25}$$

The rate of change of the daughter atoms, dN_2/dt, is due to the buildup caused by the decay of the parent less the decay of the daughter with its own decay constant, λ_2.

$$\frac{dN_2}{dt} = -\frac{dN_1}{dt} - \lambda_2 N_2$$

$$= \lambda_1 N_{10}e^{-\lambda_1 t} - \lambda_2 N_2 \tag{3.26}$$

Rearranging yields

$$\frac{dN_2}{dt} + \lambda_2 N_2 = \lambda_1 N_{10}e^{-\lambda_1 t} \tag{3.27}$$

which is an equation of the form

$$\frac{dy}{dx} + a_1(x)y = h(x) \tag{3.28}$$

This first-order ordinary differential equation can be solved through the use of an integrating factor, p.

$$p = e^{\int a_1(x)\,dx} \tag{3.29}$$

The solution for Eq. (3.28) is

$$y = \frac{1}{p}\int ph(x)\,dx + \frac{C}{p} \tag{3.30}$$

where C is a constant of integration.

Since $a_1(x) = \lambda_2$ and $p = e^{\int a_1(x)\,dx} = e^{\int \lambda_2\,dt} = e^{\lambda_2 t}$, then

$$N_2 = \frac{1}{e^{\lambda_2 t}} \int e^{\lambda_2 t} \lambda_1 N_{10} e^{-\lambda_1 t}\,dt + \frac{C}{e^{\lambda_2 t}}$$

$$= \frac{\lambda_1}{\lambda_2 - \lambda_1} N_{10} e^{-\lambda_1 t} + C e^{-\lambda_2 t} \qquad (3.31)$$

The constant of integration can be evaluated with the initial conditions, $t = 0$, $N_2 = 0$.

$$C = \frac{-\lambda_1 N_{10}}{\lambda_2 - \lambda_1}$$

Thus, the number of daughter atoms can be expressed as

$$N_2 = \frac{\lambda_1 N_{10}}{\lambda_2 - \lambda_1}(e^{-\lambda_1 t} - e^{-\lambda_2 t}) \qquad (3.32)$$

If the granddaughter element is stable, it will have a number of nuclei present, N_3, equal to

$$N_3 = N_{10} - N_1 - N_2 \qquad (3.33)$$

Figure 3.7 shows an example of this type of compound decay process for ^{140}Ba which by beta emission decays to ^{140}La, which in turn decays by beta

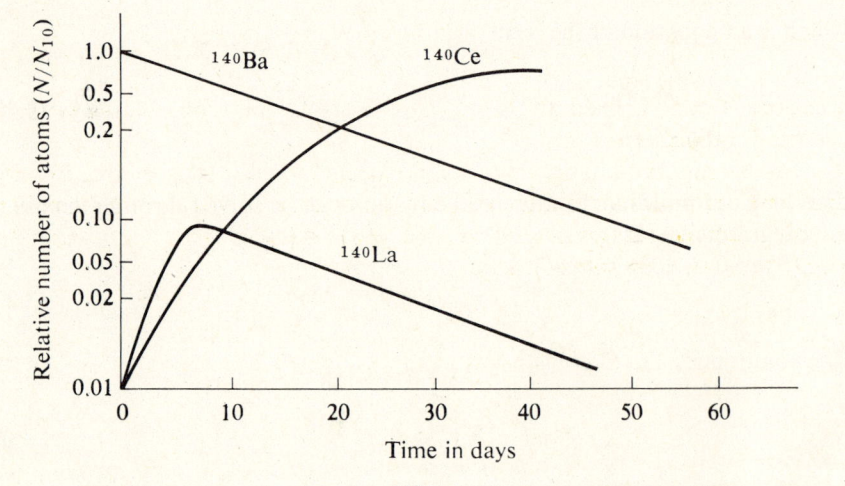

Fig. 3.7   Decay of radioactive ^{140}Ba.   After approximately 15 days the ^{140}Ba and ^{140}La will be in transient equilibrium.

emission to stable ^{140}Ce.

$$^{140}\text{Ba} \xrightarrow[\text{12.8 days}]{\beta^-} {}^{140}\text{La} \xrightarrow[\text{40.2 h}]{\beta^-} {}^{140}\text{Ce} \tag{3.34}$$

Notice here that the decay constant for the ^{140}Ba is smaller (0.0024 h^{-1}) by a considerable amount than that for the ^{140}La (0.0172 h^{-1}). After the ^{140}La passes through its maximum it will essentially decay at a rate equal to that of the parent ^{140}Ba. When $\lambda_1 < \lambda_2$ and the daughter and parent decay at essentially the same rate, they are in transient equilibrium. As t becomes large the $e^{-\lambda_2 t}$ term is considerably less than the $e^{-\lambda_1 t}$ term and contributes little to the solution. Thus,

$$N_2 \approx \frac{\lambda_1}{\lambda_2 - \lambda_1} N_{10} e^{-\lambda_1 t} \tag{3.35}$$

The ratio of the numbers of atoms of the two species which are in transient equilibrium is then

$$\frac{N_2}{N_1} = \frac{\lambda_1}{\lambda_2 - \lambda_1}$$

Multiplying both numerator and denominator by the appropriate decay constant gives the *activity ratio*.

$$\frac{A_2}{A_1} = \frac{N_2 \lambda_2}{N_1 \lambda_1} = \frac{\lambda_2}{\lambda_2 - \lambda_1} \tag{3.36}$$

The activity ratio is fixed and is only slightly larger than unity for the case of transient equilibrium.

Secular equilibrium occurs when the parent has an extremely long half-life. The daughter builds up to an equilibrium amount and decays at what amounts to a constant rate. In this situation $\lambda_1 \ll \lambda_2$ and the rate of formation of daughter atoms balances their rate of decay.

$$\frac{dN_2}{dt} = 0 = \lambda_1 N_1 - \lambda_2 N_2 \tag{3.37}$$

and thus,

$$\lambda_1 N_1 = \lambda_2 N_2 \tag{3.38}$$

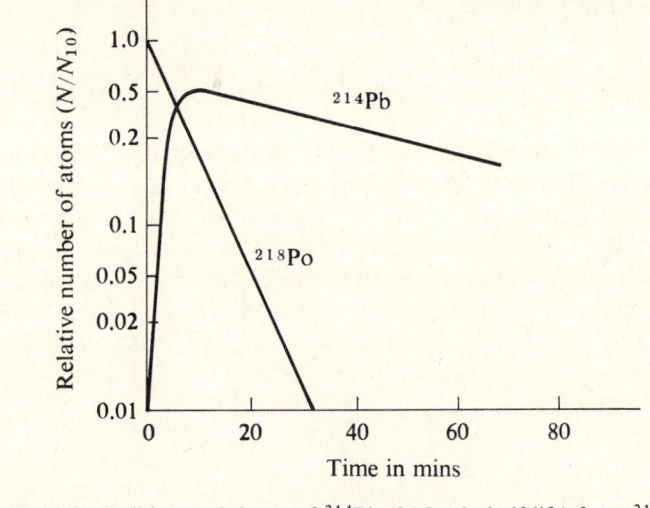

Fig. 3.8   Buildup and decay of ^{214}Pb (26.8-min half-life) from ^{218}Po (3.05-min half-life).

In the long decay chain for a naturally radioactive element such as ^{232}Th, where all the elements in the chain are in secular equilibrium, each of the descendents has built up to an equilibrium amount and all decay at the rate set by the original parent. The only exception is the final stable nth element on the end of the chain. Its number of atoms is constantly increasing. In such a chain

$$\lambda_1 N_1 = \lambda_2 N_2 = \lambda_3 N_3 = \lambda_4 N_4 = \cdots = \lambda_{n-1} N_{n-1} \qquad \textbf{(3.39)}$$

A third case of particular interest occurs when the parent is shortlived in comparison with its daughter ($\lambda_1 > \lambda_2$). In this instance no equilibrium will be established between the two species. Figure 3.8 shows the decay of 3.05-min ^{218}Po to 26.8-min ^{214}Pb. The daughter builds up to a maximum as the parent is disappearing. After the parent has all but vanished, the daughter decays according to its own half-life.

Example 3   For the ^{218}Po–^{214}Pb system, starting with freshly separated ^{218}Po, how long will it take for equal amounts of the two species to exist? At such time what fraction of the original parent atoms will be Po and Pb? Setting $N_1 = N_2$, we obtain

$$N_{10} e^{-\lambda_1 t} = \frac{\lambda_1 N_{10}}{\lambda_2 - \lambda_1} (e^{-\lambda_1 t} - e^{-\lambda_2 t})$$

By rearranging and taking logs, we get

$$t = \frac{\ln\left(\frac{2\lambda_1 - \lambda_2}{\lambda_1}\right)}{\lambda_1 - \lambda_2}$$

$$\lambda_1 = \frac{0.693}{3.05} = 0.227 \text{ min}^{-1}$$

$$\lambda_2 = \frac{0.693}{26.8} = 0.0258 \text{ min}^{-1}$$

$$t = \frac{\ln(2 \times 0.227 - 0.026)/0.227}{0.227 - 0.026} = 3.16 \text{ min}$$

$$\frac{N_1}{N_{10}} = \frac{N_2}{N_{10}} = e^{-\lambda_1 t} = e^{-0.227 \times 3.16} = 0.484$$

Thus, after 3.16 min, Po and Pb each account for 48.4 percent of the original Po atoms.

Often, more than two elements in a decay chain are radioactive. For example, the fission fragment ^{135}Te has a decay chain involving four unstable nuclei before eventually reaching ^{135}Ba, which is stable.

$$^{135}_{52}\text{Te} \xrightarrow[2\text{ min}]{\beta^-} {}^{135}_{53}\text{I} \xrightarrow[6.7\text{ h}]{\beta^-} {}^{135}_{54}\text{Xe} \xrightarrow[9.2\text{ h}]{\beta^-} {}^{135}_{55}\text{Cs} \xrightarrow[3\times10^6\text{ yr}]{\beta^-} {}^{135}_{56}\text{Ba} \qquad \textbf{(3.40)}$$

This chain is of particular interest because ^{135}Xe has an enormous thermal neutron absorption cross section (2.72×10^6 barns),* causing it to act as a neutron poison in the core of a reactor. This problem is discussed in Chapter 10.

If a sample of freshly separated parent isotope (^{135}Te above) decays, the rate of change in the number of atoms of the third isotope in the chain (^{135}Xe above) will be

$$\frac{dN_3}{dt} = \lambda_2 N_2 - \lambda_3 N_3 \qquad \textbf{(3.41)}$$

When Eq. (3.32) for N_2 is substituted into Eq. (3.41) and solved for N_3 with the initial conditions that when $t = 0$, $N_1 = N_{10}$, and $N_2 = N_3 = 0$, the solution is

$$N_3 = \lambda_1\lambda_2 N_{10}\left[\frac{e^{-\lambda_1 t}}{(\lambda_2 - \lambda_1)(\lambda_3 - \lambda_1)} + \frac{e^{-\lambda_2 t}}{(\lambda_1 - \lambda_2)(\lambda_3 - \lambda_2)}\right.$$

$$\left. + \frac{e^{-\lambda_3 t}}{(\lambda_1 - \lambda_3)(\lambda_2 - \lambda_3)}\right] \qquad \textbf{(3.42)}$$

---

* See p. 195 for definition of barn.

The *Bateman equation* gives a generalized expression for the number of atoms of the ith radioactive member of a generalized decay chain at any time, t. This assumes that initially there are only N_{10} atoms of the parent species present.

$$N_i = \lambda_1 \lambda_2 \cdots \lambda_{i-1} N_{10} \sum_{j=1}^{i} \frac{e^{-\lambda_j t}}{\prod_{k \neq j} (\lambda_k - \lambda_j)} \tag{3.43}$$

The validity of Eq. (3.42) may be verified by the application of Eq. (3.43).

Problems

1. ^{213}Po decays by alpha emission. What will be the daughter product? What will be the kinetic energies of the alpha particle and the recoiling nucleus?

2. ^{53}V emits a negative electron and an antineutrino with a combined energy of 2.53 MeV plus a 1.01-MeV gamma. What will be the daughter product? If the average electron energy is one-third of the total shared by the electron and its antineutrino, find the ratio of the velocity at this average electron energy to the speed of light. What is the atomic mass of the ^{53}V atom?

3. ^{48}V can decay by either electron capture or positron emission. Write an equation describing each process.

4. After a neutron capture by $^{78}_{34}$Se, some of the resultant $^{79}_{34}$Se nuclei are left at an elevated energy state from which they decay (3.89-min half-life) by internal conversion ($Q = 0.096$ MeV). This is followed by an isometric transition by beta-decay (6.5×10^4-yr half-life, $Q = 0.14$ MeV). No gammas accompany this beta decay.
   (a) Determine the daughter of the beta decay.
   (b) What was the mass of the unstable atom prior to each of the decay processes?
   (c) What is the electron velocity in each case? Comment.
   (d) When the nucleus recoils after the internal conversion, what will be its energy (eV)?

5. A "free" neutron is unstable and will undergo a negative beta decay having a 10.6-min half-life. Determine the maximum possible velocities for the electron and the recoiling proton. What will be the maximum kinetic energy of the proton?

6. An unstable nuclide decays by electron capture to stable $^{37}_{17}$Cl. The value of Q for the reaction is 0.814 MeV.
   (a) What was the unstable parent nucleus, and what was its mass?
   (b) Determine the velocity of the recoiling Cl nucleus. Is its kinetic energy significant?
   (c) What will be the energy of the characteristic X-ray given off if an outer shell electron falls into the K shell?

7. Carbon dating for archeological materials of organic origin is based on the fact that the organism at its death stops absorbing radioactive carbon 14 as $^{14}CO_2$ from the atmosphere. This radioactive carbon in living materials accounts for about 0.10 percent of the total carbon content. At death the absorption ceases and the ^{14}C decays

at its characteristic rate. Wood in an excavated ruin shows a ^{14}C content of 0.082 percent. Estimate the age of the structure.

8. How long will it take the activity of a 5-μCi ^{60}Co source to decrease to 1.0 μCi? What mass of ^{60}Co will be required for a 5-μCi source? What is the final activity in Bq?

9. A sample of ^{56}Mn is to have an activity of 10^5 Bq at the end of a 48-h experiment. What mass of ^{56}Mn was required at the start of the experiment, and what was its initial activity?

10. A rock returned from the surface of the moon has a ratio of ^{87}Rb to stable ^{87}Sr atoms equal to 14.45. Assuming that this was all ^{87}Rb upon the formation of the solar system, estimate the age of the solar system.

11. During the shutdown of the Connecticut Yankee Reactor, a solution of ^{58}Co in the flooded refueling cavity produced a maximum activity of 55.5 kBq/ml compared to a normal coolant activity of 37 Bq/ml. The use of flow-through filters and ion exchangers reduced the activity level by a factor of 14 in about 4 days, permitting refueling to proceed. How long would a similar attenuation take if radioactive decay of the cobalt were the only factor reducing the radiation level? If replacement power costs $50 000 per day, compute the total replacement power cost during the extra time required to wait for attenuation by radioactive decay.

12. How many grams of ^{210}Po will be required to supply a heat source for a thermoelectric generator that will produce 15 W (th) at the end of a one-year space mission?

13. From the equation for the number of atoms of a radioactive daughter product, develop an expression for the time at which this daughter product will be a maximum.

14. A sample of initially pure 4.7-day ^{47}Ca decays to 3.3-day ^{47}Sc, which decays to stable ^{47}Ti. How long will it take the ^{47}Sc to reach a maximum? What will be the relative amount of each element at this time?

15. If ^{234}U and ^{226}Ra both occur in the decay chain for ^{238}U, compute the percent of each which is present in the ore of natural uranium. How many grams of ^{226}Ra will be present in ore containing 1 metric ton of natural uranium?

16. Starting with a freshly separated sample of ^{140}Ba, plot a curve of the relative amounts of ^{140}Ba, ^{140}La, and ^{140}Ce versus time. Use a Fortran program with a DO loop to calculate these amounts for each day during a 50-day period. Calculate and print out the amounts for the time at which the ^{140}La is a maximum.

17. For a freshly separated nuclide having an unstable daughter and a stable granddaughter, derive an expression for the length of time to have the number of parent and granddaughter nuclei be equal. In the case of ^{140}Ba decay, after how many days will there be equal amounts of ^{140}Ba and ^{140}Ce?

18. Solve Eq. (3.41) for the number of atoms, N_3, the third radioactive isotope in the decay chain at time t. For the ^{135}Te decay chain, determine the fraction of the original atoms that will be ^{135}Xe at the end of one day of decay.

References

1. Lapp, R. E., and H. L. Andrews, *Nuclear Radiation Physics*. Englewood Cliffs, N.J.: Prentice-Hall, Inc., 1963.
2. Kaplan, I., *Nuclear Physics*. Reading, Mass.: Addison-Wesley Publishing Co., Inc., 1955.
3. Halliday, D., *Introductory Nuclear Physics*. New York: John Wiley & Sons, Inc., 1955.
4. Wilson, V. H., W. N. Bishop, and M. Hillman, "Production of Carrier-Free ^{132}Cs and ^{127}Cs," *Nuclear Applications* **1**, No. 6 (December 1965), pp. 556–559.
5. El Wakil, M. M., *Nuclear Power Engineering*. New York: McGraw-Hill Book Company, 1962.
6. Liverhant, S. E., *Elementary Introduction to Nuclear Reactor Physics*. New York: John Wiley & Sons, Inc., 1960.
7. Semat, H., *Introduction to Atomic and Nuclear Physics*, 4th ed. New York: Holt, Rinehart and Winston, 1962.
8. "Fourth Transformation Mode: Proton Decay," *Nuclear News* **13**, No. 12 (December 1970).
9. Jackson, K. P., et al., "^{53}Co^m: A Proton-Unstable Isomer," *Physics Letters* **33B**, No. 4 (October 26, 1970), pp. 281–283.
10. Cerny, J., et al., "Confirmed Proton Radioactivity of ^{53}Co^m," *Physics Letters* **33B**, No. 4 (October 26, 1970), pp. 284–286.
11. Graves, R. H., "Coolant Activity Experience at Conn. Yankee Reactor," *Nuclear News* **13**, No. 11 (November 1970), pp. 66–67.
12. Ray, J. W., "Tritium in Power Reactors," *Reactor and Fuel-Processing Technology* **12**, No. 1 (Winter 1968–1969), pp. 19–26.
13. Andrews, H. L., *Radiation Biophysics*. Englewood Cliffs, N.J.: Prentice-Hall, Inc., 1974.
14. Garlid, K. L., "Radioactive Waste: A Basic Issue in the Safety of Nuclear Power," University of Washington, Seattle, *The Trend in Engineering* **26**, No. 4 (October 1974), pp. 12–17, 21.
15. *The Nuclear Industry—1974*, Wash-1174-74, U.S. Atomic Energy Commission, 1974.
16. Walton, R. D., Jr., "The Management of Commercial Radioactive Waste (High Level and Transuranic Contaminated Solid Wastes)," *Proceedings of the AEC/ANS*, Nuclear Engineering Department Heads Workshop on Research in Nuclear Power Systems, October 24–25, 1974, pp. 88–112.
17. Benedict, M., and Pigford, T. H., *Nuclear Chemical Engineering*. New York: McGraw-Hill Book Company, 1957.
18. Burcham, W. E., *Elements of Nuclear Physics*. New York: Longman, Inc., 1979.
19. Chopping, G. R., and J. Rydberg, *Nuclear Chemistry*. Elmsford, N.Y.: Pergamon Press, 1980.

Nuclear Reactions

The Compound Nucleus

In a nuclear reaction an incident particle (α, d, p, n, or γ) is absorbed into a target nucleus to form a highly excited compound nucleus. This compound nucleus then ejects a particle and/or gamma radiation to drop to a lower-energy state.

If the particle being absorbed to form the compound nucleus is positively charged, it must overcome a potential barrier due to the repulsion of like charged bodies (see Fig. 4.1). As it approaches the nucleus the kinetic energy of a positively charged particle is converted to potential energy until the short-range nuclear forces balance the repulsive force. The particle then enters the nucleus, dropping into the potential well caused by the contribution of its binding energy. Classical mechanics would forbid a particle with less than the energy of the potential barrier from entering the nucleus. However, quantum mechanics predicts that by a process known as *tunneling*, there is a small probability of some particles having an energy less than that of the potential barrier entering the nucleus. Since neutrons and gammas carry no charge, they may enter the nucleus freely without having to surmount any potential barrier.

The compound nucleus may last only $\sim 10^{-14}$ s, but the binding energy of the incoming particle is quickly shared by the nucleons of the compound nucleus. During the lifetime of the compound nucleus the nucleons exchange energy among themselves until one of them attains an energy level sufficient for particle ejection. This concept is valid for particle energies below 20 MeV for the lightest nuclei and below 80 MeV for the heaviest ones. For higher particle energies individual nucleons, or groups of nucleons, may be knocked from the nucleus without the formation of a compound nucleus.

A useful example of a nuclear reaction is found in the plutonium–beryllium neutron source. These may be used to provide neutrons for subcritical assemblies or to provide a monitorable neutron flux level on startup of a critical reactor. The plutonium is an alpha emitter. These alphas then have a high

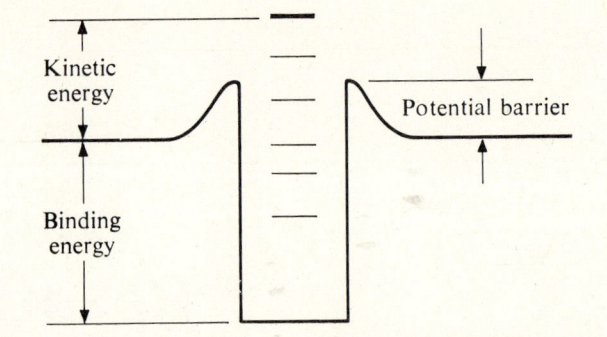

Fig. 4.1   Energy levels for a compound nucleus.   The energy of the compound nucleus is equal to the binding energy contributed by the incident particle plus its kinetic energy.

probability of being absorbed by the beryllium.   The excited compound nucleus formed is $^{13}_{6}C^*$, which subsequently ejects a neutron to produce stable $^{12}_{6}C$.

$$^{9}_{4}Be + ^{4}_{2}\alpha \rightarrow ^{13}_{6}C^* \rightarrow ^{12}_{6}C + ^{1}_{0}n \qquad (4.1)$$

In examining nuclear reactions one must allow for conservation of charge. In reaction (4.1) there are six positive charges among the incident particles in the compound nucleus and in the emergent particles (in this particular case all six are carried by the $^{12}_{6}C$).   There also must be conservation of nucleons. Again, there are 13 nucleons at each stage of reaction (4.1).

Momentum and energy must also be conserved in nuclear reactions.   As stationary observers, we view reactions in the laboratory frame of reference. The energy of an incident particle is usually given for the laboratory system. It is often convenient to consider reactions in the center-of-mass (COM) system, where the observer must be imagined to be traveling at the velocity to be attained by the compound nucleus.

Figure 4.2 shows a reaction occurring in the laboratory system.   Consider first the formation of the compound nucleus.   The incident particle of mass m_1 and velocity v_0 strikes the stationary ($V_0 = 0$) target nucleus of mass M_1 to form the compound nucleus of mass $(m_1 + M_1)$ and velocity V_c.   A momentum balance gives an expression for the compound nucleus velocity in terms of the original particle velocity.

$$m_1 v_0 + M_1(0) = (m_1 + M_1)V_c$$

$$V_c = \left(\frac{m_1}{m_1 + M_1}\right)v_0 \qquad (4.2)$$

To convert to the center-of-mass system, the compound nucleus velocity must be subtracted from each of the laboratory system velocities.   The observer

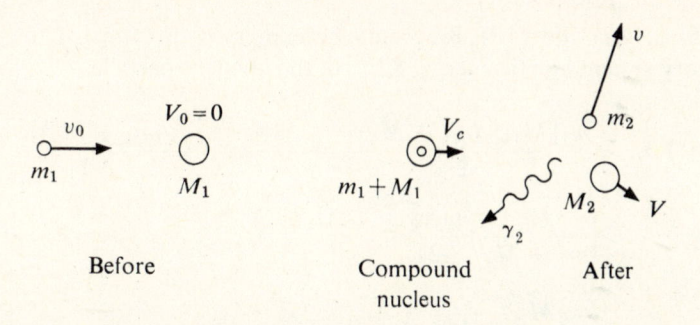

Fig. 4.2 Nuclear reaction in the laboratory system showing the formation and decay of a compound nucleus. In this frame of reference the reaction appears as it would to a stationary observer.

would then see the incident particle traveling with a reduced velocity, v_1, approaching the struck nucleus which has a velocity V_1 equal to $-V_c$ (Fig. 4.3).

$$v_1 = v_0 - V_c = v_0 - \left(\frac{m_1}{m_1 + M_1}\right)v_0$$

$$\left(v_1 = \left(\frac{M_1}{m_1 + M_1}\right)v_0 \right) \tag{4.3}$$

When they meet, the compound nucleus will appear to hang stationary, possessing no kinetic energy. All of the kinetic energy of the incident particles in the COM system will therefore contribute to the excitation of the compound nucleus.

$$\left(\text{KE}_1 = \frac{m_1 v_1^2}{2} + \frac{M_1 V_1^2}{2} \right) \tag{4.4a}$$

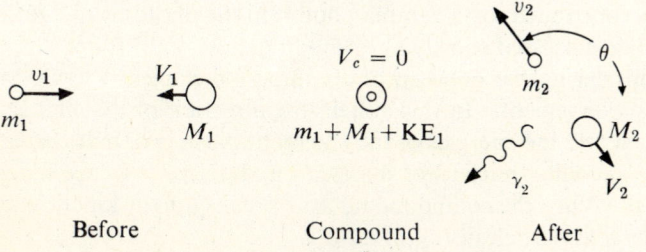

Fig. 4.3 Nuclear reaction in the center-of-mass (COM) system showing the formation and decay of a compound nucleus. In this system the observer must imagine that he or she is traveling at the velocity of the compound nucleus.

Using Eqs. (4.2) and (4.3), KE_1 may be expressed in terms of the original laboratory system kinetic energy, KE_0, of the incident particle.

$$KE_1 = \frac{m_1[M_1 v_0/(m_1 + M_1)]^2}{2} + \frac{M_1[-m_1 v_0/(m_1 + M_1)]^2}{2}$$

$$= \left(\frac{M_1}{m_1 + M_1}\right) \frac{m_1 v_0^2}{2}$$

$$\left(= \left(\frac{M_1}{m_1 + M_1}\right) KE_0 \right) \qquad \text{(4.4b)}$$

Thus, not all of KE_0 shows up as excitation energy in the compound nucleus. If the laboratory system kinetic energy of the compound nucleus is subtracted from KE_0, the result will be identical with Eq. (4.4b).

A mass–energy balance may be written for the overall reaction in the COM system.

$$m_1 c^2 + KE_1 + M_1 c^2 = m_2 c^2 + KE_2 + M_2 c^2 + \gamma_2 \qquad \text{(4.5a)}$$

The difference in the rest mass energies is often denoted as Q.

$$Q = [(m_1 + M_1) - (m_2 + M_2)]c^2$$
$$= KE_2 + \gamma_2 - KE_1 \qquad \text{(4.5b)}$$

A decrease in mass indicates an exothermic reaction and an increase in mass says that the reaction is endothermic. For endothermic reactions KE_1 must be at least large enough to offset the negative value of Q. For such endothermic reactions the threshold energy is the laboratory system kinetic energy of the incident particle (KE_0) necessary to offset the mass deficit, thus making the reaction energetically possible. Note that KE_1 may be considered to include the energy contributed by a gamma photon in the photodisintegration process, which will be discussed shortly.

In considering the breakup of a compound nucleus the COM system is again most convenient. In this system the direction of the emergent particle, θ, is random and the energies of the emergent particle of mass m_2 and velocity v_2 and the recoiling nucleus of mass M_2 and velocity V_2 are independent of direction, θ. When the compound nucleus breaks up, the kinetic energy shared by the emergent particles is

$$KE_2 = Q + KE_1 - \gamma_2 \qquad \text{(4.5c)}$$

A momentum balance ignores any small momentum of a gamma (with its zero rest mass) and indicates that particle velocities and masses are inversely pro-

portional, with the particles traveling in opposite directions.

$$(m_1 + M_1)0 = m_2 v_2 + M_2 V_2$$

$$\left(V_2 = -\left(\frac{m_2}{M_2}\right) v_2 \right) \tag{4.6}$$

An expression may then be developed for the COM kinetic energy, KE_2', of the emergent particle, m_2, in terms of the laboratory system kinetic energy, KE_0, of the incident particle. Combining Eqs. (4.4b), (4.5c), and (4.6), we get

$$KE_2 = Q + \left(\frac{M_1}{m_1 + M_1}\right) KE_0 - \gamma_2$$

$$= \frac{m_2 v_2^2}{2} + \frac{M_2(m_2 v_2/M_2)^2}{2} \tag{4.7a}$$

Rearranging yields

$$Q + \left(\frac{M_1}{m_1 + M_1}\right) KE_0 - \gamma_2 = \left(1 + \frac{m_2}{M_2}\right) \frac{m_3 v_2^2}{2}$$

$$= \left(\frac{m_2 + M_2}{M_2}\right) KE_2'$$

$$KE_2' = \frac{M_2}{m_2 + M_2}\left(Q + \frac{M_1}{m_1 + M_1} KE_0 - \gamma_2 \right) \tag{4.7b}$$

If the compound nucleus velocity is added vectorially to the emergent particle velocity, v_2, and the recoil nucleus velocity, V_2, the laboratory system velocities may be determined. Both the velocities and the kinetic energies in the laboratory system are direction dependent. This is discussed more fully for the elastic scattering of neutrons in Chapter 8.

Example 1 A 0.5-MeV neutron is absorbed by a $_3^6$Li nucleus and causes an alpha particle to be ejected without any gamma radiation. Compute the COM kinetic energy of the alpha.

$$_3^6\text{Li} + _0^1\text{n} \rightarrow _3^7\text{Li}^* \rightarrow _1^3\text{H} + _2^4\alpha + Q$$

$$Q = [(m_{\text{Li}} + m_\text{n}) - (m_\text{H} + m_\alpha)]c^2$$
$$= [(6.015\,13 + 1.008\,665) - (3.016\,05 + 4.002\,60)]931$$
$$= 4.80 \text{ MeV}$$

$$KE_2' = \frac{M_2}{m_2 + M_2}\left(Q + \frac{M_1}{m_1 + M_1} KE_0 - \gamma_2 \right)$$

$$= \frac{3}{7}\left(4.80 + \frac{6}{7} 0.5 - 0 \right)$$

$$= 2.24 \text{ MeV}$$

A compound nucleus may be formed from a variety of events. Its decay is independent of its mode of formation. An example of this is the formation of $^{15}_{7}N$ by alpha, deuteron, proton, or neutron capture. The decay of this compound nucleus may then occur by ejection of one of a variety of particles and/or by gamma emission, as illustrated by the following.

$$\left.\begin{array}{r} ^{11}_{5}B + ^{4}_{2}He \\ ^{13}_{6}C + ^{2}_{1}D \\ ^{14}_{6}C + ^{1}_{1}p \\ ^{14}_{7}N + ^{1}_{0}n \end{array}\right\} \rightarrow {}^{15}_{7}N^* \rightarrow \left\{\begin{array}{l} ^{11}_{5}B + ^{4}_{2}He \\ ^{13}_{6}C + ^{2}_{1}D \\ ^{14}_{6}C + ^{1}_{1}p \\ ^{14}_{7}N + ^{1}_{0}n \\ ^{13}_{7}N + 2^{1}_{0}n \\ ^{15}_{7}N + \gamma \end{array}\right.$$

If decay occurs by ejection of the same type of particle as that which initiated the reaction [(n, n), (p, p), etc.], the process is called *scattering*. If the emerging particle and recoiling nucleus share all of the available kinetic energy, it is known as *elastic scatter*. If some of the energy is carried off by gamma radiation, thus reducing the KE to be shared by the emergent particle and the recoiling nucleus, the process is *inelastic scatter*. If the compound nucleus drops to the ground state solely by gamma emission, the process is called *radiative capture.*

Figure 4.4 shows the energy levels of $^{15}_{7}N$ with some of the various modes of formation and decay. Note that the first excited state exists at 5.28 MeV above the ground level. The binding energy of a neutron is 10.834 MeV. Unless the energy level exceeds the binding energy of one of the particles, decay is only possible by gamma emission. The probabilities (cross sections) of neutron-induced reactions, $^{14}N(n, n)^{14}N$, $^{14}N(n, p)^{14}C$, $^{14}N(n, \alpha)^{11}B$, are shown above this binding energy for the incident neutrons.

In the (n, γ) capture process, a thermal or low-energy neutron is absorbed by a nucleus, which then will become stable after the emission of a gamma (or gammas) which drops the nucleus to its ground state. ^{14}N, for example, can absorb a thermal neutron and then emit gammas whose total energy is 10.834 MeV as the excited ^{15}N drops to its ground state, $^{14}N(n, \gamma)^{15}N$. If a single gamma is emitted, it has the total 10.834 MeV. Gamma energies in the range 2.5 to 10 MeV are sufficiently great that the recoil of a metal atom will produce a lattice vacancy and the recoil subsequently may induce further displacements, resulting in radiation damage to the material. Also, the capture gamma spectrum emitted by an unknown sample undergoing neutron irradiation may be used to identify the various isotopes present in the sample.

Photodisintegration occurs when an incident gamma provides enough excitation energy to make particle ejection possible. The binding energy of the last neutron is normally 5 to 13 MeV, but deuterium and beryllium are exceptions. In the case of beryllium this threshold energy is 1.66 MeV. The value for $^{2}_{1}D$ is calculated in the following example. If monoenergetic gammas exceeding the threshold energy are used, neutrons of a single energy will be produced.

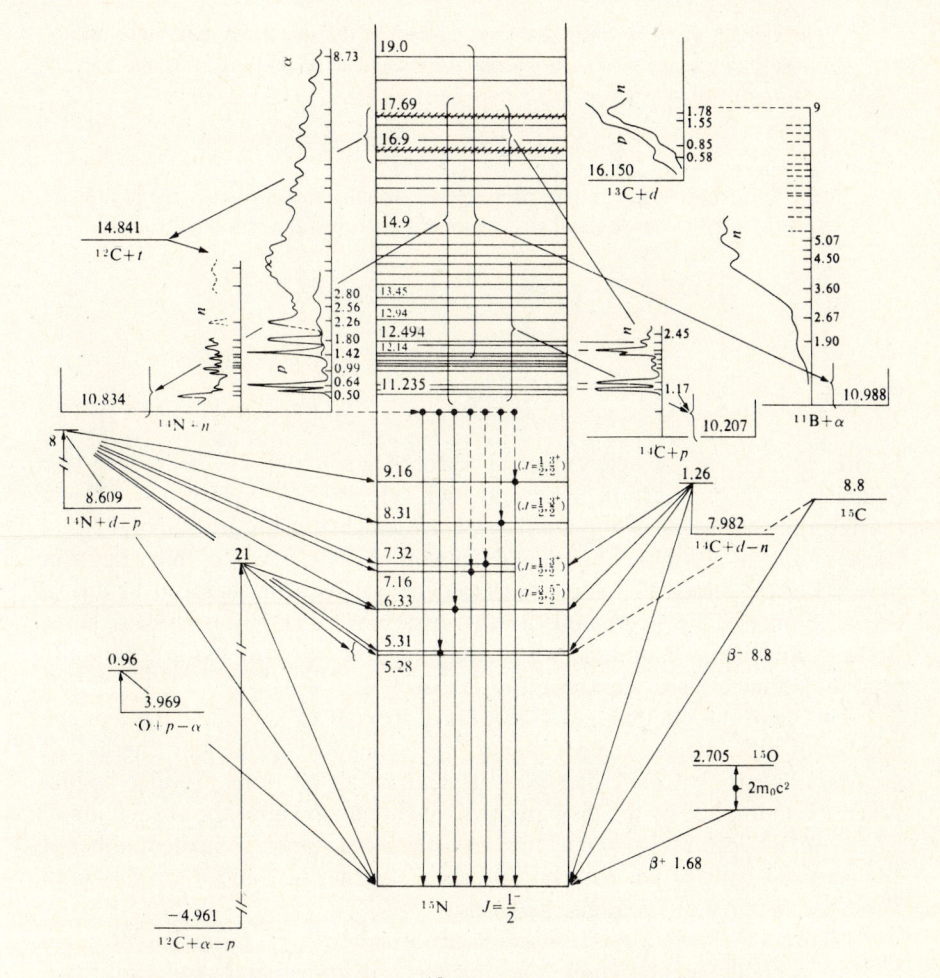

Fig. 4.4 Energy levels in the ^{15}N nucleus. [From F. Ajzenberg and T. Lauritsen, "Energy Levels in Light Nuclei, IV," *Reviews of Modern Physics* **24** (1952), p. 321.]

Example 2 A ^{24}Na–D_2O combination is used as a monoenergetic source of neutrons produced by the photodisintegration of the deuterium.

$$\gamma_1 + {}_1^2D \rightarrow {}_1^1H + {}_0^1n$$

$$
\begin{aligned}
Q &= (m_D - m_H - m_n)c^2 \\
&= (2.014\,10 - 1.007\,825 - 1.008\,665)(931) \\
&= -2.225 \text{ MeV}
\end{aligned}
$$

The threshold energy for this reaction is 2.225 MeV. The radioactive sodium gives off gammas of 2.75 MeV, 1.37 MeV, and so on. Only the 2.75-MeV

gamma provides enough energy to exceed the threshold and cause photo-disintegration. The energy shared by the neutron and the proton as they fly apart will be

$$KE_2 = Q + \gamma_1 = -2.225 + 2.75 = 0.525 \text{ MeV}$$

Since the neutron and the proton have virtually the same mass, they will share the kinetic energy equally. The kinetic energy of the photoneutrons is

$$KE_n = \frac{1}{2} KE_2 = 0.263 \text{ MeV}$$

Fission

Fission occurs when a heavy nucleus absorbs a neutron and splits into two fragments with the ejection of several high-velocity (fast) neutrons. Among the naturally occurring isotopes only ^{235}U fissions by the absorption of thermal (slow) neutrons. Its more abundant sister isotope, ^{238}U, requires that a neutron have a kinetic energy of 1 MeV or better for fast neutron–induced fission to occur. Similarly, fast neutrons can cause fission of ^{232}Th. Both of these latter isotopes are fertile materials, which may be converted to fissile (thermally fissionable) nuclei, as discussed in the next section.

The neutrons emitted per fission vary from 0 to 7 or 8. The average number of neutrons emitted per fission, v, is dependent on neutron energy. It increases linearly with energy. Not all neutron absorptions produce fission. When v is multiplied by the probability of having fission after the absorption of a neutron, the result is known as the *reproduction factor*, η. It is the number of fast neutrons emitted per neutron absorbed. Values of v and η are shown in Table 4.1 for three different energy levels.

Figure 4.5 shows a neutron being absorbed by a ^{235}U nucleus to form ^{236}U as a highly excited compound nucleus. It immediately splits into two fission fragments, say ^{140}Cs and ^{93}Rb, plus three neutrons.

$$^{235}\text{U} + {}_0^1\text{n} \rightarrow {}^{236}\text{U}^* \rightarrow {}^{140}\text{Cs} + {}^{93}\text{Rb} + 3\,{}_0^1\text{n} \qquad (4.8)$$

Table 4.1 Fast Neutrons Emitted Per Fission (v) and Per Neutron Absorbed (η)

Neutron Energy	^{233}U		^{235}U		^{238}U		^{239}Pu	
	v	η	v	η	v	η	v	η
Thermal (0.025 eV)	2.50	2.30	2.43	2.07	—	—	2.89	2.11
1 MeV	2.62	2.54	2.58	2.38	—	—	3.00	2.92
2 MeV	2.73	2.57	2.70	2.54	2.69	2.46	3.11	2.99

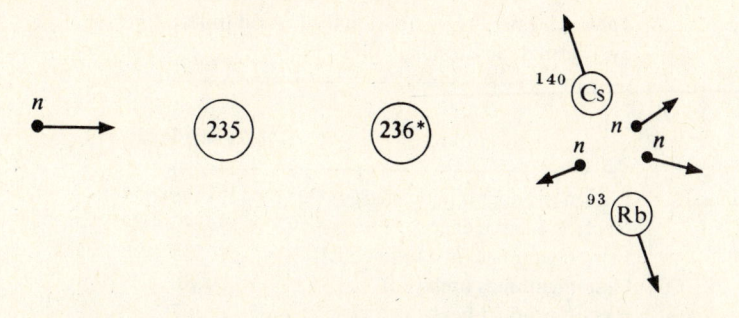

Fig. 4.5 Fission of a ^{235}U nucleus by neutron absorption. The particular fission fragments shown are ^{140}Cs and ^{93}Rb, with three neutrons being ejected.

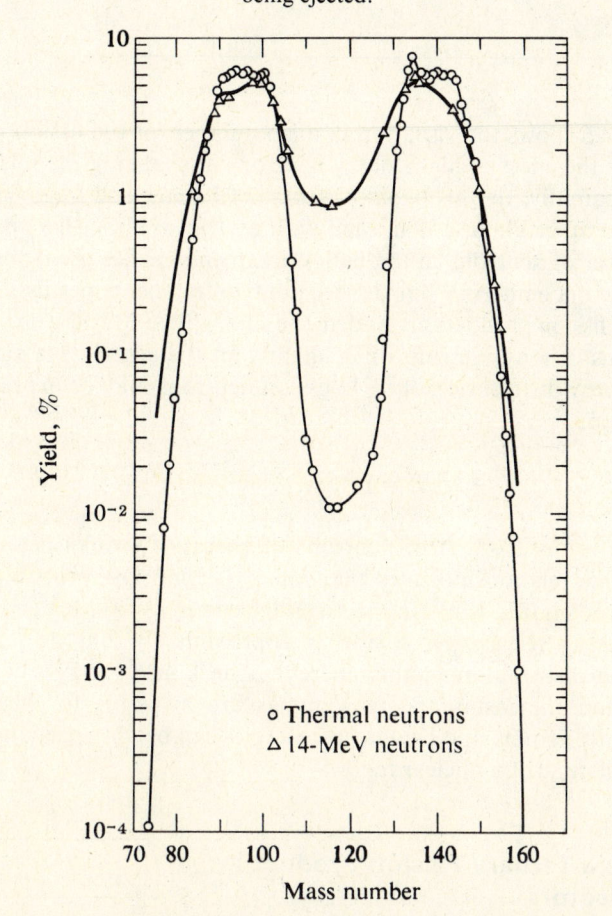

Fig. 4.6 ^{235}U fission yield versus mass number. Curves are shown for fission by thermal neutrons and by 14-MeV fast neutrons. [ANL 5800, p. 11.]

Table 4.2 Energy Distribution for Fission Induced by Thermal Neutrons in ^{235}U

Source	Energy MeV
Fission product kinetic energy	168
Neutron kinetic energy	5
Fission gammas (instantaneous)	5
Fission gammas (delayed)	6
Fission product betas	7
Total available as heat	191
Neutrino energy (not available as heat)	11
Total	202

Figure 4.6 shows the variation in mass number for the fission products of ^{235}U. Note the effect of high-energy neutrons in changing the relative fission yields. Apparently, the probability is best of having one fragment of about 95 u and the other about 139 u, each with a yield of about 6.5 percent. The excess number of neutrons in the fission fragments makes them very unstable beta and gamma emitters. Since there are two fragments per fission when all the yields of fission products are added, the sum will be 200 percent. The yields of the various fission fragments differ slightly for the other fissile nuclei.

The energy distribution of ^{235}U fission neutrons may be expressed by the *Watt equation:*

$$N(E) = 0.453e^{-E} \sinh \sqrt{2E} \qquad (4.9)$$

where $N(E)$ is the fraction of neutrons of energy E (MeV) per unit energy interval. This relation indicates that for ^{235}U the most probable energy of fission neutrons is 0.72 MeV and the average energy is 2.0 MeV.

The energy release per fission is approximately 200 MeV. Table 4.2 indicates the distribution of this energy. The kinetic energy of the fission fragments and the instantaneous gammas are available for heating. The fission product gamma and beta decay energies can be only partially recovered during the lifetime of a nuclear fuel.

Tritium as a Ternary Fission Product in Power Reactors

Tritium occurs in power reactors as a ternary fission product. It is produced at a rate of 8.7×10^{-5} triton per fission in ^{235}U-fueled reactors. The energy

of these ternary fission product tritons will be less than 14 MeV, with the most probable value being 7.5 MeV. The range of tritons in zirconium is estimated at 5 mils for 7.5-MeV tritons and 14.5 mils for 15-MeV tritons. The range is similar in UO_2 and somewhat smaller for stainless steel. Since cladding thicknesses of 16.5 to 33 mils are in use, it would appear that none of the ternary tritons will recoil through the clad. Also, there is evidence of little diffusion of this tritium through the cladding into the reactor coolant. Thus, the fission-induced tritium is delivered to the fuel reprocessing plant in the spent fuel elements. Here it is transferred to the liquid streams in the plant. It can be released after dilution in liquid water, or, after vaporization, be discharged to the atmosphere from a tall stack as a vapor.

Conversion of Fertile Nuclei

Since 99.3 percent of natural uranium is the 238 isotope, which cannot be fissioned by thermal neutrons, it would be highly desirable to convert this to fissile material. This can be accomplished by neutron capture, which produces ^{239}U, which decays by β^- emission to ^{239}Np, and subsequently to ^{239}Pu by another β^- decay.

$$^{238}_{92}U + ^{1}_{0}n \rightarrow ^{239}_{92}U^* \rightarrow ^{239}_{93}Np + ^{0}_{-1}e \qquad \textbf{(4.10a)}$$

$$^{239}_{93}Np \rightarrow ^{239}_{94}Pu + ^{0}_{-1}e \qquad \textbf{(4.10b)}$$

In a similar manner thorium-232 absorbs a neutron and by subsequent beta decays goes to protactinium-233 and uranium-233. Materials that can thus be converted to fissionable nuclei are said to be *fertile*.

Thermal fission of a ^{235}U nucleus produces an average of 2.43 neutrons. If one of these is used to produce the next fission in a steady-state chain reaction, there is a balance of 1.43 neutrons left. These are divided among:

(1) Leakage from the core
(2) Capture by nonfuel or nonfertile materials (parasitic capture)
(3) Nonfission capture in the fuel
(4) Capture by the fertile nuclei

If capture by fertile nuclei produces on an average less than one new fissionable nucleus from the available 1.43 neutrons, the reactor is said to be a *converter*. If it produces more than one new fissionable nucleus, it is a *breeder*. For a breeder, items 1, 2, and 3 listed above must consume less than 0.43 neutron per fission. In a breeder reactor, more new fissile nuclei are being produced than are consumed by fission. To achieve this end successfully in a power reactor is

a challenge to the ingenuity of reactor designers. Breeder reactors hold the promise of extending the supply of fissile material by several orders of magnitude. Because of the high value of η (2.99 at 2 MeV) for ^{239}Pu, this fuel looks to be the best prospect for use in fast breeder reactors. Because of the relatively greater number of nonfission captures in fuel, thermal breeders look less feasible. However, the η of 2.30 for ^{233}U makes it look more promising than the other two isotopes. The Molten Salt Reactor Experiment (MSRE) at Oak Ridge was a step toward the eventual development of a large thermal breeder reactor. It was the first reactor designed to be completely fueled with ^{233}U. Also, the Shippingport reactor has been converted to an experimental thermal breeder operating on the Th–^{233}U fuel cycle.

The nonproliferation issue has sparked interest in the use of *denatured fuels*. These are mixtures of ^{233}U or ^{235}U diluted with ^{238}U to have a low enough isotopic content of the fissile component to be uninteresting as a weapons material without further isotopic enrichment. On the other hand, the plutonium in a PuO_2–UO_2 mixed-oxide fuel can be separated more easily by chemical means to produce weapons-grade plutonium. The use of thorium in various fuel cycles to produce ^{233}U is under active study. The ^{233}U to initiate and sustain such a program could be produced at highly secured sites, possibly under international control, using either ^{233}Th or ^{238}U as the fertile material in the core and thorium as the fertile material in the axial and radial blanket regions. Production of fresh fissile material (^{233}U or ^{239}Pu) is enhanced by surrounding the core with a large mass of fertile material (^{232}Th or ^{238}U) known as a *blanket*. Bred ^{233}U would be chemically separated and denatured at the secure site before being shipped for use in other thermal or fast reactors at less secure sites. Some of the possible fast reactor fuel cycles are discussed more fully in Chapter 13.

Fusion

Fusion reactions form $^4_2$He by the combination of lighter nuclei. Unfortunately, the probability (cross section) for light hydrogen combining with itself or its heavier isotopes is too small to give hope of containment at reasonable temperatures. However, deuterium has a larger cross section for reaction with itself, as well as with tritium ($^3_1$T) and light helium ($^3_2$He). Figure 4.7 shows these cross sections for D-D and D-T reactions. Two deuterons may interact by either of two processes. Both have similar probabilities of occurring, hence only one D-D curve is shown.

$$^2_1D + {}^2_1D \rightarrow {}^3_2He + {}^1_0n + 3.27 \text{ MeV} \qquad \textbf{(4.11a)}$$

$$^2_1D + {}^2_1D \rightarrow {}^3_1T + {}^1_1p + 4.03 \text{ MeV} \qquad \textbf{(4.11b)}$$

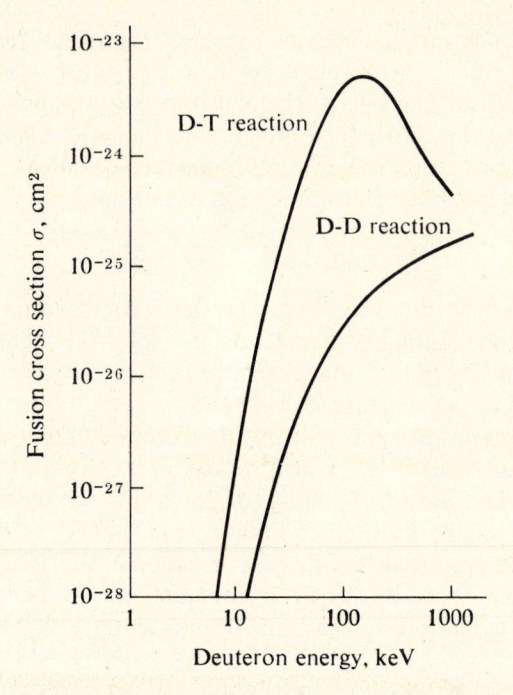

Fig. 4.7 Fusion cross section versus deuteron energy. The variation in fusion cross section is shown as a function of deuteron energy for both the D-D and D-T reactions. [From A. S. Bishop, *Project Sherwood: The U.S. Program in Controlled Fusion*. Reading, Mass.: Addison-Wesley Publishing Co., Inc., 1958.]

The tritium that is formed in the proton branch of the D-D reaction has an even greater probability of interaction with a deuteron and gives off several times as much energy.

$$\,_1^3T + \,_1^2D \rightarrow \,_2^4He + \,_0^1n + 17.6 \text{ MeV} \qquad (4.12)$$

The energy imparted to the neutron in the D-T reaction is approximately 14 MeV, which is sufficient to make the neutron relativistic (see Prob. 4.11).

The light ^{3}He formed in the neutron branch [Eq. (4.11a)] releases slightly more energy than the D-T when it interacts with a deuteron.

$$\,_1^2D + \,_2^3He \rightarrow \,_2^4He + \,_1^1p + 18.3 \text{ MeV} \qquad (4.13)$$

At temperatures where the D-T reaction would be self-sustaining, the probability of this reaction is very small. However, at temperatures three to four times as great, the D-^{3}He reaction could provide significant energy release. It is of interest, as there would be much less tritium present, resulting in less radiation hazard for such a reactor.

The 14-MeV neutrons from the D-T reaction are uncharged and, therefore, cannot be contained by a magnetic field, which must envelop the reaction volume to contain the plasma. If the neutrons pass through the wall of the containment vessel, they can enter a surrounding moderator blanket containing ^{6}Li. On absorption by the ^{6}Li they can produce an additional source of tritium to be separated and fed back into the fusion chamber.

$$^6_3\text{Li} + ^1_0\text{n} \rightarrow ^7_3\text{Li*} \rightarrow ^3_1\text{T} + ^4_2\text{He} \qquad (4.14)$$

In a fast neutron flux there will also be a contribution to the tritium production from the ^{7}Li(n, n'α)^{3}T reaction. In a blanket containing 0.3 m of molten lithium and 0.2 m of graphite, 46.5 percent of the tritium breeding ratio can be attributed to this reaction (Reference 32).

The energy given up as the neutrons slow down and that produced by the n–^{6}Li reaction can be used as the heat source for a conventional steam or gas turbine cycle. The transport medium for conveying the thermal energy to a steam generator might be molten lithium, LiF, LiO_2, or Li_7Pb_2. Extra neutrons could be produced by either Be or fission of thorium or a subcritical amount of uranium. The Be has an (n, 2n) reaction. As mentioned earlier, it can also undergo photodisintegration, if sufficiently energetic photons are available.

In order for fusion to occur, tremendous temperatures are required. Two colliding nuclei must have sufficient energy to overcome the electrostatic forces of repulsion due to their like charges. Figure 4-8 shows the power density attained for both D-D and D-T reactions as a function of kinetic temperature. Since the kinetic energy at the most probable velocity for a group of particles with a Maxwellian energy distribution is the product of the Boltzmann constant, k, and the temperature, T, temperature is sometimes expressed in electron volts, with 1 keV being equal to 1.16×10^7 K. A temperature of 100 000 000 K (8.6 keV) will permit fusion. However, for the D-D reaction the radiation losses from the plasma exceed the energy released and thus there is a net power loss. The D-T reaction only requires 4 keV (4.6×10^7 K) to attain a balance between energy released and energy lost, as compared to 36 keV (4.1×10^8 K) for D-D. The major energy loss is due to X-rays induced by bremsstrahlung, that is, collisions between electrons and the positive ions. The temperature at which the reaction becomes self-sustaining is called the *ignition temperature*. At such temperatures the hot plasma must be kept from physical contact with the container walls. This is accomplished by containing the plasma within a magnetic field.

When a particle with a velocity, v, moves across a magnetic field of strength, B, as shown in Fig. 4.9, there is a mutually perpendicular force, F, set up which is equal to the product of the field strength in gauss, the electronic charge, e, and the velocity of the particle.

$$F = Bev \qquad (4.15)$$

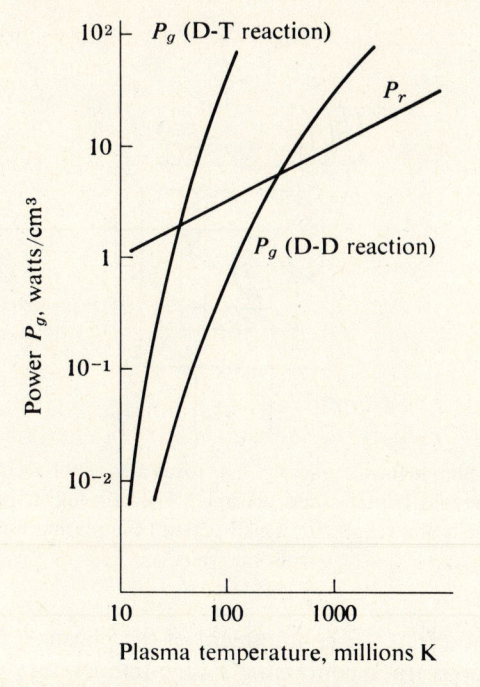

Fig. 4.8 Power generated versus plasma temperature. The total power generated (P_g) per cm^3 is shown for both the D-D and D-T reactions as a function of temperature. The power radiated (P_r) is also shown. The curves are for plasma densities of about 10^{15} particles per cm^3. [From A. S. Bishop, *Project Sherwood: The U.S. Program in Controlled Fusion.* Reading, Mass.: Addison-Wesley Publishing Co., Inc., 1958.]

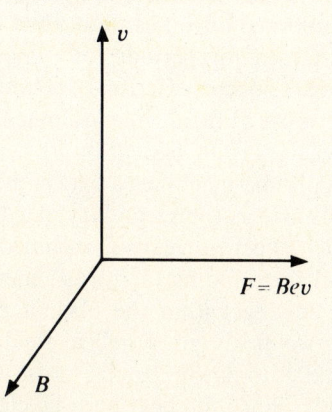

Fig. 4.9 Vector diagram showing the relationship among particle velocity, v, magnetic field strength, B, and force on the particle, F.

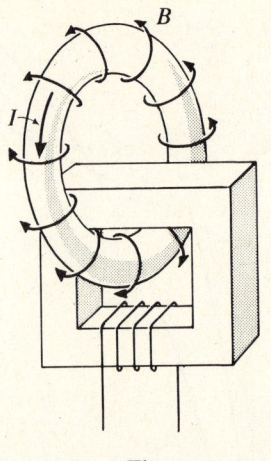

Fig. 4.10 Simple fusion reactor. The torus acts as the secondary of a transformer and contains the plasma. The current, *I*, induces the magnetic field of strength, *B*, which pinches the plasma inward away from the walls.

It is this force that affects the containment of the plasma. If a heavy current flows in a continuous track containing a high-temperature plasma, such as a torus, just the current flow will induce a field around itself. This *pinch effect* squeezes the plasma into the center of the track. A simple fusion reactor might take the form of a torus containing the plasma (see Fig. 4.10). The torus acts as the secondary of a transformer. The energy pumped into the torus from the primary windings induces a high current and produces the enormous temperatures necessary for fusion. This current also causes a magnetic field which pinches the plasma.

It is interesting to note that the *mean free path*, or the average distance a deuteron must travel to undergo fusion, is thousands of miles. The stability of the plasma must be sustained for periods long enough to allow significant numbers of fusion reactions to occur. Control or elimination of the instabilities in plasmas presents a most difficult technological challenge. Successful development of fusion reactors is dependent on its being overcome.

The most successful plasma machine to date is the Soviet Tokamak design (Fig. 4.11). A large torus acts as the secondary of a transformer and contains the plasma with the pinch effect preventing expansion in the direction of the minor axis. A heavy coil surrounding the torus generates a toroidal magnetic field, while the current flowing within the plasma produces the pinch field. Vector addition of the two fields gives helical field lines that lie on closed magnetic surfaces (nested toroids of circular cross section). This is shown schematically in Fig. 4.12. Since the field lines lie closer together at the inner radius of the torus than at the outermost radius, there is a tendency for the plasma to be pushed toward the outermost wall. To counteract this tendency,

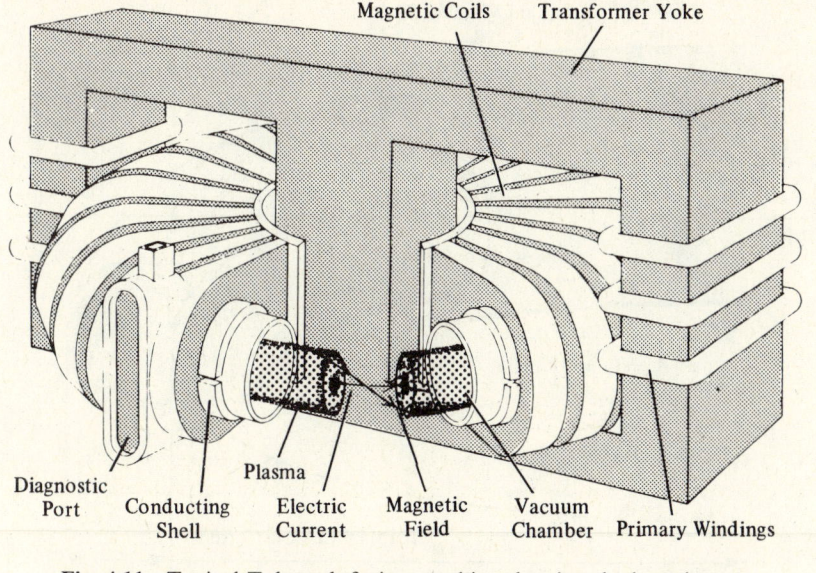

Magnetic Coils   Transformer Yoke

Diagnostic Port   Conducting Shell   Plasma   Electric Current   Magnetic Field   Vacuum Chamber   Primary Windings

Fig. 4.11 Typical Tokamak fusion machine showing the hot plasma confined by helical magnetic field lines (only one shown). The toroidal component is produced by the circular array of wedge-shaped coils enveloping the toroidal plasma chamber. The poloidal component of the field is caused by the heavy current flow in the plasma, which, in turn, is induced by current flowing through the primary windings around the iron yoke. [From B. Coppi and J. Rem, "The Tokamak Approach in Fusion Research." Copyright © July 1972 by *Scientific American, Inc.* All rights reserved.]

a vertical field is applied to control the position of the plasma column within the toroidal shell.

A kink instability will occur when the *Kruskal–Shafranov limit* is exceeded. This limit occurs when the ratio of the toroidal magnetic field strength to the poloidal field strength becomes equal to the aspect ratio of the torus (the ratio of the length of the major axis to that of the minor axis). At this limit the magnetic field line fits just once on the length of the plasma column and once around it, closing on itself. The column then is given a growing helical displacement that destroys the equilibrium configuration. Thus, the stronger the toroidal field is, the higher the current (and hence the higher poloidal field) that is possible. Usually, a safety factor of 2.5 to 3 is allowed to ensure stability.

Another type of instability is known as the *banana regime.* The field varies inversely as the distance from the toroidal axis, so that field lines have a maximum strength at the inner radius and a lesser value at the outermost radius—as was mentioned previously. Low-velocity particles may be trapped in the minimum-field region. On the cross section of the toroid, the circulating particles follow a nearly circular path, but the trapped ones follow a banana-like

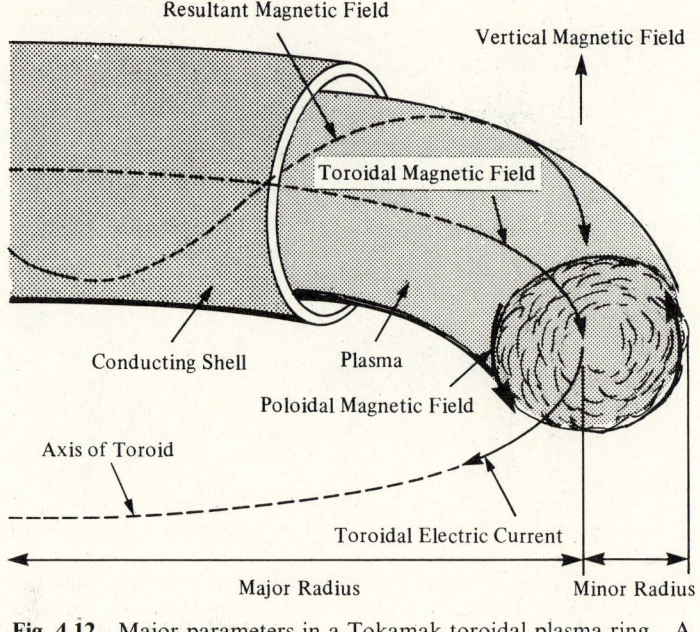

Fig 4.12 Major parameters in a Tokamak toroidal plasma ring. A combination of the toroidal magnetic field set up by external coils and the poloidal magnetic field set up by current flow in the plasma gives a resultant set of helical field lines. Expansion in the direction of the minor axis is prevented by the pinch forces produced by the current flow through the plasma. Expansion in the direction of the major radius is prevented by the conducting magnetic shell surrounding the plasma or by the vertical magnetic field. [From B. Coppi and J. Rem, "The Tokamak Approach in Fusion Research." Copyright © July 1972 by *Scientific American, Inc.* All rights reserved.]

curve. It is feared that these trapped particles will leave the system. There is as yet no clear-cut evidence of this.

Success of the Soviet design has led to a number of Tokamak-like machines being built in the United States. The Model C Stellarator at Princeton was converted to become the Symmetric Tokamak. Before its dismantling, the Symmetric Tokamak demonstrated a doubling of the ion temperature through the application of radio-frequency power to the plasma. At Princeton this has been followed by the Adiabatic Toroidal Compressor and the Princeton Large Torus. At the Hollifield (Oak Ridge) National Laboratory, there is a Tokamak machine known as ORMAK. Figure 4.13 shows the ALCATOR, a powerful Tokamak-type machine built at MIT's National Magnet Laboratory. Very strong toroidal currents heat the plasma to elevated temperatures. It is hoped to attain plasma densities and temperatures where the banana regime will become evident.

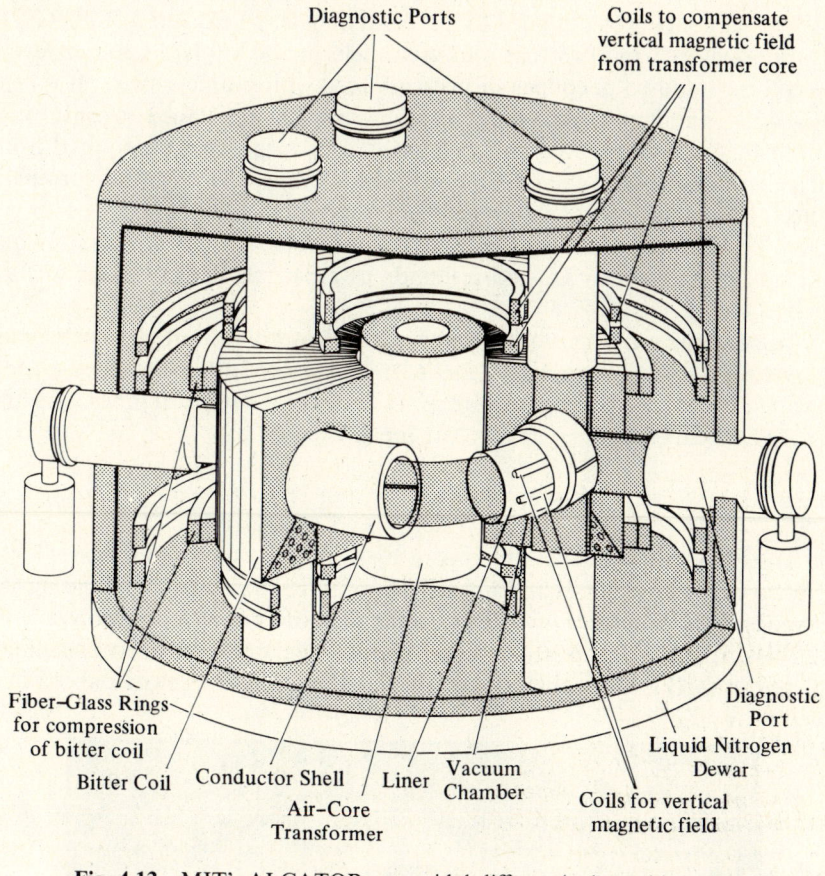

Diagnostic Ports

Coils to compensate
vertical magnetic field
from transformer core

Fiber–Glass Rings
for compression
of bitter coil

Diagnostic
Port

Bitter Coil   Conductor Shell   Liner   Vacuum
Chamber

Liquid Nitrogen
Dewar

Air–Core
Transformer

Coils for vertical
magnetic field

Fig. 4.13 MIT's ALCATOR, a toroidal-diffuse-pinch machine. This Tokamak version is designed to produce and confine a plasma with properties similar to those of a thermonuclear plasma. The large current in the plasma will cause it to be heated. This is expected to produce microturbulence and hence enhance the electrical resistivity. A toroidal plasma column that carries such a large current tends to be destroyed by macroscopic fluid-like instabilities unless a large enough toroidal magnetic field is applied. Thus, the ALCATOR is designed for magnetic fields as large as 120 kG in a plasma chamber having a minor radius of 12.5 cm and a major radius of 54 cm. This should allow ALCATOR to carry currents up to 600 kA with a reasonable safety factor against instabilities. [From B. Coppi and J. Rem, "The Tokamak Approach in Fusion Research." Copyright © July 1972 by *Scientific American, Inc.* All rights reserved.]

The ALCATOR is designed for toroidal magnetic fields up to 120 kG. It was reported early in 1975 that a toroidal field of 100 kG had been achieved. The coil is composed of copper sheets reinforced with stainless steel. It is held at 77 K by liquid nitrogen, reducing the electrical resistance to only one-seventh that at room temperature. Twenty-four megawatts is required from four generators to hold the maximum field for 0.25 s. The plasma current is induced by an air-core transformer inside the bitter coil. It is hoped to get current densities of 1400 A/cm^2, a fivefold improvement over the best previous Tokamak experiments. This corresponds to a particle density of 2.5×10^{14} particles/cm^3 at 5 keV.

The *Lawson criterion* states that the particle density (particles/cm^3) times the average containment time (seconds) must exceed 10^{14} to produce power. Previous Tokamaks had come within 10^{-1} of the Lawson number, and the ALCATOR will come very close to the break-even point.

$$2.5 \times 10^{14} \times 0.25 = 0.625 \times 10^{14}$$

The Tokamak-like machines are improving our ability to contain dense, high-temperature plasmas for significant lengths of time, thereby understanding the dynamics of the processes involved in controlled thermonuclear fusion.

Fusion devices of another class are the open-field magnetic mirror machines such as the 2X II and Baseball II, both at the Lawrence Livermore Laboratory.

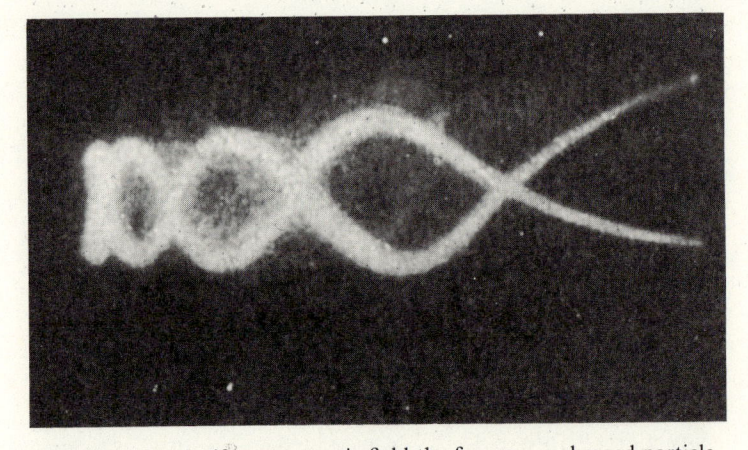

Fig. 4.14 In a uniform magnetic field the force on a charged particle bends its path into a spiral. When the field is made nonuniform at, say, the end of an open tube, the lines curve so that the particle is turned back from what is effectively a magnetic mirror. This photograph shows the tracks of electrons being turned back from such a mirror (at right). The light for the photo comes from the excitation of the background gas. [From H. Roderick and A. E. Ruark, "Thermonuclear Power," *International Science and Technology* **45** (September 1965).]

A straight tube, open at the ends, uses a stronger field at the ends of the tube to make the ions spiral more tightly until they are reflected back, as illustrated by Fig. 4.14. The ions then just spiral back and forth, and, unfortunately, some escape from the ends. It has yet to be proven that plasmas can be contained in mirror machines long enough and at high enough densities to provide any net power.

Inertial Confinement Fusion

It has been shown that fusion may be accomplished by an intense beam of monochromatic and coherent light energy from a laser. A neodymium glass laser with suitable amplification can produce, when focused, a power density of 10^{17} W/cm². Figure 4.15 shows how a single pulse may be extracted from a train of pulses and be shaped. The shaping permits the low-power leading edge to vaporize a solid deuterium–tritium fuel pellet prior to the heating of the main pulse that induces fusion. The energy release will far exceed that required to drive the laser and to recycle and condense (or freeze) the unused fuel together with any bred tritium.

Some concepts would try to restrain the plasma by a magnetic field. However, it has been shown that fusion can be induced in a plasma expanding freely in a vacuum.

The laser pulse must be absorbed in the expanding plasma (inverse bremsstrahlung), where the absorbed energy is transferred to electrons. The electrons then transfer energy to the heavier positive ions present in the expanding plasma, which takes the order of 10^{-11} s. Also, the pulse must not exceed the time for the plasma to expand to the point where it is too dilute to absorb the radiation (a few nanoseconds). Thus, a pulse with a duration of 10^{-10} s will be bracketed by these two times. Figure 4.16 shows how a pulse of laser energy is absorbed in an expanding plasma.

The outlook for laser-induced fusion appeared momentarily bright when in 1974 KMS announced its first production of fusion neutrons via laser. This novel application for the high-technology laser seemed much simpler than the massive magnetic confinement schemes. It led to funding of ambitious programs for development of the neodymium glass laser at the Lawrence Livermore Laboratory and gas laser development at the Los Alamos Scientific Laboratory. The initial enthusiasm has waned somewhat, as it has been shown that the Nd–glass laser must cool for a matter of minutes between pulses, making it unsuited for continuous power production. The CO_2 laser is 8 to 10 percent efficient, but it has not been shown to be capable of a sufficiently high pulse rate or a high enough energy input to the pellet. The KrF gas laser operates at a higher frequency, possibly high enough to produce energy gains greater than unity.

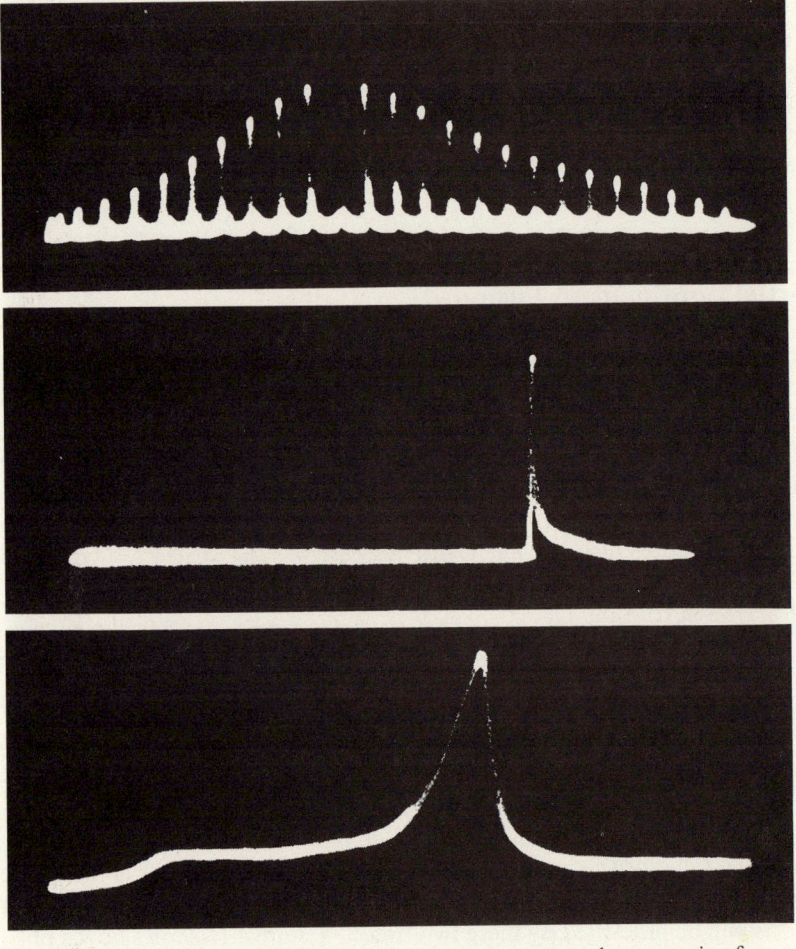

Fig. 4.15 Tailoring of a laser pulse. The upper trace shows a train of
laser pulses with one pulse removed. The extracted pulse (center) is
shown after amplification. The pulse is then shaped (bottom) with a
leading edge to vaporize the fuel prior to heating by the main pulse to
produce fusion. [By permission of the Laboratory for Laser Ener-
getics, University of Rochester.]

   Much interest for inertial confinement has shifted away from laser energy
to vaporize the D-T pellet to the use of ion pulses produced by particle accelerators. Pulses of light ions (protons or deuterons) have a 20 to 30 percent
efficiency for energy delivery and initially appear to have lower cost than heavy
ions, which are only 15 to 20 percent efficient. The Sandia Electron Fusion
Beam Accelerator has been cleverly adapted to provide beams of either light
ions or electrons. The heavy-ion driver pulses are being approached by two

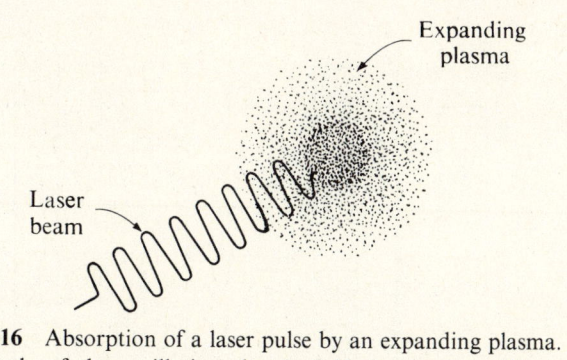

Fig. 4.16 Absorption of a laser pulse by an expanding plasma. The amplitude of the oscillating electric field decreases rapidly as it is absorbed in the surface layer of the plasma.

methods. The Lawrence Berkeley Laboratory would accelerate its 3-MJ heavy-ion pulses with Liniac accelerators, while the Argonne National Laboratory would use synchrotrons and storage rings for the production of 1-MJ pulses.

One problem with laser pulses is that the late-arriving energy carried by the back end of the pulse may arrive too late to be of use in producing fusion. The initial part of the pulse vaporizes the shell of the pellet and absorption of the late-arriving energy by the plasma may prevent it from reaching the pellet core. The charged ion beam may be able to prevent this problem, as the pulse can be magnetically manipulated in flight to become bunched up. As the pulse passes through the magnetic field, it becomes stronger, causing the latter part to catch up with the first part. Thus, the energy is delivered in a shorter burst to reduce the plasma losses.

It would appear at the present time that inertial confinement will probably not be as energetically pursued as magnetic confinement. It is doubtful that a commercial system could go into operation until well after the magnetic confinement system, well into the twenty-first century.

Transuranium Elements

Elements with an atomic number greater than that for uranium do not exist in nature but can be produced by bombardment of uranium or thorium nuclei with neutrons, deuterons, alphas, or even heavier nuclei. Figure 4.17 shows how elements between plutonium ($Z = 94$) and rutherfordium ($Z = 104$) can be produced by a combination of intense thermal neutron irradiation and beta decays.

The new element hahnium ($Z = 105$), named in honor of Otto Hahn, who received a Nobel prize for discovering nuclear fission, was produced by the bombardment of ^{249}Cf with 84-MeV ^{15}N nuclei to produce hahnium-260 accompanied by the emission of several neutrons.

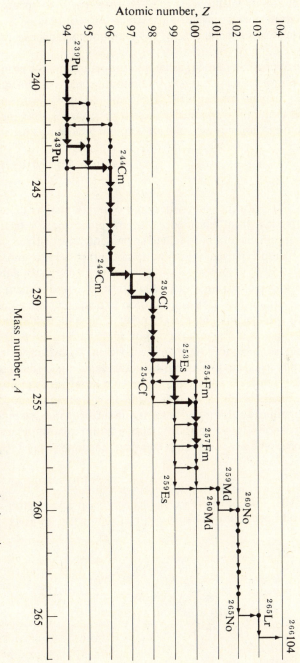

Fig. 4.17 Nuclear reactions for the formation of transuranium elements by intensive thermal neutron irradiation. [From G. T. Seaborg, *Nuclear Applications & Technology,* **9**, No. 6 (December 1970), p. 843.]

An even more exciting development has been the possible formation of a superheavy element, eka-mercury ($Z = 112$). This was accomplished by the bombardment of pure tungsten by high-energy (24-GeV) protons. Occasionally, a recoiling tungsten nucleus will be energetic enough to cause a fusion with another similar nucleus. The resulting isotope has a long-lived alpha decay (6.73 MeV) different from any known isotope. It also fissions spontaneously with a half-life of approximately 500 yr. Such superheavy elements have very interesting fission properties in that the energy release per fission is about 50 percent greater than that for U or Pu isotopes. Also, the number of neutrons released per fission (v) is estimated to be about 10.5.

Among the transuranium elements ^{239}Pu and ^{241}Pu are established as fissionable fuels, while ^{238}Pu is an important heat source due to alpha decay. ^{242}Cm is not too useful as a heat source because of its 163-day half-life; however,

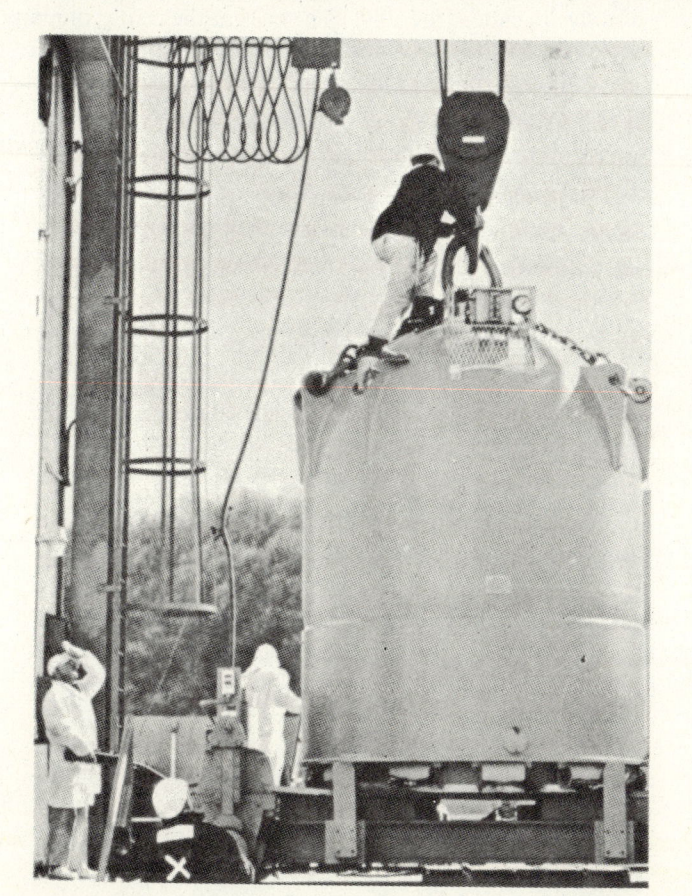

Fig. 4.18 Fifty-ton shipping cask built at Oak Ridge National Laboratory for transporting up to 1 g of ^{252}Cf. [From G. T. Seaborg, *Nuclear Applications & Technology* **9**, No. 6 (December 1970), p. 848.]

it does decay to ^{238}Pu. ^{244}Cm should ultimately be produced at less cost than ^{238}Pu and be used as a heat source. Because of its spontaneous fission it will require more shielding.

Americium-241 emits a 60-keV gamma with 35 percent of the alpha decay events. It has a 433-yr half-life and probably has more applications than any of the other actinide isotopes. As a gamma emitter it can be used in thickness gages, to determine sediment concentration in a flowing stream, to measure soil compaction, or as a location-sensing device. As an alpha emitter it can be mixed with beryllium as a neutron source for oil-well logging, measuring moisture content of soils, or determining water content in chemical process streams. The ionization of air by its alphas is useful in static eliminators and fire detectors.

Californium-252 is a 2.65-yr alpha emitter and also spontaneously fissions with a half-life of 85.5 yr. It is useful as an intense source of neutrons and it is particularly effective in cancer therapy. The cells at the center of many cancers are starved for oxygen. These oxygen-starved cells are much more resistant to X-rays than are ordinary cancer cells, whereas the effectiveness of the ^{252}Cf is much less effected by the oxygen deficiency. Figure 4.18 shows a 50-ton shipping cask built at Oak Ridge to transport up to 1 g of ^{252}Cf, giving some indication of the intensity of its radiation.

High-Energy Reactions

As more energetic projectiles have become available with the advent of more powerful accelerators, a whole menagerie of strange particles has been observed. These result from energy-to-mass conversions similar to electron pair production at lower energy levels. Additional species appear during the decay of other unstable strange particles.

Bubble Chambers

A powerful experimental tool for studying high-energy reactions is the bubble chamber. In 1952, D. A. Glasser reported the operation of the first such unit, a 3 cm by 1 cm glass tube filled with diethylether. These since have grown to the bubble chamber at the Fermi National Accelerator Laboratory (Fermilab), which has a dimension of 4.57 m (15 ft) in the direction of the entering beam of neutrinos and contains 30 000 liters of liquid hydrogen, deuterium, or neon, depending on the particular experiment (Fig. 4.19). When filled with liquid hydrogen, the temperature will be ~27 K. The pressure of the subcooled liquid is suddenly reduced below the saturation pressure. The passage of a charged particle causes ionization along its track in the now-superheated liquid. This ionization produces a train of bubbles that mark the path of the particle, which then can be photographed. The chamber is operated in a strong magnetic

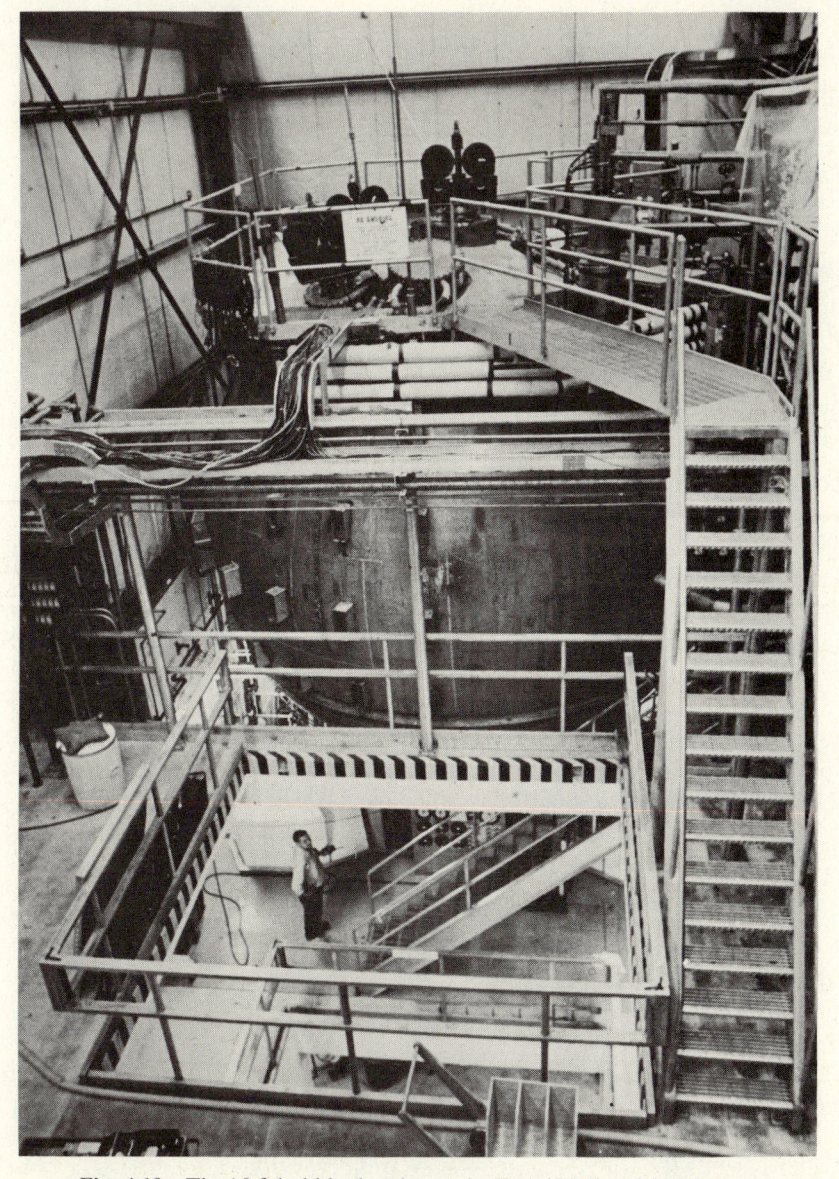

Fig. 4.19 The 15-ft bubble chamber at the Fermi National Accelerator Laboratory, Batavia, Illinois. High-energy particles from the accelerator enter at the right. The interactions of those particles with liquid hydrogen in the chamber form "tracks" that may be photographed by seven cameras located in the top of the chamber. [Courtesy of the Fermi National Accelerator Laboratory.]

field so that the paths of charged particles are curved. Particles with opposite charges have paths with opposite directions of curvature, the degree of curvature being a function of the particle's momentum and the magnetic field strength.

Particle–Antiparticle Pairs

At low energy levels, gammas can interact in the field of a nucleus to create an electron pair, one being positively charged and the other being negatively charged. A threshold energy of 1.02 MeV is required to produce the rest mass of the two particles in this pair production process. Any excess energy is shared nearly equally as the kinetic energy of the electrons.

$$h\nu = {}_{-1}^{0}e + {}_{+1}^{0}e + 2KE \qquad (4.16)$$

Looking ahead at Fig. 4.25, one can see two examples of pair production. The dashed lines in the right half of this illustration represent the paths of unchanged particles that leave no track in the actual photograph on the left. The dashed lines are marked γ_1 and γ_2 with the sudden appearance of a pair of divergent tracks. The opposite curvature of the tracks indicates the opposite charges of the electrons as they move through the magnetic field in the bubble chamber.

Pair production is but one form of mass–energy conversion. Proton–antiproton and neutron–antineutron pairs have been produced by accelerators able to produce particles with energies in the range of 1 billion electron volts (GeV). The particle–antiparticle pairs have the following characteristics:

(1) Masses equal and positive
(2) Charges opposite
(3) Spins equal
(4) Magnetic moments opposite
(5) Equal lifetimes
(6) Creation and annihilation in pairs

To produce a proton–antiproton pair, a high energy of 5.6 GeV is required when a high-energy proton strikes a stationary proton. The reaction is

$$p + p \rightarrow 3p + \bar{p} \qquad (4.17)$$

where $\bar{p}$ represents the antiproton. Figure 4.20 is of interest since it shows the disappearance of an antiproton. It may be surmised that it has produced a neutron–antineutron pair (being uncharged, there are no tracks). After traveling 9.5 cm, the antineutron is annihilated, producing an annihilation star by interaction with a neutron in a nucleus. The star is formed by the bubble tracks due to several charged particles produced in the collision.

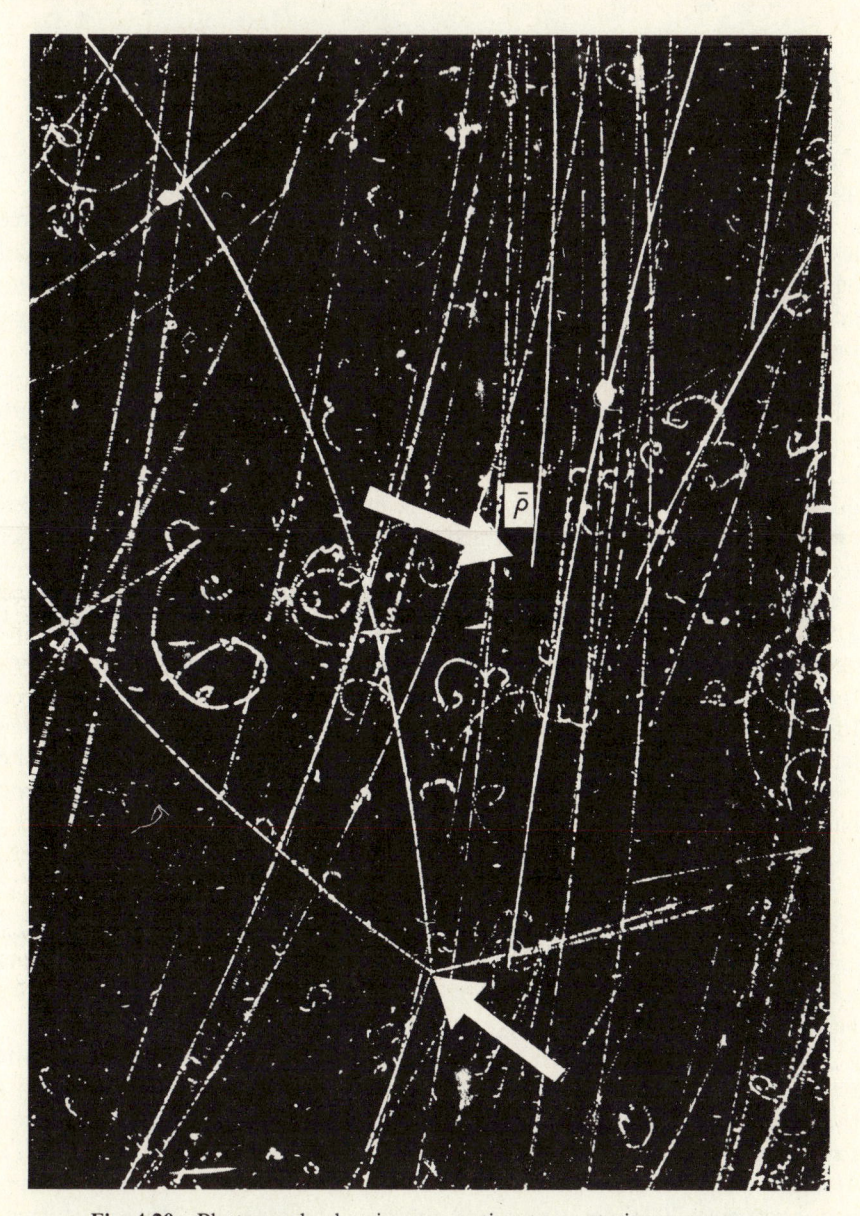

Fig. 4.20 Photograph showing an antiproton entering a propane bubble chamber. It undergoes a charge exchange with a proton to form a neutron–antineutron pair (upper arrow). The antineutron being uncharged leaves no track. After traveling 9.5 cm it interacts with a nucleus to form an annihilation star. [Photograph courtesy of E. Segré, Lawrence Radiation Laboratory, University of California, Berkeley.]

Classes of Particles

Photons, which include both X-rays and gammas, are uncharged, massless quanta of electromagnetic radiation, traveling at the speed of light.

Those weakly interacting particles not subject to the nuclear strong force are known as *leptons*. These include electrons, muons, and two types of neutrinos. They are considered to be truly elemental, having no substructure nor any measurable size. Not being affected by the strong internucleon forces, the leptons make excellent probes to study nucleons and their substructure.

All particles other than photons and leptons are known as *hadrons*. These particles are subject to the nuclear strong force and are considered to be composed of various combinations of *quarks* and *antiquarks*. *Mesons* consist simply of a quark and an antiquark; *baryons* consist of three quarks. The various quarks have been classified by rather whimsical names, sometimes called "flavors." The first four are named *up*, *down*, *strange*, and *charm*. Rather recently, a fifth quark has been confirmed, known as *truth*, with the name *beauty* reserved for an expected sixth quark. It was the discovery of the 3.1-GeV J or psi particle in 1974 which indicated the existence of charm, followed by the 9.5-GeV upsilon particle, which indicated the fifth quark, truth.

An unusual characteristic of quarks is their possession of fractional charges. In the case of the neutron and the proton, their constituent up and down quarks have charges of $\frac{2}{3}$ and $-\frac{1}{3}$, respectively. Their antiquarks have opposite charges. Thus, a proton consists of two up and one down quarks and a neutron has two down and one up.

Virtual particles

Virtual particles have such short lifetimes and distances of travel that their existence may be inferred only indirectly. They are, however, thought to serve as intermediaries for transmission of energy and various other characteristics between interacting particles.

For interactions between high-energy positive and negative electrons, the intermediary is considered to be a virtual photon. In Fig. 4.21a it is shown that the intermediary virtual photon may be transformed into another $\pm$ electron pair. A second possibility (Fig. 4.21b) is the formation of a muon pair ($\pm\mu$). *Muons* are leptons with the same characteristics as electrons except that they have a rest mass 207 times as great. A third possibility (Fig. 4.21c) is the formation of a pair of mesons, known as *pions*. The positive pion (π^+) consists of an up quark (u) and a down antiquark (d) while the negative pion (π^-) is composed of a down quark (d) plus an up antiquark ($\bar{u}$).

In the elastic scattering of protons (see Fig. 4.22a) the interaction is between protons acting as a whole. In this case the intermediary is thought to be an uncharged pion, composed of an up quark and its antiquark. For a hard-

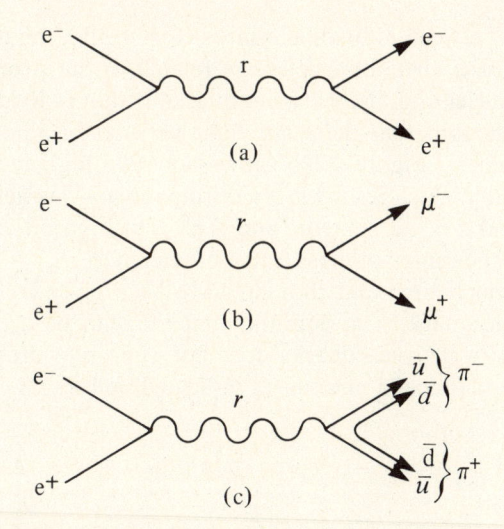

Fig. 4.21 Positron–electron collisions where the electrons are annihilated to form a virtual photon which may decay to (a) another electron pair ($\pm$e), (b) a muon pair ($\pm\mu$), or (c) a pair of pions ($\pm\pi$).

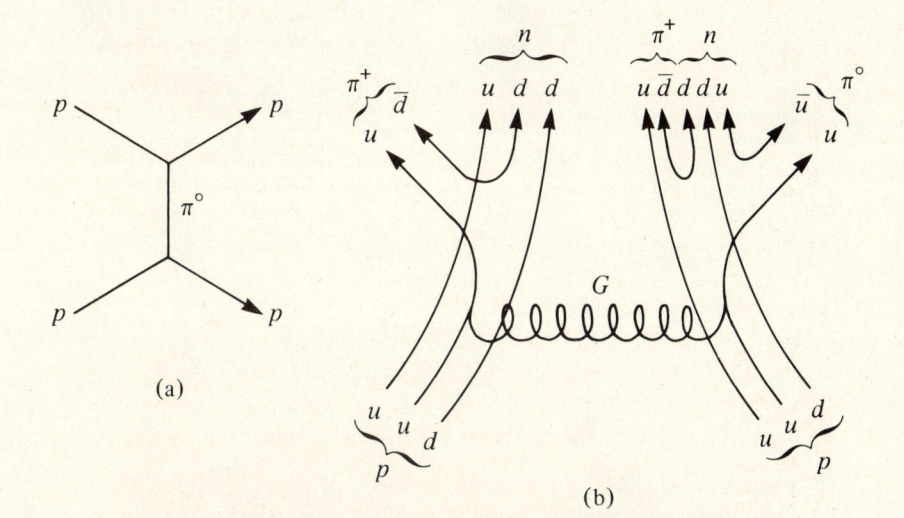

Fig. 4.22 (a) An elastic scattering process between protons where the interaction between the two protons is mediated by an uncharged pion (π^0). In (b) there is a hard scatter between two up quarks. This results in two jets; one consists of a π^+ and a neutron and the other produces a π^+, a π^0, and another neutron.

97

scattering event between individual quarks (Fig. 4.22b), the mediating particle is the *gluon*. These virtual particles are related to the strong binding forces within atomic nuclei and also between quarks within individual nucleons. It has been hypothesised that there are eight varieties of gluons, which "glue" the quarks together. Figure 4.22b shows how two high-energy protons may undergo a collision which results in a neutron and a π^+, which go off as a jet in one direction and a π^+, a neutron, and a π^0, which go off as an opposing jet when viewed in the center-of-mass frame of reference.

Although much time and thought have been given to the possibility of isolating a single quark, it is doubtful that this can be accomplished. The strong force acting between quarks does not appear to drop off at distances greater than the radius of a nucleus, as indicated by Fig. 2.8.

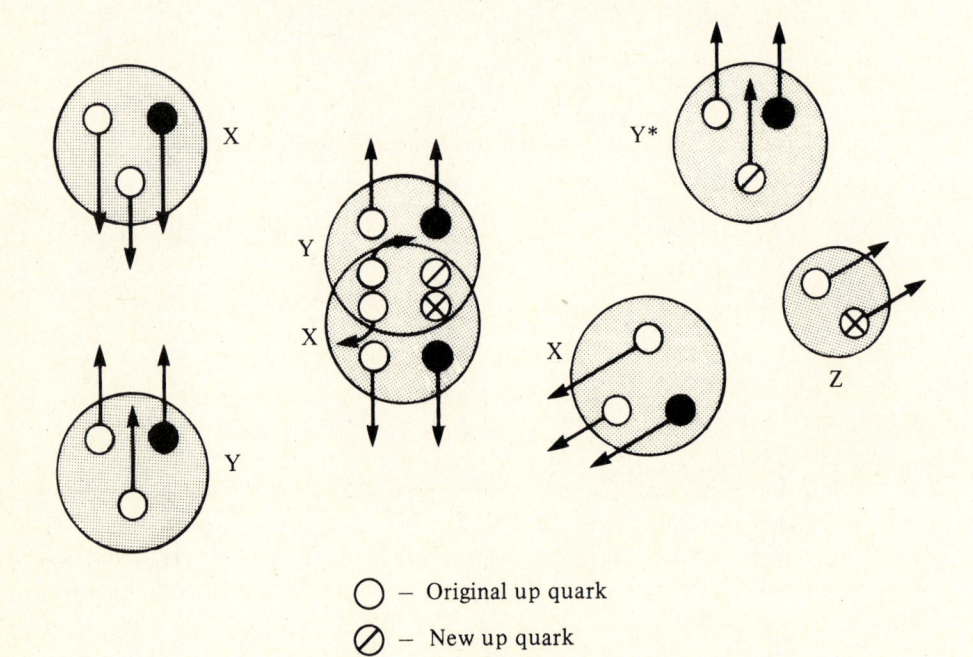

○ — Original up quark

⊘ — New up quark

⊗ — New anti-up quark

● — Down quark

Fig. 4.23 Simplified quark model in which two protons approach one another, each with two up quarks and one down quark. The gray areas represent the strong force field binding the subparticles together. As one proton passes through the other, only two up quarks interact, creating a new $u\bar{u}$ pair. The new quark is appended to Y* and its antiquark $\bar{u}$ is joined by an original u to fly off at a large angle as a π^0 meson, Z. The proton X is deflected to conserve momentum.

Table 4.3 Properties of Elementary Particles

Class of Particle	Particle and Antiparticle	Charge	Mass m_e	Mass (MeV)	Mean Lifetime (s)
Photon	γ	0	0	0	Stable
Lepton	$\nu_e, \bar{\nu}_e$	0	0	0	Stable
	$\nu_\mu, \bar{\nu}_\mu$	0	0	0	Stable
	e^-, e^+	$\pm e$	1	0.511	Stable
	μ^-, μ^+	$\pm e$	207	105.7	2.2×10^{-6}
Hadrons					
Mesons	π^-, π^+	$\pm e$	273.2	139.6	2.602×10^{-8}
	π^0	0	264.2	135.0	0.84×10^{-16}
	K^-, K^+	$\pm e$	966.1	493.7	1.24×10^{-8}
	K^0	0	974.0	497.7	0.882×10^{-10}
	η	0	1074.0	548.8	2.5×10^{-17}
	J or ψ	0	6084	3109	1.8×10^{-17}
	Υ	0	18 600	9500	
Baryons	$p, \bar{p}$	$\pm e$	1836	938.26	Stable
	$n, \bar{n}$	0	1838.6	939.55	918
	$\Lambda, \bar{\Lambda}$	0	2183	1115.6	2.52×10^{-10}
	$\Sigma^+, \bar{\Sigma}^+$	$\pm e$	2327	1189.4	0.8×10^{-10}
	$\Sigma^0, \bar{\Sigma}^0$	0	2333	1192.5	$< 10^{-14}$
	$\Sigma^-, \bar{\Sigma}^-$	$\pm e$	2343	1197.3	1.48×10^{-10}
	$\Xi^0, \bar{\Xi}^0$	0	2573	1314.7	3.0×10^{-10}
	$\Xi^-, \bar{\Xi}^-$	$\pm e$	2586	1321.3	1.7×10^{-10}
	$\Omega^-, \bar{\Omega}^-$	$\pm e$	3272	1672	1.3×10^{-10}

Figure 4.23 gives a somewhat different simple model for a proton–proton collision. Here two quarks collide to form a new $u\bar{u}$ pair. The new up quark joins one of the original protons, while one of the original u's combines with the $\bar{u}$ to form a new meson, an uncharged pion (π^0). Both momentum and energy must be conserved during this reaction.

Table 4.3 shows various characteristics of some of the currently known particles, both leptons and hadrons. Note that the heavy J and upsilon particles are classed as mesons, as they are merely a quark and its antiquark. In the case of the J particle it is a charmed quark with its antiquark; with the upsilon it is the fifth quark, truth, and its antiquark.

Production and Decay of Strange Particles

The bombardment of a carbon target with 380-MeV alphas will produce pi mesons (pions). An energy of 139.6 MeV is required to create a pion whose ·

charge can be of either sign. They are unstable and decay with a mean lifetime of only 2.55×10^{-8} s.

$$p + p \rightarrow p + n + \pi^+ \tag{4.18a}$$

$$p + n \rightarrow p + p + \pi^- \tag{4.18b}$$

$$p + n \rightarrow n + n + \pi^+ \tag{4.18c}$$

An uncharged pion (π^0) can be formed by the interaction of a negative pion and a proton.

$$\pi^- + p \rightarrow n + \pi^0 \tag{4.19}$$

Observe that in Table 4.3 the uncharged pion weighs less than its charged sisters by the amount of the proton–neutron mass difference. This pion decays to two gammas in a mean lifetime of 1.8×10^{-16} s.

$$\pi^0 \rightarrow \gamma + \gamma \tag{4.20}$$

The charged pions, however, decay to a muon and a neutrino.

$$\pi^+ \rightarrow \mu^+ + \nu_\mu \tag{4.21a}$$

$$\pi^- \rightarrow \mu^- + \bar{\nu}_\mu \tag{4.21b}$$

The muons, in turn, decay with a mean lifetime of 2.2×10^{-6} s. They form an electron of the same sign and a neutrino–antineutrino pair.

$$\mu^\pm \rightarrow e^\pm + \nu_\mu + \bar{\nu}_\mu \tag{4.22}$$

It has been determined that there are two types of neutrino pairs: those that accompany beta decay (ν_e and $\bar{\nu}_e$) and those that accompany pion decay (ν_μ and $\bar{\nu}_\mu$). When undergoing an inverse process, they will produce either electrons or muons, depending on their ancestry. The muons have properties identical to the electrons, except for their mass, which is 207 times as great.

The strange particle baryons of mass between one and two nucleons are also known as hyperons. They are designated as Λ (lambda), Σ (sigma), Ξ (xi), and Ω (omega) particles. Figure 4.24 is exceptionally fine in that it shows the formation of a positive sigma particle, Σ^+, and a positive kaon, K^+, by the interaction of a positive pion, π^+, and a proton.

$$\pi^+ + p \rightarrow \Sigma^+ + K^+$$
$$\qquad\qquad \big\downarrow$$
$$\qquad\qquad \hookrightarrow \pi^+ + n \tag{4.23}$$

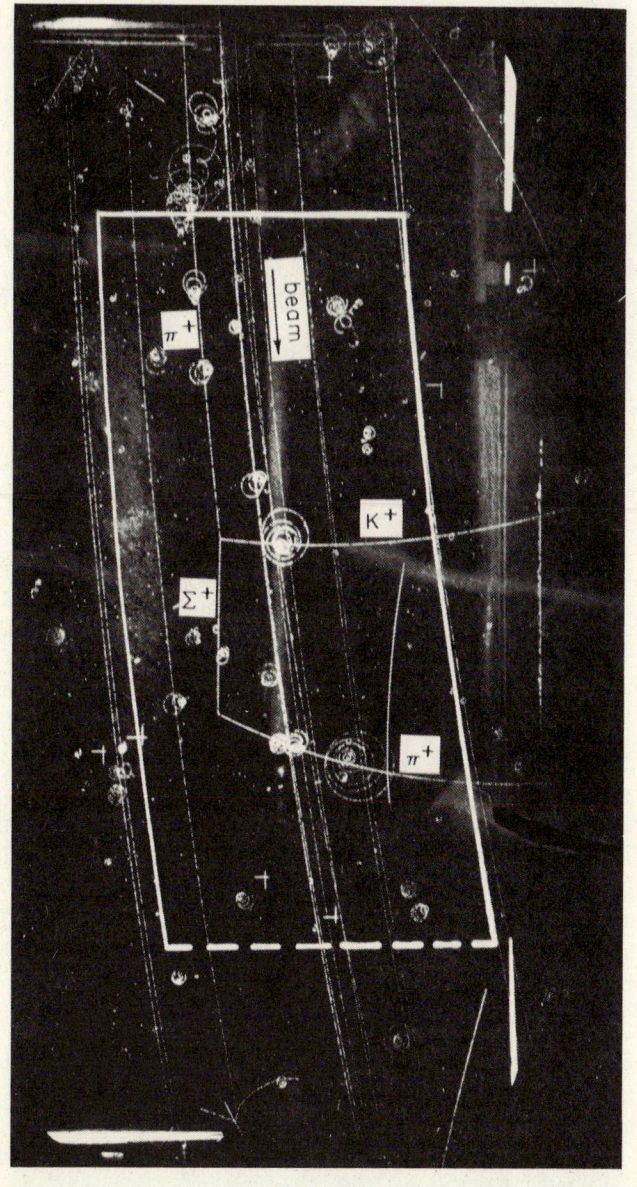

Fig. 4.24 Photograph of the tracks of particles in a liquid hydrogen bubble chamber showing the production of a positive sigma particle by the interaction of a positive pion and a proton. The initial momentum of the pion is 1.23 GeV/c the bubble chamber is in a magnetic field of 17 000 G. [From C. Baltay et al., *Rev. Mod. Phys.* **33** (1961), p. 374.]

The positive kaon is a heavy meson with a mass of $466.7m_e$. It leaves the bubble chamber before decaying, but the positive sigma particle decays into a positive pion and an uncharged neutron.

The series of events shown in Fig. 4.25 confirms the existence of the Ω^- particle, which had been predicted previously by a gap in the array of known particles. About 100 000 pictures were taken to find the first example of this event. At the bottom of the photograph is shown a stream of K^- mesons entering the bubble chamber with a momentum of 5.0 GeV/c. A collision between one of these negative kaons and a proton produces the Ω^-, as well as a K^+ and a K^0. The Ω^- undergoes the following sequence of decay events:

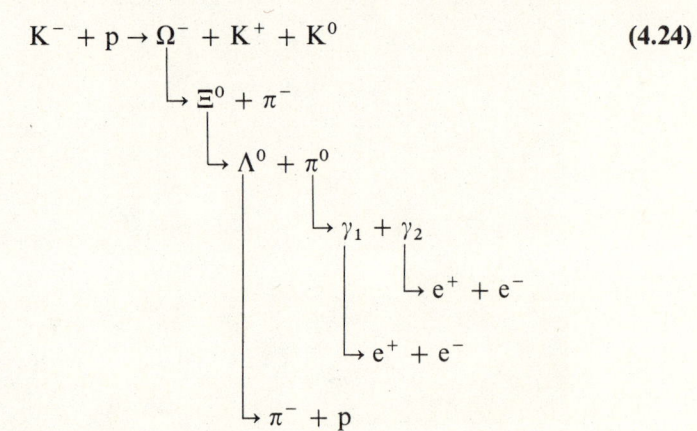

$$K^- + p \rightarrow \Omega^- + K^+ + K^0 \tag{4.24}$$

The study of these strange particles has led to their first medical application. At Stanford's Linear Accelerator Center, π^- mesons have been used for cancer therapy. These π^- mesons have the unique ability to penetrate deeply with little damage to outer tissues. When focused on the diseased area, oxygen atoms capture the negative pions, causing star explosions that destroy the adjacent malignant cells.

Fermi National Accelerator Laboratory

Located at Batavia, Illinois, the Fermi National Accelerator Laboratory (Fermilab) is designed to produce protons with energies as high as 400 GeV, and with the addition of a ring of superconducting magnets it is hoped to reach 1000 GeV. Figure 4.26 is an aerial photograph of Fermilab, and Fig. 4.27 shows a map of the laboratory.

Protons are accelerated to 0.75 MeV in a Cockcroft–Walton generator, and then are directed to a linear accelerator 145 m long and 1 m in diameter, where they are boosted up to 200 MeV. From this energy level a synchrotron boosts the proton energy to 8 GeV after which they are injected into the 6.3-km-circumference main ring. After 200 000 revolutions in the main ring, the protons will acquire an energy of 400 GeV.

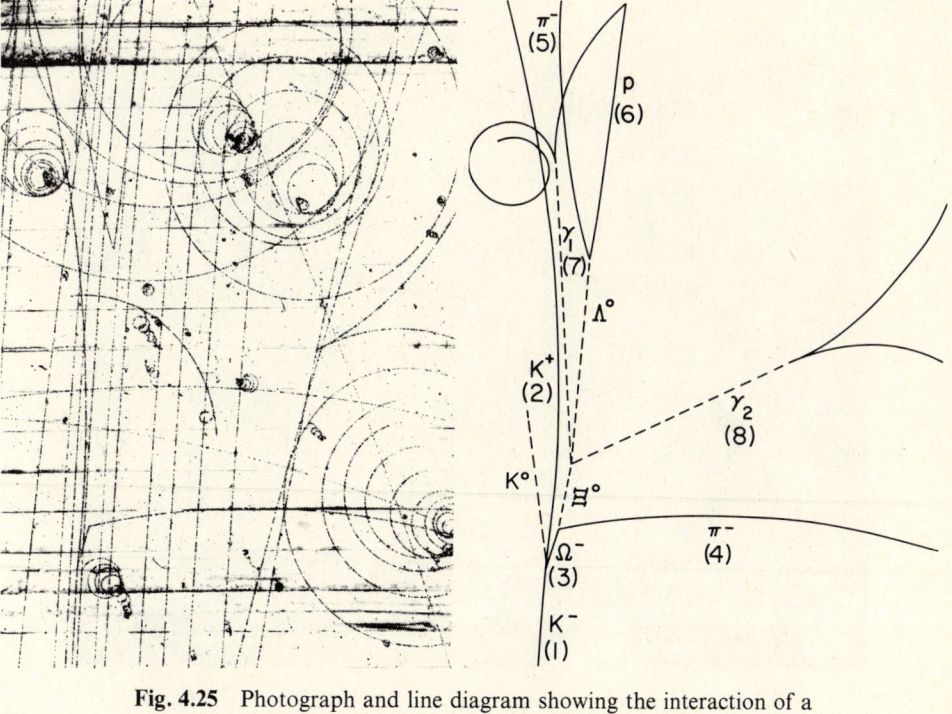

Fig. 4.25 Photograph and line diagram showing the interaction of a negative kaon with a proton to form an omega minus particle together with a positive kaon and an uncharged kaon. The series of events due to the decay of the Ω^- is also shown. [Photograph courtesy of Brookhaven National Laboratory Bubble Chamber Group.]

The high-energy protons are deflected by magnets and directed to the various experimental areas (proton, meson, or neutrino). When heading for the neutrino area the beam travels 1000 m before striking a target (usually steel) to produce a shower of particles, mostly pions and kaons. These pass into a 400-m decay tube and most decay to a muon and a neutrino. At the end of the decay tube, there are 1000 m of dirt fill, which filter out pions, kaons, and protons, leaving only neutrinos and antineutrinos to arrive at the bubble chamber. It is interesting to note that Pauli proposed the neutrino in 1931 to account for the apparent loss of momentum and energy in beta decay processes. It was not until 1956 that the first neutrino-induced event was observed experimentally through inverse beta decay.

$$\nu + p \rightarrow n + \beta^+ \tag{4.25a}$$

$$\beta^+ + \beta^- \rightarrow \gamma + \gamma \tag{4.25b}$$

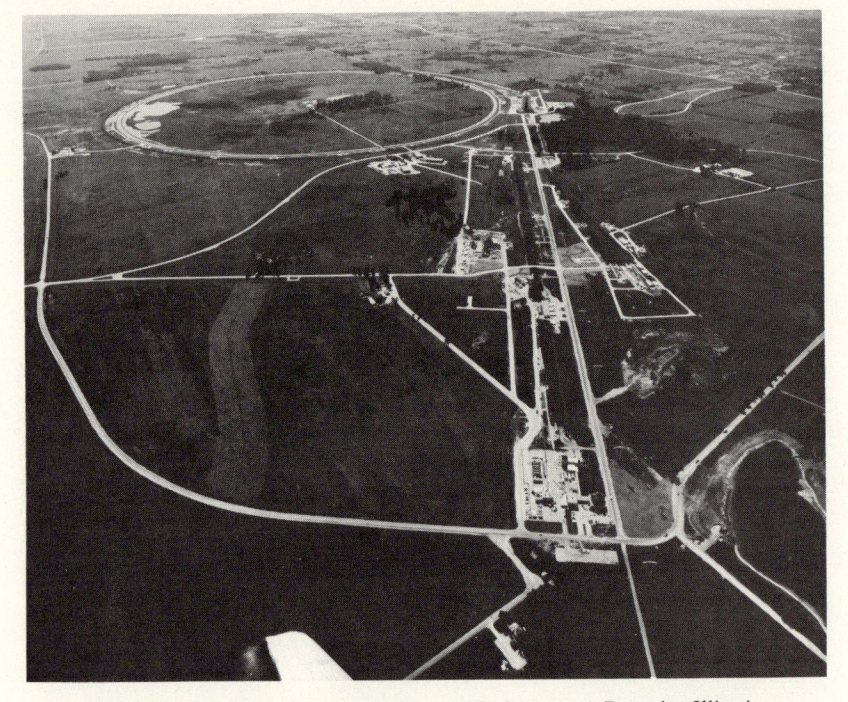

Fig. 4.26   Fermi National Accelerator Laboratory, Batavia, Illinois.
The main ring of the world's largest accelerator is housed in a circular
tunnel 2 km in diameter and buried some 6 m underground. [Courtesy
of the Fermi National Accelerator Laboratory.]

Figure 4.28 shows a bubble chamber photograph of a neutrino–neutron
interaction that produces another neutrino, an uncharged lambda, and an
uncharged kaon.

$$v_\mu + n \rightarrow v'_\mu + \Lambda + K^0$$

$$\qquad\qquad\qquad \hookrightarrow \pi^+ + \pi^- \qquad\qquad (4.26)$$

$$\qquad\quad \hookrightarrow p + \pi^-$$

This photograph was taken in the 3.66-m (12-ft) bubble chamber at the ANL
Zero Gradient Synchrotron from amoung 580 000 photographs taken with the
tank filled with deuterium.

   A more complete picture of the work being done on strange particles can
be obtained from the references at the end of the chapter.

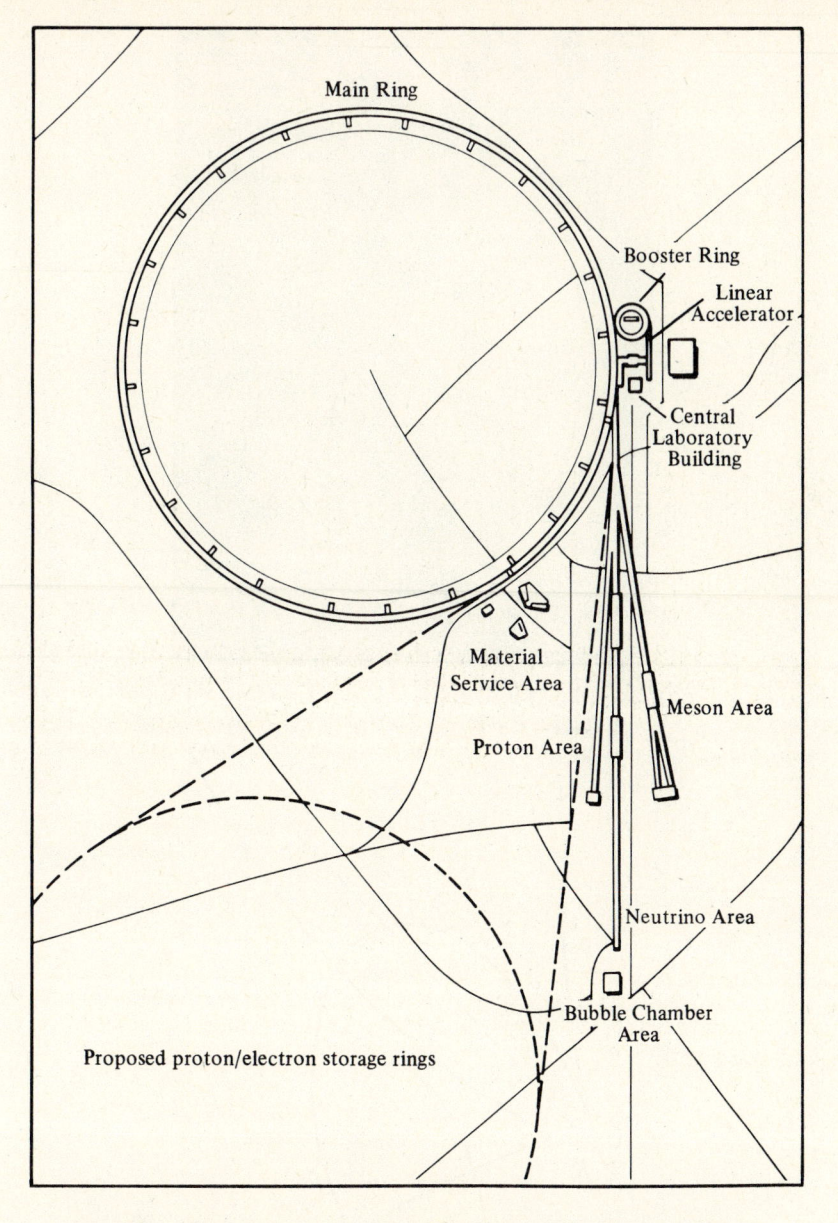

Main Ring

Booster Ring

Linear Accelerator

Central Laboratory Building

Material Service Area

Proton Area

Meson Area

Neutrino Area

Bubble Chamber Area

Proposed proton/electron storage rings

Fig. 4.27 Map of the National Accelerator Laboratory showing the main ring and the various experimental areas (Proton Area, Meson Area, and the Neutrino Area, with its enormous bubble chamber.) Also shown is a proposed proton/electron storage ring where energetic protons would be diverted from the main ring to run head on into protons or electrons stored in the ring. Head-on collision of two 1000-GeV protons would correspond to the energy released when protons from a single 2.2×10^6-GeV beam strike target protons that are at rest. [Courtesy of the Fermi National Accelerator Laboratory.]

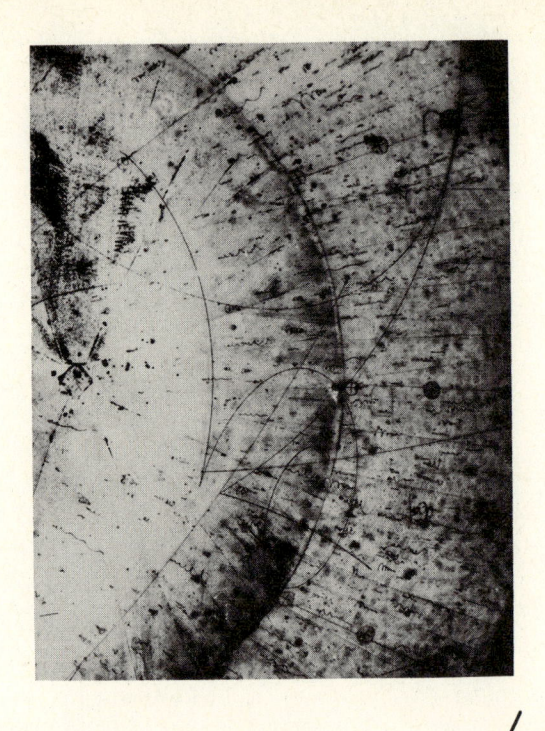

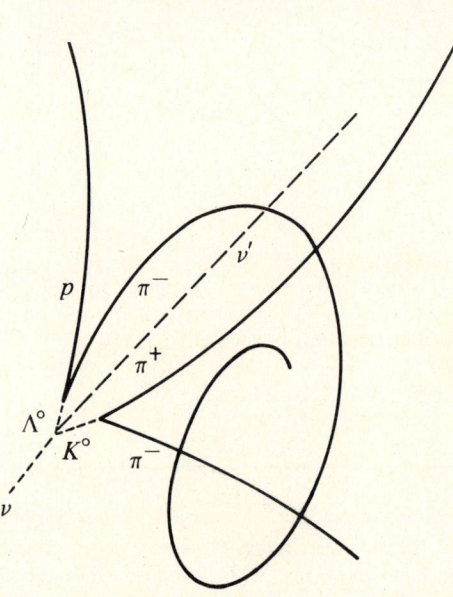

Fig. 4.28 Photograph of the $\nu n \nu \Lambda^0 K^0$ event. Below the charged particles are represented by solid lines, and the neutral particles by dashed lines. [From S. J. Barish et al., *Physical Review Letters* **33**, No. 24 (December 9, 1974), p. 1448.]

Problems

1. ^{37}Cl is struck by a proton and emits a 1-MeV neutron (COM). Write the reaction equation and compute the energy of the incident proton in the laboratory system.

2. A 2-MeV deuteron strikes a $^6_3$Li nucleus. The compound nucleus splits into identical nuclides. Identify the resultant particles and calculate the COM kinetic energy of each.

3. A 2-MeV neutron strikes a $^{14}_7$N nucleus, causing an alpha particle to be ejected. What element is the recoil nucleus? What will be the kinetic energy (COM) of the recoil nucleus? What will be the maximum kinetic energy of the alpha in the laboratory system?

4. A ^{54}Fe nucleus absorbs a thermal neutron and gives off a single very energetic capture gamma. Calculate the energy of the capture gamma and the kinetic energy for the recoiling iron nucleus.

5. Compute the threshold energy for photoemission of a neutron by a $^9_4$Be nucleus. The resultant $^8_4$Be nucleus is in turn unstable and splits into two alphas. What will be the velocity of the two emergent alphas if the original gamma had just the threshold energy?

6. Show that in the laboratory system, when the kinetic energy of the compound nucleus is subtracted from the kinetic energy of the incident particle, the difference is identical to Eq. (4.4b).

7. A light hydrogen atom absorbs a thermal neutron in a radiative capture process. Compute the energy of the gamma emitted. Calculate the recoil velocity and kinetic energy of the deuteron.

8. Thermal fission of a ^{233}U nucleus produces four neutrons and two fission fragments, one of which is ^{143}La (142.9157 u). What is the other fission fragment? Compute the energy released if the second fragment has a mass of 86.9224 u.

9. If the Watt equation (4.9) defines the energy spectrum of the fission neutrons for ^{235}U, show that the most probable fission energy is 0.72 MeV.

10. In a fusion reactor the uncharged neutrons produced in some of the reactions will escape the reaction chamber. Their energy can be transformed to heat by a surrounding moderator. If the moderator contains $^6_3$Li, neutron absorption by this isotope will produce an extra source of tritium. Compute the energy available from the reaction per neutron absorbed.

11. Show that the deuterium–tritium fusion reaction produces 17.6 MeV. Determine the kinetic energy of the emergent neutron. Is the neutron relativistic?

12. Show that a laboratory energy of 5.6 GeV is required to produce a proton–antiproton pair when a proton strikes another stationary proton.

13. In the laboratory system, what is the minimum energy required for a proton striking a stationary proton to create a positive pion (π^+), as shown by Eq. (4.18a)?

14. Proton decay of ^{53}Co was mentioned in Chapter 3. To form the ^{53}Co nucleus ^{40}Ca can be bombarded by an energetic ^{16}O nucleus. An alternative method of forming the ^{53}Co is by proton bombardment of ^{54}Fe. What particles are emitted in each case?

15. In the Li-cooled blanket of a D-T Tokamak-type fusion reactor, why can the (n, n'α) reaction for ^{7}Li take place only with fast neutrons?

References

1. Liverhant, S. E., *Elementary Introduction to Nuclear Reactor Physics*. New York: John Wiley & Sons, Inc., 1960.
2. Semat, H., *Introduction to Atomic and Nuclear Physics*. New York: Holt, Rinehart and Winston, 1962.
3. Kaplan, I., *Nuclear Physics*. Reading, Mass.: Addison-Wesley Publishing Co., Inc., 1955.
4. Lapp, R. E., and H. L. Andrews, *Nuclear Radiation Physics*. Englewood Cliffs, N.J.: Prentice-Hall, Inc., 1963.
5. Roderick, H., and A. E. Ruark, "Thermonuclear Power," *International Science and Technology* **45** (September 1965), pp. 18–29.
6. Murray, R. L., *Introduction to Nuclear Engineering*. Englewood Cliffs, N.J.: Prentice-Hall, Inc., 1961.
7. Bishop, A. S., *Project Sherwood—The U.S. Program in Controlled Fusion*. Reading, Mass.: Addison-Wesley Publishing Co., Inc., 1958.
8. Glasstone, S., and R. H. Loveberg, *Controlled Thermonuclear Reactions*. Princeton, N.J.: D. Van Nostrand Company, 1960.
9. Katkoff, S., "Fission Product Yields from U, Th, and Pu," *Nucleonics* **16**, No. 4 (April 1958), p. 78.
10. Jensen, J. E., *The Cryogenic Bubble Chamber—An Accelerator Research Facility*, ASME Paper No. 66-WA/NE 22, 1966.
11. Hill, R. D., *Tracking Down Particles*. New York: W. A. Benjamin, Inc., 1964.
12. Barnes, V. E., et al., "Observation of a Hyperon with Strangeness Minus Three," *Physical Review Letters* **12** (February 1964), pp. 204–206.
13. Alvarez, L. W., et al., "1660-MeV Y_1^* Hyperon," *Physical Review Letters* **10**, No. 5 (March 1963), pp. 184–188.
14. Segré, E., *Nuclei and Particles*. New York: W. A. Benjamin, Inc., 1965.
15. Seaborg, G. T., "The Synthetic Actinides—from Discovery to Manufacture," *Nuclear Applications & Technology* **9**, No. 6 (December 1970), pp. 830–850.
16. "Final Report of the IAEA Panel on International Co-operation in Controlled Fusion Research and Its Applications," *Nuclear Fusion* **10**, No. 4 (December 1970), pp. 413–421.
17. "Enter Element 105," *Nuclear News* **13**, No. 6 (June 1970), p. 20.
18. "Pi Mesons in Cancer Treatment," *Nuclear News* **13**, No. 6 (June 1970), p. 34.
19. Butler, J. W., et al., "Report of the Argonne Senate Subcommittee on Controlled Thermonuclear Research," *Argonne Reviews* **6**, No. 1 (July 1970), pp. 30–34.
20. "Eka-Mercury, Evidence for Element 112," *Science News* **99**, No. 8 (February 20, 1971), pp. 127–128.

21. Kendall, H. W., and W. K. H. Panofsky, "The Structure of the Proton and the Neutron," *Scientific American* **224**, No. 6 (June 1971), pp. 60–77.
22. Andrews, H. L., *Radiation Biophysics*. Englewood Cliffs, N.J.: Prentice-Hall, Inc., 1974.
23. Wilson, R. R., "The Batavia Accelerator," *Scientific American* **230**, No. 2 (February 1974), pp. 72–83.
24. Weinberg, S., "Unified Theories of Elementary Particle Interaction," *Scientific American* **231**, No. 1 (July 1974), pp. 50–59.
25. Coppi, B., and Rem, J., "The Tokamak Approach in Fusion Research," *Scientific American* **229**, No. 2 (July 1972), pp. 61–75.
26. Longo, M. V., *Fundamentals of Elementary Particle Physics*. New York: McGraw-Hill Book Company, 1973.
27. Burcham, W. E., *Elements of Nuclear Physics*. New York: Longman, Inc., 1979.
28. Steiner, D., "Fusion Power Development: Status and Prospectus," *Mechanical Engineering* **102**, No. 6 (June 1980), pp. 48–53.
29. Lederman, L. M., "The Upsilon Particle," *Scientific American* **239**, No. 4 (October 1978), pp. 72–80.
30. Jacob, M., and P. Landshoff, "The Inner Structure of the Proton," *Scientific American* **242**, No. 3 (March 1980), pp. 66–75.
31. 't Hooft, G., "Gauge Theories of the Forces Between Elementary Particles," *Scientific American* **242**, No. 6 (June 1980), pp. 104ff.
32. Jung, J., "A Comparative Study of Tritium Breeding Performance of Lithium, LiO_2, and Li_7Pb_2 Blankets in a Tokamak Power Reactor," *Nuclear Technology* **50**, No. 1 (mid-August 1980), pp. 60–82.
33. Choppin, G. R., and J. Rydberg, *Nuclear Chemistry*. Elmsford, N.Y.: Pergamon Press, 1980.

Radiation Detection

Radiation detection results from ionization of the medium through which radiation passes. The charge collected is a measure of the radiation, either in the form of pulses or current. Four general types of radiation which are of interest are discussed in this chapter.

Heavy Charged Particles

The following comments apply to alpha particles and, generally, to fission fragments and protons as well. Of the four interactions that alpha particles have with matter, scattering and nuclear transmutation are rare. Excitation and ionization, therefore, are almost exclusively used as the processes for alpha detection. An alpha particle loses an average of 32 to 35 eV per ion pair produced. This characteristic is utilized while the alpha particle moves along its straightline path. Ionization can be said to be a property of the particle's passing through matter. The charged particle transfers some of its kinetic energy to the electrons it encounters. The electrons are either raised to an excited state or removed from the atom. The heavy particle (α) loses energy by gradually transferring small amounts of energy to atomic electrons of absorbing material (32 to 35 eV per ion pair produced). Finally, the alpha is stopped and acquires two electrons to become an uncharged helium atom.

Each alpha (or other particle) has a definite range. The range depends on the initial energy of the alpha and the properties of the absorbing material. Since the mass of electrons is small, they cannot deflect a heavy particle from its path. Occasionally (usually toward the end of its track), an alpha may collide with a nucleus and be deflected through a large angle (see Fig. 5.1). The range for a group of alphas of the same initial energy has a distribution over a small limit, also shown by Fig. 5.1.

The Bragg curve, Fig. 5.2, shows that the specific ionization (energy loss) increases along the particle track to a maximum and then rapidly drops to

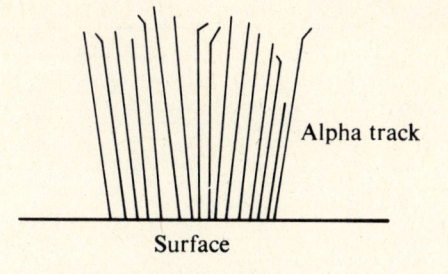

Fig. 5.1 Sketch of cloud chamber tracks of alpha particles.

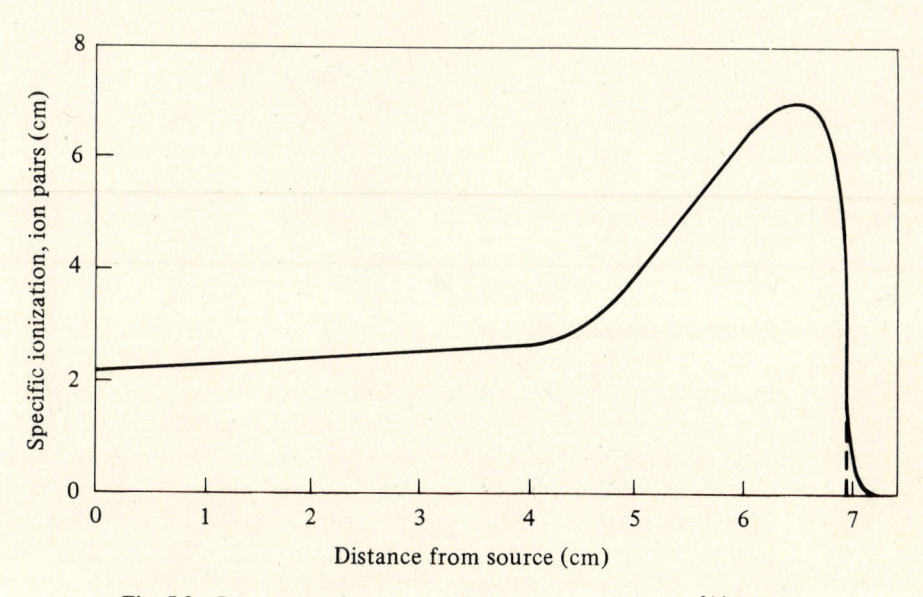

Fig. 5.2 Bragg curve for single alpha track due to decay of ^{214}Po(7.68 MeV), having an extrapolated range of 6.95 cm in air at 1 atm and 15°C.

zero. Another relationship, Fig. 5.3, shows that the alpha range is energy sensitive.

Specific ionization is the number of ion pairs produced per centimeter of track. *Total ionization* is the total number of ion pairs produced. The stopping power of an absorber, dE/dx, is related to the specific ionization, I, by

$$de/dx = -wI \tag{5.1}$$

The minus sign occurs because energy, E, decreases as x increases; w is the mean energy expended per ion pair produced. Fortunately, w is nearly independent of the energy of the primary particle. In the case of argon, w is the same

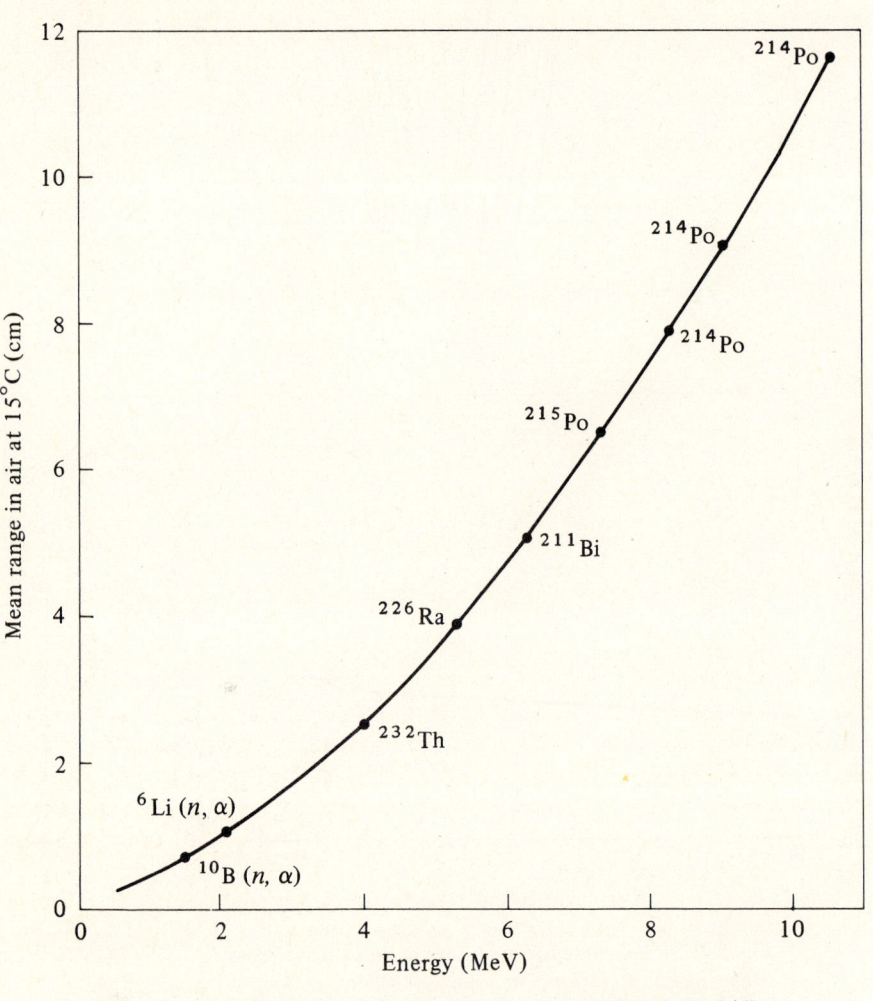

Fig. 5.3   Mean range for alpha particles in air at 15°C and 101.3 kPa.

for alphas, protons, electrons, and other light particles. Thus, argon is particularly well suited as a gas for use in ion chambers and other counting instruments. Both w and I, as yet, have proven difficult to calculate theoretically; therefore, experimental measurement is relied on to determine their values. The range of alpha particles can be measured experimentally by measuring the intensity of monoenergetic alphas at different distances from a thin source. The intensity of alpha plotted versus distance from source appears as in Fig. 5.4.

    The extrapolated range, R_e, is the range at which the tangent to the curve at its inflection point intersects the horizontal axis. The mean range, $\bar{R}$, is the

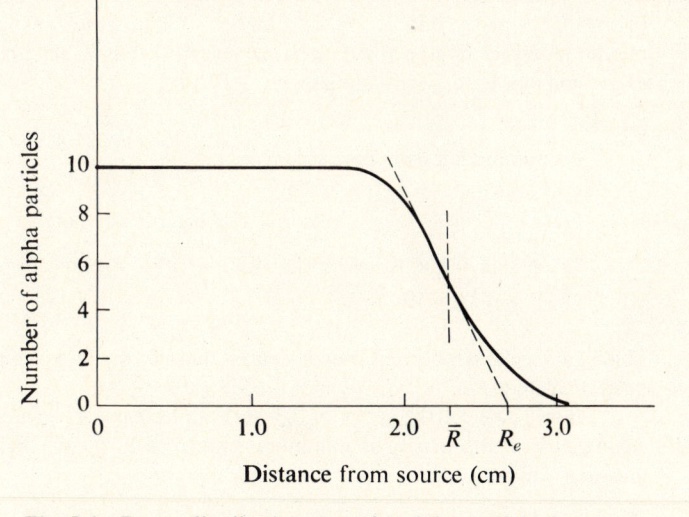

Fig. 5.4   Range distribution curve for alpha particles in a gas.

range at which half of the particles have a greater range and half a smaller range.   This occurs at the inflection point, and the alphas have a Gaussian distribution about this point.   The difference between the mean and extrapolated ranges is the straggling.   It occurs because of the random nature of the collisions. If there were no straggling, the curve would drop sharply to zero at the range of the alpha.   The mean range and mean energy expended in air of some alpha particles are shown in Table 5.1.

Table 5.1 Range of Alpha Particles

Energy (*MeV*)	Range in Air (*cm*)	*w* (*eV/ion pair*)	Range in Aluminum (*mg/cm²*)	Range in Lead (*mg/cm²*)
2	1		1.5	3.7
3.5	2		3.1	6.7
5.3	3.7	35.6	5.6	13.7
6.3	5		7.6	18.0
7.8	7.3	35.1	10.8	25.2
9.7	10		14.8	34.5

Data from W. A. Aron, B. D. Hoffman, and F. C. Williams, AECU Pamphlet 663, 1949.

Example 1 Determine the number of ion pairs produced and the ionizing current of 100 alphas per second of 5.3-MeV energy. What is the thickness of aluminum necessary to stop these alpha particles? From Table 5.1, $w = 35.6$ eV per ion pair for 5.3-MeV alphas.

$$\text{number ion pairs produced} = \frac{(5.3 \times 10^6 \text{ eV/}\alpha)(100 \text{ }\alpha/s)}{35.6 \text{ eV/ion pair}}$$

$$= 1.489 \times 10^7 \text{ ion pairs/s}$$

$$I = (1.489 \times 10^7 \text{ electrons/s})(1.6 \times 10^{-19} \text{ C/electron})$$

$$= 2.38 \times 10^{-12} \text{ Å}$$

This is a very small current, but it can be detected by fairly conventional microammeters.

From Table 5.1 the range of 5.3-MeV alpha particles in aluminum is 5.6 mg/cm^2. If the density of aluminum is taken as 2.7 g/cm^3, the necessary thickness will be

$$x = \frac{5.6 \times 10^{-3} \text{ g/cm}^2}{2.7 \text{ g/cm}^3}$$

$$= 2.1 \times 10^{-3} \text{ cm}$$

Fission fragments have more nuclear collisions and produce less ionization than alpha particles. The fission fragments are initially highly ionized; the ionization is gradually reduced by picking up electrons. The state of ionization decreases therefore, causing less ionization in the material through which it passes. An alpha may also pick up some electrons but will lose them in succeeding collisions. Fission fragments lose most of their energy near the beginning of their track. whereas alpha energy loss increases along the track (Bragg curve). The total range of an average fission fragment is 6×10^{-4} cm in uranium and 12×10^{-4} cm in aluminum. (For aluminum cladding the thickness is usually 50 to 100 times this range.) Since few fission fragments can escape, most of the energy loss occurs inside the fuel element; this causes the fuel and cladding to heat up considerably.

Light Charged Particles (β)

Light charged particles are not as easy to analyze as heavy charged particles because:

(1) Their path is very irregular.
(2) An electron has a smaller mass, so it suffers many abrupt deflections. It is difficult, therefore, to associate a range with it (see Fig. 5.6).

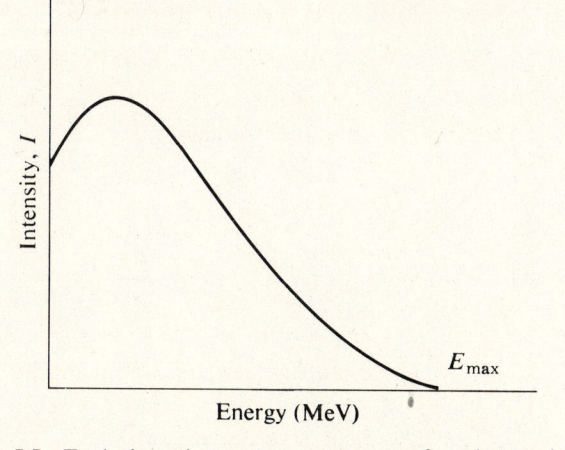

Fig. 5.5 Typical continuous energy spectrum for a beta emitter.

(3) A mechanical effect is involved since a beta particle and an electron are identical.
(4) The high speed of a beta particle necessitates a relativistic treatment of the collision process.
(5) Beta radiation shows a continuous initial energy distribution. Figure 5.5 is a typical energy spectrum.

While different isotopes emit beta radiation with different spectra, the general shape appears as in Fig. 5.5.

The processes by which a beta particle interacts with matter are as follows:

(1) *Elastic collision.*
(2) *Energy conversion.* Mass in the form of an electron and a positron is converted to radiation, resulting in two oppositely directed photons.
(3) *Inelastic collision* (ionization, excitation). This process generally predominates for beta energies of 1 MeV or less. The collision is similar to the ionizing effect of heavy charged particles. The energy loss varies approximately as

$$-\frac{dE}{dx} \propto \frac{NZ}{v^2} \tag{5.2}$$

where Z is the atomic number and N the atomic density of the absorber. The ions produced in the primary ionization often produce further or secondary ionization as they release their excitation energy. The total ionization is the sum of the primary and secondary ionization.
(4) *Bremsstrahlung* (braking radiation). When electrons have an inelastic collision with nuclei, they radiate energy in the form of a continuous X-ray emission. The electron path is bent as it nears the nucleus. This path change results in an acceleration of the electron. The radiation emitted is directly proportional to the

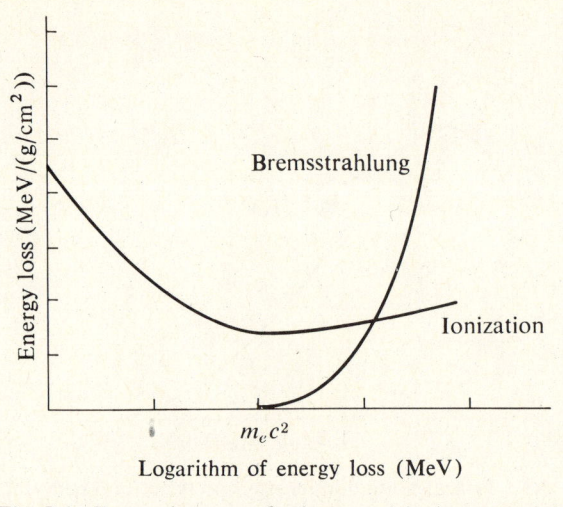

Fig. 5.6   Energy-loss rates for beta particles in a material.

acceleration squared.   With heavier particles the acceleration as a result of inelastic collision is small and consequently the *bremsstrahlung* is negligible.   *Bremsstrahlung* accounts for the spectrum emitted by X-ray tubes.   The energy loss can be stated as

$$-\frac{dE}{dx} \propto NZ^2E \qquad (5.3)$$

where E is the electron energy and N and Z are as before.

By combining the energy loss from *bremsstrahlung* and ionization one can plot the continuous energy loss noted previously (see Fig. 5.6).

Since beta particles suffer so many deflections while they lose energy, it is difficult to associate a definite range with a given beta energy.   Figure 5.7 is a plot of the number of beta particles versus their range (expressed in g/cm^2). Note that there is a definite upper limit, but that the curve is not flat up to this range.   The extrapolated range can be used to take the place of an actual range of beta particles.   (*Extrapolated range* is the thickness of material required to reduce the intensity to the background rate.)   Fortunately, this relationship also holds for a continuous beta spectrum if the range is associated with the maximum beta energy (refer to Fig. 5.5).   If the range is extrapolated as in Fig. 5.7, then for the most probable beta energy, $E_p = \frac{1}{3}E_{max}$, the intensity of radiation follows an exponential curve

$$I = I_0 e^{-\mu x} \qquad (5.4)$$

where μ is the linear absorption coefficient and has dimensions of reciprocal centimeters.   A mass absorption coefficient may be defined by

$$\mu_m = \mu/\rho \qquad (5.5)$$

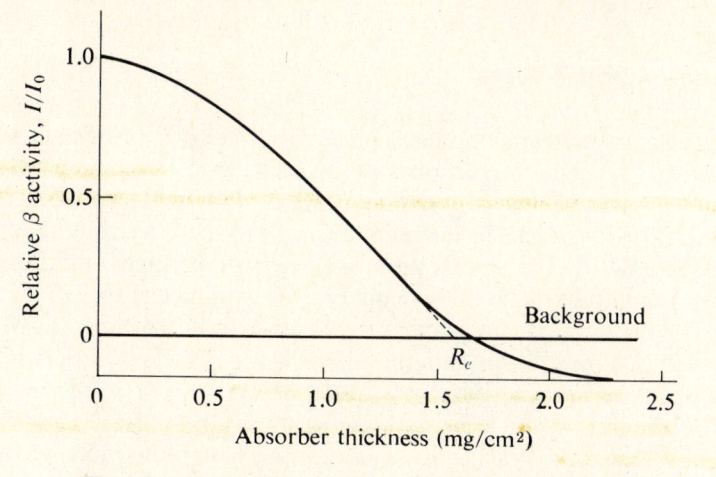

Fig. 5.7   Absorption curve for a single-energy beta particle.

Equation (5.4) then becomes

$$I = I_0 e^{-\mu_m d} \tag{5.6}$$

where d is the absorber thickness in g/cm^2.

The result of all this is that one finds that for a given material, beta particles have a greater range than alpha particles.

Example 2   Find the thickness of aluminum absorber necessary to absorb 99 percent of the 5.3-MeV maximum beta particles striking it.   An empirical relationship (see Reference 8)

$$\mu_m = \frac{22}{E_m^{1.33}} \tag{5.7}$$

gives mass absorption coefficients for 0.5 MeV $< E_m <$ 6 MeV.

$$\mu_m = \frac{22}{(5.3)^{1.33}} = 2.39 \ cm^2/g$$

$$I/I_0 = 0.01 = e^{-2.39d}$$

$$-4.59 = -2.39d$$

$$d = 1.92 \ g/cm^2$$

$$x = \frac{1.92 \ g/cm^2}{2.7 \ g/cm^3} = 0.710 \ cm$$

Thus, it is seen that a beta particle of 5.3 MeV maximum energy will have a greater range than a corresponding alpha particle.

Gamma (γ) and X-Rays

Almost all gamma interaction takes place with electrons. Normally a gamma ray has only a single interaction with an electron. The photoelectric effect predominates when low-energy gammas interact with tightly bound atomic electrons. It is believed that the interaction is with K-shell electrons about 80 percent of the time. In the process the atomic electron becomes detached from its atom and acquires kinetic energy equal to the gamma energy less the binding energy for the electron; the gamma ray is used up in the process. When the outer-shell electrons fill the gap, they emit X-rays. The photoelectric effect is predominant for energies of about 0.1 MeV or less.

The *Compton effect* is an inelastic scattering between a photon and an individual electron. Practically speaking, the Compton effect becomes important for gamma energies of about 0.1 MeV and up. Since the gamma radiation is absorbed by the interacting medium, a linear absorption coefficient can be defined for gamma radiation. Figure 5.8 shows that the Compton coefficient consists of two parts: one portion caused by scattering of the photon and the remainder due to true absorption. The photon energy is reduced by scattering and eventually the photoelectric process will take place. There is essentially no gamma scattering in the photoelectric effect, so there need be no division of the absorption coefficient. However, Compton scattering occurs between a gamma ray and an electron, so the linear absorption coefficient for Compton scattering must be dependent on the number of electrons present (Z of absorber).

Pair production results in complete absorption of the gamma ray and production of a positron–electron pair. This energy-to-mass conversion takes place normally in the vicinity of a nucleus. Since the rest energy of the positron–negatron pair is $2m_0 c^2$, the incident photon must have at least this energy (1.022 MeV). Any excess energy will be shared between the positron and electron.

Combining the linear absorption coefficients for a given material results in a graph similar to Fig. 5.9. It should be noted that the curves will be different

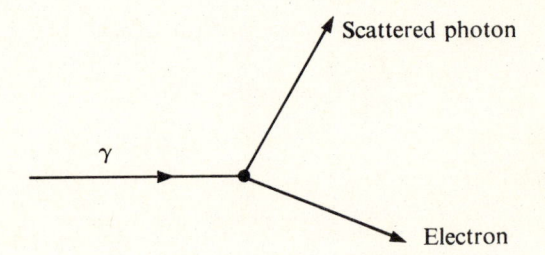

Fig. 5.8   Compton collision between a gamma ray and an electron.

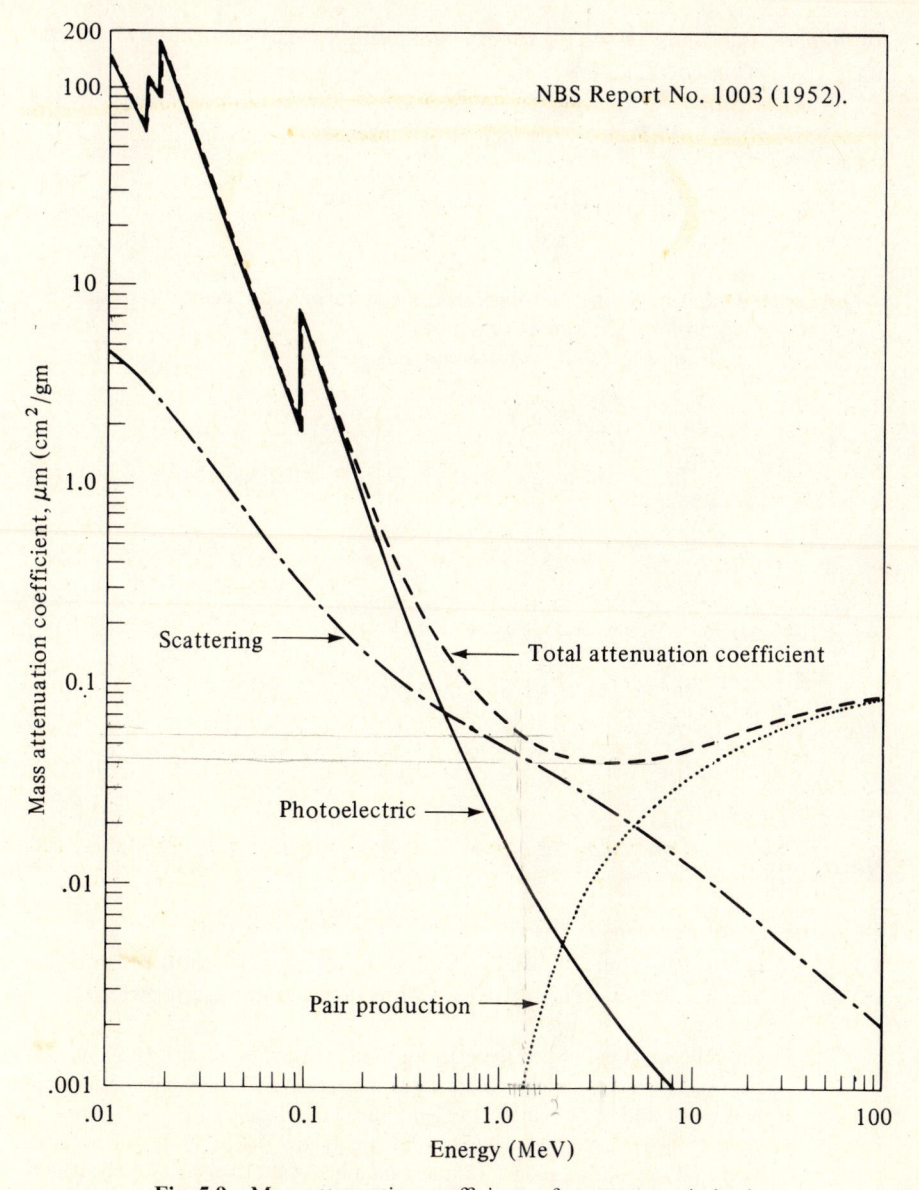

Fig. 5.9 Mass attenuation coefficients of gamma rays in lead.

for different materials.   As in the case of beta radiations, we can define linear and mass absorption coefficients.

A mean range, $\bar{R}$, for gamma penetration in matter can be defined as the average distance a photon travels before it is absorbed.

$$\bar{R} = \frac{\int_{I_0}^{0} x \, dI}{\int_{I_0}^{0} dI} = \frac{\int_{0}^{\infty} x e^{-\mu x} \mu \, dx}{\int_{0}^{\infty} e^{-\mu x} \mu \, dx} = \frac{1}{\mu} \qquad (5.8)$$

Example 3   Calculate the half-thickness, mean range, and mass absorption coefficients for ^{60}Co gamma rays in lead.
From Fig. 5.9, $\mu_m = 0.058$ cm^2/g.

$$\mu = \mu_m \rho = 0.058(11.32) = 0.66 \text{ cm}^{-1}$$

The *half-thickness* is the thickness of lead that will reduce the intensity by one-half.

$$I/I_0 = 0.5 = e^{-\mu x_{1/2}}$$

$$x_{1/2} = \frac{0.693}{\mu} = \frac{0.693}{0.66}$$

$$= 1.05 \text{ cm}$$

$$\bar{R} = \frac{1}{\mu} = \frac{1}{0.66} = 1.52 \text{ cm}$$

Neutrons

The detection of neutrons is based on neutron interactions with matter.   Since a neutron carries no charge, nuclear forces rather than Coulomb forces act between it and a nucleus.   The six neutron interactions are as follows:

(1) Elastic collision occurs when the neutron shares its kinetic energy with a nucleus without exciting the nucleus.   This is the primary mode of energy loss for neutrons as they are slowed to thermal by the light nuclei of a moderator.
(2) Inelastic collision usually occurs with fast neutrons.   Here the target becomes excited, emits a gamma, and shares the remainder of the available kinetic energy with the scattered neutron.
(3) Radiative capture (n, γ) takes place when a neutron is absorbed to produce an excited compound nucleus which attains stability by emission of a gamma.   These reactions are more probable with thermal and epithermal neutrons.
(4) Ejection of a charged particle occurs normally with fast neutrons.   This process is frequently used for detection of neutrons (both fast and thermal).   An (n, p) reaction is used for fast neutrons, and an (n, α) reaction is used for thermal detectors such as ^{10}B + n → ^{7}Li* + α.

(5) Fission reactions occur with both fast and thermal neutrons. Certain fission reactions are energy sensitive. These are also used in detectors.

(6) Shower occurs for very high neutron energies (greater than 100 MeV). The neutron energy appears as a shower of photons and light particles. The total shower energy produced by the neutrons equals the sum of all energies of photons, light particles, and so on, produced.

It should be noted that all six interactions can occur, but the probability of a given interaction occurring depends on the neutron energy. A more complete treatment of neutron interactions is given in Chapter 8.

Electrostatic Charge Accumulating Instrument (Dosimeter)

One of the simplest means of detecting (and measuring) ionizing radiation is with an electroscope (illustrated schematically in Fig. 5.10). The support and quartz fiber are positively charged to about 150 V dc. This is done by depressing the charging spring into contact with the support and simultaneously applying the dc voltage between the wall (ground) and the charging spring.

The support and quartz fiber are repelled and assume an equilibrium position. The electrostatic force balances the spring force of the quartz fiber. When an ionizing particle (γ) ionizes the gas, the negative ion that is formed discharges the quartz fiber and the deflection diminishes. When the deflection is observed with a microscope, it can be calibrated in units of radiation dose.

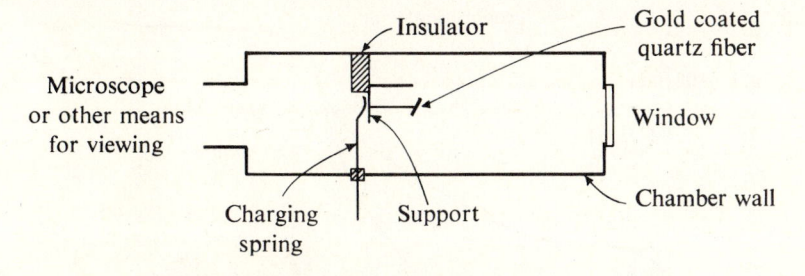

Fig. 5.10   Sketch of a typical electrostatic dosimeter.

Simplest Ionization-Type Detector

Incoming radiation ionizes gas in the gas-filled tube. The ions produced are collected at the electrodes. A negative ion collected on the anode drives the vacuum tube grid negative and causes the meter to deflect (Fig. 5.11). If the voltage, V, on the tube is gradually increased, a characteristic curve (Fig. 5.12) can be plotted.

Radiation

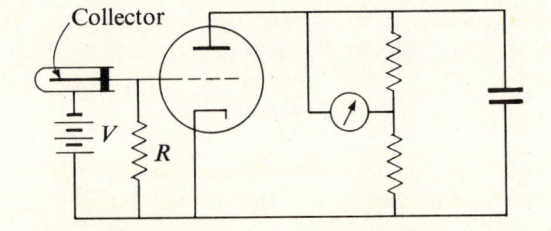

Fig. 5.11   Simple ionization-type detector.

Below V_1 some ions recombine.   As V increases, the drift velocity of ions increases and recombination decreases.   At saturation voltage, V_1, the recombination is at a minimum.   Region II is called the *ionization region* and extends over a few hundred volts.   In this region all ion pairs are collected.   The voltage range is normally about 100 to 300 V.   Region III is the *proportional region.*   Here the primary ions acquire enough energy (from V) to cause

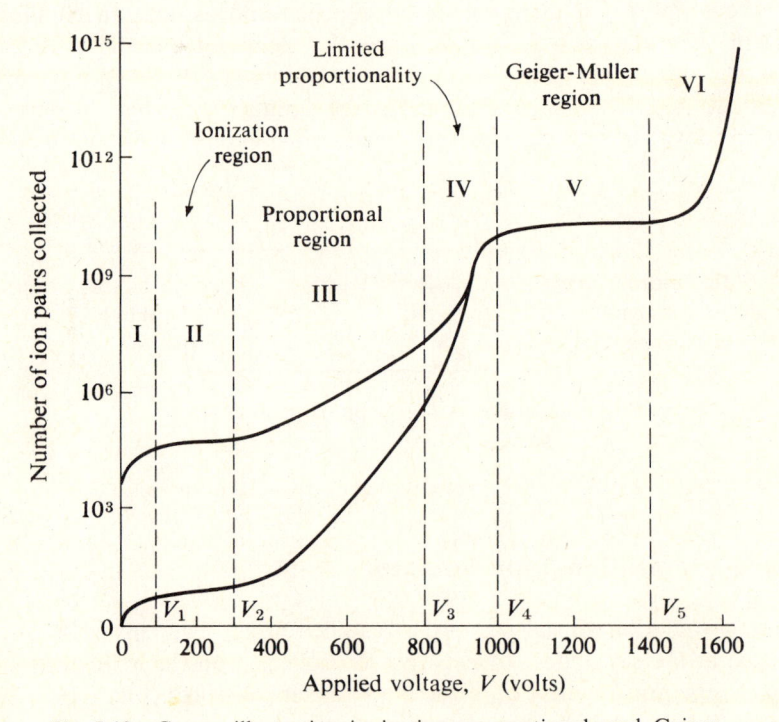

Fig. 5.12   Curves illustrating ionization, proportional, and Geiger–Müller regions.

secondary ionization (gas amplification) and increase the charge collected. The secondary ionization may cause further ionization. In this region there is a linear relationship between number of ion pairs collected and voltage V. In region IV no linear relationship exists between ions collected and primary ions. An avalanche of ions starts and increases as V increases. In region V there is a complete discharge in the vicinity of the central wire. This region is called the *Geiger–Müller region*. The complete discharge is independent of the initial ionizing radiation and an avalanche of electrons develops all over the central wire and in the gas. V_4 is called the *threshold voltage*. Here the number of ion pairs levels off and remains relatively independent of the applied voltage V. This leveling-off is called the *Geiger plateau*. The plateau extends over a region of 200 to 300 V; the threshold is normally about 1000 V. The operating voltage of the tube is normally about 1200 V. It should be recalled that the ionization in this region is independent of the nature and energy of the original radiation. An increase in V above V_5 produces a continuous discharge in the tube and will very quickly destroy the tube.

The different methods of detecting ionizing radiation (ion chamber, G-M counter, etc.) depend on the varying behavior of ions produced by radiation in their passage through the tube. Thus, one type of detector may have advantages over another in certain applications (i.e., proportional counter as neutron counter, Geiger tube for small specific ionization).

Ion Chambers

An ion chamber can be used as either a pulse- or a rate-type detector. In *pulse operation* the output of the chamber is indicated as a series of signals (usually voltage) separated in time. In *rate operation* no attempt is made to resolve individual actions. Instead, the output is the time average of many interactions. Measuring the output current of the ion chamber is an example of a mean-level system. When the total dose measured is averaged over the time interval of measurement, the electrostatic dosimeter of the last section becomes an example of a mean-level-type ionization detector.

In a pulse-type ionization chamber the magnitude and duration of the voltage pulse produced by the tube become important. If the detector is to count only the number of particles, the voltage pulses must be short enough to be distinguished from one another; they only need to be large enough to be picked up by the amplifier. However, if energy is to be measured, the pulse amplitude is important. When current flows in the external circuit of Fig. 5.11, the ions begin to drift to the collectors. When all charge is collected, the current ends. Curve a of Fig. 5.13 shows the voltage output pulse of the ionization tube for a long time constant. The electrons are normally collected within 10^{-6} s and the positive ions within 10^{-3} s. There is no multiplication. This type of ionization chamber is, therefore, limited to low counting rates (usually

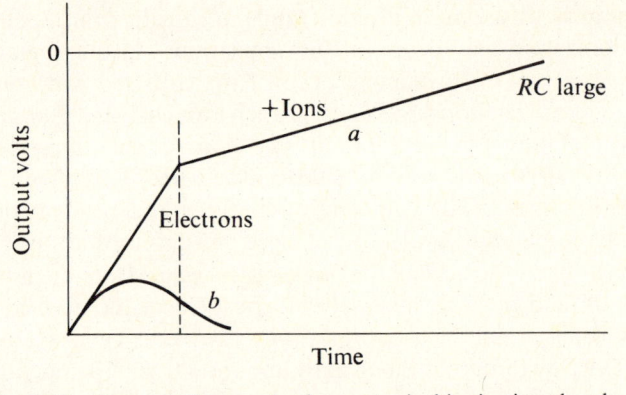

Fig. 5.13 Output voltage pulse from a typical ionization chamber.

less than 100 counts/s). To remove this restriction of low counting rates, the time constant of the chamber is usually made less than the maximum collection time. A typical value is 0.2 times the maximum electron collection time. This gives a uniform pulse height but a reduced amplitude. Curve *b* of Fig. 5.13 illustrates the shorter time constant.

The gas filling is usually air, at or near atmospheric pressure. Almost any gas can be used if it has a small electron affinity. The insulators (between + and − electrodes) are commonly made of aluminum oxide, quartz, polystyrene, or Teflon. Organic materials are generally not used for high-dose-rate applications because of their susceptibility to radiation damage. Teflon is least susceptible to water absorption, which may reduce surface leakage resistance. Stress currents appear across an insulator after it receives electrical or mechanical stress. Aluminum oxide and quartz have very low stress currents and are used for low-current measurements. Guard rings are used in one type of ionization chamber (parallel plate). The guard ring defines the active volume of the chamber and reduces the leakage current through the insulators. The guard rings and electrodes are usually made from tungsten or platinum.

Proportional Counters

Proportional counters can be used either as pulse-type or rate-type detectors, but pulse-type operation is the most frequent. Proportional counters are useful in beta radiation measurements because ionization chambers do not have high enough sensitivity. They also offer an advantage because the voltage pulse retains a proportional relationship with energy but produces a larger pulse than an ionization chamber. In the proportional region shown in Fig. 5.12 ionization is produced in the region surrounding the anode. The resulting pulse (Fig. 5.14) is independent of the exact region where primary ionization is

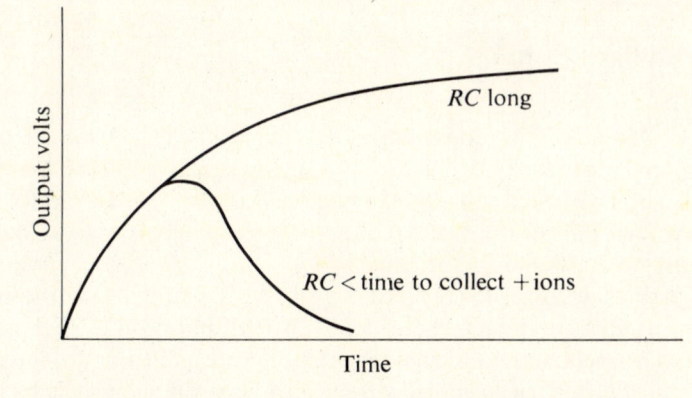

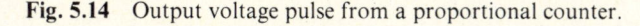

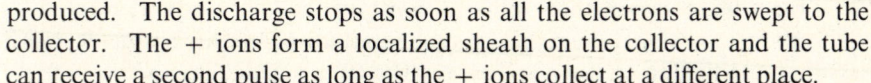

Fig. 5.14 Output voltage pulse from a proportional counter.

produced. The discharge stops as soon as all the electrons are swept to the collector. The + ions form a localized sheath on the collector and the tube can receive a second pulse as long as the + ions collect at a different place.

If the multiplication is too large, the + ion sheath enlarges and the sensitive time of the tube decreases. If the amplitude of the pulses must be measured, the time between pulses must be great enough for the effect of collecting the + ions on the cathode to be negligible. In most cases this can be less than 100 μs.

Multiplications resulting from secondary ionization of 10^4 are normal and 10^6 is a possibility. The multiplication must be independent of the position of the primary ionization. This is made possible by making the electrode of fine platinum wire 0.001 or 0.002 in. in diameter. Consequently, the secondary ionization is produced close to the wire. The gas filling is normally a mixture of 90 percent argon and 10 percent methane at atmospheric pressure. The methane absorbs photons, reducing photoemission of electrons and limiting multiplication. The photons are produced with the secondary electrons.

A proportional counter is a fast counter and pulse rates as high as 10^6 per second can be counted with it. A proportional counter can also be used to discriminate energies. The multiplication is linearly dependent on operating voltage. If the counts at lower operating voltages are compared with those at higher voltages, the contribution of highly ionizing particles is compared with those of low ionization. An electronic differential pulse-height analyzer is used for this type of discrimination. These analyzers measure the slope of the energy distribution curves. The discriminator allows pulses of height between set limits, say H and $H + \Delta H$, to be passed to a counter. The height, H, and $H + \Delta H$ can be varied as desired. As the difference ΔH is decreased, the result approaches a true differential pulse-height distribution. The instruments that perform these differential energy-distribution measurements are commonly known as *spectrometers*.

Geiger–Müller Counters

The Geiger–Müller (G–M) counter has been the most widely used detector. It has been popular primarily because (a) it is highly sensitive to even the smallest radiations; (b) it can be used with many different types of radiation; (c) it has a very high voltage pulse output; and (d) its cost is reasonable. The G–M counter is frequently used as a pulse-type detector.

The applied voltage, V (refer to Fig. 5.12), is so high that the primary ionization produces an avalanche of ions. A multiplication as high as 10^8 is common. The avalanche is independent of the nature and energy of the original radiation; in addition, electrons may be ejected from the tube walls by gamma interaction. The immediate region around the anode becomes insensitive to more radiation very quickly. This occurs because the free electrons are collected rapidly, leaving a sheath of + ions all along the anode. The output voltage pulse from the tube resembles that of the proportional counter, shown in Fig. 5.14. As in the proportional counter, the time constant of the tube is made much less than the time to collect the + ions. The tube remains insensitive to more radiation until the + ions move away from the anode. The + ions finally collect on the cathode and quenching must be supplied to prevent an undesirable afterpulse. The afterpulse would take the form either of an electron emitted from the cathode or a photon radiated from the cathode. Either of these could start another avalanche. Quenching may be provided by: (a) lowering the tube voltage for a few microseconds (external quenching); (b) introducing an organic gas into the tube; or (c) introducing a halogen gas. External quenching is very seldom used. The + ions transfer their energy to quench gas molecules and when the quench gas molecules reach the cathode they dissociate rather than produce an electron. Alcohol-quenched tubes have a lifetime of about 10^{10} events, whereas halogen ions recombine, thereby extending the lifetime of the tube almost indefinitely. The gas is generally about 90 percent argon at a total pressure of 10 cm Hg.

The ionization depends on the physical characteristics and construction of the tube. Normally, the tube output pulse is high (on the order of volts). The tube is usually cylindrical with a 0.003- or 0.004-in.-diameter center electrode made of tungsten. The cathode is usually the wall of the tube and is made of glass coated with a stainless steel or nickel conductor.

When electrons collect on the anode (collector), the positive ions originating near the anode reduce the electric field intensity too low to support discharges, and there is no output. The dead time, t_d, is the time for a subsequent small pulse to appear. The recovery time, t_r, is the time for a full-amplitude pulse to appear. The effective dead time, D, depends on the electronics of the system and the tube characteristics. It is generally less than $t_d + t_r$ (Fig. 5.15). Dead time is normally introduced electronically after each count. This enables each operator to accurately determine the true count. The dead time

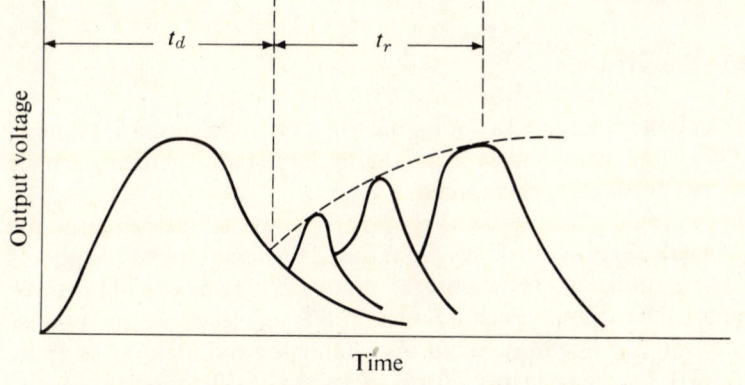

Fig. 5.15 Dead time in a G-M tube. [From H. G. Stever, *Physical Reviews* **61**, No. 38 (1942). Reprinted by permission.]

also limits the maximum number of counts. The true counting rate is determined as follows:

$$n = \text{observed counts per second}$$

$$N = \text{true counts per second}$$

$$D = \text{effective dead time, seconds per count}$$

Then

$$nD = \text{total time not counted}$$

$$1 - nD = \text{total time counted}$$

$$n = N(1 - nD)$$

$$N = \frac{n}{1 - nD} \tag{5.9}$$

Thus, $1/(1 - nD)$ is the factor by which the observed count rate may be multiplied to give the true count rate.

Example 4 What is the counting error for a G-M counter with a dead time of 200 μs and an observed counting rate of 1500 counts/min?

The counting error may be expressed by the factor

$$\frac{1}{1 - nD} = \frac{1}{1 - (1500/60)(200 \times 10^{-6})}$$

$$= \frac{1}{1 - 5 \times 10^{-3}} = 1.005\ 03$$

The true count would then be

$$N = \frac{n}{1 - nD}$$

$$= 1500(1.005\ 03) = 1508 \text{ counts/min}$$

Scintillation Counters

Scintillators can be used for any radiation but are most widely used for gamma counting. The scintillators discriminate energy levels and therefore can be used to detect unknown gamma emitters.

Radiation entering the crystal (Fig. 5.16) causes luminescence in the scintillation crystal. This light is reflected through the light pipe to the photo-multiplier. The scintillation crystal absorbs gamma energy by one of the three means of gamma interaction. Each interaction lifts an electron to an excited state. The electron is brought to rest in the scintillator and in doing so emits light [for NaI(Tl), one photon per 50 eV of energy]. The intensity of the emitted light is directly proportional to the energy lost by the incident particle. The focusing of emitted photons onto the photocathode is very critical. MgO or Al_2O_3 (aluminum foil) is used to reflect photons into the light pipe and to prevent light transmission to the outside. The reflector normally covers all but one side of the scintillator. Impurities are added to inorganic crystals to "soften" them (i.e., make them more transparent to photons).

Scintillators are made in many shapes, including disks, right circular cylinders, and cylinders with holes. The light pipe is commonly made from glass, Lucite, or plexiglass. Its purpose is to provide a large critical angle of incidence so that radiation will not be trapped within the crystal. It also serves to separate the crystal from the photomultiplier (if desired), to spread the light over a large cathode area, and to shape the surface to a flat photo-cathode if necessary. The photocathode emits electrons when excited by the photons. The electrons are collected and focused from one dynode to another. Each dynode is at a higher potential than the previous one and emits several electrons for each electron incident on its surface. Thus, by using several dynodes (~ 10), amplifications of 10^7 to 10^{11} can be obtained. The photo-cathode is usually a thin coating of antimony–cesium on the inside of the

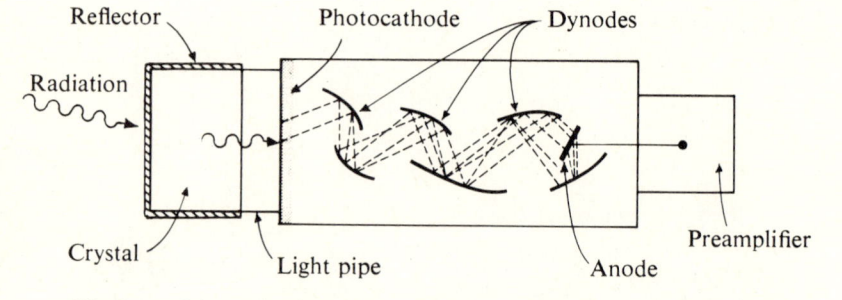

Fig. 5.16 Schematic diagram of scintillation crystal and photomultiplier.

photomultiplier; the dynodes are either antimony–cesium or silver–magnesium.

A good scintillator emits light in less than 10^{-8} s. Since all the light flashes occur so nearly simultaneously, they are integrated into a single pulse. The time for the emission of 63.2 percent of the photons is called the *time constant* (τ). For

NaI (Tl)	$\tau = 0.25 \times 10^{-6}$ s
Anthracene	$\tau = 0.27 \times 10^{-7}$ s
ZnS	$\tau = 10^{-5}$ s
Plastics	$\tau = 5 \times 10^{-9}$ s

[handwritten: γ of specific discrete energies given off — can discriminate γ's]

Scintillators have several advantages over other methods of counting gammas. Among them are: (a) a pulse height proportional to energy; (b) an efficiency for gammas which may be as high as 50 percent since there are more atoms in a solid than a gas; (c) a short dead time $(10^{-9}$ s); and (d) the probability of absorbing a scattered gamma which increases with crystal size, thereby enhancing the photopeak. Table 5.2 lists some scintillators with their applications.

One of the most useful applications of a scintillation crystal is the scintillation spectrometer. A typical gamma spectrometer is illustrated in Fig. 5.17. In this setup the scintillator–photomultiplier output combination is fed into a single-channel pulse-height analyzer. The lower discriminator will only pass pulses from the linear amplifier which are greater than its setting (45 V). The upper discriminator will only pass pulses greater than its setting (55 V). The anticoincidence circuit will only pass pulses which do not arrive in coincidence from the two discriminators. Therefore, only pulses falling within a preset value (window) will be passed on to the counter (45 to 55 V in this case). This ensures that only pulses between the two discriminator settings will be passed

Table 5.2 Applications of Scintillators

Scintillator	Most Common Use
Anthracene	β, neutron
trans-Stilbene	β
ZnS(Ag)	Alpha particles
NaI(Tl)	γ-ray
LiI(Sn)	neutrons
p-Terphenyl	$\beta(^{14}C, {}^3H)$
Diphenylorazole	β, neutrons
Tetraphenyl butadiene	β
Terphenyl in polystyrene	
Xenon	Neutrons

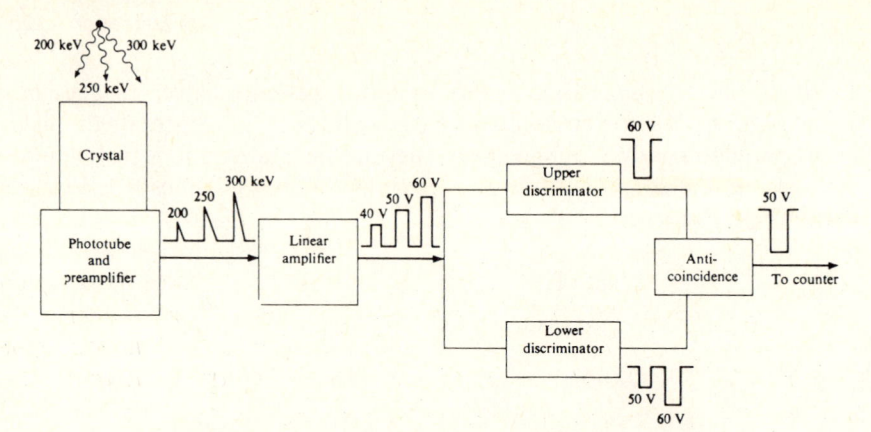

Fig. 5.17 Schematic arrangement of a scintillation spectrometer.

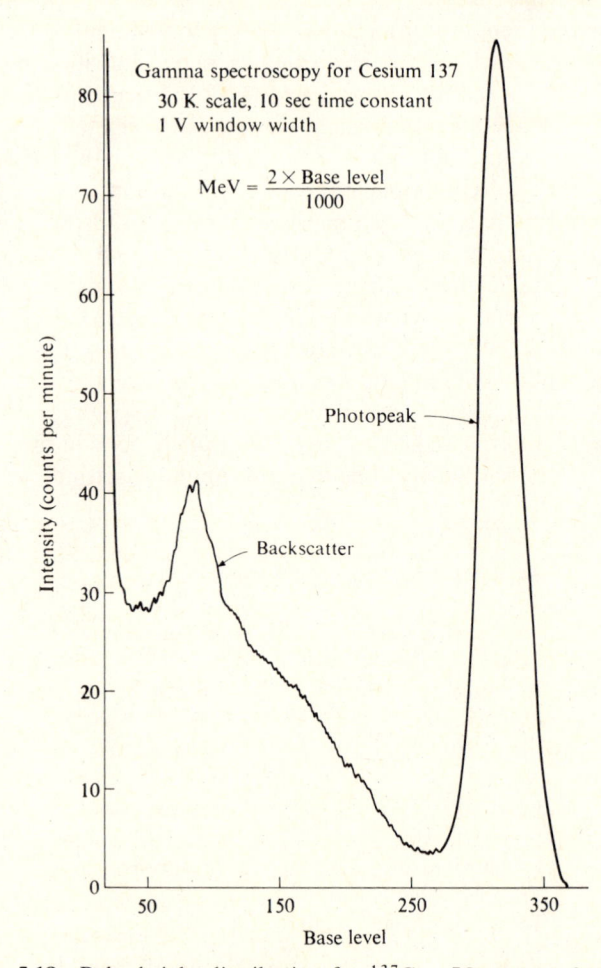

Gamma spectroscopy for Cesium 137

30 K. scale, 10 sec time constant

1 V window width

$$MeV = \frac{2 \times \text{Base level}}{1000}$$

Photopeak

Backscatter

Intensity (counts per minute)

Base level

Fig. 5.18 Pulse-height distribution for ^{137}Cs. [Courtesy of North-eastern University Nuclear Engineering Laboratory.]

to the counter. The discriminator settings and/or the window width may be varied over any desired range. Figure 5.18 illustrates a pulse-height curve for ^{137}Cs gammas taken with a single-channel pulse-height analyzer and a NaI(Tl) crystal. Notice that the photopeak appears very sharp, whereas the scatter peak is spread over a wider range. At this energy level the gammas interact by both the photoelectric process and Compton scatter. All of the photoelectric interactions contribute to the photopeak, as well as those Compton events where both the scattered electron and the attenuated gamma dump their energy in the scintillator during its resolving time. If the attenuated gamma interacts elsewhere, then only the scattered electron converts its energy to light. These events cause a significant number of counts to occur at energies below that of the photopeak.

Neutron Detectors

The products of neutron interactions are measured because neutrons themselves cannot be detected. The most useful neutron interactions are as follows: (a) transmutations (n, α), (n, p), (n, γ), (n, fission); (b) elastic collision (n, p); and (c) foil activation.

Transmutations are generally energy sensitive. For example, the reaction ^{10}B + n $\rightarrow$ ^{7}Li + α is very sensitive to thermal neutrons (0.025 eV) but relatively insensitive to neutrons above 100 keV.

Gas chambers, scintillators, and thermopiles are used for neutron detection. Gas chambers are lined with B_4C or filled with BF_3 gas. A chamber may be lined with ^{235}U (for thermal neutrons) or ^{238}U (for fast neutrons). In a ^{10}B chamber the ^{10}B must be extremely pure to prevent capture of neutrons and thereby reduce the ionization present.

A BF_3-filled counter is normally cylindrical in shape with a 0.002-in.-diameter center anode and a cathode about 0.8 in. in diameter and is 4 to 6 in. long. The BF_3 filling is at about 10 cm Hg pressure. The detector must not disturb the neutron flux. As long as the product $\Sigma_a d$ is very small, the absorption will be negligible. Σ_a is the macroscopic absorption cross section for neutrons and d is the distance a neutron can travel through the BF_3.

One type of boron-lined counter is sketched in Fig. 5.19. The polarity of alternate plates is the same; therefore, the chamber is in effect one large parallel-plate chamber. It is sensitive to both gamma and neutrons, but by making the surface large in comparison to volume, the sensitivity to gamma can be kept at a minimum. The alpha and ^{7}Li produced by the reaction in turn produce ionization which is measured in a current-type ionization chamber. The boron is applied in thin layers to make it transparent to neutrons from outside and to produce reactions such that the particles can escape into the chamber. ^{10}B chambers are usually used for neutron fluxes in the order of 10^4 to 10^{10} neutrons/cm^2 s.

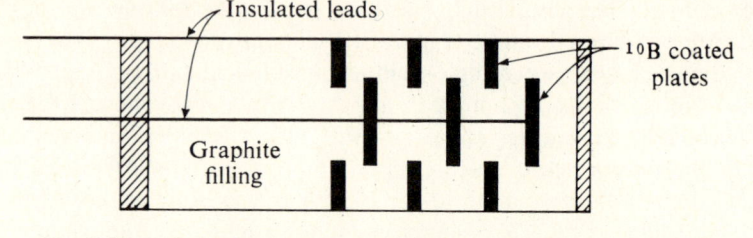

Fig. 5.19 Schematic diagram of a parallel circular plate neutron chamber (uncompensated ion chamber).

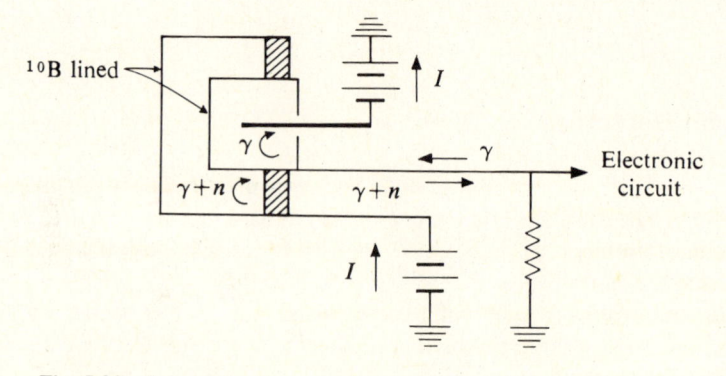

Fig. 5.20 Schematic diagram of a compensated ion chamber.

When a pulse-type proportional chamber is used to measure the neutron flux, the gamma radiation accompanying the neutrons can be discriminated by the circuit electronics. One type of compensated ion chamber which isolates the neutron-induced ionization from the gamma-induced ionization is sketched in Fig. 5.20. With this compensated ion chamber an accurate measure of the neutron flux (and power level) in a reactor can be made. The net current flow is proportional to current due to $\gamma + n$ minus current due to γ, or the neutron current alone. The internal pressure, volume, and voltage are adjusted so that the γ just balances the γ contribution in the $\gamma + n$.

Fission Chambers

Fission chambers are used during reactor startup because of the large energy released during fission relative to the flux of γ-rays present. Neutrons are absorbed and the fission fragments produce ionization either through kinetic energy or radioactivity. The resulting ionization is measured either in a pulse-type ion chamber or a proportional counter. One type of fission chamber has

enriched ^{235}U electroplated on the inside surface; the chamber is filled with high-pressure oxygen-free argon. The fission fragments lose their energy in the gas. If fast neutrons are desired, ^{10}B or Cd is used to screen out the thermal neutrons and the fast neutrons react with the ^{238}U. Other materials such as ^{232}Th, which are fissionable only by fast neutrons, are also used as the lining in fast-neutron fission chambers.

Fast-Neutron Detectors

The elastic scattering (n, p) is the method most used for detecting fast neutrons. One type of chamber is lined with polyethylene and filled with hydrogenous gas such as methane. Thermal neutrons will not impart an appreciable amount of energy to the protons. On the other hand, fast neutrons will impart energy to the hydrogen nuclei through elastic scattering. The recoil protons will then induce ionization or excitation by inelastic collision. The chamber can be either a pulse- or rate-type detector, but is usually a pulse-type proportional counter. If the pulse heights are made proportional to the proton recoil energies, the fast-neutron energies can be determined.

Neutron-induced foil activation as a means of detection is discussed in Chapter 8. The general use of neutron detectors in reactor instrumentation is shown in Fig. 5.21.

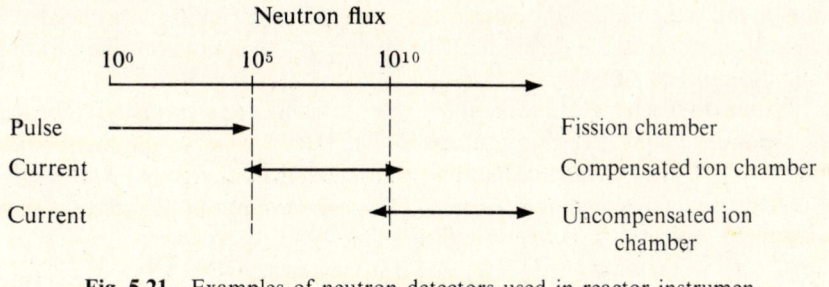

Fig. 5.21 Examples of neutron detectors used in reactor instrumentation.

Solid-State Detectors

There are two principal types of solid-state radiation detectors: bulk and barrier layer. They both offer some extremely attractive advantages over other methods of detection.

They have, for example, high energy resolution, linear output with particle energy regardless of the nature of the particle, almost 100 percent detection efficiency, very fast pulse rise time, and no apparent dead time. In addition, the detectors are quite stable, low in cost, and very small in size.

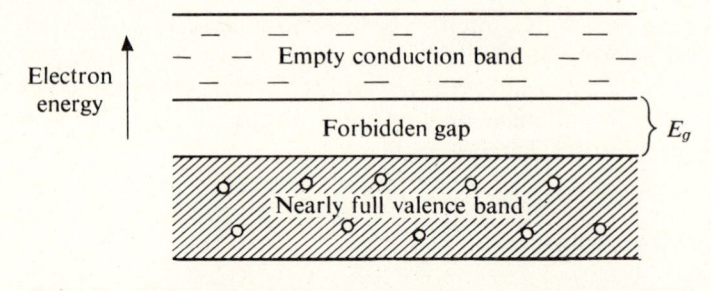

Fig. 5.22   Schematic representation of semiconductor energy bands.

Unfortunately, all the manufacturing problems have not yet been overcome; the detectors have a small output (millivolts), and they are sensitive to high temperatures. Their life is also quite limited under strong radiation.

The theory and operation of semiconductor-type detectors depend on an understanding of quantum mechanics, a specialized and complex field. The action of radiation in semiconductors, therefore, will only be described qualitatively. A schematic diagram of the electron energy bands of semiconductors (Fig. 5.22) will aid in the description of particle interaction in a semiconductor. The uppermost filled band is separated from a higher empty band by a small forbidden zone. At sufficiently high temperatures (100 to 300 K) some electrons cross the forbidden zone. These electrons are now able to conduct, and the empty band is then called the *conduction band*. The minimum energy needed for an electron to cross the forbidden gap and enter the conduction band ranges from about 0.1 to 3.0 eV.

When an electron leaves the filled valence band it leaves a *hole*. This hole corresponds to a positive charge carrier. The electrons in the conduction band and these holes in the valence band become carriers of charge and will respond to an electric field placed across them. The concentration of electrons (n) and holes (p) is given by

$$n = p = 10^{19} e^{-E_g/2KT} \qquad (5.10)$$

where E_g is the minimum energy for an electron to enter the conduction band. Table 5.3 lists some properties of semiconductors used in particle detection.

Example 5   What is the concentration of charge carriers in silicon at room temperature? Compare this with the concentration in a metal conductor.

$$n = p = 10^{19} e^{-E_g/2KT}$$
$$= 10^{19} e^{-1.08/2(293.6)(8.65 \times 10^{-5})}$$
$$n = 2 \times 10^{10} \text{ cm}^{-3}$$

The carrier concentration of a metal is about 10^{22} cm^{-3}. Thus, the carrier concentration of a semiconductor is much less and is temperature dependent.

Table 5.3 Properties of Some Semiconductors Used in Radiation Detection

Semiconductor	Energy, E_g (eV)	Electron Carrier Lifetime (s)	Density (g/cm^3)
Silicon	1.08	10^{-3}	2.33
Germanium (77 K)	0.75	10^{-3}	5.32
Germanium (300 K)	0.66	10^{-3}	5.32
Gallium arsenide	1.39	10^{-8}	5.3

If impurities (dope) are introduced into the crystal, it is possible for an electron to reach the conduction band without originating in the valence band. In this situation one of the valence electrons remains bonded to its parent nucleus. It can, therefore, move into the conduction band with less than the necessary E_g. If this occurs, there will be no hole left by the electron, the impurity is called a *donor*, and the crystal is called an *n-type semiconductor*. If the opposite effect occurs with the introduction of an impurity, it is called an *acceptor*, and the crystal is a *p-type semiconductor*.

When radiation enters a bulk detector it creates hole–electron pairs. These hole–electron pairs, acting with the electric field applied to the detector (Fig. 5.23), produce an output pulse which can be counted. The number of carriers released is proportional to the incident energy of the particle, making solid-state detectors useful for spectrometry. The energy necessary to create a hole–electron pair is about one-tenth that necessary to create an ion pair in gas. Thus, the particles can be stopped in a small volume (~ 1 cm^3). The

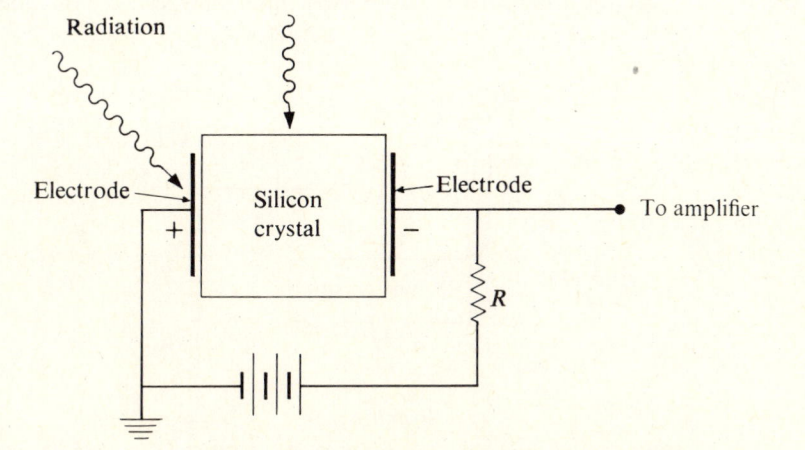

Fig. 5.23 Bulk-type solid-state radiation detector.

problem is that there is no multiplication on the semiconductor, which necessitates a very sensitive counting circuit. On the other hand, the statistical variation is not multiplied, either. The bulk detector requires a high electric field to collect the carriers in a short time. This means that the resistivity must be high to prevent leakage current. One way to increase the resistivity is to "dope" silicon with gold, but this increases carrier collection time. Another method has been to dope the silicon with boron. A typical carrier collection time is 10^{-7} s.

The barrier-layer detectors are of two types; diffused junction and surface barrier. The two are very similar; they differ mainly in the method of obtaining a surface junction of p- and n-type semiconductors. In the junction type, practically the whole crystal is p-type but there is an extremely small portion of n-type material on the surface (see Fig. 5.24). The n-type material is usually about 1 μm or less. The radiation enters the crystal through the n-face. The carriers on the two sides of the junction can cross the potential existing across the junction if they are given energy. The incoming radiation does this (3.6 eV per pair of silicon and 3.0 eV per pair for germanium). When the holes and electrons cross the junction, they produce an output pulse which can be counted. The pulse is small and is proportional to the energy of the incident particle but is independent of the nature of the particle. The barrier layer detectors are extremely small (<1 cm^3) and the collection time is about 10^{-9} s. Silicon detectors of the p-type are used as diffused junction types while n-type silicon and n-type germanium have been used for surface barrier-type detectors.

Barrier layer-type detectors can be used as neutron detectors. One such detector employs the reaction ^{6}Li(n, α)^{3}H. Lithium fluoride is coated onto the surfaces of two silicon-surface barrier-type detectors. The two detectors are placed so that the lithium fluoride is sandwiched between the sensitive surfaces of the detectors. The detectors are placed about 0.5 mm apart. The products of the neutron reaction are recorded on a pulse-height analyzer. If a fissionable

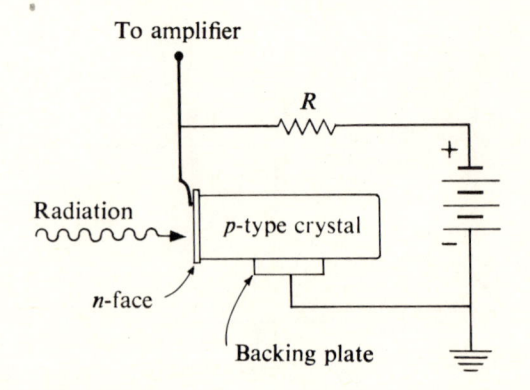

To amplifier

Fig. 5.24 Diffused junction barrier-type solid-state radiation detector.

material is substituted for the lithium fluoride, a fission product spectrum can be obtained.

Thermoluminescent Dosimetry

A growing development in the field of radiation measurement is the use of thermoluminescent materials. Certain materials will release stored energy in the form of visible light. When the stored energy is produced as a result of ionizing radiation and is measured quantitatively, the technique is commonly known as *thermoluminescent dosimetry* (TLD). Although the theory is not well known, the phenomenon is well understood. In an insulating material, ionizing radiation may release an electron from the valence band into the conduction band. Frequently, the electron holes do not recombine but are trapped between the conduction and valence bands. If the material is subsequently heated, the electron or the hole may gain enough energy to recombine with the other. When this occurs, a thermoluminescent photon is emitted. The energy possessed by the electrons determines the temperatures at which the photons will be released. A curve of light emitted in relation to time heated is called the *glow curve*. Figure 5.25 is a typical glow curve for a thermoluminescent material.

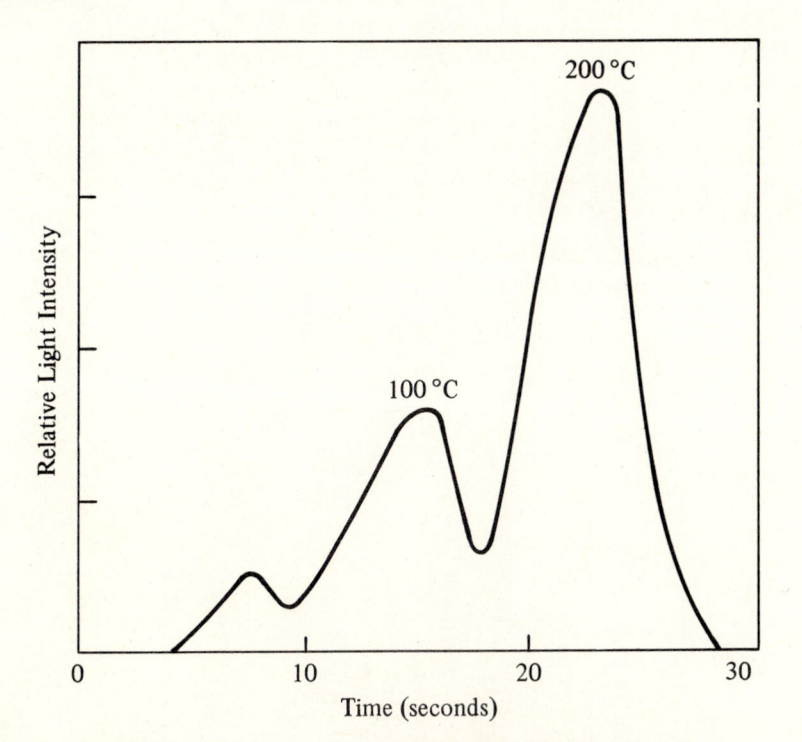

Fig. 5.25 Typical glow curve for a thermoluminescent phosphor.

Table 5.4 Characteristics of TLD Materials

Characteristic	LiF	CaF$_2$:nat.	CaF$_2$:Mn	Li$_2$B$_4$O$_7$:Mn	BeO	Al$_2$O$_3$
Temperature of useful glow peak	190°C	260°C	260°C	200°C	180–220°C	220°C
Relative response to ^{60}Co	1.0	23	3	0.3		3
Useful range	1 mR–10^5 R	mR–10^4 R	1 mR–3 × 10^5 R	mR–10^6 R	1 mR–10^5 R	10 R–up
Sensitivity to light	None	High	None	None	None	None
Peak fading	<5% per week	None	10% (first month)	8% (first month)	None	Some
Physical forms	Powder, glass capillaries, embedded, extruded		Powder, glass capillaries, embedded chips	Powder embedded	Powder	
Energy response	Essentially flat	Flat above 300 keV	Flat above 300 keV	Flat		Flat

The lower-temperature peaks generally disappear a short time after irradiation, leaving the higher peaks suitable for measurement. Annealing of the crystal after irradiation also removes these low-temperature peaks. Before reuse, the crystals must also be annealed in order to restore the normal glow curve characteristics.

Several materials are suitable for use as TLDs, but LiF is used most widely. The materials are used in both powder and solid forms, making their use very flexible. Table 5.4 lists the characteristics of some TLD materials.

In order to read a TLD it is necessary to heat the material and measure the emitted light. Measurement is done with a photomultiplier or photodiode. The output from the phototube is amplified and observed as a voltage pulse or current. For commercial applications the glow curve (Fig. 5.25) is not measured. Instead, either the area under the glow curve is measured or the peak height is measured. The area under the glow curve is proportional to the integral of the photomultiplier current, and is relatively insensitive to the heating rate. It is usually used with LiF and $Li_2B_4O_7$ TLDs. The measurement of peak height is sensitive to heating rate, but it is preferable for measurement of low dose measurements. CaF_2 materials, which have very good low-exposure response, are often measured using the peak-height technique.

Heating (which is necessary to produce the thermoluminescence) is accomplished by placing the dosimeter in a pan and electrically heating the pan. The heating systems, sample, and phototube are enclosed in a light-tight box. Some commercial personnel dosimeters have a small heating element built into the instrument. They can be inserted into a specially built reader, and the exposure evaluated and recorded very quickly. TLD has also been used successfully for radiation therapy measurements and for archeological dating.

Statistics

The emission of radioactive particles is a random process. In addition, there is always a small amount of background radiation present from cosmic sources and natural radiation from nearby sources. Radiation from nuclear reactors or X-ray machines may also affect the background radiation. The effect of background radiation is reduced by taking a separate measurement of the background count and subtracting it from the measured count. It is also helpful to place shielding around the counting equipment to minimize the effect of radiation from outside the apparatus.

The number of counts recorded on a counting apparatus will be different each time a measurement is made. It is desired to approach as close as possible to a true average counting rate. To do this, the deviation to be expected due to the random fluctuations must be known.

Since each radioactive decay is an independent event, the average disintegration rate can be determined exactly only by counting for an infinite

time. The deviation from this true average will, however, obey the laws of random processes. In the decay process the binomial distribution law

$$p(n) = \frac{N_0!}{m!\, n!} p^n q^m \qquad (5.11)$$

may be applied. In this expression $p(n)$ is the probability that n disintegrations out of a total of N_0 radioactive atoms will occur, and q is the probability that n disintegrations will not occur: $m = N_0 - n$. The average value n then becomes pN_0. Several deviations from the average are used to indicate the precision of the resulting n disintegrations. These are (a) standard deviation, (b) probable error, and (c) average error. The most frequently used is the standard deviation, σ.

$$\sigma = \left[\frac{1}{N} \sum_0^{N_0} (n - \bar{n}) \right]^{1/2} \qquad (5.12)$$

Only a few of the disintegrations, n, that occur in a fixed time interval are measured by counting apparatuses, so for radioactive decay processes it is virtually impossible to measure the quantities of Eq. (5.11). In the case of radioactive decay, only two events are possible—either the atom decays or it does not decay—so that the binomial distribution can be stated in a more convenient form. The probability that an atom will not decay in time t is $e^{-\lambda t}$, and the probability that it will decay is $1 - e^{-\lambda t}$. Also, for radioactive decay $\lambda t \ll 1$, $N_0 \gg 1$, and $n \ll N_0$. Making these substitutions into Eq. (5.11) gives us the Poisson distribution.

Since it is impossible to determine the exact parameters of the entire population, it is normal to select sample evaluations on a random basis. Such is the case in counting radioactive samples. For a normal distribution the standard deviation is a measure of the width of distribution to be expected.

$$p(n) = \frac{(\bar{n})^n e^{-\bar{n}}}{n!} \qquad (5.13)$$

The Poisson distribution is applicable for λt as large as 0.01 and N_0 as small as 100. The Poisson distribution is in reality the limiting value of the binomial distribution when the probability of decay is small and the number of atoms present is very large.

In the case of radioactive decay, the binomial distribution can be approximated by a normal distribution. This approximation is possible because N_0 is very large. The normal distribution is a limiting form of the binomial distribution when $p = \frac{1}{2}$ and N_0 approaches infinity. It takes the familiar bell shape. In a normal distribution the standard deviation of a single measurement becomes

$$\sigma = (pN_0)^{1/2} = (\bar{n})^{1/2} \qquad (5.14)$$

The standard deviation from the mean then equals $\bar{n}^{1/2}$. For the population as a whole, 68.26 percent of measurements have a smaller deviation than the standard, and 95.44 percent of measurements have smaller deviation than two times the standard.

However, for a set of measurements, which may include additional random fluctuations, the standard deviation of a single measurement should be calculated from Eq. (5.12).

If several sets of measurements are made, each will have its own mean (also random) and its own standard deviation. The standard deviation of the mean is given by

$$\sigma_m = \frac{\sigma}{\sqrt{n^n}} \cong \sqrt{\bar{n}/\text{no. count sets}} \qquad (5.15)$$

The results of counting measurements should be reported as follows:
If several measurements are made,

$$\bar{R} = \frac{\bar{n} \pm \sqrt{\bar{n}/\text{no. count sets}}}{t} \qquad (5.16)$$

If only one measurement is made,

$$R = \frac{n \pm \sqrt{n}}{t} \qquad (5.17)$$

where t is the time interval of counting and n is the number of counts. It can be seen that longer counting periods increase the probability of obtaining a true average. In the case of a single measurement, the larger the sample size (n), the more accurate is the estimate.

The probable error, p, is

$$p = 0.6745\sigma \qquad (5.18)$$

For a normal distribution 50 percent of all measurements taken will have a deviation of less than p. Obviously, 50 percent will also have a deviation of more than p.

Probable error is not used often, but when it is used results are usually reported as

$$R = \bar{n} \pm \frac{0.6745\sigma}{\sqrt{\text{no. count sets}}} \qquad (5.19)$$

Example 6   A radioactive sample was counted for 9 min. The average counting rate was 3476 counts/min. The background was counted for 5 min and was 562 counts. What is the counting rate with its standard deviation?

The standard deviation of the resulting count squared is the sum of the standard deviations squared of the sample counting and the background counting.

← corrected count rate.

$$for (R_s - R_B)$$

$$\sigma^2 = \sigma_s^2 + \sigma_B^2$$

$$\sigma^2 = \sigma_s^2 + \sigma_b^2$$

$$\sigma_s = \sqrt{3476 \times 9}/9 = 19.65$$

$$\sigma_b = \sqrt{562}/5 = 4.74$$

$$\sigma^2 = 19.65^2 + 4.74^2$$

$$\sigma = 20.2$$

The counting rate due to the source alone, therefore, is

$$R = (n_s - n_h) \pm \sigma$$
$$= (3476 - 112) \pm 20$$
$$= 3364 \pm 20 \text{ counts/min}$$

The preceding material can be used to aid in the design of counting experiments for minimum error. There are three situations of interest to be considered when specifying a given standard deviation:

(1) *Minimum total counting time.* In this case the differential $d(t_s + t_h) = 0$, and

$$t_s = \frac{R_s + \sqrt{R_s R_b}}{\sigma^2} \tag{5.20}$$

(2) *Equal counts.* In this case $n_s = n_h$, and

$$n_s = \frac{R_s^2 + R_b^2}{\sigma^2} \tag{5.21}$$

(3) *Equal counting times.* In this case $t_s = t_h$, and

$$t_s = \frac{R_s + R_b}{\sigma^2} \tag{5.22}$$

Problems

1. Determine the thickness of lead and aluminum necessary to absorb 4.0-MeV alpha particles.

2. Compare the ionization loss of beta particles of maximum energy 1.7 MeV in aluminum and lead.

3. Find the approximate range of $^{90}_{38}$Sr beta particles in aluminum. What is the half-thickness for these particles in the aluminum?

4. Using Fig. 5.9, calculate the half-value thickness for 1.25-MeV gamma radiation from a source. What is the mass absorption coefficient for lead at this gamma energy? What is the mean range of gammas?

5. Find the half-thickness, mean range, and absorption coefficients for $^{110}_{47}$Ag in lead.

6. Calculate the maximum thickness of aluminum window in a G-M tube to permit counting of 2.7-MeV beta particles.

7. The capacitance of an electroscope is 15 $\mu\mu$F, and it is charged to 150 V. If it is subjected to a count rate of 29.5 counts/s of 1.60-MeV gammas for 1 h, how much will the dosimeter deflect? Assume deflection proportional to charge.

8. How many ions would be required to neutralize all the quench gas in a 100-cm^3 G-M tube if it did not recombine? The quench gas is 0.1 percent Cl by volume of argon at atmospheric pressure.

9. A G-M counter has an operating voltage of 1200 V. The Geiger plateau has a slope of 2.5 percent. What is the maximum permissible voltage fluctuation if the count rate fluctuation is not to be more than 0.1 percent?

10. A G-M tube has an operating voltage of 1000 V and the counting circuit has a sensitivity of 5 V. If the count rate-fluctuation is not to be more than 0.5 percent, what must be the slope of the Geiger plateau?

11. Determine the counting error for a counter dead time of 150 μs at an observed counting rate of 100 counts/s.

12. Plot a curve of observed counting rate versus percent error for errors from 0 to 10 percent at a dead time of 180 μs.

13. The background counting rate is 25 ± 4 counts/min. The sample counting rate is 126 ± 10 counts/min. Determine the net counting rate with its standard deviation.

14. Before counting a certain sample the background was found to be 12 ± 4 counts/min. The sample was counted for 12 min and an average of 3276 counts was observed. Determine the net counting rate and its standard deviation.

15. It is desired to make a measurement of counting rate with the largest possible statistical accuracy. The requirements limit the total counting time to 15 min. A test count gave a background of 37 counts/min and a sample including a background of 1012 counts/min. What should be the counting schedule for minimum error?

References

1. Birks, J. B., *Scintillation Counters*. New York: McGraw-Hill Book Company, 1953.
2. Goldstein, H., *Fundamental Aspects of Reactor Shielding*. Reading, Mass.: Addison-Wesley Publishing Co., Inc., 1959.
3. Halliday, D., *Introductory Nuclear Physics*. New York: John Wiley & Sons, Inc., 1955.

4. Evans, R., *The Atomic Nucleus*. New York: McGraw-Hill Book Company, 1955.

5. Lapp, R. E., and H. L. Andrews, *Nuclear Radiation Physics*. Englewood Cliffs, N.J.: Prentice-Hall, Inc., 1965.

6. Price, W. J., *Nuclear Radiation Detection*, 2d ed. New York: McGraw-Hill Book Company, 1964.

7. Taylor, J. M., *Semiconductor Particle Detectors*. London: Butterworth & Company (Publishers), Ltd., 1963.

8. Goodman, C., ed., *The Science and Engineering of Nuclear Power*, Vol. 1. Reading, Mass.: Addison-Wesley Publishing Co., Inc., 1947.

9. Hurst, G. S., and J. E. Turner, *Elementary Radiation Physics*. New York: John Wiley & Sons, Inc., 1969.

10. Cameron, J. R., N. Suntharalingam, and G. N. Kenney, *Thermoluminescent Dosimetry*. Madison, Wis.: The University of Wisconsin Press, 1968.

11. Knoll, G., *Radiation Detection and Measurement*. New York: John Wiley & Sons, Inc., 1977.

Health Physics and Biological Radiation Effects

Radiation is either a form of energy or a particle carrying energy with it. As we have seen, this radiation can have certain interactions with matter; both the matter and the radiation will be altered. If the matter undergoing the reaction happens to be a living organism, the effects of the interaction can damage the living tissue to the point of severe illness or death. The exact effects and mechanism of radiation damage to biological tissue are not known. *Health physics* is concerned not only with the effects of radiation on living tissue but also with the detection of radiation that may be injurious to human beings. Health physicists evaluate permissible exposure levels of radiation and devise procedures and methods to protect individuals from excess exposure to radiation.

It was partly by accident that people became aware of the damaging nature of radiation on living tissue. Early workers suffered "burns" on their skin after exposure to X-rays. This naturally led to the study of the biological effects of radiation. Unfortunately, there were some deaths and severe injuries as a result of overexposure to radiation before the serious nature of radiation damage was known. The classic example of this is the case of workers who painted watch dials. They died as a result of accidentally ingesting some of the radium that the paint contained.

It was not until the 1920s (over 25 years after X-rays were discovered) that safety measures for handling radioactive materials were proposed. In the 1930s maximum permissible levels for exposure were set. The general acceptance of the maximum permissible exposure levels has kept down the incidence of radiation injuries.

Units and Measurements

An amount of radiation is usually referred to as a *dose.* It is important to make a distinction between an exposure and a delivered or absorbed dose. For personnel protection, it is the exposure which is of interest, whereas for biological (or structural) damage the absorbed dose would be of concern. The first unit of radiation, the roentgen, was adopted in 1928 and modified in 1937. It is still the most prevalent unit in use. The *roentgen* (R) is "that quantity of X- or gamma radiation such that the associated corpuscular emission per 0.001 293 g of air produces, in air, ions carrying 1 esu of quantity of electricity of either sign." This means that 1 roentgen, when ionizing 1 cubic centimeter of air at STP, will produce ions such that the total charge on all the ions will be 1 esu. Notice that only the primary ionization need be produced in the 1.0 cm^3, but secondary ionization can be produced in any air. The secondary ionization contributes to the 1 esu of charge. Note also that a roentgen expresses a charge density and applies only to X- or gamma rays in air and, therefore, is considered a unit of exposure. Exposure expressed in roentgens does not depend on time, so an exposure rate can be expressed in terms of roentgen per unit time.

> **Example 1**   What energy is imparted to air when it completely absorbs 1 R of radiation?
>
> The charge on an electron is 4.8×10^{-10} esu, and it requires 34 eV to produce one ion pair. The latter unit of 34 eV per ion pair was adopted by the International Commission on Radiological Units in 1956.
>
> $$1 \text{ R} = \frac{1 \text{ esu}}{0.001\ 293 \text{ g air}} \times \frac{1 \text{ ion pair}}{4.8 \times 10^{-10} \text{ esu}} \times \frac{34 \text{ eV}}{\text{ion pair}} = 54.8 \times 10^{12} \frac{\text{eV}}{\text{g}}$$
>
> $$1 \text{ R} = \frac{54.8 \times 10^{12} \text{ eV}}{\text{g air}} \times \frac{1.6 \times 10^{12} \text{ erg}}{\text{eV}} = 87.7 \frac{\text{ergs}}{\text{g air}}$$

Even though 1 R is a reasonably large quantity of radiation, it is trivial from a physical energy standpoint. For example, it takes about 85.5 ergs to move a new sharpened No. 2 lead pencil one-sixteenth of an inch. As we will see later, the amount of energy delivered in a lethal dose of radiation is also trivial.

It seems desirable to have a more general unit of radiation which is independent of both the type of radiation and the irradiated material. This unit would then express an absorbed or delivered dose. The unit has been named the *gray* (Gy) and can be defined as being equal to the absorption of 1 J/kg (100 rads) at the point of interest. The gray therefore applies to any material and any radiation. Obviously, an exposure of 1 R will deliver a dosage to a material in Gy's dependent on its absorption coefficient. The absorption coefficient is related to the probability of absorption and is expressed as cm^2/cm^3 or cm^2/g.

The absorbed dose does not allow for severity or probability of harmful health effects since not all radiation has the same effect on body tissue even though it may dissipate the same energy in the tissue. A unit termed a *dose-equivalent* has been introduced to correlate the effect of different radiations on body tissue. The "dose-equivalent" is named a *sievert* (Sv). Dose-equivalent in sieverts is

$$H = DQ \qquad (6.1)$$

where D is the absorbed dose (Gy) and Q is the *quality factor*. A sievert is equivalent to 100 rems (roentgens equivalent man). The quality factor allows for the relative biological effect of different ionizing radiation. Table 6.1 gives recommended values of Q. The quality factor is based on relative biological effectiveness and extrapolations from higher absorbed doses at which effects in humans have been determined.

Probably the most significant characteristic of passage of radiation through matter relating to biological damage is the *linear energy transfer* (LET). LET expresses a rate of energy loss per micron of particle track.

$$LET = wS = \frac{dE}{dL} \qquad (6.2)$$

where S is the specific ionization, number of ion pairs produced per micron of tissue, and w is the energy (eV) necessary to produce one ion pair. Neither S nor w is well known, but dE/dL can be predicted from theoretical considerations. The Q would then depend on the values of LET for the particular type of damage and particular radiation. For most biological changes Q increases as LET or S increases, from small values for secondary electrons to largest values for protons or alpha particles.

An advantage in using dose-equivalent is that dosages of different radiations (when expressed in sieverts) are additive. Thus, maximum permissible exposure limits for persons can be expressed in sieverts without regard to the type of radiation present. However, this implies a simplifying assumption that there is a linear relationship between delivered dose and effect and that the severity of each type of radiation is independent of the other. It must be remembered, though, that the values of Q are not exact. Furthermore, the sievert should not

Table 6.1 Approximate Q for Various Radiations

Radiation	Q
X- and gamma rays	1
Electrons	1
Neutrons, protons, and singly charged particles of rest mass 1 amu of unknown energy	10
Alpha particles and multiply charged particles of unknown energy	20

be used for acute exposure (accidents) because the Q's may be different from those for chronic exposure.

Unfortunately, radiation exposure causes no bodily sensation (except possibly at high intensities). This means that instruments must be used to survey the areas where radiation hazards may exist. A good survey meter should respond in a manner proportional to tissue ionization produced.

All areas around a reactor are monitored frequently to determine the radioactivity present on surfaces and in the air. There is usually an instrument that continuously monitors the radiation present in the air. In addition, all wastes discharged from a plant are monitored to ensure that they do not cause an environmental hazard. Personnel are provided with dosimeters to measure the radiation to which they are exposed. The dosimeter can be read as often as desired in an area where radiation levels are high. Workers may be taken off a project before they receive more than a permissible exposure. Anyone exposed to an above-normal dose can be identified and measures must be taken to prevent further injury.

In the United States the National Council on Radiation Protection and Measurement recommends maximum levels of radiation exposure for individuals. The Environmental Protection Agency sets the standards for radioactive protection in the United States and the NRC (for nuclear activities) makes and enforces the regulations. It is the responsibility of the health physics organizations to adhere to these recommendations and regulations and to conduct radiation monitoring. It is well to remember that, despite the efforts of health physicists, an individual is ultimately responsible for his or her own safety.

Radiation monitoring instruments must be calibrated regularly and kept in perfect operating condition. Ionization chamber instruments are widely used for quantitative surveys since they respond to a wide range of energies. This means, of course, that their response can be designed to be proportional to tissue ionization. Geiger–Muller counters are also used for survey measurements; since they respond to the number of ionizing events rather than energy, however, they should not be used for quantitative measurements. A Geiger counter is a radiation detector rather than an instrument for radiation measurement. Preferably a Geiger counter used for radiation monitoring should read in counts rather than in milliroentgens per minute. The milliroentgen per minute reading would apply only to the radiation for which the instrument was calibrated. Figures 6.1 and 6.2 show available radiation monitoring instruments.

To calculate a dose rate the flux of radiation (photons/cm^2 s) must be multiplied by the rate at which energy is delivered to the absorbing material (MeV/cm^3). The product can be converted to Gy/g h. It is general procedure to use the energy absorption coefficient, μ, as the total coefficient minus the Compton scattering coefficient. Figure 6.3 shows variation of μ with gamma energy for air, water, and tissue. The density of tissue is taken as 1.0 g/cm^3.

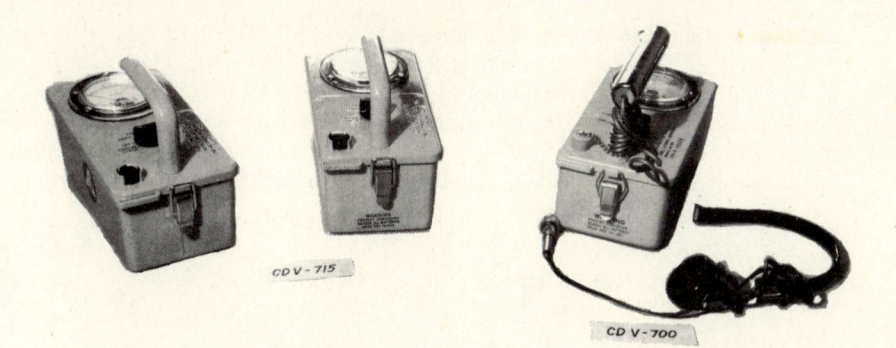

Fig. 6.1 Radiation monitoring instruments. The CDV-715 is a gamma survey meter, the CDV-700 a Geiger counter.

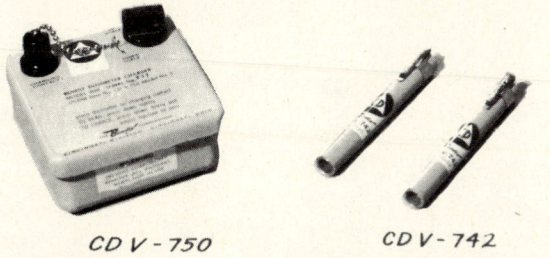

Fig. 6.2 Radiation monitoring instruments. The CDV-750 is a dosimeter reader-charger and the CDV-742 is a dosimeter.

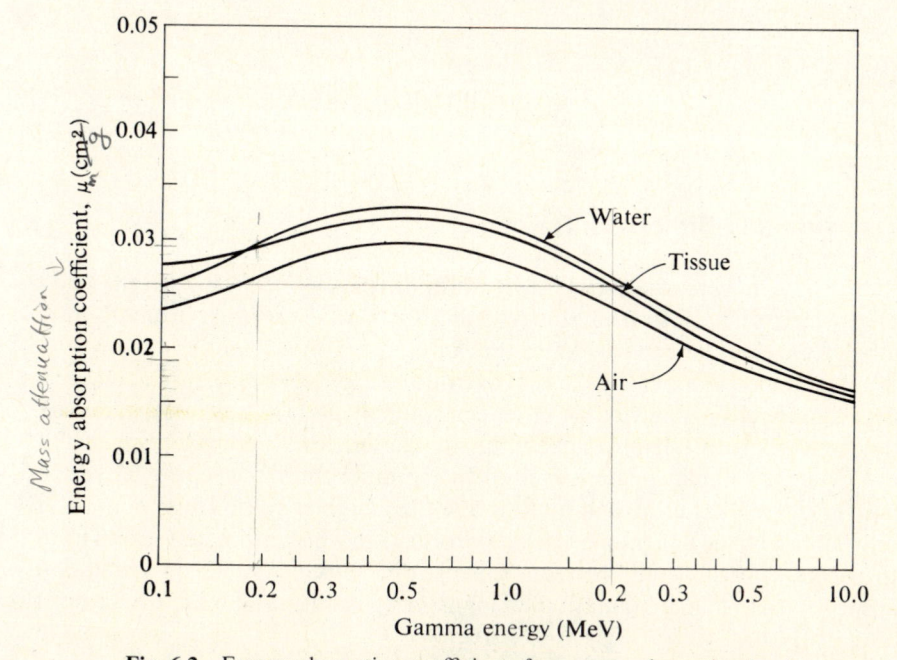

Fig. 6.3 Energy absorption coefficients for gamma absorption in air, water, and tissue. [Data from ORNL-421, *Absorption of Gamma Rays.*]

149

Example 2   Calculate the dose rate in tissue at a distance of 10 cm from a 1-mCi source of ^{60}Co. ^{60}Co emits gamma rays of energies 1.17 MeV and 1.332 MeV. Since μ is a function of energy, the dose rate will be different for each of the two gammas; therefore, each must be calculated separately and added to give the total dose rate. The absorption of radiation by air in the 10-cm interval will be neglected. The flux can be calculated from a point source using the inverse-square law.

$$\phi = \frac{\text{source strength}}{4\pi R^2}$$

$$= \frac{3.7 \times 10^7}{4(100)\pi} = 2.94 \times 10^4 \text{ photons/cm}^2 \text{ s}$$

From Fig. 6.3,

$$1.17 \text{ MeV} \qquad \mu = 0.029 \text{ cm}^2/\text{g}$$

$$1.332 \text{ MeV} \qquad \mu = 0.0275 \text{ cm}^2/\text{g}$$

$$\text{dose rate} = \phi\mu E$$

$$D_{1.17} = 0.029 \text{ cm}^2/\text{g} \times 2.94 \times 10^4 \text{ photons/s cm}^2 \times 1.17 \text{ MeV/photon}$$
$$= 997 \text{ MeV/g s}$$

$$D_{1.332} = 0.0275 \qquad \times 2.94 \times 10^4 \qquad \qquad \times 1.332 = 1080$$

$$\text{Total dose rate} \qquad\qquad\qquad\qquad\qquad\qquad = 2077 \text{ MeV/g s}$$

This is equivalent to

$$D = \frac{2077 \text{ MeV/g s} \times 1.6021 \times 10^{-13} \text{ J/MeV}}{1 \text{ J/kg Gy} \times 10^{-3} \text{ kg/g}} = 3.328 \times 10^{-7} \text{ Gy/s}$$
$$(0.02 \text{ mSv/min})$$

Exposure and Biological Damage

Radiation exposure can occur in two ways: (a) external exposure to the body or clothing directly from a source and (b) internal exposure from sources inhaled, ingested, or absorbed into the body. Irradiation of a cell (by internal or external exposure) can cause damage to the cell nucleus or to any of the other cell components. It is generally believed that cells can repair damage unless the interaction is with the cell nucleus. In this case, the cell may be destroyed or the chromosomes in a reproductive cell may be altered, resulting in a mutation in the daughter cell. These mutations are retained by the daughter cells and the organism which ultimately grows from these cells will be different (better or worse) from one which suffered no radiation absorption. In humans it may take a long time (generations) for these mutations

to become evident. Thus, many of the long-term effects of radiation are still unknown

The potential results of these genetic effects cause great concern to health physicists. Many feel that mutations will be disadvantageous to humans because we are so highly developed. This means that offspring will be more or less defective and these defects may be passed on to succeeding generations. It seems possible that a general *large* increase in level of radiation may increase the number of members of society who are unable to contribute their proper share to the general welfare. It is interesting to note that offspring of the Hiroshima and Nagasaki survivors have *not* shown an abnormal rate of genetic defects.

More specifically, all radiation effects stem from ionization of the cells. The ionization can be produced in any of the interaction processes described in Chapter 5. It should be noted, however, that the LET of a particle plays an important part in the cell damage. The radiation produces ionization and excitation in the cells which interfere with their normal functioning. This is evidenced by (a) the breaking apart of cells, (b) the displacement of cells or the swelling of cell nuclei beyond their normal position, (c) the increasing permeability of cell membranes, and (d) the forming of radioactive isotopes which themselves decay. Almost all the effects of radiation result from secondary ionization by the electron and free radical of the original cell. Large amounts of radiation destroy many cells and a large quantity of tissue is damaged.

The effects on the body of exposure to radiation are heavily dependent on the nature of the exposure. *Acute exposure* is a large dose of radiation received over a short period of time. For example, radiation absorbed due to an accident would be acute. *Chronic exposure* means relatively low dosages received over a long period of time, such as that experienced in normal everyday handling of radioactivity. In general, chronic exposure is less serious than acute exposure because the body can make continuing repairs to the damage.

The most striking and most significant characteristic of biological radiation exposure is the *latent period.* This is observed as a delay in appearance of symptoms. It is comparable to sunburn in which the redness and soreness usually become more marked several hours after one has come out of the sun. Radiation exposure produces an effect similar to sunburn but is much more serious since the radiation can penetrate to tissue beneath the skin. The latent period applies as well to other effects of radiation exposure. In some cases the latent period may even be longer than the individual's lifetime.

If a large dose is absorbed, most of the apparent changes will be observed within a few weeks and apparent recovery will take place within a few months. These relatively short-term effects show themselves as reddening of the skin (erythema), nausea, vomiting, decrease in both white and red blood cell count, loss of appetite, diarrhea, and formation of ulcers. The time lag in the appearance of these symptoms depends on the absorbed dose. As little as 1 Gy received as an acute dose will cause vomiting and fatigue in most persons, but

some persons will show no effects for larger doses. Thus, it is nearly impossible to assign strict levels that cause specific biological damage.

In an emergency or an accident an individual might suffer a large acute dose of radiation. We can estimate the probable effects of an acute whole-body dose. These whole-body effects are shown in Table 6.2.

If exposure occurs daily for a long period, no symptoms will be immediately observed (possibly for years). The damage, however, worsens as long as exposure continues. If the exposure is stopped, the damage may be repaired or long-term effects may occur years later. It is difficult to tell when recovery is completed because of these long-term effects. With a moderate dose there will be only a few such effects, but the delay may be as long as 25 years. In the interval between recovery from short-term effects and appearance of long-term

Table 6.2 Effects of Acute Whole-Body Radiation Doses[a]

| Effect | Dose (Gy) | | | | |
	0–1	1–2	2–6	6–10	10–50
Vomiting	None	5–50%	100% at 3 Gy	100%	100%
Latent period	None	3 h	2 h	1 hr	30 min
Characteristic sign		Leukopenia	Leukopenia, purpura, hemorrhage	Diarrhea, fever	Convulsions, tremor, lethargy
Therapy	Reassurance	Reassurance	Blood transfusion, antibiotics	Possible bone marrow transplant	Sedatives
Convalescent period	None	Several weeks	1–12 months	Long	
Incidence of death	None	None	0–80%	80–100%	90–100%
Time of death			2 months	2 months	2 weeks
Cause of death			Hemorrhage, infection	Hemorrhage, infection	Circulatory collapse

[a] This table is based on doses received by large numbers of individuals and on extrapolations from animal studies. For up to 2 Gy the information is based on reliable information, but beyond 2 Gy the data from exposed humans decrease rapidly. Above 6 Gy the information is based almost entirely on observations made on animals exposed to radiation.

Data from Reference 1.

effects, the victim may be completely free of symptoms. As the dose is increased, the interval between short- and long-term effects becomes shorter.

There may also be some synergism; that is, irradiation from a mixture of two types of radiation and/or exposure of two or more body organs may produce damage greater than the sum of individual components when received separately.

Some organisms are more sensitive to radiation than others. The most sensitive cells are those which are constantly reproducing or growing. Examples are the blood-forming organs, intestines, reproductive organs, lenses of the eye, the skin, and the thyroid. The most insensitive cells are those which are not growing or reproducing themselves such as adult brain cells and muscles. Recovery can take place in cells and tissue that can grow new ones; no recovery is possible in cells that do not repair or reproduce themselves. For a comprehensive discussion of radiation effects, see Reference 1.

External Effects, Internal Effects, and Treatment

The following outline lists the noted biological effects of radiation. The single most significant effect is the LD 50–30. This is the acute dose that would prove fatal to some 50 percent of the population within 30 days of exposure.

(1) *LD 50–30 acute dose in roentgens (measured in air).* Examples:

Humans	450	Guinea pig	250
Monkey	500	Dog	300–430
Rat	590	Sheep	520
Rabbit	790–875	Donkey	580–780
Chicken	1000	Turtle	1500

(2) *Erythema.* A reddening of the skin similar to sunburn, but different because of the more penetrating effect of radiation.

(3) *Growth rate and stunting of growth.*

(4) *Growth abnormalities.* Various localized abnormal growths, such as tumors, may be produced.

(5) *Blood count.* The most reliable indication of degree of exposure. First indication is a decrease in the white blood count followed somewhat later by a loss in red blood count. Blood count by itself, however, is not used as a measure of overexposure to radiation.

(6) *Mitotic index.* A quantitative and rapidly responding measure of a radiation exposure. The mitotic index is a ratio of the number of cells in a particular phase of their development to the total number of cells of the same kind. The mitotic index changes normally and continually, but exposure to radiation will cause an abnormal change.

(7) *Life span*. So far this is only a statistical measure of the effects of radiation. It is fairly certain, however, that chronic exposure will reduce the normal life span slightly. Examples:

General population	65.6 yr
Some exposure	63.3 yr
Radiologists	60.7 yr

(8) *Cataracts*. The eye is not particularly sensitive to gamma rays but is quite sensitive to fast neutrons. It is estimated that about 3×10^9 neutrons will cause an eye cataract.

(9) *Cancer and leukemia*. The latent period for the incidence of cancer and leukemia is very long (about 25 yr). For this reason the usual practice is to limit exposure to levels that show no permanent deleterious changes which might lead to cancer.

(10) *Genetic effects*. The complete genetic effects on human beings cannot yet be given, because it will take about 10 generations for them to be produced. However, it has been shown that radiation-induced mutations are about the same as those occurring in nature. Therefore, it is desirable to limit the exposure to as few persons as possible and, among those persons, to limit exposure through the reproductive years.

Overexposure of the reproductive cells to radiation in both men and women may produce temporary or permanent sterility. This reduction in fertility or sterility may be passed on to descendants. A genetic injury of this type does not affect the health of the exposed person. It only appears in the descendants in the form of inherited disorders, diseases, and increase of stillbirths, malformations, and so on. It is impossible, at present, to predict accurately the amount of genetic injury resulting from overexposure to radiation. The sensitivity of the reproductive cells, however, is not as high as that of the blood-forming organs, so that the level of permissible exposure for the blood-forming organs is also satisfactory for the reproductive cells.

Hands almost inevitably will receive a higher dose rate than other parts of the body. It is improbable, therefore, that the maximum permissible dose rate of the rest of the body will be exceeded if the hands do not receive it. It is difficult to measure the exposure of hands, so the usual practice is to measure the dose rate to the whole body by a film badge and/or pocket dosimeter similar to that of Fig. 5.10. Ring-type film badges have become available and are in use.

The effects of exposure from internal sources are no different except that the source presents more concentrated exposure near vital organs. This is particularly important in the case of alpha and beta particles. Once the radioactive source has been introduced into the body through (a) wounds, (b) the gastrointestinal tract, or (c) the lungs, there is no way to control the damage. The duration of internal exposure depends on the half-life of the source and body elimination. Some materials are rapidly eliminated in body wastes. Others, such as plutonium, strontium, and radium, concentrate in the bone. Uranium

concentrates in the kidneys, but tritium distributes itself throughout the body. In this regard the critical organ refers to the most vital organ to which the isotope goes. In evaluating internal hazards two concepts are useful: effective half-life and maximum permissible body content.

$$\frac{1}{t_{1/2\text{eff}}} = \frac{1}{t_{1/2\text{biol}}} + \frac{1}{t_{1/2\text{rad}}} \tag{6.3}$$

where $t_{1/2\text{eff}}$ = effective half-life
$t_{1/2\text{biol}}$ = biological half-life or the length of time to eliminate one-half the isotope biologically
$t_{1/2\text{rad}}$ = radioactive half-life

The effective half-life cannot be predicted exactly for all persons because of normal variation in biological half-life. The effective half-life for ^{90}Sr, for example, is 3000 to 5000 days. The maximum permissible body content is the amount of radioactive isotope that gives the maximum permissible internal dose rate to a critical organ. The rate of intake into the body, which eventually results in a specified body content, is inversely proportional to the effective half-life in the critical organ.

Treatment of radiation sickness can only minimize the effects of exposure to radiation. The usual treatment is bed rest for long periods of time. Antibiotics are given to increase body resistance and blood transfusions are given to counteract the blood damage. In some recent serious cases bone marrow has been transplanted, but transplants have not met with complete success because the body tends to reject foreign tissue.

Low-Level Effects

Data are inadequate to evaluate fully the risk of injury to the general population from low levels of radiation. Low-level ionizing radiation does not produce previously unknown effects in human beings, but the effects, if any, have not been measured in population groups of existing people. The most important of the long-term effects from low-level radiation is the potential production of cancer. The risks of low-level radiation can be predicted only by statistical methods. For example, it is not possible to attribute a case of cancer specifically to low-level radiation exposure, but through statistical modeling it is possible to estimate the number of cancer deaths among a large segment of the population. The latest estimates of the NAS BEIR Committee (Reference 17) estimate a 0.5 to 1.4 percent increase in cancer deaths per million persons exposed to a 0.1-Gy (10-rad) single whole-body dose. Based on an accepted rate of 160 000 cancer deaths per million persons, this means 800 to 2250 increased deaths or an

increase in rate of 0.08 to 0.22 percent. Such a small increase cannot be detected by normal methods, considering the variability in cancer frequency. To pinpoint the relationship between low-level radiation dose and effects, large numbers of people would have to be studied over several generations under laboratory conditions. The risks of long-term, low-level radiation are estimated by extrapolating high-level, short-term doses received by Japanese atomic bomb survivors, animal experiments, radiation therapy patients, industry workers, and accidental exposures.

The International Committee on Radiation Protection (ICRP) and the National Council on Radiation Protection (NCRP) were formed in the 1920s to provide guidelines for occupational exposures to radiation. However, after development of the atomic bomb, interest in more and better information on the hazards of radiation arose. Several national committees were formed to study radiation risk. When the EPA was formed in 1970, The National Academy of Sciences formed its Committee on the Biological Effects of Ionizing Radiations (BEIR). At the request of the EPA the BEIR Committee (as well as other organizations) has studied the biological effects of low-level radiation. The BEIR III report (Reference 17) represents (at the time of its publication) the latest statement of the knowledge of the risks of low-level radiation.

The human body has adapted itself to small continuous doses of radiation because it has always been exposed to natural background radiation. This background radiation is cosmic rays and radioactive material in the soil, air, and water, and ^{14}C and ^{40}K in the body. Now small additional amounts of radiation from atomic weapons tests, operating reactors, medical procedures, flying, and other sources have been added to daily natural exposures. Natural background radiation is by far the largest contributor to low-level radiation exposure. It varies from 1.0 to 2.5 mSv (100 to 250 mrem) per year, depending on location. A national average dose is about 1.3 mSv (130 mrem) per year. Most persons receive an additional dose of 0.70 to 0.75 mSv (70 to 75 mrem) per year from medical X-rays. Exposures from other sources are not significant (~ 11 mrem).

The difficulty in assessing risk of development of cancer from low-level radiation arises in extrapolating the high-dose, high-dose-rate exposures on record to low-dose, low-dose-rate exposures. The prevailing assumption has been that the low-dose effects will be proportional to those at high doses (curve *A* in Fig. 6.4). Some biologists believe that the low-dose effects should follow a quadratic model (curve *B* in Fig. 6.4) and some believe that a threshold exists below which no risk exists (curve *C* in Fig. 6.4). There are also some who believe the risks are higher than predicted by the linear theory. The consensus of the BEIR Committee as reported in Reference 17 is that the extrapolation should be linear-quadratic at low doses (curve *D* in Fig. 6.4). Although not all scientists agree on the precise model to be used (including members of the BEIR Committee), they do generally agree on the risk estimates.

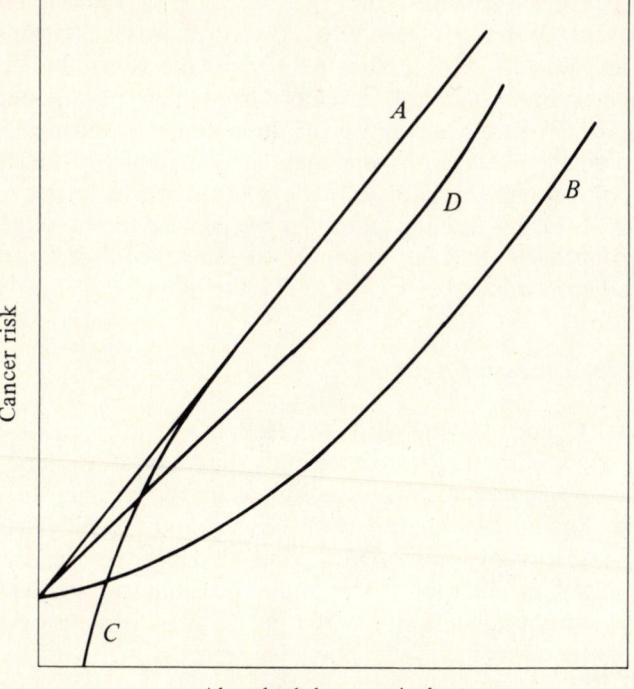

Fig. 6.4 Forms for dose–response data of low-level radiation. [Adapted from "Radiation and Human Health," *EPRI Journal* **4**, No. 7 (September 1979).]

Some major conclusions in the BEIR III report are:

(1) Cancers arising in organs and tissues are the principal long-term somatic effects of radiation exposure.
(2) There is a 0.5 to 1.4 percent increase in risk of cancer deaths above the naturally occurring rate from a single whole-body absorbed dose of 0.1 Gy (10 rad).
(3) There is a 3 to 8 percent increase in risk of cancer deaths above the naturally occurring rate from a continuous lifetime exposure of 0.01 Gy (1 rad) per year.
(4) Effects of low-level radiation on the embryo and fetus are strongly related to the development stage at which exposure occurs. There may be a threshold below which malformations or functional impairments are not produced.
(5) Data do not show an increased risk of other somatic effects from low-level radiation.
(6) A parental exposure of 0.01 Sv (1 rem) throughout the population will produce 5 to 75 additional serious genetic disorders per million liveborn children. Note that the normal incidence of serious genetic disorders is about 10 percent of liveborn children.

In the late 1970s several studies (References 21–23) were published that have conclusions contrary to generally accepted estimates. These studies are important and do foster the desire to improve the generally accepted risk estimates and exposure standards developed from them, even though they may not be accepted. After the publicity given these studies by the media, they were subjected to detailed analysis by the scientific community. The studies cited have not been considered reliable enough to be used in estimates of radiation risk. Principal reasons include questionable statistical techniques, inadequate sample sizes, inconsistent data, unconfirmed estimates of exposure, and unorthodox statistical methods.

Exposure Protection Guides

The National Council on Radiation Protection and Measurement (NCRP), with the guidance of the ICRP, recommends standards for radiation exposure protection for occupational workers as well as for members of the public. In addition, the Environmental Protection Agency and the NRC are active in establishing radiation exposure guides. The guides, of course, are subject to continuing study and revision. The limitations imposed by the guides are recommended as upper limits, not as routine permissible doses. All radiation exposures should be kept as low as possible.

The ICRP Radiation Protection Guides for individuals exposed to radiation in their occupations are based on certain simplifying assumptions and a procedure which takes into account the risk attributable to exposure of various irradiated tissues. Risk factors for different tissues are based on the likelihood of inducing fatal disease in the individual or substantial genetic disease in descendants. Exposure risk is recommended in terms of an annual dose-equivalent upper limit for uniform irradiation of the whole body. For occupational exposure the ICRP recommends that the risk be comparable to other occupations having a high level of safety. This is generally acknowledged to be an average annual death rate of 1 in 10 000.

The ICRP has divided risks into stochastic and nonstochastic. For nonstochastic (i.e., effects where the severity of effect varies with dose, such as cataracts, nonmalignant damage to skin, hematological deficiencies, and so on), a limit of 0.5 Sv/yr (50 rem) to all tissues except the lens of the eye is recommended. A limit of 0.3 Sv/yr (30 rem) is recommended for the lens. For stochastic risks (i.e., effects where probability of an effect occurring is a function of dose, such as cancers and hereditary effects), the recommended dose limitation for tissue is 50 mSv/yr (5 rem) for the whole body. For tissues, the dose limitation is based on a weighting factor representing the proportion of risk for the tissue in question to the whole body. The annual dose-equivalent limiting relationship is

$$\sum_T w_T H_T \leq H_{wb,L} \tag{6.4}$$

Table 6.3 Dose-Equivalent Weighting Factors

Tissue	w_T
Gonads	0.25
Breast	0.15
Red bone marrow	0.12
Lung	0.12
Thyroid	0.03
Bone surface	0.03
Five remaining highest	0.06 (each)

where w_T is a weighting factor for a given tissue, H_T is the dose-equivalent in that tissue, and $H_{wb,L}$ is the annual dose-equivalent limit for the whole body (50 mSv/yr). Values of w_T are given in Table 6.3.

For individual members of the public, dose-equivalent limitation is a concept intended to ensure that operation of radioactive facilities are not likely to expose individual members of the public to a specified dose-equivalent. The ICRP recommends an annual dose-equivalent limit from all sources of 5 mSv (0.5 rem) as applied to a critical group. A critical group is a representative group expected to receive the highest dose-equivalent. This limitation would then restrict the lifetime dose to an individual to less than 1 mSv (0.1 rem) per year, whole-body exposure.

Shielding

Chapter 1 indicated that shielding must be provided around a reactor to protect both personnel and material. Shielding that is adequate for neutrons and gamma rays will also stop alpha and beta particles. Previous sections of this chapter have indicated that neutrons are potentially more dangerous than gamma rays in causing biological damage. Even so, the contribution of neutrons to total dose is much less than that of gamma rays. The problem, then, is basically one of providing protection from (a) primary gamma radiation, (b) radiation due to n–γ reactions in the shield, and (c) fast neutrons.

The weight of shielding to be used is almost independent of the shielding material itself. It is usually advantageous to use dense material, particularly concrete, lead, and water. Water can be used where heating (due to absorption of gamma energy) is important and for storage purposes. Because of its low cost and structural characteristics, concrete is the most commonly used shielding material.

In its simplest form shielding involves interposing distance and materials between the source and recipient of radiation. Design considerations and the calculation of the resultant dose complicate the problem.

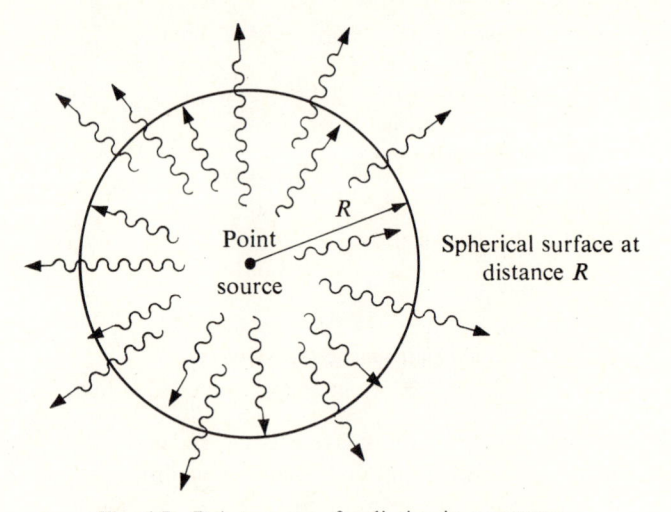

Fig. 6.5   Point source of radiation in a vacuum.

To gain some insight into shielding calculations we shall consider an oversimplified situation which involves a point source of radiation. Figure 6.5 shows a point source radiating isotropically through a vacuum. According to the inverse-square law, the intensity of radiation on the surface of a sphere of radius R will be

$$I = \frac{S}{4\pi R^2} \qquad (6.5)$$

where S is the source strength (number of particles or rays per unit time). If we place enough distance between ourselves and the source, the intensity of radiation will be reduced to safe levels. However, if we place material between ourselves and the source, we can take advantage of attenuation provided by the material. For the case of a collimated beam of gammas, Eq. (5.4) predicts the intensity at a point in the material where μ is the total linear absorption coefficient. Equation (6.6),

$$I = I_0 e^{-\mu R} \qquad (6.6)$$

applies strictly to a collimated beam and only when scattered radiation is removed from the beam. In a thick shield such as that in Fig. 6.6, Eq. (6.6) would give a low result because some of the radiation is backscattered into the path. This is the result of Compton scatter in the shield; the gammas decrease in energy. To correct for this backscattering an experimental *buildup factor*, $B(\mu R)$, is employed. Equation (6.6) becomes

$$I = B(\mu R)I_0 e^{-\mu R} \qquad (6.7)$$

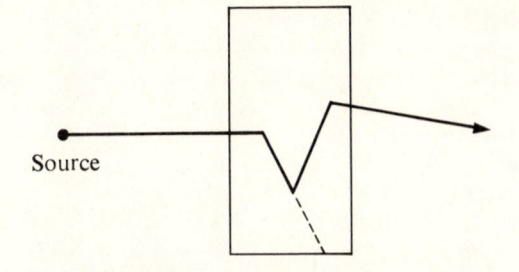

Fig. 6.6 Backscattering of radiation in a shield.

The buildup factor, a function of linear absorption coefficient and shield thickness, is commonly given in the form

$$B(\mu R) = 1 + a\mu R \qquad\qquad (6.8)$$

where a is a value depending on gamma energy and material. For example, the buildup factor for ^{60}Co gammas in air is (see Reference 12)

$$B = 1 + 0.55R \qquad\qquad (6.9)$$

Combining Eqs. (6.5) and (6.7) gives the intensity from an isotropic point source of radiation:

$$I = \frac{SB(\mu R)e^{-\mu R}}{4\pi R^2} \qquad\qquad (6.10)$$

Example 3 What is the intensity of gamma rays from a 10-Ci source of ^{22}Na surrounded by 5 cm of lead? The buildup factor is 2.1.

$$I = \frac{SB(\mu R)e^{-\mu R}}{4\pi R^2}$$

From Appendix D the ^{22}Na gamma energy is 1.28 MeV. The total linear absorption coefficient (Fig. 5.9) is 0.66 cm^{-1}.

$$I = \frac{10(3.7 \times 10^{10})(2.1)e^{-(0.66)(5)}}{4\pi(5)^2}$$

$$= 9.03 \times 10^7 \text{ photons/cm}^2 \text{ s}$$

This can be converted to a dose rate in the manner of Example 2.

Equation (6.10) is applicable only to a point source of radiation. With the introduction of a buildup factor, it is seen that intensity in a shield is not an

Table 6.4   Relaxation Lengths

Material	Density (g/cm^3)	Fast Neutrons (cm)	8-MeV Gamma (cm)
Water	1.00	10	40
Concrete	2.30	12	18
Lead	11.30	9	2
Beryllium	1.85	9	30
Graphite	1.65	9	25

exponential function. However, it would be convenient if we could devise an expression that retains the exponential character. This can be done by representing intensity as

$$I = I_0 e^{-a/\lambda} \tag{6.11}$$

where λ is defined as the *relaxation length*. In other words, it is the length in which the intensity decreases by a factor of $1/e$. Equation (6.11), therefore, includes a contribution for scattering in the material. Relaxation length is frequently used in rough shielding calculations. If the radiation variation is not exponential, this means that the relaxation length varies from point to point in the material. For thick shields variation is very nearly exponential. Table 6.4 gives relaxation lengths for some common shielding materials. Notice that the relaxation length for both neutrons and gammas is about the same in concrete.

Almost all sources of radiation are larger than point sources. Fortunately, gamma rays do not interact with each other, so total radiation can be considered as the sum of radiations from point sources. For example, consider a circular plane uniform isotropic source, Fig. 6.7. If the source is considered to be made up of a number of point sources having a total strength S, the dose (intensity) at point P is

$$D = \int_0^R \frac{Se^{-\mu a}B(\mu a)}{4\pi a^2} 2\pi r \, dr \tag{6.12}$$

In terms of the distance from P to the plane, Eq. (6.12) becomes

$$D = 2\pi S \int_z^{(z^2 + R^2)^{1/2}} G(a)a \, da \tag{6.13}$$

$G(a)$ is called the *point attenuation kernel*. It is the radiation intensity observed at distance a from a unit isotropic point source. In our case

$$G(a) = \frac{B(\mu a)e^{-\mu a}}{4\pi a^2} \tag{6.14}$$

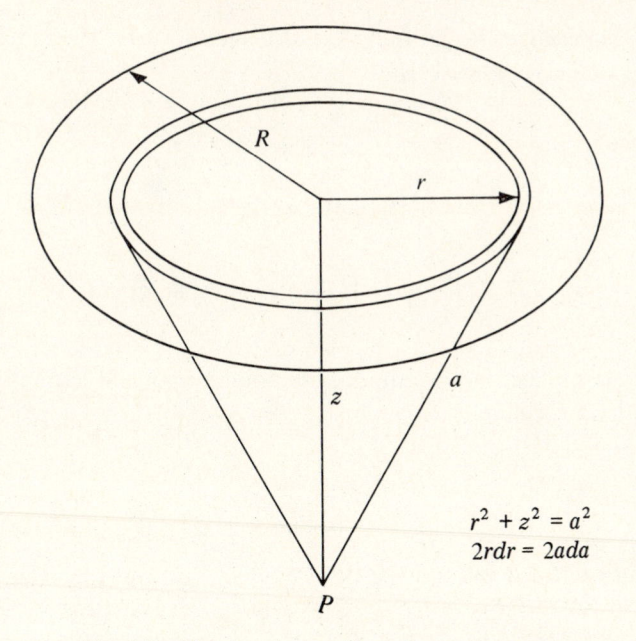

$$r^2 + z^2 = a^2$$
$$2rdr = 2ada$$

Fig. 6.7   Circular plane uniform isotropic source.

Integration of Eq. (6.13) for a plane surface leads to an exponential integral of the first kind.   By employing this concept of a point kernel the intensity for other geometries can be evaluated.

A volume source can be related to a surface source as follows.   Let S_v represent a uniform source strength within the volume element dV of Fig. 6.8. The source strength due to the volume element is $S_v\, dV$.   The width of the

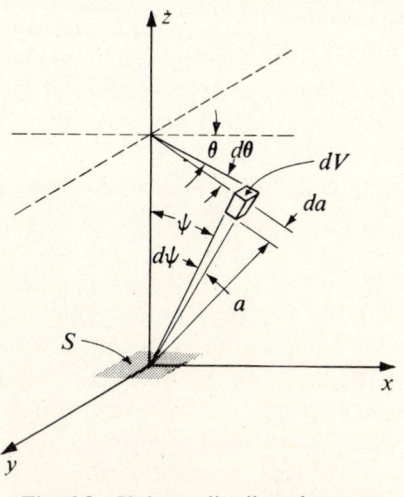

Fig. 6.8   Volume-distributed source.

element is $r \sin \psi \, d\theta$, its height is dr, and its length is $r \, d\psi$. The radiation at the surface due to the volume dV is

$$dS = S_v \, G(a)a^2 \sin \psi \, d\theta \, d\psi \, da \qquad (6.15)$$

and the total intensity at the surface due to the entire volume is

$$S = S_v \int_0^{2\pi} \int_0^{\pi/2} \int_0^D G(a)a^2 \sin \psi \, d\theta \, d\psi \, da \qquad (6.16)$$

Introducing the relaxation length into the point kernel and letting the buildup factor equal 1, we obtain

$$G(a) = \frac{e^{-a/\lambda}}{4\pi a^2} \qquad (6.17)$$

Evaluating Eqs. (6.16) and (6.17) gives

$$S = \tfrac{1}{2}S_v \lambda \qquad (6.18)$$

and since Eq. (6.16) was written for sources from one direction only, Eq. (6.18) must be multiplied by 2 to obtain the total surface source equivalent of the volume source.

If S_v can be represented by a simple function, Eq. (6.18) can be combined with Eq. (6.13) and integrated. In the more general case, an average S_v is used and assumed constant throughout the core. This leads to a conservative result.

The shielding materials used in reactor systems are dictated not only by neutrons and primary and secondary gammas but also by the purpose of the reactor. For example, in a submarine reactor the weight of the shielding is of extreme importance, but in a power reactor it is mostly an economic consideration. Radiation level outside the reactor is of prime importance, of course.

Frequently, shielding is designed by a comparison method. An existing reactor is used as the basis for predicting the shielding characteristics of the reactor being designed.

When the physical layout of the reactor system has been determined, the maximum radiation exposure at various points is computed. Of course, some knowledge of the reactor power and core materials is necessary to determine the distribution of radiation. If this information is known, a comparison can be made. A mockup of the design can then be tested in a facility such as the Bulk Shielding Facility at Oak Ridge, Tennessee.

If no similar system is available, analytic calculations are made. Because of the complexity of radiation interactions, the calculations cannot be exact,

and normally, recourse must be made to experimental models. Finally, as the reactor design is finalized, the shielding determination becomes more and more refined.

Problems

1. Is there a difference between the contributions of dose rate and total dose to the biological damage done by radiation? Explain.

2. How long will it take to receive an annual dose-equivalent of 2-MeV gammas to the hands and forearms from a 1-μCi source at 10 cm?

3. Calculate the total energy dissipated if a 160-lb man received an LD 50–30.

4. Calculate the dose rate in tissue given by a 10-mCi source of ^{137}Cs gammas. Neglect any absorption of gammas in the air between tissue and source. The source distance is 100 cm.

5. How long will it take a worker to receive his annual dose-equivalent of neutrons if he is exposed to a flux of 10^8 thermal neutrons/cm^2 s and a flux of 10^4 fast neutrons/cm^2 s?

6. If a worker was exposed to 50 mR of ^{60}Co gammas and 20 mrads of fast neutrons, what would be her total exposure?

7. In cleaning up a radioactive spill a worker received a dose of 2 mSv of gamma radiation and 1.5 mSv of unknown fission product radiation. What was his dose-equivalent?

8. A worker is suspected of receiving a dose-equivalent of 12 mSv to her thyroid and 22 mSv to her lungs. What is her whole-body dose-equivalent?

9. The radiation protection guides are intended to prevent what injuries during the lifetime of the individual?

10. A man exposed to radiation receives a dose of 40 mrads of fast neutrons. What is his whole-body dose? Did he exceed the recommended weekly dose on any part of his body?

11. What should be done in the case of the man exposed to radiation in Prob. 10?

12. If a test tube containing 400 mg of ^{90}Sr was dropped and broken in a room 40 ft square and 10 ft high, would there be a dangerous concentration of ^{90}Sr in the air in the room?

13. What distance must an experimenter remain from a 3-Ci ^{226}Ra source if she is not to receive more than the recommended maximum weekly dosage?

14. Find the thickness of lead necessary to reduce the dose rate from a 100-Ci ^{60}Co source to safe levels for continuous exposure. (*Note*: Use a buildup factor of 5.8 for both the ^{60}Co gammas.)

15. What is the dose rate from a 5-Ci source of ^{60}Co surrounded by a 10-cm-thick lead shield?

16. By completing this table, you will get an idea of the amount of low-level radiation you are exposed to every year.   The average dose to Americans is 1.3 mSv.

Source	Annual Dose (mSv)
Cosmic radiation reaching the earth.	0.28

Add 0.01 for every 100 ft above sea level.
   Pittsburgh is 1200 ft; add 0.12.
   Denver is 5300 ft; add 0.53.
   Atlanta is 1050 ft; add 0.10.
   Chicago is 600 ft; add 0.06.
	.08

If your house is brick, stone, or concrete, add 0.07.	.07
Ground radiation (average).	0.26
Water, food, air, and internal sources.	0.28
Nuclear weapons testing fallout.	0.04
If you have had a chest X-ray, add 0.09 for each.	——
If you have had your teeth X-rayed, add 0.03.	.03
If you have had intestinal X-rays, add 1.0.	——
If you wear a luminous wristwatch, add 0.01.	——
If you watch color TV, add 0.0015 for each hour of daily watching.	.003
For each 1500 mi you have flown in a jet airplane, add 0.01.	.1
If you live 1 mi from a nuclear plant, add 0.0002 for each hour you are at home during the day.	——
If you live from 1 to 5 mi from a nuclear power plant, add 0.00002 for each hour you are at home during the day.	——
If you live within 5 mi of a coal-fired power plant, add 0.001 for each hour you are at home during the day.	.001
If you sleep with your spouse (or any other person), add 0.01.	——
Total	1.169 mSv/yr

References

1. Glasstone, S., ed., *The Effects of Nuclear Weapons*, U.S. Atomic Energy Commission, 1964.
2. *Nuclear Safety*, U.S. Atomic Energy Commission, quarterly publication.
3. *Proceedings of the Second United Nations International Conference on the Peaceful Uses of Atomic Energy*, Vol. 21, Geneva, 1958.

4. Glasstone, S., and A. Sesonske, *Nuclear Reactor Engineering*. New York: D. Van Nostrand Company, 1963.

5. Etherington, E., *Nuclear Engineering Handbook*. New York: McGraw-Hill Book Company, 1958.

6. Blatz, H., ed., *Radiation Hygiene Handbook*. New York: McGraw-Hill Book Company, 1959.

7. International Commission on Radiation Units and Measurements, *ICRU Report 19*, Washington, D.C., 1973.

8. *Radiation Protection*, ICRP Publication 26, Annals of the ICRP 1, No. 3 (January 1977).

9. E. L. Saenger, ed., *Medical Aspects of Radiation Accidents*, U.S. Atomic Energy Commission, 1963.

10. *Radiological Health Handbook*, U.S. Department of Health, Education and Welfare, 1962.

11. *Occupational Radiation Protection*, U.S. Department of Health, Education and Welfare, 1965, Chap. 5.

12. Batter, J. F., "Cobalt and Iridium Buildup Factors near the Ground/Air Interface," *Transactions of the American Nuclear Society* 6, No. 1 (1963).

13. Spencer, L. V., *Structure Shielding Against Fallout Radiation from Nuclear Weapons*, NBS Monograph 42, 1962.

14. Eisenhauer, C., *An Engineering Method for Calculating Protection Afforded by Structures Against Fallout Radiation*, NBS Monograph 76, 1964.

15. *Basic Radiation Protection Criteria*, NCRP Report 39, 1971..

16. "Radiation Effects on Man," *Nucleonics* 21, No. 3 (March 1963).

17. *The Effects on Populations of Exposure to Low Levels of Ionizing Radiation*, National Academy of Sciences, July 1980.

18. *Radiological Factors Affecting Decision-Making in a Nuclear Attack*, NCRP Report 42, 1974.

19. *The Effects on Populations of Exposure to Low Levels of Ionizing Radiation*, NAS/NRC Report, November 1972.

20. Sagan, L., "Radiation and Human Health," *EPRI Journal* 4, No. 7 (September 1979), pp. 6–13.

21. Mancuso, T. F., A. Stewart, and G. Neale, "Radiation Exposures of Hanford Workers Dying from Cancer and Other Causes," *Health Physics* 33 (1977), pp. 369–385.

22. Bross, I. D., M. Ball, and S. Falen, "A Dosage Response Curve for the One Rad Range: Adult Risks from Diagnostic Radiation," *American Journal of Public Health* 69 (1979), pp. 130–136.

23. Najarian, T., and T. Colton, "Mortality from Leukaemia and Cancer in Shipyard Nuclear Workers," *Lancet* 1 (1978), pp. 1018–1020.

Radioisotope Application

Categories of Radioisotopes

Almost every industry and service organization in the world has a potential use for radiation. This chapter aims to acquaint the reader with some examples of current uses of radioisotopes. Because of the magnitude of present and future uses of radioisotopes, the discussion is limited; however, the reader can gain some knowledge and, hopefully, will be stimulated to explore other uses in detail.

It is convenient to classify the usage of radioisotopes in four categories:

(1) Irradiation of a target material to make a change in its physical properties. The change may enhance the usefulness of the material or destroy it. Examples are production of wood–plastic materials and destruction of cancerous tissue.

(2) Injection of a small amount of radioisotope with normal material in order to trace it through some process. Examples are wear studies and tracing water flow to locate water supplies.

(3) Fixed sources of radiation used as gages. Examples are thickness gages and radiographic inspection.

(4) Fixed sources of radiation used for power generation, heat, or illumination.

The use of radioisotopes, although it does not now represent a large capital outlay by the nuclear industry, is increasing rapidly. In addition to the users of radioisotopes, the industry encompasses producers of primary isotopes, processors who prepare radioactive chemicals and specialized sources for medical applications, fabricators who prepare large sources such as those in radiography and medical therapy, and equipment manufacturers who make the radiography equipment, power generators, process irradiators, and so on. The number of companies involved in these activities is small.

Industrial Applications

Industrial companies use radioisotope gages to measure thickness, density, and level of materials. These measurements are frequently combined with other mechanical measurements, such as speed, to determine other properties of the material.

Radioisotope gages have a big advantage over other gages in that they can continuously monitor without contacting the material to be measured. Also, since the radioisotopes are used in conjunction with an electronic detector, it is relatively easy to amplify the output signal and use it to control a process.

The radioisotope sources are sealed and emit beta or gamma radiation. ^{192}Ir, ^{90}Sr, and ^{60}Co are popular sources. The gages are set up either as transmission or backscatter types. In the transmission gage the source and detector are on opposite sides of the material to be gaged. The attenuation depends on the thickness and density of material between source and detector. The detector is commonly an ionization chamber. A backscatter gage has the source and detector on the same side of the material to be gaged. Shielding prevents any direct radiation from entering the detector. Only the reflected radiation is detected. The degree of reflection, or backscatter, depends on the thickness of the material. Thickness measurement by backscatter requires access to only one side of the sample. When a beta particle is incident on a surface it will undergo scattering with the nuclei of the material. It is probable that at some time the beta may be scattered back out from the surface. Figure 7.1 shows the intensity of backscattered radiation in relation to material thickness. The intensity of this radiation is a function of beta energy and the scattering characteristics of the material being measured.

At the saturation thickness, t_s, the intensity reaches a limiting value. Thus, to measure thickness it is only necessary to have a large enough difference in

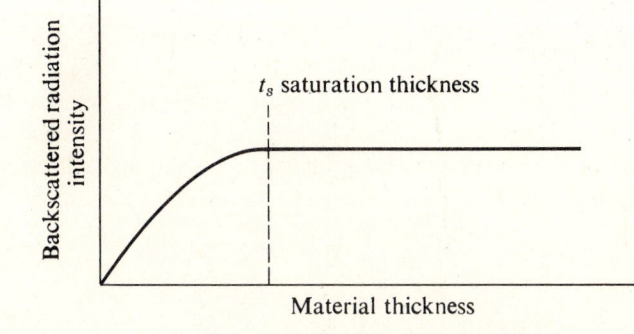

Fig. 7.1 Intensity of backscattered radiation with thickness of material.

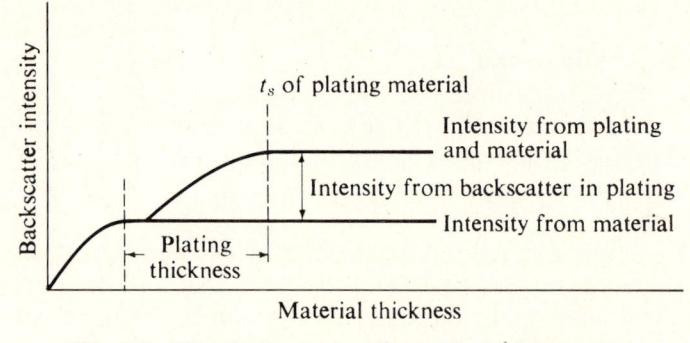

Fig. 7.2   Film thickness measurement by backscattering.

backscattered intensity from the two materials.   An example of this measurement is shown in Fig. 7.2.

Coating thicknesses in the range 5 to 100 μin. can be measured in this manner as long as the geometry of the detection system is fixed, the film thickness is less than the saturation thickness of the plating material, and the plated material has a thickness greater than its saturation thickness.

Use of thickness gages has been successful in the manufacture of sandpaper and emery cloth.   Gages are placed before and after successive coating operations.   The difference in detector responses represents the difference in weight; the response automatically adjusts the coating applicators.   This technique results in an extremely uniform product and may save up to $35,000 per year for a large manufacturer.

Example 1   Figure 7.3 shows a schematic arrangement for measuring the thickness of a material using a transmission gage.   Suppose that such a gage is to be used for monitoring the thickness of rolled steel plate.   The source is ^{192}Ir which has a half-life of 74.5 days and emits several gammas, 0.316 MeV being the most abundant.   What should be the reduction in intensity if the steel thickness is to be 0.100 in.?   The reduction in gamma ray intensity can be represented as an exponential function.

$$I = I_0 e^{-\mu_m \rho x} \tag{7.1}$$

The mass attenuation coefficient for 0.316-MeV gammas in steel is 0.107 cm^2/g.

$$I = I_0 e^{-(0.107)(7.80)(0.100)(2.54)}$$
$$= I_0 e^{-0.212} = 0.808 I_0$$

The reduction in intensity is therefore 19.2 percent of the intensity with no specimen present for 0.316-MeV gamma rays.

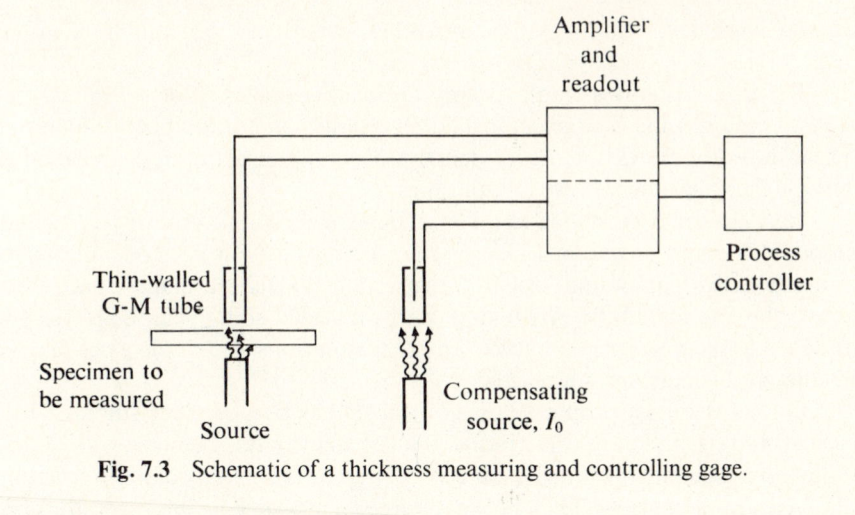

Fig. 7.3 Schematic of a thickness measuring and controlling gage.

Probably the best known use of radioisotope gages is in the control of cigarette density. It has been estimated that 80 to 90 percent of cigarette production is controlled by radioisotope gaging (see Reference 1). In a typical application a 20-μCi source of ^{90}Sr is used in the density gage. The gage measures the density of cigarette rod before it is cut and assembled into packages. The output of the detector controls the rate of tobacco feed, maintaining a constant weight of tobacco from pack to pack. This technique enables an automatic production of about 2000 cigarettes per minute.

Some of the newer industrial uses of radioisotopes include using ^{32}S to study the constituents of stack gas, ^{60}Co treatment to break down detergents, in pesticides, and to treat bacteria in sewage treatment plants. Activation analysis of trace elements in air and in polluted waterways to pinpoint polluters and measurement of contaminants in uranium mines are also interesting applications of radioisotopes.

Radiography using radioactive isotopes amounts to about 60 to 70 percent of the total radiography done in the United States. Most of the radiography is done using ^{60}Co with a specific activity of about 160 Ci/g. In some applications ^{192}Ir with a specific activity of about 300 Ci/g is used. The tendency is to make specific activity larger, since this results in a higher sensitivity. The sensitivity is akin to the shadow a light source makes. In other words, the smaller the source size, the more sensitive the radiography.

The most valuable use of radioisotopes is in portable radiography work, where it would be impossible to transport X-ray machines. For example, radioisotopes are used extensively for weld inspection in pipelines, bridges, boilers, and other custom-built structures. Radioisotopes are frequently used in routine maintenance inspections. Airlines use a ^{60}Co inspection for the core of jet engines. With routine radiographic inspection flaws can be detected inexpensively. Faulty boiler tubes with insulation and furnace walls corroded

can be identified.   Ordinarily, some tubes would not be seen in a visual inspection.

Research programs employ many radioactive tracer studies.   In a tracer study a radioisotope is attached to a larger quantity of nonradioactive material. The radioisotope does not lose its identity during chemical or physical reactions. Thus, its path can be followed at all times.

Radioisotopes are added to material being traced in several ways.   A small amount of liquid or gaseous isotope may be mixed with a larger amount of nonradioactive compound.   In some cases, chemicals are made from radio-active starting materials.   With solid parts, a sealed source can be physically attached to the object to be traced.   In other situations, the whole piece may be irradiated by neutrons in a reactor.

One of the most profitable uses of tracers has been in wear studies.   In a typical study, a pinion gear is irradiated in a reactor.   It is then run with a non-radioactive gear and the rate of wear is measured by the activity of the lubricating oil.   An additional benefit can be gained from a test such as this, by autoradio-graphing the initially nonradioactive gears.   The radiograph will show any areas that have picked up radioactive ^{59}Fe from the pinion.   The gears can then be redesigned easily to prevent this wear.   In internal combustion engine wear tests the cylinder is frequently plated with ^{51}Cr (0.32-MeV gamma) and the piston rings are irradiated to produce ^{59}Fe (1.11- and 1.13-MeV gammas). Using a pulse-height analyzer it is possible to separate the ring and cylinder wear.   Alternatively, the rings could be labeled with different isotopes to com-pare the wear rates of those rings.   To make studies of piston ring rotation a small amount of another isotope (such as ^{60}Co) could be embedded in the ring. Wear tests can also be made by impregnating the surface with ^{14}C or ^{32}P. The surface can be scanned to determine the time variations in radioactivity.   This method is most valuable in cases where wear particles are difficult to collect.   In any wear tests, precaution must be taken to obtain accurate measurements, such as by screening the engine from the detector.   The significant difference between the wear tests using more conventional physical measurements and the radioactive tracers methods lies in the fact that the radioisotope experiments can be made in days while conventional tests take about six months.

A unique tracer study was made in studying the rate of tool wear in a pro-duction machining operation.   Instead of activating the tool, as is done in laboratory experiments, the chips were collected and irradiated to activate tungsten from the cutting tool.

The Atomic Energy Commission sponsored development of a tracer technique that should find wide applicability in studies of solid surfaces.   In this method the material is either bombarded with positive radioactive krypton ions or radioactive krypton ions are diffused into the substance at high pressures and temperatures.   The krypton is a mixture of approximately 5 percent ^{85}Kr (half-life 10.7 yr) and 95 percent stable krypton isotopes.   Some of the ^{85}Kr gas is substituted in the lattice structure and some is located interstitially.   Almost

any solid material can be made radioactive by "kryptonation." An important side effect of kryptonation allows determination of the surface temperature distribution. Kryptonates lose krypton at high temperatures in a reproducible and controlled way. That is, when the temperature of a kryptonate is raised to a certain point, a fixed percentage of ^{85}Kr is lost. If the temperature is held at this point, or lowered, no further loss of krypton occurs. Surface temperature distributions of jet engine blades have been measured using this technique. Kryptonates have also been proposed for use in corrosion studies, phase-change studies, and stress–strain studies.

The chemical and electronic industries are making use of radiation for processing. The most successful radiation processing has been with polyethylene film and electrical insulation. The radiation used for these products is mainly electron beams from accelerators. The irradiated polyethylene film or insulation has a tensile strength up to six times as great as the conventional product and it possesses a "thermal memory." In the case of polyethylene, a sheet or tube is irradiated, heated, and stretched to double its original size, then cooled under tension. When the polyethylene is used (for packaging, say) it is again heated. The second application of heat causes it to shrink to its normal size.

Wood–plastic combinations promise an interesting use of radiation processing. In this process, the wood fibers are impregnated with a plastic compound. The impregnated wood is then exposed to a high-dose-rate ^{60}Co source (up to 10^4 Gy total dose) which polymerizes the plastic compound within the wood. Most of the impregnation has been done with styrene, methyl methacrylate, polymethyl methacrylate, or vinyl acetate.

The resulting wood–plastic combination has the natural appearance of wood, but is much more durable. The combination has increased strength, dimensional stability, and hardness. It can be fabricated by normal methods used for other woods; the only finishing required is sanding to the desired surface characteristics. The major difficulties in making the process economically feasible have been a lack of quantity markets for the product, difficulty in producing uniform impregnation, and fabrication problems. Recently published research indicates that flooring and veneer modification are the two most likely applications for this material.

Studies in the United States and Germany indicate that production of detergents in the presence of gamma radiation enhances their biodegradable characteristics. The original detergents are composed of alkyl benzene sulfonates. By reacting oxygen and sulfur dioxide with hydrocarbons in the presence of gamma radiation an alkane sulfonate compound is produced. The alkyl benzene sulfonates are only about 25 percent decomposed in sewage plants. Tests of the alkane sulfonate compounds showed them 100 percent degraded after two days, whereas the alkyl benzene sulfonate was degraded only 10 percent in two days. The physical properties and detergent action were nearly the same.

Power Generators

In 1956, the U.S. Atomic Energy Commission began developing direct-conversion devices to convert the energy of decaying isotopes into electricity. The impetus for this program was the potential use of the generators in space and in isolated locations such as Antarctica, and in their use as navigational buoys. The first isotopic power generator was produced in 1959. It employed a thermoelectric generator, produced 2.5 W of power, and used ^{210}Po as fuel. The whole generator weighed only 4 lb. Contrast this with a conventional nickel–cadmium battery (weighing about 700 lb) producing equivalent energy. The first commercial isotopic power generator became available in 1966. It was produced through the joint efforts of the Martin Company and the Atomic Energy Commision. The unit is a thermoelectric device powered by ^{90}Sr. Initial prices for the unit ranged from $53 587 to $63 230 for a unit guaranteed to produce 25 W of electricity for five years. Figure 7.4 is a picture of a Martin–Marietta Corporation isotopic power generator.

Two types of direct conversions have been proposed for isotopic power generators: thermoelectric and thermoionic. Figure 7.5 is a simplified sketch of a thermoelectric radioisotope generator. The outer shell serves as the heat radiator as well as the container for the generator itself. Theoretically, the radiation shield could be placed on either side of the thermoelectric converters. Placing it inside the converters reduces the hot junction temperature and the generator efficiency. Placing it outside the converters, as in the figure, increases the volume of shield material and the cost. Uranium is a common shield; consequently, the mass and cost of the shield may be the major portion of the whole generator. In fact, the shield may go as high as 90 percent of the generator mass.

The Apollo missions have included some of the more spectacular applications of radioisotope generators. The first use was in the Apollo-11 mission, where a ^{238}Pu-fueled heater in the seismometer keeps the equipment warm. The Apollo-12 mission left a SNAP-27 generator as the sole source of power on the moon. The SNAP-27 generator produces 63 W of electricity. It is fueled by ^{238}Pu. The ^{238}Pu capsule was encased in graphite and carried on the side of the LEM. This permitted a simpler design and the graphite protected the ^{238}Pu in the event that the mission was aborted and the LEM reentered the earth's atmosphere.

In 1969, the first Nimbus weather satellite using a radioisotope generator was launched. The power supply was PuO_2 fuel. The heat from the decaying ^{238}Pu is converted into electrical power in a bank of PbTe thermocouples. The thermocouples are wired in series and parallel to provide the optimum current–voltage relationship. The heat rejected is radiated into space by four fins. The Nimbus satellite has two 25-W generators each about 9 in. high × 20 in. in diameter.

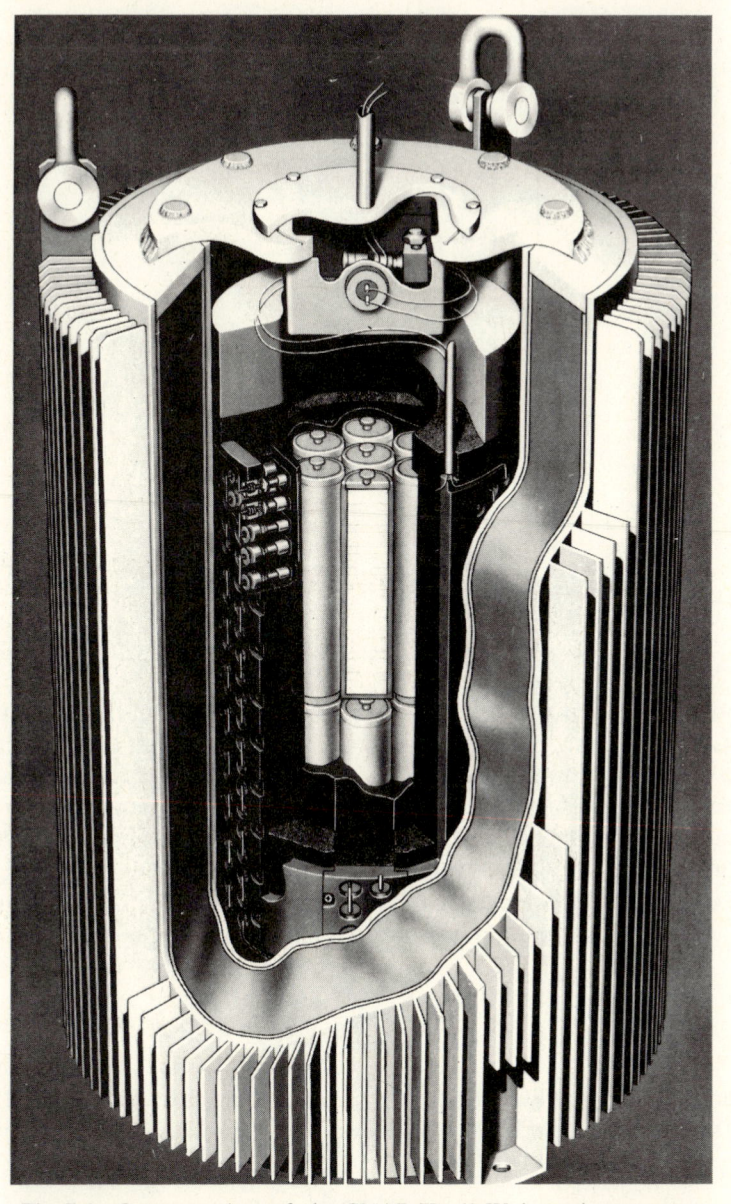

Fig. 7.4 Cutaway view of the SNAP-7B 60-W isotopic generator showing 7 of the 14 tubular fuel capsules as its center which contain strontium titanate pellets providing the source of power. Around the capsules are 120 pairs of lead telluride thermocouples, which convert heat from the decaying radioisotope into electricity. [Courtesy of Teledyne Energy Systems.]

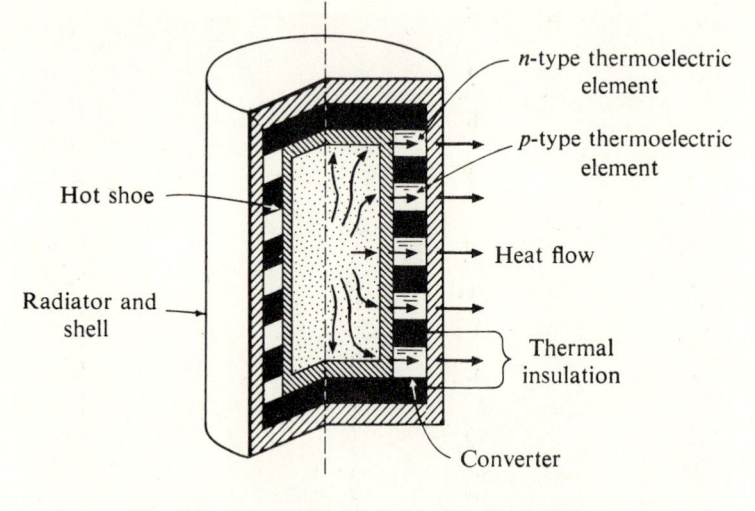

Fig. 7.5   Thermoelectric isotopic power generator.

Thermoelectric Converters

Ordinary metal wire thermocouples were used originally, but their low efficiencies (<1 percent) made them impractical.   All thermoelectric isotopic power generators now use doped semiconductors as the conversion device (Fig. 7.6). Unfortunately, there are many problems with semiconductor thermoelectric elements, not the least of which is efficiency.   Several doped materials have been used for the n- and p-type semiconductors.   Either n- or p-type semiconductors can be used alone for conversion, but a couple makes a much better converter.
The efficiency of a thermoelectric generator can be expressed as

$$\eta_t = \frac{I^2 R_L}{\alpha_{pn} I T_{\mathrm{H}} + K\,\Delta T - \frac{1}{2} I^2 R} \tag{7.2}$$

The current I is determined by the thermoelectrical potential, $\alpha_{pn}\,\Delta T$, divided by the sum of the couple resistance, R, and the load resistance, R_L.

$$I = \frac{\alpha_{pn}\,\Delta T}{R + R_L} \tag{7.3}$$

The useful output is the energy dumped in the load, $I^2 R_L$.   The input energy is made up of the Peltier effect, which is the product of the Seebeck coefficient, α_{pn}, the current, and the hot junction temperature, plus the thermal heat leak from the hot side to the side through the semiconductor elements, minus one-half the

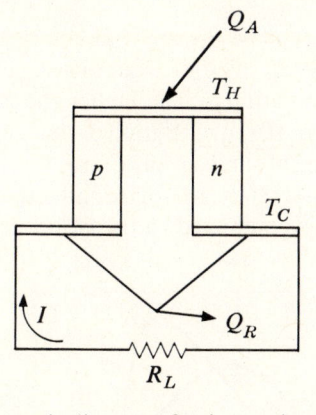

Fig. 7.6 Schematic diagram of a thermoelectric generator.

joule internal heating effect which is returned to the hot junction.

$$Q_A = \alpha_{pn} I T_{\mathrm{H}} + K\,\Delta T - \tfrac{1}{2} I^2 R \qquad (7.4)$$

If the p and n legs have their area ratio as follows:

$$\frac{A_p}{A_n} = \left(\frac{\rho_p k_n}{\rho_n k_p}\right)^{1/2} \qquad \begin{array}{l} \rho = \text{electrical resistivity} \\ k = \text{thermal conductivity} \end{array} \qquad (7.5)$$

the KR product will be a minimum and the figure of merit becomes

$$Z = \frac{\alpha_{pn}}{[(\rho_n k_n)^{1/2} + (\rho_p k_p)^{1/2}]^2} \qquad (7.6)$$

It can be seen that for a high figure of merit, the Seebeck coefficient should be as large as possible, while the thermal conductivity and electrical resistivity should be as low as possible.   Using this optimized geometry, and using the value of R to give the maximum efficiency, an expression depending only on the temperatures and the figure of merit will result:

$$\eta_{\mathrm{opt}} = \frac{\Delta T}{\dfrac{\sqrt{1 + ZT_m} + 1}{\sqrt{1 + ZT_m} - 1}\, T_m + \dfrac{\Delta T}{2}} \qquad (7.7)$$

where $T_m = (T_{\mathrm{H}} + T_{\mathrm{C}})/2$. The ratio of the load resistance to the internal couple resistance is

$$\frac{R_L}{R} = \sqrt{1 + ZT_m} \qquad (7.8)$$

When the load resistance equals the internal couple resistance, the maximum output results, but the efficiency is somewhat lower. Figure 7.7 shows the figure of merit for various combinations. The higher the Z value, the better will be the thermoelectric material. Even with good design and placement of insulation only about 5 to 10 percent of the heat produced by the decaying elements can be converted into electricity. If the thermal conductivity of the converters is too high and too much heat is conducted to the cold junction, not enough will be

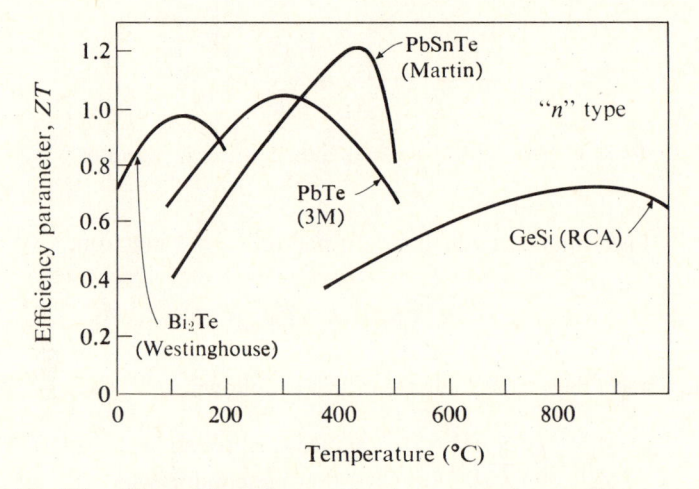

Fig. 7.7a   Figure of merit as a function of temperature for n-type materials.   [From F. Schulman, *Nucleonics* **21**, No. 9 (1953), p. 56. Reprinted by permission.]

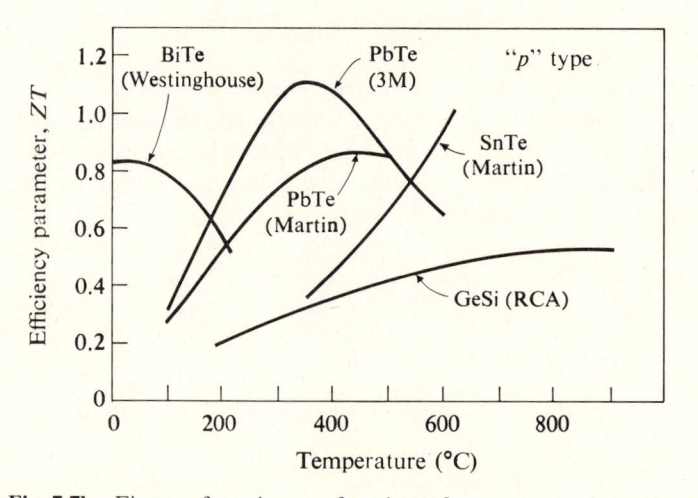

Fig. 7.7b   Figure of merit as a function of temperature for p-type materials.   [Ibid. Reprinted by permission.]

converted to electricity. On the other hand, if the electrical resistance is too high, too much of the power converted appears as I^2R loss—thus the importance of the figure of merit, Z.

Example 2 A thermoelectric generator is to be designed to operate between a high-side temperature of 327°C and a sink temperature of 27°C. The n-type couples are to be made of 75 percent Bi_2Te_3–25 percent Bi_2Se_3, while the p-type material is to be 25 percent Bi_2Te_3–75 percent Bi_2Se_3 with 1.75 percent excess Se. The properties of these semiconductor materials at the average couple temperature are given as follows:

$$\text{Seebeck coefficient, } \alpha_{pn} = 430 \ \mu\text{V/C}$$

$$\text{Electric resistivity, } \rho_p = 0.001 \ 75 \ \Omega\text{-cm}$$

$$\rho_n = 0.001 \ 35 \ \Omega\text{-cm}$$

$$\text{Thermal conductivity, } k_p = 0.012 \ \text{W cm}^{-1}\text{C}^{-1}$$

$$k_n = 0.014 \ \text{W cm}^{-1}\text{C}^{-1}$$

The elements are to be 0.25 cm in length and the diameter of the n-type elements is to be 0.3 cm. The diameter of the p-type units should produce the minimum value of the RK product. The load is to match the generator so as to produce the optimum value of thermal efficiency. There are to be 20 couples in series electrically and in parallel thermally. Find the heat that would have to be supplied by an isotopic heat source, the watts of power provided to the load, and the thermal efficiency.

The diameter of the positive leg should be determined first.

$$\frac{A_p}{A_n} = \left(\frac{\rho_p k_n}{\rho_n k_p}\right)^{1/2} = \left(\frac{0.001 \ 75 \times 0.014}{0.001 \ 35 \times 0.012}\right)^{1/2} = 1.23$$

$$d_p^2 = 1.23 d_n^2 = 1.23 \times 0.3^2 = 0.110 \ 8$$

$$d_p = 0.333 \ \text{cm}$$

The electrical resistance and the thermal conductance for the individual couples must be calculated.

$$R = \frac{\rho_p l_p}{A_p} + \frac{\rho_n l_n}{A_n} = \frac{0.001 \ 75 \times 0.25}{(\pi/4)(0.333)^2} + \frac{0.001 \ 35 \times 0.25}{(\pi/4)(0.30)^2}$$

$$= 0.005 \ 05 + 0.004 \ 77 = 0.009 \ 82 \ \Omega/\text{couple}$$

$$K = \frac{A_p k_p}{l_p} + \frac{A_n k_n}{l_n} = \frac{(\pi/4)(0.333)^2(0.012)}{0.25}$$

$$+ \frac{(\pi/4)(0.30)^2(0.014)}{0.25} = 0.004 \ 2 + 0.003 \ 95$$

$$= 0.008 \ 15 \ \text{W C}^{-1}$$

The optimized figure of merit is

$$Z = \frac{\alpha_{pn}^2}{(\sqrt{\rho_p k_p} + \sqrt{\rho_n k_n})^2}$$

$$= \frac{(430 \times 10^{-6})^2}{(\sqrt{0.001\ 75 \times 0.012} + \sqrt{0.001\ 35 \times 0.014})^2}$$

$$= 0.002\ 32 \text{ K}^{-1}$$

The load ratio that will result in the best efficiency is

$$\frac{R_L}{R} = \sqrt{1 + ZT_m} = \sqrt{1 + 0.002\ 32 \times 450}$$

$$= 1.43$$

The load resistance is then

$$R_L = 1.43 \times 0.009\ 82 = 0.014\ 03\ \Omega/\text{couple}$$

The current developed by the unit will be

$$I = \frac{\alpha_{pn}\ \Delta T}{R + R_L} = \frac{430 \times 10^{-6} \times 300}{0.009\ 82 + 0.014\ 03} = 5.41 \text{ A}$$

The heat input required from an isotopic heat source is

$$Q_1 = \alpha_{pn} T_1 I + K\ \Delta T - \tfrac{1}{2} I^2 R$$

$$= (430 \times 10^{-6} \times 600 \times 5.41) + (0.008\ 15 \times 300)$$

$$- (\tfrac{1}{2} \times 5.41^2 \times 0.009\ 82)$$

$$= 1.395 + 2.44 - 0.143\ 5$$

$$= 3.69 \text{ W/couple}$$

$$= 73.8 \text{ W for the assembly of 20 couples}$$

The power generated by the thermoelectric unit is

$$W = I^2 R_L = (5.41)^2 \times 0.014\ 03 = 0.411 \text{ W/couple}$$

$$= 8.22 \text{ W for the total unit}$$

The thermal efficiency is then

$$\eta_t = \frac{W}{Q_1} = \frac{8.22}{73.8} = 11.13 \text{ percent}$$

This could also have been calculated using Eq. (7.7).

$$\eta_t = \frac{\Delta T}{\dfrac{\sqrt{1 + ZT_m} + 1}{\sqrt{1 + ZT_m} - 1} \times T_m + \frac{1}{2}\Delta T}$$

$$= \frac{300}{(2.43/0.43)(450) + 300/2} = 11.13 \text{ percent}$$

There are other mechanical difficulties in the development of thermoelectric converters. For example, it is extremely difficult to bond the thermoelectric converters to the hot junction. The delicately doped semiconductors must be protected from any chemical reactions with their surroundings. Finally, the thermoelectric materials are very fragile.

Thermionic Converters

Thermionic converters convert heat directly into low-voltage dc. They offer the advantage of higher efficiency than thermoelectric converters, but at the expense of higher temperatures. The basic configuration consists of a refractory metal emitter spaced about 0.025 cm from a collector. Cesium vapor forms a conductive plasma gap between emitter and collector. Figure 7.8 is a schematic sketch of an isotopic power generator utilizing thermionic conversion. Heat from the fuel raises the temperature of the emitter to about 1500°C. Electrons are then boiled off from the emitter and captured by the collector. The flow of electrons provides current flow through an external load. Low-pressure

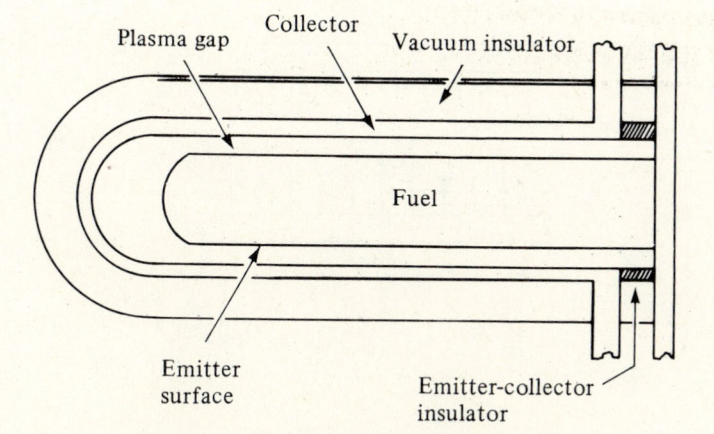

Fig. 7.8 Thermionic isotopic power generator.

cesium vapor in the interelectrode gap neutralizes any space charge in the gap. Collector temperatures are generally fixed by the surroundings at about 650°C, the temperature of conventional superheated steam cycles. Ideally, the efficiency can be made to approach Carnot cycle efficiency by adjusting the collector and emitter energy levels. Materials are a limiting factor in increasing emitter temperatures and, therefore, efficiency. However, since thermionic converters operate near complete vacuum, the problems associated with combined high temperature and pressure are reduced.

In practice the Carnot efficiency will not be reached because of inherent losses. Figure 7.9 shows these losses. The efficiency of this device can be expressed as

$$\eta = \frac{J_n(V_1 - V_2 - \Delta V)}{J_1(V_1 + ZkT_1) - J_2(V_1 + ZkT_2) + \Sigma Q} \tag{7.9}$$

J_n represents the net current flow and ΔV accounts for electrical leakages. The Q terms in the denominator represent an increase in input energy to allow for thermal conduction and radiation in the gaseous plasma gap and losses in the electrical leads connected to the electrodes. The kinetic energy of the plasma gas electrons also contributes to a decrease in the efficiency. Practical thermionic devices have attained efficiency of just over one-third their Carnot efficiencies.

High power and current densities require very high emitter temperatures to produce a satisfactory flow of electrons. To work, the emitter and collector

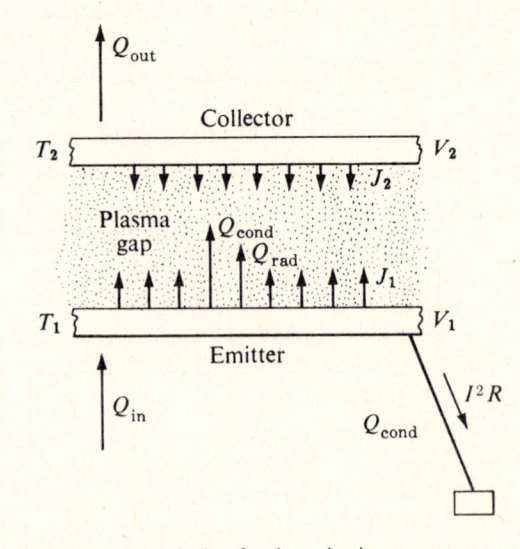

Fig. 7.9   Schematic of a thermionic converter.

must have different work functions. The electron that binds other electrons to the metal surface is measured by its work function. Thus, the greater the work function, the more electrons emitted. To provide useful current densities the work function of the emitter must be kept low. In thermionic converters this is done by enclosing the emitter surface with cesium gas. The presence of a partially ionized gas between electrodes produces electron scattering which reduces the passage of electrons across the space. Some of the energy of the emitted electrons must be used to produce ions in the space. These ions make up a large fraction of the current flow. While collecting electrons the collector also collects contaminants present. These contaminants may come from emitter and collector materials or possibly the fuel clad. Even small amounts of contaminants reduce collector absorption. The effectiveness of the collector is affected by the plasma and it is necessary to optimize the conditions of the emitter and collector. The high-temperature emitter surfaces typically have been tungsten with a silicon carbide envelope for protection. A difficulty in producing an isotope-fueled thermionic power generator will be in insulating the high-temperature emitter. As much as one-half of the heat produced may escape the converter. If the temperature becomes too high the electron kinetic energy will be converted to waste heat in the collector, raising its temperature. Lower temperatures will, of course, reduce the output and efficiency.

Isotope Fuels

The selection of a fuel for a radioisotope generator is of critical importance. Of special concern is safety. The fuel must not subject anyone to unnecessary exposure to radiation and it must be safe from possible nuclear reaction. Extensive tests have been made of all fuel capsule configurations to ensure that they are completely safe under all possible operating or accident conditions. Other equally important considerations are reliability, weight (for space applications), power density, and cost.

There are over 1300 possible radioisotopes from which to choose. However, a suitable isotope must have a long half-life, low gamma emission, power density of at least 0.1 W/g; it must be cheap to produce, easy to shield, and it must have desirable physical properties. These restrictions narrow the selection to nine practical isotope fuels. Four of these are beta-emitting fission products which can be recovered in fuel reprocessing. Four are alpha emitters which must be produced in reactors and, consequently, are more expensive. The ninth, ^{60}Co, is available in plentiful supply but is a gamma emitter. Table 7.1 lists the nuclear and physical properties of available isotopic power sources.

So far the most successful fuel has been ^{90}Sr, but the only fuel used so far in space applications has been ^{238}Pu. Plutonium has been used because it is easily shielded, and there is no danger of nuclear reaction.

^{90}Sr is cheap and plentiful. There are millions of curies available in the DOE's waste facility in the state of Washington. It has a half-life of 27.7 years

Table 7.1 Properties of Available Radioisotopic Power Sources

	^{144}Ce	^{90}Sr	^{137}Cs	^{147}Pm	^{60}Co	^{242}Cm	^{244}Cm	^{210}Po	^{238}Pu
Compound	Ce_2O_3	$SrTiO_3$		Pm_2O_3		Cm_2O_3	Cm_2O_3	GdPo	PuO_2
Half-life	284.5 days	27.7 yr	30 yr	2.67 yr	5.26 yr	162.5 days	18.1 yr	138 days	86 yr
Activity (Ci/g)	440*	33	16	742	360 max.	3044	72.6		
Specific power (W/g)	2.84	0.223	0.0774	0.41	5.32	44.1	2.53	140	0.4
Thermal energy (Ci/W)	126	148	207	2440	65.1		29.2	31.2	
Melting point (°C)	2680	1910		2350	1480	1950	1950	590	
Strength		Fair	Brittle	Good	Excellent	Fair	Fair		Good
Stability	Good in air	Good	Decreases above 1000°C	Good	In inert gas				Good
Shielding†	3.5	1.0	3.6	Little	5.7	Neutron	Neutron	Neutron	Neutron
Capsule compatability	Reacts above 1400°C	Excellent	Excellent	Excellent	Excellent				

* After a 1-year decay.

† Number of centimeters of uranium necessary to attenuate radiation to 0.1 Gy/h at 100-cm distance with 100 W of power.

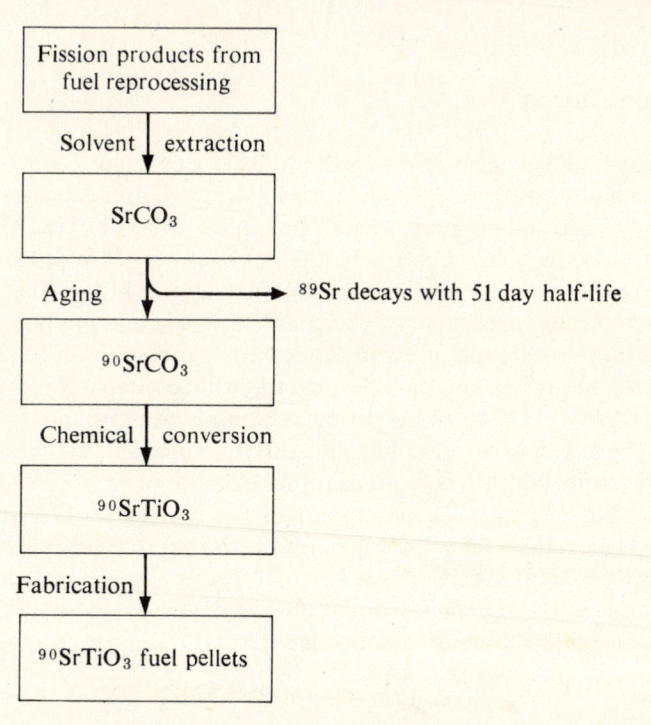

Fig. 7.10 Flow diagram for production of ^{90}Sr fuel.

and has a fairly high power density; its disadvantage is its affinity for bone and consequent damage to blood production. Strontium titanate was chosen as the fuel form because $SrTiO_3$ is insoluble in water, resistant to shock, and has a high melting point. Furthermore, ^{90}Sr is a beta emitter and does not require extremely heavy shielding. Figure 7.10 traces schematically the flow of ^{90}Sr from fission products to final fabrication into small cylindrical pellets. ^{90}Sr has found its most publicized use in the SNAP-7 series of power generators. The SNAP-7 series is used for navigational and weather station purposes. Presently, ^{90}Sr produces more electric power than any other isotope.

Example 3 Calculate the theoretical initial power density of $SrTiO_3$. The composition of the Sr is 55 percent ^{90}Sr, 43.9 percent ^{88}Sr, 1.1 percent ^{86}Sr.

$$A_0 = \frac{0.693(6.023 \times 10^{23})}{27.7(8760)} = 1.72 \times 10^{18} \frac{\text{disintegrations}}{\text{hour-atom } ^{90}\text{Sr}}$$

power density $= 1.72 \times 10^{18}$ dis./hour-atom ^{90}Sr$(0.55$ atom ^{90}Sr/atom

$SrTiO_3$)(1 atom $SrTiO_3$/185 g$SrTiO_3$)

$(0.54$ MeV/dis.$)(4.44 \times 10^{-17}$ W-h/MeV)

$= 0.1223$ W/g $SrTiO_3$ or 0.457 W/g ^{90}Sr

^{14}Carbon Dating

Cosmic ray bombardment of the earth produces a constant supply of neutrons in the earth's atmosphere. The neutrons react with atmospheric nitrogen to produce ^{14}C, ^{3}H, and probably a small amount of ^{4}He with ^{11}Be. The ^{14}C and ^{3}H are radioactive, ^{14}C having a half-life of 5568 yr. It is assumed that the radioactive carbon reacts with oxygen to form CO_2. It is possible, therefore, to say that absorption of cosmic ray neutrons is equivalent to production of radioactive carbon dioxide mixture with atmospheric carbon dioxide. Since plants and animals live off carbon dioxide, they too will be radioactive.

Extensive theorizing and experimentation reveal that there is equilibrium between the rate of decay of radioactive carbon atoms and the rate of assimilation of the radioactive carbon atoms for all living organisms. When the living organism dies, the assimilation stops and the radioactive ^{14}C in the tissue decays. The equilibrium value has been determined to be 15.3 disintegrations per minute per gram of carbon (see Reference 2). This value provides a method for determining the age since death of organic materials. The equation for the specific activity of ^{14}C in an organic material is

$$\text{sp. act.} = 15.3\, e^{-0.693t/5568} \tag{7.10}$$

where t is the number of years since death. When age is to be measured, the specimen must be very carefully sampled. It must contain only carbon atoms originally present at the time of death; chemical changes cannot have caused replacement of the carbon atoms.

> **Example 4**   A charred log recovered from ruins being excavated in the central United States was found to have an activity of 0.510×10^{-3} μCi. The sample weighed 525 g after separation and processing. What is the approximate age of the civilization existing at the time the tree was cut?
>
> The specific activity of the tree log in dis./min g is
>
> $$\text{sp. act. } ^{14}C = \frac{0.510 \times 10^{-9} \text{ Ci}}{525 \text{ g}} \frac{3.7 \times 10^{10} \text{ dis./s}}{\text{Ci}} \frac{60 \text{ s}}{\text{min}}$$
>
> $$\times\ 2.16 \text{ dis./min g}$$
>
> $$\text{sp. act.} = 15.3 e^{-0.693t/5568}$$
>
> $$2.16 = 15.3 e^{-0.693t/5568}$$
>
> $$t = 16\,000 \text{ yr}$$

Normally, the ^{14}C dating method is used only to corroborate other archaeological dating methods. It is difficult to separate ^{14}C specimens; since specimens are susceptible to many changes over the thousands of years following their death, ^{14}C dating in itself is not reliable.

Carbon dating techniques were recently used (Reference 18) in the Reactor Hazards Program of the U.S. Geological Survey. The program is aimed at identifying geologic processes that might affect siting of nuclear power reactors. Movement along a fault near Augusta, Georgia, was seen when a trench was cut across one of the faults.

The trench cut exposed several scattered areas of organic material. A composite sample revealed a radiocarbon age of 400 ± 300 yr. Clayey sand containing organic material was incorporated into the fault gouge. A composite sample of this material yielded a radiocarbon age of 2450 ± 1000 yr. At least 3 ft of movement must have taken place since deposition of the material dated 400 ± 300 yr. Up until this discovery and its related radiocarbon dating, fault movements in the southeastern United States were believed to be many millions of years old. However, the conclusion that movement has taken place within the last few hundred years does not mean that the area is seismically active.

The reason that errors are also large is due to the very small samples of carbon. The samples had to be diluted with gas of a known age in order to count the radioactivity. The dilution reduces the reliability of the data. Ordinarily, 10 g of organic material is a large enough sample, but in this case up to 1300 g of material was burned to obtain enough CO_2.

Food Processing

Radiation can be used to aid in the preservation of foods. Small doses can be used in pasteurization and large doses can be used to sterilize foods. Sterilization requires 2 to 4.5 million rads, while a dose of 200 000 to 500 000 rads is used to pasteurize. Both gamma radiation and electrons produced by accelerators are used.

In addition to retarding spoilage, small doses of radiation can tenderize vegetables and will inhibit maturing of vegetables and fruits, thereby increasing shelf life.

The major problem in radiation-processed foods has been a slight alteration of taste and color. As in all preservation or sterilization of foods, there will be some slight alterations. The problems of undesirable tastes can be partially eliminated by controlling the irradiation conditions. For example, irradiating meats at $-26°C$ ($-78°F$) virtually eliminates undesirable tastes.

The Department of Energy (DOE) and the Food and Drug Administration (FDA) continue to make lengthy and exhaustive tests of radiation-processed foods to ensure wholesomeness and quality. No radiation-processed food may be offered for public consumption without the FDA's specific approval of both the food and its packaging. Packaging of preserved food is a considerable problem because the food must be packaged before it is processed. The radiation

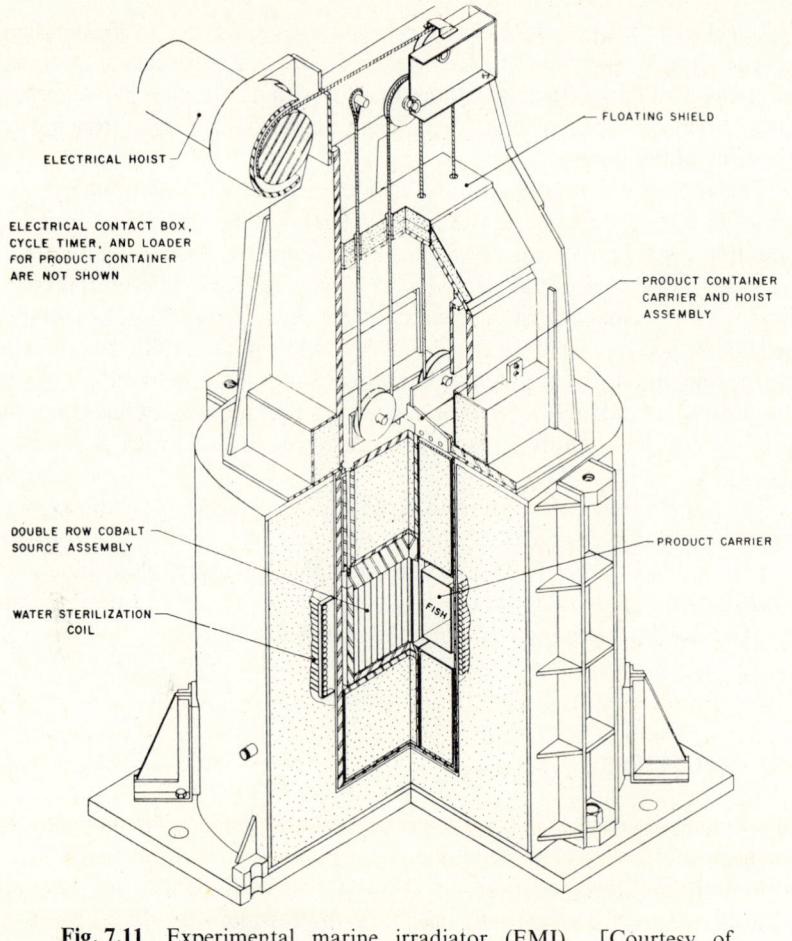

ELECTRICAL HOIST

ELECTRICAL CONTACT BOX,
CYCLE TIMER, AND LOADER
FOR PRODUCT CONTAINER
ARE NOT SHOWN

FLOATING SHIELD

PRODUCT CONTAINER
CARRIER AND HOIST
ASSEMBLY

DOUBLE ROW COBALT
SOURCE ASSEMBLY

PRODUCT CARRIER

FISH

WATER STERILIZATION
COIL

Fig. 7.11 Experimental marine irradiator (EMI). [Courtesy of ARCO Radiation Center, Karthaus, Pa.]

(particularly the doses used in sterilization) may be harmful to the packaging; there could possibly be some interaction between the food and packaging. The ordinary metal cans used in heat processing have been found most satisfactory for radiation-sterilized foods. Radiation pasteurizing does not present as great a packaging problem as does radiation sterilizing.

One of the earliest and most successful programs utilizing radiation processing has been in the fish industry. This program has been a cooperative venture between the DOE and private industry. It has been shown that irradiation can easily double the refrigerated shelf life of fish and shellfish. Increasing shelf life in this fashion will permit much wider distribution of fish and fish products. It is hoped that this wider distribution will make the process more attractive economically. The first instrument designed specifically for fish irradiation

was the Marine Products Development Irradiator at Gloucester, Massachusetts. An alternative to this is the Experimental Marine Irradiator, Fig. 7.11. These irradiators are small and are capable of irradiating to 1000 Gy(100 000 rads) 150 lb. of fish per hour after the catch. The irradiated fish are given a second dose on land after processing and packaging.

Agriculture

Only about 1 percent of the energy that plants receive from the sun is used to manufacture food. The remaining 99 percent is wasted. Agriculturists are using radioisotopes to study photosynthesis. In one study CO_2 and H_2O were labeled with ^{15}O. The labeled compounds were supplied separately to plants. The result was that no ^{15}O was liberated when the plants were fed labeled CO_2, but ^{15}O was liberated when labeled H_2O was supplied. Thus, through a relatively simple test using radioisotopes, an important discovery concerning photosynthesis was made.

Labeled fertilizers are also used successfully in agriculture. The labeled fertilizers proved conclusively that plants absorb fertilizer through leaves, bark, and roots. In fact, about 95 percent of fertilizer applied to leaves is absorbed, whereas only about 10 percent applied to soil is absorbed.

Almost all food used by animals as well as by plants can be tagged with radioactive compounds to trace its utilization. For example, calcium and phosphorous have been labeled to show exactly how bone is formed and destroyed. Vitamin B_{12} has been tagged with ^{60}Co and fed to sheep and chickens. The results help show the effect of B_{12} and why it makes the chicken and sheep grow faster.

Radiation is also used to produce mutations in plants. Much of this work was done at Brookhaven National Laboratory. Plants are set out in circular rows in a 1-acre plot. The source can be lowered into the ground to limit exposure and to permit personnel to harvest and examine the plants, vegetables, fruit, and shrubs. If a mutation is produced, cuttings or seeds are taken from it and planted in a normal environment for test of hardiness and fruitfulness. Through this "gamma field" and other irradiating fields many spectacular developments have been made in a very short time.

Irradiation has been successfully used for insect control. Radiation-sterilized male insects are released into a controlled wild population. Mating by the sterilized males results in a decline in size of successive generations. Advantages of this procedure include avoidance of insecticides and increasing efficiency as population of the pest is reduced. To be successful, however, an economic method of raising large numbers of insects must be available, the insects must be easily dispersed, and the females must mate only once.

The screw-worm fly was irradicated from southeastern United States in about 18 months after the initial release of sterile screw-worm flies. At the

height of the project about 50 million flies per week were raised, and a total of 2.5 billion sterile flies were released. The initial cost was about $10 000 000.

Medicine

In the medical profession the use of radioisotopes is focused into three categories: diagnosis, therapy, and research.

Physicians have used radium in therapy for many years, but it has always been very expensive. When radioisotopes became readily available ^{60}Co replaced radium as the chief source for therapeutical applications. It is estimated that the activity of $100 of ^{60}Co is equivalent to $20 000 of radium. ^{60}Co, however, is limited to external applications. In many instances it is desirable to have the isotope inside the body. Phosphorous-32 (a beta emitter) has been used to treat polycythema vera, a disease of the blood. A patient is fed ^{32}P, which concentrates in bone marrow. The radioactive ^{32}P slows the production of red blood cells in the marrow, arresting the polycythema vera. ^{32}P is also being used to control rejection of transplanted organs. It helps destroy some of the blood factors that cause rejection.

As a treatment for cancer, ^{198}Au has been physically implanted in the diseased cells. The gamma radiation destroys the cells and the gold eventually decays to negligible activity. An important side benefit is the fact that the radioactive gold inside the body can be used as a radiography source to locate its position accurately.

In a normal heart an electrochemical pulse starts the heart contraction. The contraction begins in the sinus node near the top of the heart and spreads downward. In some people, for various reasons, the heart does not beat in synchronism with the sinus node impulse or does not beat at all. A pacemaker placed beneath the skin stimulates the heart with an electrical pulse that passes through a catheter wire implanted in the heart.

One such isotopic powered pacemaker is illustrated in Fig. 7.12. It contains 0.25 g of ^{238}Pu in the form of PuO_2, and is contained, together with the electronics, in a hermetically sealed titanium alloy capsule. The entire unit weighs only 61 g and occupies 33 cm^3. This is much smaller than more conventional battery-operated pacemakers.

The decay heat of the ^{238}Pu is converted to electricity by Tophel–cupron thermocouples connected in series. Tophel–cupron was selected because of its properties and shock resistance. At the beginning of life, the thermal output is 141 mW. The electrical output is a pulse with amplitude of about 8 mA at a basic rate of 72 beats per minute. The 87.8-yr half-life of the ^{238}Pu causes a decrease in power to 121 mW after 20 yr. This results in a drop in output to approximately 70 beats per minute and 7 mA. The average life expectancy of this pacemaker is calculated as 38.9 yr. and the median life is considered 34.0 yr.

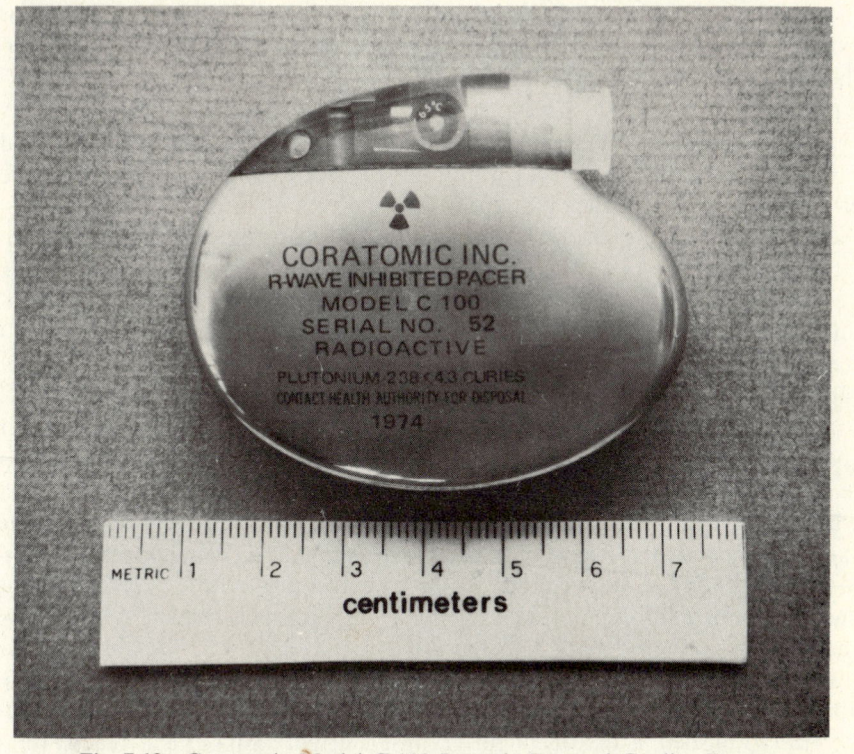

Fig. 7.12 Coratomic Model C-100 Isotopic Powered Cardiac Pacemaker. [Courtesy of Coratomic, Inc.]

During development of the pacemaker, it was tested under a variety of extreme accident conditions. The purpose was to ensure that no fuel release could occur as a result of an accident. To make sure the unit would not release fuel in the event of a large building fire, tests subjected the pacemaker to 800°C for 30 min followed by quenching with water and the application of a 1000-kg load. Another test involved the effect of accidently leaving the pacemaker in a body that might be cremated. The pacemaker was placed in a furnace at 1380°C for 90 min. Other tests simulated falls from an airplane, explosion, and sinking in the ocean.

^{238}Pu is an alpha and gamma emitter. In addition, neutrons are produced in other elements as the Pu decays. The patient is exposed to a dose rate of about 350 mrem/yr, well below that from chest X-rays or dental X-rays. The exposure to the patient's spouse, assuming separation of 50 cm for 3000 h/yr, is about 20 mrem/yr. Thus, there is negligible radiation exposure and virtually no possibility of the Pu being released to the environment.

The cost of an isotopic pacemaker is about five times that of a battery-operated device; however, its long life may make it more economical.

As important or more important than therapy uses are tracers. For example, a small quantity of NaCl containing ^{24}Na ($t_{1/2} = 15$ h) is injected into the blood. By following the radioactive buildup with a Geiger counter the circulation can be gaged. If there is a restriction to blood flow, the activity may even drop sharply at the restriction, pinpointing its exact location. ^{99}Tc and ^{67}Ga have been used to detect certain kinds of brain tumors. The radioactive elements have a preference for certain kinds of tumors and are absorbed into them. By tracing this absorption externally, the location and type of tumor can frequently be diagnosed.

^{252}Cf is being investigated as a treatment for cancer. ^{252}Cf is a high-intensity neutron source. Therefore, it may be implanted to give a highly localized radiation dose to diseased tissue while healthy tissue is exposed to lesser radiation.

Brookhaven National Laboratory and the Nassau County Medical Center (New York) have collaborated in developing a method to analyze calcium in bone structure. It is done by using whole-body neutron activation and whole-body counting techniques. The natural calcium in the body is made slightly radioactive by exposure to neutrons. The patient is quickly transferred to a whole-body radiation counter where 54 detectors read the calcium content and distribution throughout the skeleton. The radioactive calcium has a very short life, so small that there is no residual activity. The radiation dose is about the same as that for a routine medical X-ray. The technique is used in studying and treating bone embrittlement. After the patient's natural calcium distribution is established, a treatment program is started. Subsequent reactivation and counting is used to monitor the effectiveness of treatment.

The method is used chiefly in diagnosing and detecting osteoporosis. Osteoporosis is a disease that deprives the body skeleton of its natural calcium. By the time clinical symptoms appear, bones usually have lost their structural strength. This may lead to spontaneous fractures that are very difficult to heal.

Of course, in medical research and diagnosis the isotope must be carefully chosen. Both its decay energy and its effective half-life must be minimal to prevent any harmful effects to the patient. Even so, the use of radioisotopes is limited only by the enthusiasm and ingenuity of the experimenter.

Problems

1. What phenomena must be taken into account for the interpretation of measurements made with tracers?

2. A transmission gage is to be used to control the thickness of coating on a certain paper. The source of radiation is ^{90}Sr. The paper is composed of material with density 1.0 g/cm^3 and mass attenuation coefficient 0.224 cm^2/g. The coating material has density of 1.38 g/cm^3 and mass attenuation coefficient of 0.235 cm^2/g. What should be the reduction in intensity if the paper is 0.092 in. thick and the coating 0.019 in. thick?

3. In controlling the weight of sugar in a refining process a transmission gage is used. The gage is set to control the limits of flow between 0.70 and 0.75 of initial intensity. For a ^{137}Cs source, what is the mass flow rate of sugar if the 6-in. conveyor travels at a speed of 2.25 cm/s? The sugar density is 0.308 g/cm^3.

4. Carbon believed to be from a cremation was found in the northeastern United States. After separation and processing, 12 g of carbon remained. The carbon had an activity of 94.3 counts/min above background. What is the age of the cremation?

5. In the southwestern United States several corncobs were found in a cave. After processing, 5.54 g of carbon with an activity of 66 counts/min remained. The background was 1.7 counts/min. How many years ago was the corn eaten?

6. Repeat Example 2, but make the load resistance equal to the resistance of the generator. Assume that it is to work between the same temperature limits. Determine the required heat input, power output, and thermal efficiency of the unit.

7. The generator in Prob. 6 is to be fueled with ^{210}Po. The source at the start of the mission is adequate to provide the necessary amount of heat. After 138 days, what will be the heat supply temperature (T_1), the power output, and the generator efficiency? Assume that the sink temperature is maintained at 27°C.

8. What is the minimum ^{147}Pm that must be produced to have an initial power of 100 W?

9. What amount of $^{90}_{38}$Sr would be necessary initially to power a radioisotope generator of 200 W? How much $SrTiO_3$ is required? What is the power density after two years of use?

10. What is the initial theoretical power density of $^{238}_{94}PuO_2$?

11. Theoretically, how many curies of ^{242}Cm would be required initially to fuel the generator of Prob. 9? Assume all conditions the same except the fuel.

12. An engine piston weighing 1 kg was made radioactive by neutron activation. The resulting initial activity was 8.5 mCi. After a few hours of running the lubricating oil was indicating a true count rate of 13 000 counts/min corrected for background. Estimate the piston wear (in grams) that has taken place.

References

1. Stone, E. W., et al., *Isotopes in Industry*, U.S. Atomic Energy Commission Report 3337–16.
2. Libby, W. F., *Radiocarbon Dating*, 2nd ed. Chicago: University of Chicago Press, 1955.
3. *Isotopes and Radiation Technology*, U.S. Atomic Energy Commission Quarterly.
4. *Proceedings of the Third United Nations International Conference on the Peaceful Uses of Atomic Energy*, Vol. 15, Geneva, 1964.
5. Angrist, S., *Direct Energy Conversion*. Boston: Allyn and Bacon, Inc., 1965.
6. Urbain, W. M., "Food Irradiation," *Nuclear News* 9, No. 7 (September 1966).
7. Wagner, H. M., "Radiopharmaceuticals—Their Use in Nuclear Medicine," *Nucleonics* 24, No. 7 (July 1966).

8. *Strontium-90 Fueled Thermoelectric Generator Power Source for Five Watt U.S. Coast Guard Light Buoy*, U.S. Department of Commerce Report MND–P–2720, 1962.

9. Bormat, M., et al., "SNAP–III Electricity from Radionuclides and Thermoelectric Conversion," *Nucleonics* **17**, No. 166 (May 1959).

10. "Radionuclide Power for Space," Parts 1, 2, and 3, *Nucleonics* **21**, No. 3 (March 1963), No. 4 (April 1963), No. 9 (September 1963).

11. *Shipboard* ^{60}*Co Radiopasteurizer*, BNL-808, U.S. Atomic Energy Commission, 1963.

12. Desrosier, N. W., *The Technology of Food Preservation.* Westport, Conn.: AVI Publishing Company, 1959.

13. Goldsmid, H. J., *Applications of Thermoelectricity.* London: Methuen & Co. Ltd., 1960.

14. Chleck, D., R. Maehl, and O. Cucchiare, *Development of* ^{85}*Kr as a Universal Tracer*, NYO-2757, U.S. Atomic Energy Commission, February 1966.

15. Coombe, R. A., *An Introduction to Radioactivity for Engineers.* New York: Macmillan Publishing Co., Inc., 1968.

16. Lewis, J., and Henley, E. J., eds., *Advances in Nuclear Science and Technology*, Vol. 5. New York: Academic Press, Inc., 1969.

17. Purdy, D. L., et al., "A New Radioisotope-Powered Cardiac Pacemaker," *Journal of Thoracic and Cardiovascular Surgery* **69**, No. 82 (January 1975).

18. Prowell, D. C., B. J. O'Connor, and M. Rubin, *Preliminary Evidence for Holocene Movement Along the Belair Fault Zone near Augusta, Ga.*, U.S. Geological Survey Report 75–680, 1975.

19. *Thermionic Topping Converter for a Coal-Fired Power Plant*, U.S. Office of Coal Research and Development, Department of the Interior. Report No. 52.

8

Neutron
Interactions

Fission neutrons are born with an average energy of 2 MeV. These fast neutrons interact with the core materials in absorption and scattering reactions. Collisions that result in scatter are useful in slowing neutrons to thermal energies. Thermal neutrons can be absorbed by fissionable nuclei to produce more fissions or can be absorbed in fertile material for conversion to fissionable fuel. Unfortunately, parasitic absorption by structural material, moderator, or coolant removes some neutrons without their having fulfilled any useful purpose.

Cross Sections

The cross section (σ) for a reaction is an indication of the probability that a particle (neutron) will interact with a nucleus. Loosely, this can be considered as the target area through which the particle must pass if an interaction is to occur. A square centimeter is tremendous in comparison to the effective area of a nucleus; hence, it is convenient to express cross sections in barns, where

$$1 \text{ barn} = 10^{-24} \text{ cm}^2$$

Nuclear folklore suggests that a physicist referred to a particularly large cross section as being "big as a barn door." The name has persisted.

Neutron Interactions

Scattering reactions are important because they allow the moderation or slowing-down process to occur. When neutrons reach thermal energy levels they have much higher cross sections for fission. The mechanics of neutron energy loss during slowing down is discussed later in this chapter.

Elastic scatter occurs when both the kinetic energy and momentum of the incoming and emergent particles are conserved. The neutrons may bounce

off the struck nucleus, as when one billiard ball strikes another (potential scatter). They may also be absorbed into the nucleus and subsequently be expelled. This resonant scatter accounts for the resonance peaks in the elastic scattering cross section when it is plotted versus energy.

Inelastic scatter occurs when the nucleus is raised to an excited state and emits a gamma ray in addition to a neutron. Since light elements commonly have excited energy levels 1 MeV or more apart, the slowing-down process is usually considered as taking place by elastic scatter only. With heavy nuclei the energy levels are only about 0.1 MeV apart and inelastic scatter is more prominent. Generally, only fast neutrons engage in inelastic scatter, where because of the energy of the gamma, the emergent particles have less kinetic energy and momentum than the incident particles.

Absorption processes may be of three types:

(1) *Radioactive capture (n, γ).* The nucleus absorbs a neutron and the new isotope drops to ground state by the emission of a gamma. This type of reaction accounts for the relatively high absorption cross section of light hydrogen.

$$_{1}^{1}H + _{0}^{1}n \rightarrow _{1}^{2}H^* \rightarrow _{1}^{2}H + \gamma$$

For a reaction such as this we can use a shorthand notation having the incident and emergent particles within a bracket preceded by the target nucleus and followed by the recoil nucleus.

$$_{1}^{1}H(n, \gamma)_{1}^{2}H$$

Not all neutrons absorbed by fissionable nuclei produce fission. ^{235}U, for example, has a radiative capture cross section of 112 barns at 0.025 eV.

(2) *Capture with particle emission.* The capture of a neutron may result in the compound nucleus decaying by the ejection of a charged particle. Examples of this are:

$$_{4}^{7}Be \,(n, p) \,_{3}^{7}Li$$

$$_{7}^{14}N \,(n, p) \,_{6}^{14}C$$

$$_{5}^{10}B \,(n, \alpha) \,_{3}^{7}Li$$

The last of the reactions above is important in neutron detection where a BF_3 counter measures thermal neutron flux. Uncharged neutrons produce little ionization in a counting chamber, but the alphas produced in boron trifluoride gas are highly ionizing. Each alpha emission indicates a neutron interaction within the counting chamber.

(3) *Fission.* The capture of a neutron by ^{233}U, ^{235}U, ^{239}Pu, or ^{241}Pu may result in splitting the heavy nucleus. As the fission fragments fly apart an average of between two and three neutrons are emitted. One of these must cause a subsequent fission if a steady-state chain reaction is to take place. A detailed discussion of the fission process is presented in Chapter 4.

Attenuation of a Neutron Beam

Consider a collimated beam of neutrons impinging perpendicularly on a surface of area A, as shown in Fig. 8.1.

A = surface area, cm^2

x = distance from front face, cm

I_0 = initial beam intensity, neutrons/cm^2 s

I = intensity at distance x from front face, neutrons/cm^2 s

σ = microscopic cross section, cm^2/nucleus

N = density of target atoms, nuclei/cm^3

$NA\,dx$ = number of target atoms in the differential thickness, dx

$\sigma NA\,dx$ = total effective area presented by nuclei to the neutrons, cm^2

The ratio of the total effective area in the differential slab to the full area gives the probability of interaction in the distance dx.

$\sigma NA\,dx/A$ = probability of interaction (fraction of beam that undergoes an interaction), cm^2 effective area/cm^2 total area

The decrease in intensity, dI, as the neutrons pass through the differential slab, is the intensity at that point times the probability of interaction:

$$dI = -I\sigma N\,dx \tag{8.1}$$

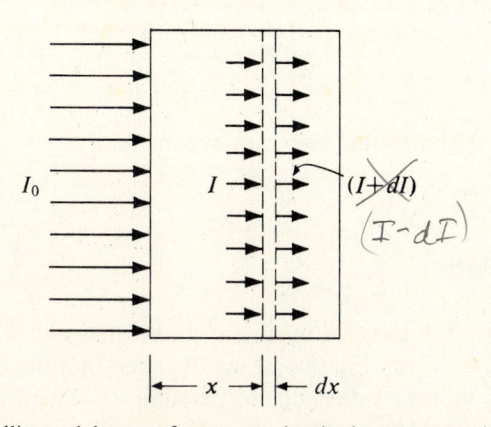

Fig. 8.1 Collimated beam of neutrons impinging on material with cross-sectional area of A (cm^2). Note that, since neutrons are being removed, dI will be negative.

Separating variables yields

$$\frac{dI}{I} = -\sigma N \, dx$$

Then, integrating for a thickness x,

$$\int_{I_0}^{I} \frac{dI}{I} = -\sigma N \int_0^x dx$$

$$\ln \frac{I}{I_0} = -\sigma N x$$

$$I = I_0 e^{-\sigma N x} = I_0 e^{-\Sigma x} \qquad (8.2)$$

[handwritten note: probability that neutron will get thru]

where $\Sigma = \sigma N$ is the macroscopic cross section, cm^2/cm^3. This represents the effective target area per unit volume of material.

The number density of the target atoms can be found by the product of the density multiplied by Avogadro's number divided by the mass number.

$$N = \frac{\rho \text{ g/cm}^3 \times 6.023 \times 10^{23} \text{ atoms/g atom}}{A \text{ g/g atom}}$$

$$= \frac{6.023 \times 10^{23} \rho}{A} \qquad (8.3)$$

Differentiating Eq. (8.2), we get

$$\frac{dI}{dx} = -I_0 \Sigma e^{-\Sigma x} \qquad (8.4)$$

Thus, the rate of absorption decreases exponentially.

Mean Free Path

The average distance traveled by a neutron before interaction is known as the *mean free path*, λ. From Eq. (8.4) it may be seen that the decrease in intensity while traveling a distance dx is due to the number of neutrons interacting and, hence, removed from the beam.

$$dI = -I_0 \Sigma e^{-\Sigma x} \, dx$$

These neutrons have traveled a distance x without interaction. The total distance traveled by all the neutrons as they interact in an infinite thickness of material is

$$-\int_{x=0}^{x=\infty} x \, dI = +I_0 \Sigma \int_0^\infty x e^{-\Sigma x} \, dx$$

This represents the summation for the infinitely thick slab of the distance traveled by neutrons absorbed in each differential thickness. The mean free path is then this total interaction distance divided by the original beam intensity.

$$\lambda = \frac{+I_0 \Sigma \int_0^\infty x e^{-\Sigma x} \, dx}{I_0} = \Sigma \int_0^\infty x e^{-\Sigma x} \, dx = \frac{1}{\Sigma} \tag{8.5}$$

Thus, the reciprocal of the macroscopic cross section (cm^2/cm^3) is the mean free path (cm). This same result is useful for considering neutron flux where the neutrons are not collimated but are traveling in random directions.

Since cross sections are probabilities of interaction, individual probabilities can be summed to give a total probability. The total cross section is the sum of the absorption and scatter cross sections:

$$\sigma_T = \sigma_a + \sigma_s \tag{8.6}$$

In turn, for a fissionable nucleus the absorption cross section is the sum of the fission and radiative capture cross sections.

$$\sigma_a = \sigma_f + \sigma_c \tag{8.7a}$$

Similarly, the scattering cross section is made up of an elastic plus an inelastic value.

$$\sigma_s = \sigma_{se} + \sigma_{si} \tag{8.7b}$$

Macroscopic cross sections can also be added:

$$\Sigma_T = \Sigma_f + \Sigma_c + \Sigma_s = \frac{1}{\lambda_f} + \frac{1}{\lambda_c} + \frac{1}{\lambda_s} = \frac{1}{\lambda_T} \tag{8.8a}$$

Thus, the total mean free path is

$$\lambda_T = \frac{\lambda_f \lambda_s \lambda_c}{\lambda_c \lambda_s + \lambda_f \lambda_s + \lambda_f \lambda_c} \tag{8.8b}$$

The relaxation length for a material is the thickness of material necessary to attenuate the neutron beam by a factor of e.   Setting $x = \lambda$, we obtain

$$I_\lambda = I_0 e^{-\Sigma \lambda} = I_0 e^{-1} = \frac{I_0}{e} \qquad (8.9)$$

Therefore, the mean free path is also the relaxation length.

Example 1   For a 95 mass percent Al–5 mass percent Si mass alloy, determine the absorption, scatter, and total macroscopic cross sections for 2200-m/s neutrons.   What are the corresponding mean free paths?   The density of the alloy is 2.66 g/cm^3.

$$N_{Al} = \frac{0.95 \times 2.66 \times 6.023 \times 10^{23}}{26.98} = 5.59 \times 10^{22} \text{ Al nuclei/cm}^3$$

$$N_{Si} = \frac{0.05 \times 2.66 \times 6.023 \times 10^{23}}{28.06} = 2.85 \times 10^{21} \text{ Si nuclei/cm}^3$$

$$\Sigma_a = N_{Al}\sigma_{aAl} + N_{Si}\sigma_{aSi}$$

$$= 5.59 \times 10^{22} \times 0.23 \times 10^{-24} + 2.85 \times 10^{21} \times 0.16 \times 10^{-24}$$

$$= 0.012\,83 + 0.000\,456 = 0.013\,29 \text{ cm}^2/\text{cm}^3$$

$$\Sigma_s = N_{Al}\sigma_{sAl} + N_{Si}\sigma_{sSi}$$

$$= 5.59 \times 10^{22} \times 1.4 \times 10^{-24} + 2.85 \times 10^{21} \times 1.7 \times 10^{-24}$$

$$= 0.0782 + 0.004\,85 = 0.0831 \text{ cm}^2/\text{cm}^3$$

$$\Sigma_t = \Sigma_a + \Sigma_s = 0.013\,29 + 0.0831 = 0.0964 \text{ cm}^2/\text{cm}^3$$

$$\lambda_a = \frac{1}{\Sigma_a} = \frac{1}{0.013\,29} = 75.3 \text{ cm}$$

$$\lambda_s = \frac{1}{\Sigma_s} = \frac{1}{0.0831} = 12.02 \text{ cm}$$

$$\lambda_t = \frac{1}{\Sigma_s} = \frac{1}{0.0964} = 10.37 \text{ cm}$$

Thus, it is seen that a 2200-m/s neutron must travel an average of 75.3 cm to interact by absorption, 12.02 cm to interact by scatter, but only 10.37 cm to have an interaction by either scatter or absorption.

Neutron Cross Sections

High scattering cross sections are imperative for effective slowing of fast neutrons to thermal energy levels.   (Values change only slowly with energy.)

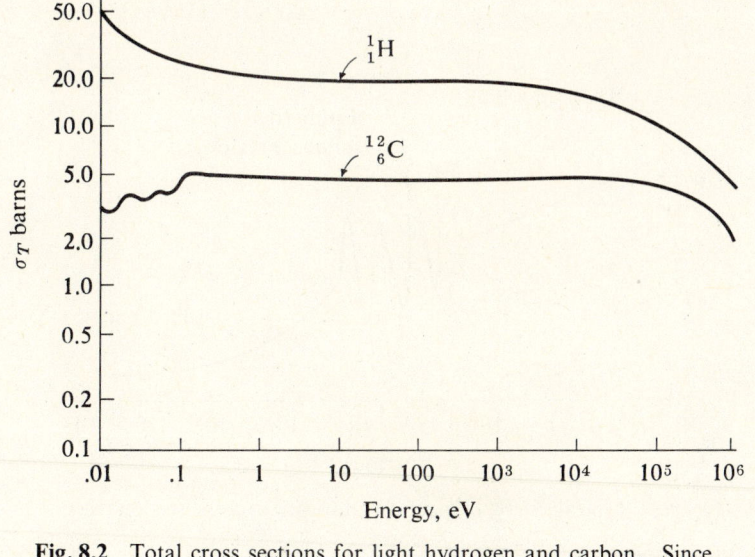

Fig. 8.2   Total cross sections for light hydrogen and carbon.   Since $\sigma_a \ll \sigma_s$, $\sigma_s = \sigma_T$ for these two isotopes.

Both thermal and epithermal cross sections, as well as 2200-m/s absorption cross sections, are listed in Appendix A.   It can be seen that sometimes the epithermal cross section is larger than the thermal scattering cross section and sometimes it is the smaller of the two.   For example, the thermal value of the scattering cross section for deuterium is 7 barns, while the epithermal value is 3.4 barns.   In the case of copper, the values are 7.2 barns thermal and 7.7 barns epithermal.   For most of the lighter elements, scattering cross sections run 2 to 6 barns, while the heaviest ones run up to 12 barns.   Light hydrogen is an exception.   At low energies a single atom has a scattering cross section of 38 barns.   In the epithermal region the value drops to 20 barns.   When in the bound state as light water, composite values of 80 barns thermal and 44 barns epithermal can be used for each molecule.   Figure 8.2 shows the total cross section for $_1^1$H and $_6^{12}$C as a function of energy.   Since absorption cross sections are small for these isotopes, the total cross section is approximately equal to the scattering cross section.

Neutron absorption cross sections are highly energy dependent.   In the thermal region they are usually inversely proportional to neutron velocity. When values in the thermal region are plotted versus energy on a log-log plot, the curve will have a slope of $-\frac{1}{2}$, as illustrated in Fig. 8.3.   As energies increase, resonance peaks are apt to be prominent in the epithermal region.   Many of these resonances are due to radiative capture.   As energies continue to increase into the fast region, cross sections drop to low values of the order of 1 barn.

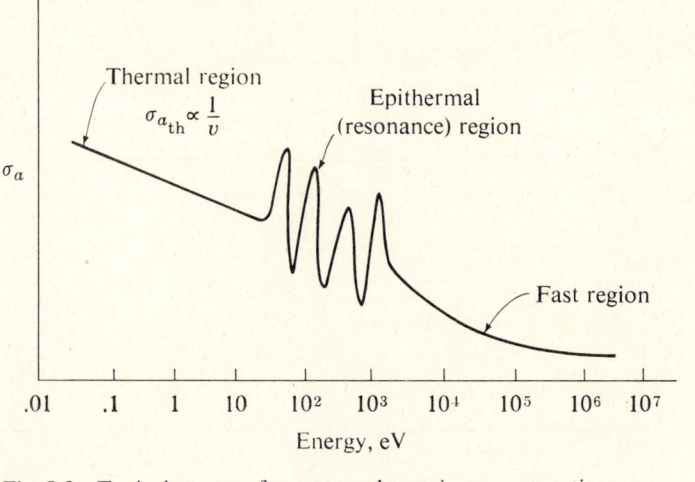

Fig. 8.3 Typical curve of neutron absorption cross section versus energy.

Thermal Neutron Velocity (Energy) Distribution

When fast neutrons have been slowed by successive collisions to thermal energy, they have as much probability of gaining energy as of losing it on any subsequent collisions. These thermalized neutrons diffuse throughout the core until they either leak out or are absorbed. If there are n_0 thermal neutrons per cubic centimeter, their population density as a function of velocity, $n(v)$, is the number of neutrons per cubic centimeter per unit velocity interval. This velocity distribution for thermal neutrons is called a *Maxwellian velocity distribution*, which is described by

$$\frac{dn}{dv} = n(v) = n_0 \frac{4\pi v^2}{(2\pi kT/m)^{3/2}} e^{-mv^2/2kT} \qquad (8.10)$$

where n_0 = thermal neutrons per cm^3

m = neutron rest mass

T = temperature, K

k = Boltzmann constant

The effect of temperature on $n(v)$ is illustrated by Fig. 8.4, where the fraction of neutrons per unit velocity interval [$n(v)/n_0$] is plotted versus neutron velocity. Note how an increase in temperature spreads out the neutrons over a wider

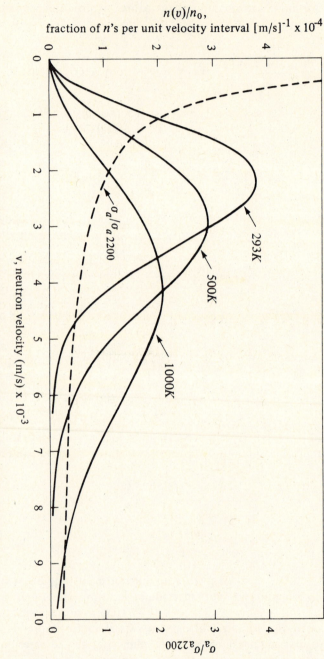

Fig. 8.4 Fraction of neutrons per unit velocity interval $[n(v)/n_0]$ versus neutron velocity at three different temperatures. The $\sigma_a/\sigma_{a\,2200}$ curve shows the $1/v$ behavior for absorption cross sections.

spectrum of velocities, thus lowering the peak of the curve. Also, with increasing temperature the most probable and average neutron velocities are increased, as will be indicated by Eqs. (8.12) and (8.15). The dashed line shows how absorption cross sections vary inversely with velocity. For a $1/v$ absorber σ_a/σ_{a2200} is plotted versus neutron velocity.

At a fixed temperature the temperature may be combined with the other constants to simplify the expression for $n(v)$. Let

$$B = \frac{4\pi}{(2\pi k T/m)^{3/2}}$$

Then

$$n(v) = n_0 B v^2 \, e^{-mv^2/2kT} \qquad (8.11)$$

The most probable neutron velocity, v_p, is found by setting the derivative of $n(v)$ with respect to velocity equal to zero.

$$\frac{dn(v)}{dv} = 2n_0 B v e^{-mv^2/2kT} + n_0 B v^2 \frac{-mv}{kT} e^{-mv^2/2kT} = 0$$

$$v_p = \left(\frac{2kT}{m}\right)^{1/2} \qquad (8.12)$$

The kinetic energy of neutrons at the most probable velocity is

$$KE_p = \frac{mv_p^2}{2} = \frac{m}{2} \frac{2kT}{m} = kT \qquad (8.13)$$

This is not the average kinetic energy, which has a value of $\frac{3}{2}kT$. Note that the kinetic energy is independent of the particle mass. For neutrons at 20°C,

$$v_p = \left(\frac{2 \times 1.38 \times 10^{-23} \times 293}{1.66 \times 10^{-27}}\right)^{1/2} 3 \text{ s}$$

$$= 2200 \text{ m/s}$$

At the most probable velocity the kinetic energy is

$$KE_p = 1.38 \times 10^{-27} \text{ J/K} \times 293 \text{ K} \times \frac{1}{1.6 \times 10^{-19}} \text{ eV/J}$$

$$= 0.025 \text{ eV}$$

It is at these conditions that neutron absorption cross sections are tabulated in Appendix A.

To find the average neutron velocity,

$$\bar{v} = \frac{\int_0^\infty n(v)v\,dv}{\int_0^\infty n(v)\,dv} = \frac{\int_0^\infty \frac{4\pi n_0 v^3 e^{-mv^2/2kT}\,dv}{(2\pi kT/m)^{3/2}}}{n_0} \qquad (8.14)$$

Let $y = v^2$; then $dy = 2v\,dv$ and $v\,dv = dy/2$. Also, let $b = -m/2kT$.

$$\bar{v} = \frac{4\pi}{(2\pi kT/m)^{3/2}} \int_0^\infty y e^{by}\,\frac{dy}{2}$$

$$= \frac{2\pi}{(2\pi kT/m)^{3/2}} \left[\frac{e^{by}}{b^2}(by - 1)\right]_0^\infty$$

$$= \sqrt{\frac{8kT}{\pi m}} \qquad (8.15)$$

Therefore, the ratio of the average velocity to the most probable velocity is

$$\frac{\bar{v}}{v_p} = \frac{\sqrt{8kT/\pi m}}{\sqrt{2kT/m}} = \frac{2}{\sqrt{\pi}} = 1.128 \qquad (8.16)$$

Neutron chopper experiment

In the neutron chopper experiment it is possible to determine the distribution of neutrons with respect to velocity. A narrow beam of neutrons is allowed to pass through a beam port in the reflector region of a core. In this region the neutrons are well thermalized. They are aimed at the edge of a spinning cadmium disk containing four slots. These allow bursts of neutrons to pass toward a BF_3 counter a fixed distance from the rotating disk. A phototube triggers the time sweep on a multichannel analyzer as a photocell receives a pulse of light as one of the slots passes the light source, as shown in Fig. 8.5. Each channel represents a discrete time interval, say 40 μs, so the thirtieth channel after the start of the sweep represents a time of 1200 μs. Since the neutrons travel a known distance from the slot to the BF_3 counter, the fastest ones will be in the earlier channels and the slower ones in the later channels.

As the sweep is repeated many times, the number of counts in each channel accumulates and may be displayed on a cathode ray tube as *a* counts per channel [$n(t)$] versus channel number (time) curve. $n(t)$ is proportional to the neutron population density with respect to time, neutrons/cm^3 s. To convert this to $n(v)$, the $n(t)$ value must be multiplied by dt/dv.

$$n(v) = n(t)\,dt/dv = dn/dt \times dt/dv = dn/dv$$

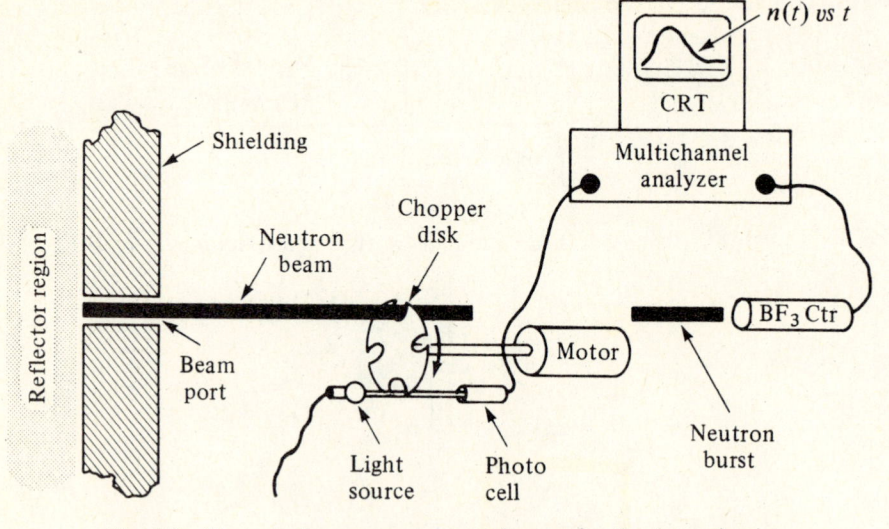

Fig. 8.5   Neutron chopper experimental setup for the determination of neutron velocity distributions.

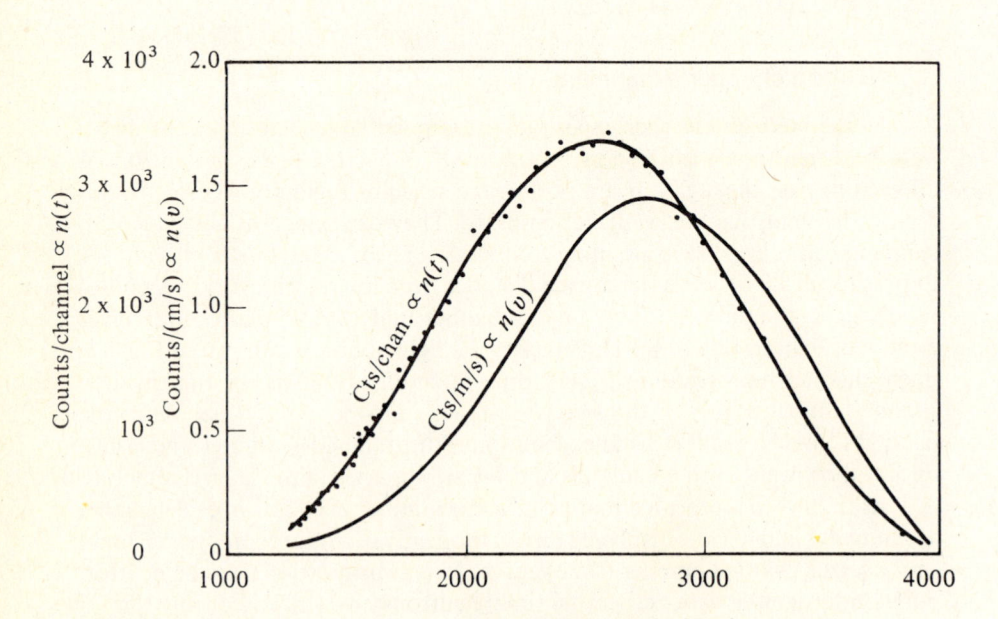

Fig. 8.6   Neutron spectrum for a beam taken from the reflector region of the MIT reactor. (Courtesy MITR.)

After making this conversion for each of the channels, $n(v)$ may be plotted versus the velocity for each channel and the curves should show approximately a Maxwellian distribution. Figure 8.6 shows results obtained at the MIT reactor. Knowing the value of the velocity where the $n(v)$ function peaks, one may calculate the effective neutron temperature for the moderator.

Example 2 During a neutron chopper experiment the neutron flight distance from the spinning disk to the BF_3 counter is 2 m. Each channel represents a 40-μs time increment. In the twentieth channel from the start of the sweep there have been accumulated 4000 counts. What is the average speed for neutrons counted by this channel, and what is $n(v)$ counts per meter per second?

The neutron speed is

$$v = s/t = \frac{2\text{ m}}{40 \times 10^{-6}\text{ s/ch.} \times 20\text{ ch.}}$$

$$= 2500\text{ m/s}$$

Since $s = vt$, it follows that

$$dv/dt = -s/t^2$$

and its reciprocal is

$$dt/ds = -t^2/s$$

The negative sign merely reflects the fact that the velocity decreases with increasing flight time (channel number). Thus, for the twentieth channel,

$$n(v) = \frac{dn}{dt} \times \frac{dt}{dv} = n(t) \times \frac{t^2}{s}$$

$$= \frac{4000\text{ counts}}{40 \times 10^{-6}\text{ s}} \times \frac{(800 \times 10^{-6})^2\text{ s}^2}{2\text{ m}}$$

$$= 32\, \frac{\text{counts}}{\text{m/s}}$$

Corrected Absorption Cross Sections

In the thermal region most materials have an absorption cross section that varies inversely with neutron velocity. Therefore, the absorption cross section will vary inversely as the square root of both the kinetic energy and the absolute temperature.

$$\sigma_a \propto \frac{1}{v} \propto \frac{1}{KE^{1/2}} \propto \frac{1}{T^{1/2}} \tag{8.17}$$

$$\sigma_a' = \sigma_{a293}\sqrt{\frac{293}{T}} \tag{8.18}$$

Since the average neutron velocity is larger than the most probable neutron velocity by the factor of $2/\sqrt{\pi}$, the absorption cross sections at these velocities are in the inverse of this ratio.   The cross section at the average neutron velocity at a temperature T is

$$\bar{\sigma}'_a = \sigma_{a_{293}} \frac{\sqrt{\pi}}{2} \sqrt{\frac{293}{T}} \tag{8.19}$$

Neutron Flux

At a given point in a reactor, neutrons will be traveling in all directions.   The flux is defined as

$$\phi = \int n(v) v \, dv \tag{8.20}$$

Dividing both sides of Eq. (8.20) by n_0, the total number of neutrons per cubic centimeter, we obtain

$$\frac{\phi}{n_0} = \frac{\int n(v) v \, dv}{\int n(v) \, dv} = \bar{v} \tag{8.21}$$

This represents the average neutron speed.

Although flux is velocity (or energy) dependent, it is often convenient to treat thermal neutrons as a monoenergetic group traveling at the average neutron speed.

$$\phi_{th} = n_0 \bar{v} \tag{8.22}$$

Thus, ϕ_{th} is equivalent to a flux of neutrons at a single speed and the velocity distribution can be ignored.   The flux can be considered as the sum of the distances traveled per second by the neutrons in 1 cm³ of volume.

Dividing the distance traveled per second, ϕ, by the mean free path for interaction, λ, results in the interaction rate, R.

$$R = \frac{\phi}{\lambda} = \phi\Sigma = \phi N\sigma \qquad \text{interactions/s cm}^3 \tag{8.23}$$

Neutron Activation

A material in a neutron flux will form new isotopes as neutrons are absorbed. If these new isotopes are unstable, they will begin to decay, as well as being removed by neutron capture. Let

N_1 = number density of original target nuclei, nuclei/cm³

σ_{a_1} = microscopic absorption cross section of the target nuclei, cm²/parent nucleus

N_2 = number of new nuclei present at any time t

σ_{a_2} = microscopic cross section of the newly formed nuclei, cm²/new nucleus

ϕ = thermal flux, neutrons/cm²-s

λ_2 = probability of radioactive decay of the activated nuclei, s⁻¹

The rate of change of the newly activated nuclei, dN_2/dt, is, then, the rate of formation by neutron capture ($\phi N_1 \sigma_{a_1}$) less the rate of their decay ($N_2 \lambda_2$), less their rate of removal by neutron capture ($\phi N_2 \sigma_{a_2}$).

$$\frac{dN_2}{dt} = \phi N_1 \sigma_{a_1} - N_2 \lambda_2 - \phi N_2 \sigma_{a_2}$$

$$= \phi N_1 \sigma_{a_1} - (\lambda_2 + \phi \sigma_{a_2}) N_2 \tag{8.24}$$

Notice that $\phi \sigma_{a_2}$ represents the probability of removal for the activated species by neutron capture, just as λ_2 is the probability of removal by decay. The sum of the two may be denoted as an effective probability of removal, λ_{2T}. Unless the new species has an absorption cross section that is quite large, and unless the fluxes are of the order found in power reactor cores, $\lambda_2 \approx \lambda_{2T}$. On the other hand, if the new species is stable, removal can only occur by neutron capture and $\lambda_{2T} = \phi \sigma_{a_2}$.

Rearranging Eq. (8.24) yields

$$\frac{dN_2}{dt} + \lambda_{2T} N_2 = \phi N_1 \sigma_{a_1} \tag{8.24a}$$

This is a first-order differential equation of the form

$$\frac{dy}{dx} + a(x)y = h(x)$$

An integrating factor, $p = e^{\int a(x)\,dx}$ gives a solution of the form

$$y = \frac{1}{p} \int ph(x)\,dx - \frac{C}{p}$$

where C is a constant of integration. For this particular case, $a(x) = \lambda_{2T}$ (constant), and $h(x) = \phi N_1 \sigma_{a_1}$ (also constant).

$$p = e^{\int \lambda_{2T}\,dt} = e^{\lambda_{2T} t}$$

$$N_2 = \frac{1}{e^{\lambda_{2T} t}} \int e^{\lambda_{2T} t} \phi N_1 \sigma_{a_1}\,dt - \frac{c}{e^{\lambda_{2T} t}}$$

$$= \frac{\phi N_1 \sigma_{a_1}}{\lambda_{2T}} - \frac{c}{e^{\lambda_{2T} t}}$$

$$= \frac{\phi \Sigma_{a_1}}{\lambda_{2T}} - ce^{-\lambda_{2T} t}$$

For a sample that is unirradiated initially, the initial conditions are when $t = 0$, $N_2 = 0$; and the constant of integration may be evaluated.

$$C = \frac{\phi \Sigma_{a_1}}{\lambda_{2T}}$$

The number of new nuclei present at any time, t, is then

$$N_2 = \frac{\phi \Sigma_{a_1}}{\lambda_{2T}} (1 - e^{-\lambda_{2T} t}) \tag{8.25}$$

Figure 8.7 shows the buildup of radioactive nuclei during irradiation in a neutron flux. When the irradiated sample is removed from the core, it will decay with its own characteristic half-life.

After 4 or 5 half-lives, the exponential contributes very little to the solution. The sample is said to be saturated, the rate of formation of new nuclei being balanced by the rate of removal by decay and/or neutron capture.

$$N_{2_{sat}} = \frac{\phi \Sigma_{a_1}}{\lambda_{2T}} \tag{8.26}$$

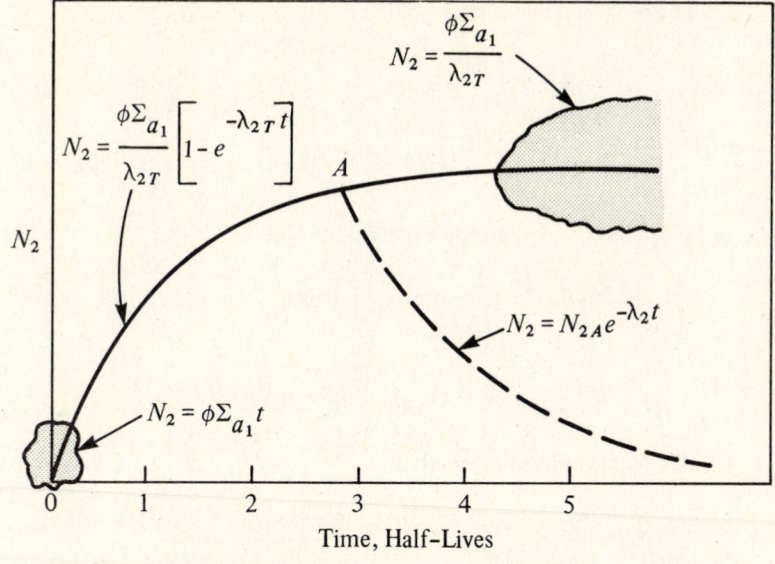

Fig. 8.7 Buildup of radioactive nuclei during irradiation. The dashed curve commencing at point A represents the decay of an activated specimen removed after being irradiated for 3 half-lives. At removal it contained N_{2A} activated nuclei.

On irradiation for a short period, say less than 0.1 half-life, the number of new nuclei is so small that their rate of removal may be ignored and

$$\left(\frac{dN_2}{dt} = \phi\Sigma_{a_1} \right) \tag{8.27a}$$

Integrating yields

$$\left(N_2 = \phi\Sigma_{a_1} \int_0^t dt = \phi\Sigma_{a_1} t \right) \tag{8.27b}$$

Example 3 A 2.5-cm-diameter gold foil 1 mm thick is irradiated in a flux of 10^8 neutrons/cm² s. The temperature during irradiation is 50°C. How many curies will it emit on removal if it is irradiated for 24 h?

The corrected cross sections are

$$\sigma_{a_1} = 98.8 \frac{\sqrt{\pi}}{2} \sqrt{\frac{293}{323}} = 83.396 \text{ barns}$$

$$\sigma_{a_2} = 26\,000 \frac{\sqrt{\pi}}{2} \sqrt{\frac{293}{323}} = 21\,946 \text{ barns}$$

The effective removal cross section is

$$\lambda_{2T} = \frac{0.693}{(t_{1/2})_2} + \phi\sigma_{a_2}$$

$$= \frac{0.693}{2.7} + 10^8 \times 21\,946 \times 10^{-24} \times 3600 \times 24$$

$$= 0.2567 + 1.896 \times 10^{-7} = 0.2567 \text{ day}^{-1}$$

Thus, at this flux level, neutron capture is not significant. The volume of the foil is

$$V = \frac{\pi \times 2.5^2}{4} \times 0.1 = 0.4909 \text{ cm}^3$$

The number of the target atoms is

$$N_1 = \frac{V\rho \times 6.02\,252 \times 10^{23}}{A} = \frac{0.4909 \times 19.3 \times 6.0225 \times 10^{23}}{196.9666}$$

$$= 2.897 \times 10^{22} \text{ atoms } ^{197}\text{Au}$$

Using Eq. (8.25), the number of activated ^{198}Au atoms may be calculated.

$$N_2 = \frac{\phi\Sigma_{a_1}}{\lambda_{2T}}(1 - e^{-\lambda_{2T}t})$$

$$= \frac{10^8 \text{ (neutrons/cm}^2 \text{ s) } 2.897 \times 10^{22} \text{ (atoms } ^{197}\text{Au)}}{0.2567 \text{ day}^{-1}}$$

$$\times 83.39 \times 10^{-24} \text{ (cm}^2/\text{atom } ^{197}\text{Au)} \times 1 \text{ (atom } ^{198}\text{Au/neutron)}$$

$$\times (3600 \times 24) \text{ s/day } (1 - e^{-0.2567 \text{ day}^{-1} \times 1 \text{ day}})$$

$$= 1.840 \times 10^{13} \text{ atoms } ^{198}\text{Au}$$

The activity is then

$$A_0 = \frac{N_2\lambda_2}{3.7 \times 10^{10}} = \frac{1.84 \times 10^{13} \text{ atoms } ^{198}\text{Au} \times 0.2567 \text{ day}^{-1}}{3.7 \times 10^{10} \text{ (atoms } ^{198}\text{Au/Ci s) } \times (3600 \times 24) \text{ s day}^{-1}}$$

$$= 1.4775 \times 10^{-3} \text{ Ci}$$

If the flux had been 10^{14} nuetrons/cm^2 s, what would be the activity on removal?

$$\lambda_{2T} = \frac{0.693}{2.7} + 10^{14} \times 21\,946 \times 10^{-24} \times 3600 \times 24$$

$$= 0.2567 + 0.1896 = 0.4463$$

In this case, removal by neutron absorption becomes quite significant.

$$N_2 = \frac{10^{14} \times 0.4909 \times 5.902 \times 10^{22} \times 83.39 \times 10^{-24} \times 3600 \times 24}{0.4463}$$

$$\times (1 - e^{-0.4463 \times 1}) = 1.6839 \times 10^{19} \text{ atoms } {}^{198}\text{Au}$$

$$A_0 = \frac{1.6839 \times 10^{19} \times 0.2567}{3.7 \times 10^{10} \times 3600 \times 24} = 1352 \text{ Ci}$$

Notice how different the probability of removal was for the high-flux case. When the sample is removed from the reactor core, there is no more neutron capture and the probability of removal, λ_2, is by radioactive decay only.

Neutron Activation Analysis

Activation analysis is a powerful tool for determining the elements present in an unknown sample. This particularly valuable technique can identify trace amounts of elements as small as a few parts per million, or, in some cases, parts per billion. Upon neutron irradiation of the unknown material, radioactive isotopes are formed. Examination of the energy spectrum of the radiation being emitted will identify the elements present. A multichannel analyzer with as many as several hundred channels can display on an oscilloscope the gamma spectrum being given off. The elements are identified by the energies at which photopeaks occur. Knowledge of such factors as the height of the photopeak, irradiation time, post-irradiation time, and counter efficiency will allow the determination of the amount of each element present. It should be noted that neutron activation analysis is insensitive to the chemical form or state of the element being bombarded.

The uses of activation analysis are myriad. It has been used to determine trace-level impurities in semiconductors such as germanium and silicon. It also can determine the dopant levels in finished semiconductor devices. It is useful in determining trace amounts of oxygen in steel, titanium, sodium, and beryllium. It can detect deleterious traces of catalyst residues in plastics. In agriculture it is useful for determining residual amounts of bromine on crops and in foodstuffs. Figure 8.8 indicates how an oil slick can be identified by comparing the spectrogram of a known sample with the spectrograms of various unknown samples.

The results of activation analysis are being used successfully as evidence in court. In criminal investigations it can trace a sample of paint, grease, tire rubber, and so on, to its manufacturer by the amounts of trace elements present. Gunpowder residues on the skin of a suspect can be identified in a reliable manner.

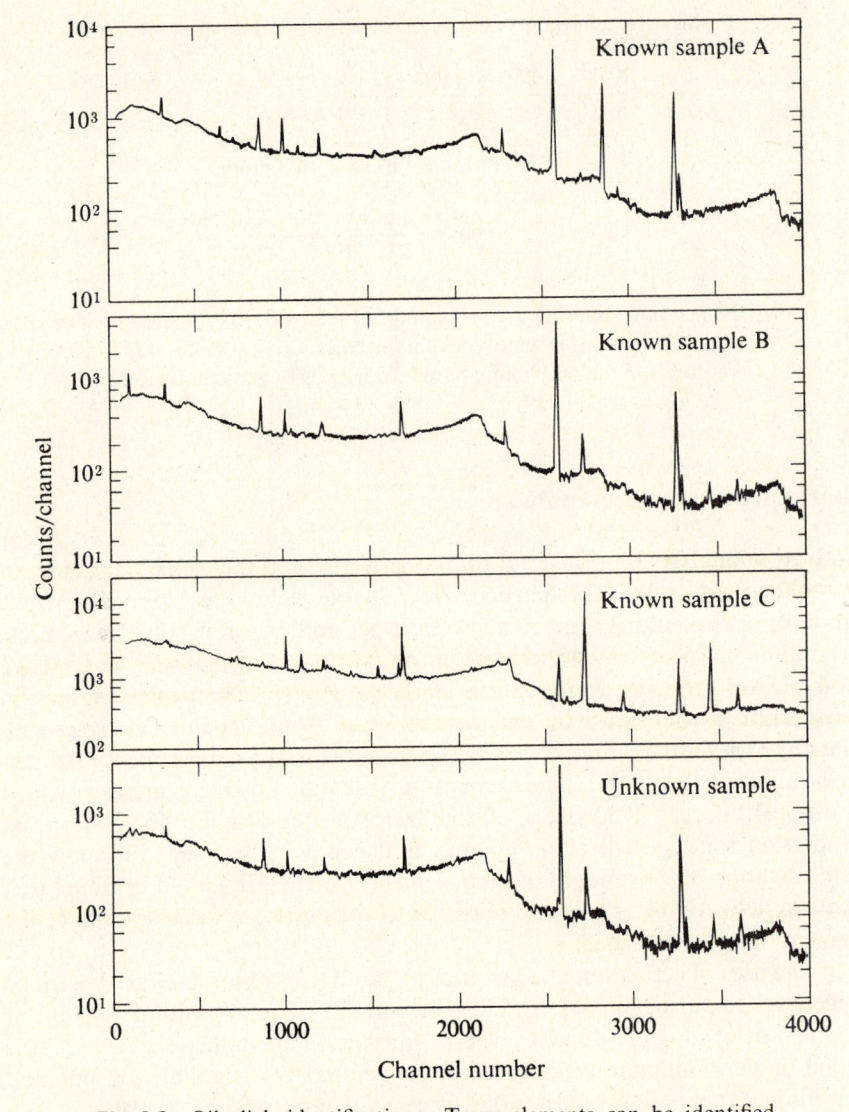

Fig. 8.8 Oil slick identification. Trace elements can be identified from the peaks in each spectrum. The trace element fingerprint of the unknown sample closely matches that of sample B. [From J. John, H. R. Lukens, and H. L. Schlesinger, *Industrial Research* **13**, No. 9 (September 1971), p. 50.]

One particularly attractive use of activation analysis is in the introduction of minor amounts of several easily activated trace elements as coding to guard against counterfeiting of the protected product. This method of identification is being considered for products as diverse as drugs, foodstuffs, cement, alkali metals, and legal tender.

Flux Determination by Foil Irradiation

Neutron flux levels can be inferred by inserting foils or wires of such metals as gold or copper into a neutron field. The gold, for example, has a fairly large neutron absorption cross section (98.8 barns at 2200 m/s). Upon neutron capture the ^{198}Au atoms formed will decay with their characteristic 2.7-day half-life. The initial count rate will be proportional to the number of activated nuclei, which is, in turn, proportional to the flux at the point of irradiation, as shown by Eq. (8.25).

Indium foil is often used to determine flux levels. Because of a strong resonance absorption of 29 400 barns at 1.44 eV, the ^{115}In will be activated by both thermal neutrons and those neutrons absorbed in the resonance region.

A second foil can be covered with cadmium and irradiated in a similar position. A large resonance exists in cadmium at the upper end of the thermal region. The cross section then falls rapidly to about 10 barns at the 1.44-eV energy level. The cadmium covers are, therefore, nearly transparent to the epithermal neutrons and opaque to the thermal neutrons. The variation in cross section with energy for the two foil materials is shown in Fig. 8.9. The

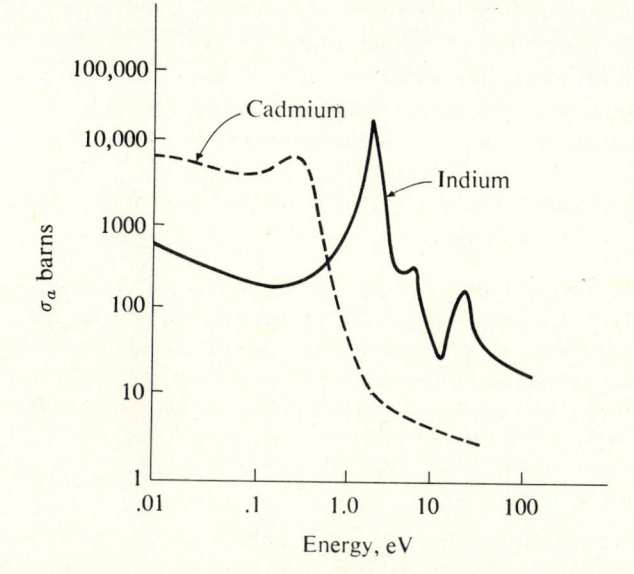

Fig. 8.9 Cadmium and indium neutron absorption cross sections.

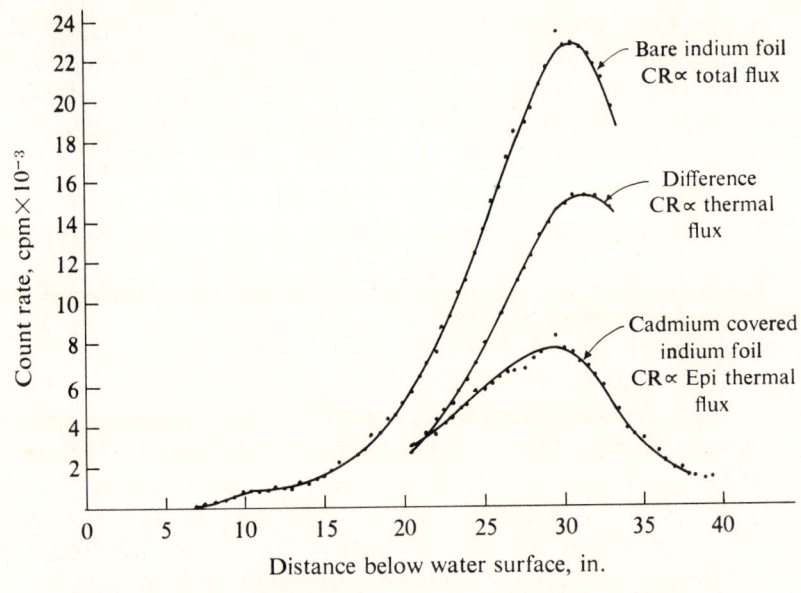

Fig. 8.10 Vertical flux distribution from data taken 4 in. from source tube in a Nuclear Chicago Student Training Reactor, Model 9000. The foil-traversing technique uses bare and Cd-covered indium foils, a Nuclear Chicago D-47 Gas Flow Counter, and a Nuclear Chicago Actigraph (foil speed $1\frac{1}{2}$ in./min, window width $\frac{1}{2}$ in.).

activity of the covered foil is due almost entirely to the epithermal neutrons. The difference between the bare and the Cd-covered foil readings indicates the activity due to absorption of thermal neutrons.

Figure 8.10 shows data taken with a bare indium foil and a Cd-covered indium foil in a vertical direction, 4 in. from the central source tube of a light water-moderated, natural uranium-fueled, subcritical assembly. The ratio of the count induced by the thermal neutrons is roughly twice that for the epithermal count rate. This ratio gives an indication as to the degree of thermalization of the flux.

Example 4 Two indium foils, each weighing 0.5 g, are placed in geometrically similar locations in the core of a subcritical reactor. The effective neutron temperature is 77°C. One foil is bare and the other is cadmium covered. The time of irradiation for both foils is 1 h. The bare foil has a count rate of 10 000 counts/min when counted 54 min after removal, and the cadmium-covered foil has a count rate of 4000 counts/min when it is counted 27 min after removal from the core. The efficiency of the G-M counter being used is 1.5 percent. Determine:

(a) The activities of both foils at the time of removal from the core
(b) The ratio of disintegrations due to thermal absorption to those resulting from capture in the epithermal resonance
(c) The thermal flux level for the position at which the foils are located in the core

A delay in counting of several minutes is desirable to allow for decay of the 72-s ^{114}In and the 14-s ^{116}In. Only the 54-min half-life of the ^{116}In need be considered, as the 49-day ^{114}In will contribute little to the count rate, owing to its long half-life and short time of irradiation.

The decay probability of ^{116}In is

$$\lambda = \frac{0.693}{t_{1/2}} = \frac{0.693}{54} = 0.01\,283 \text{ min}^{-1}$$

The activity of the foils at the time of removal is determined by dividing the count rate by both the efficiency of the counting system and the factor at attenuation between the times of removal and counting.

$$A_{bare} = \frac{(CR)_{bare}}{\eta e^{-\lambda t_{bare}}} = \frac{10\,000}{0.015 e^{-0.012\,83 \times 54}} = 1.333 \times 10^6 \text{ counts/min}$$

$$A_{Cd} = \frac{(CR)_{Cd}}{\eta e^{-\lambda t_{Cd}}} = \frac{4000}{0.015 e^{-0.012\,83 \times 27}} = 3.771 \times 10^5 \text{ counts/min}$$

The difference in these activities gives the activity resulting from thermal neutron capture.

$$A_{th} = A_{bare} - A_{Cd} = 1.333 \times 10^6 - 3.771 \times 10^5 = 9.559 \times 10^5 \text{ counts/min}$$

The ratio of activity due to thermal neutron absorption to that resulting from resonance capture is

$$R = \frac{A_{th}}{A_{Cd}} = \frac{9.559 \times 10^5}{3.771 \times 10^5} = 2.53$$

Before using Eq. (8.25) to determine the flux level, it is necessary to calculate the number of ^{115}In target atoms and the corrected cross section for the ^{115}In. Notice that the 14-s and 54-min decays are sequential, and thus the sum of the two σ_a's must be used.

$$N_1 = \frac{0.5 \times 6.023 \times 10^{23} \times 0.9577}{114.82} = 2.512 \times 10^{21} \text{ atoms } ^{115}\text{In}$$

$$\sigma_1 = (50 + 150)\frac{\sqrt{\pi}}{2}\sqrt{\frac{293}{350}} = 162.2 \text{ barns}$$

$$\phi_{th} = \frac{N_2 \lambda}{N_1 \sigma_1 (1 - e^{-\lambda t})} = \frac{A_{th}}{N_1 \sigma_1 (1 - e^{-\lambda t})}$$

$$= \frac{9.559 \times 10^5 \text{ (atoms } ^{116}\text{In/min)} \times 1 \text{ (neutron/atom } ^{116}\text{In)} \times (1 \text{ min/60 s})}{2.512 \times 10^{21} \text{ (atoms } ^{115}\text{In)} \times 162.2 \times 10^{-24} \text{ (cm}^2/\text{atom } ^{115}\text{In})(1 - e^{-0.012\,83 \times 60})}$$

$$= 7.283 \times 10^4 \text{ neutrons/cm}^2 \text{ s}$$

Neutron–Capture Gamma Analysis

Analysis of the gamma spectrum due to neutron capture may be utilized to determine the isotopic content of an irradiated sample. The gammas, many in the range 5 to 10 MeV, are available immediately and without the problems with transient decay that occur in neutron activation analysis. The development of high-sensitivity germanium lithium drifted [Ge(Li)] detectors has allowed the sharp resolution of gammas at these high energies. As the predominant mode of gamma interaction at these energies is by pair production, there will be three peaks for each gamma energy. There is a full peak (f) where the total gamma energy is dissipated in the crystal, a single escape peak (s) where one of the 0.51-MeV annihilation gammas escapes the detector, and a double escape peak (d) where both annihilation gammas leave the crystal. This is shown effectively in Fig. 8.11, which is the capture gamma spectrum for iron. Iron has full peaks at 9.298, 7.646, 7.632, 6.018, and 5.921 MeV. For each

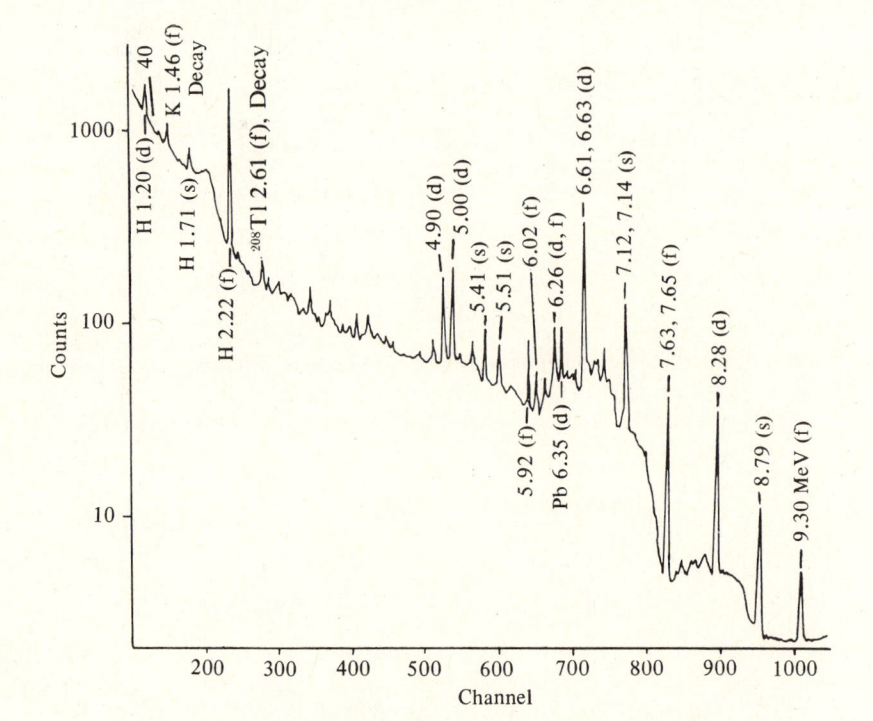

Fig. 8.11 Capture gamma spectrum for an iron plate irradiated for 133 min by a 10^7-neutron/s ^{252}Cf source. [From D. Duffey, J. P. Balogna, and P. F. Wiggins, *Nuclear Technology* **27**, No. 3 (November 1975), p. 492.]

of these there will be single and double escape peaks at 0.51 and 1.02 MeV less than the full peak values. The source of the neutrons was a small $(2 \times 10^7$ neutrons/s) ^{252}Cf source.

A 1-in.-thick block of polyethylene between the source and the sample moderates the neutrons. It also accounts for the three hydrogen peaks. Iron is a good calibration material as the peaks run from 9.3 MeV down to the double escape peak for hydrogen at 1.20 MeV.

This technique has been applied to the analysis of potential sources of hot water for geothermal power plant use. The impurities such as sodium, calcium, and chlorine can cause difficulties with corrosion or deposition of solids. A spectrum for a sample of water from Soda Dam Spring is shown in Fig. 8.12, and

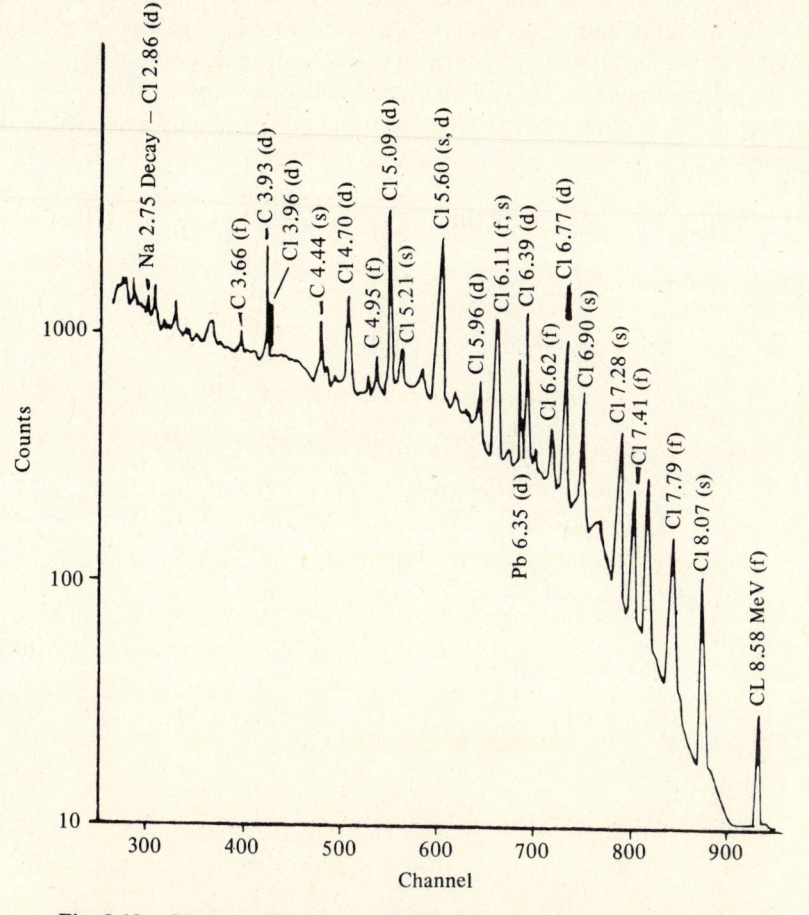

Fig. 8.12   Capture gamma spectrum for a 600-g sample of water from Soda Dam Spring irradiated 570 min by a 10^{10}-neutron/s ^{252}Cf source. [From D. Duffey, J. P. Balogna, and P. F. Wiggins, *Nuclear Technology* **27**, No. 3 (November 1975) p. 496.]

it shows the complexity that the neutron-capture spectrum may have. It indicates the impurities present and gives a general indication as to their amount. This method could also be used for monitoring flows, for process control, and for scanning pipes for solid deposits in an operating geothermal plant.

Polonium Production

The activation of a particular nuclide in itself may not be as useful as its daughter or granddaughter element. For example, ^{210}Po, an alpha emitter, is a popular isotopic heat source (see Prob. 3.12). It is the unstable daughter of the ^{210}Bi formed by activation of naturally available ^{209}Bi. In the conversion of ^{232}Th and ^{238}U to fissionable ^{233}U and ^{239}Pu, the fissile isotopes are the grand-daughters of the ^{333}Th and ^{239}U formed upon neutron activation.

The production of ^{210}Po will be used as an example of neutron activation for the production of an isotope other than the one originally activated in the target material.

$$^{209}_{83}\text{Bi} + ^{1}_{0}\text{n} \longrightarrow ^{210}_{83}\text{Bi} \xrightarrow[\text{5 days}]{\beta^-} ^{210}_{84}\text{Po} \xrightarrow[\text{138 days}]{\alpha} ^{206}_{82}\text{Pb}$$

In the following derivations,

N_1 = no. target nuclei (^{209}Bi)

N_2 = no. nuclei of activated isotope (^{210}Bi)

N_3 = no. nuclei of radioactive daughter (^{210}Po)

N_4 = no. nuclei of granddaughter (^{206}Pb)

The rate of change of activated ^{210}Bi nuclei is developed from Eq. (8.24) as

$$\frac{dN_2}{dt} = \phi N_1 \sigma_{a_1} - \lambda_{2T} N_2 \tag{8.28}$$

and Eq. (8.25) gives the number of ^{210}Bi nuclei as

$$N_2 = \frac{\phi N_1 \sigma_{a_1}}{\lambda_{2T}} (1 - e^{-\lambda_{2T} t}) \tag{8.29}$$

The rate of change of ^{210}Po nuclei will be equal to its rate of formation due to ^{210}Bi activity less its own decay rate. Note that any removal of ^{210}Bi and ^{210}Po by neutron capture could be ignored as having a probability of removal by neutron capture $\phi \sigma_a$ much less than its probability of decay λ. This effect

may not be ignored for ^{233}U and ^{239}Pu, so λ_{2T} and λ_{3T} will be retained for this derivation.

$$\frac{dN_3}{dt} = N_2 \lambda_2 - N_3 \lambda_{3T} \tag{8.30a}$$

Substituting Eq. (8.29), we obtain

$$\frac{dN_3}{dt} = \phi N_1 \sigma_{a_1} \frac{\lambda_2}{\lambda_{2T}} (1 - e^{-\lambda_2 T t}) - N_3 \lambda_{3T} \tag{8.30b}$$

Rearranging yields

$$\frac{dN_3}{dt} + N_3 \lambda_{3T} = \phi N_1 \sigma_{a_1} \frac{\lambda_2}{\lambda_{2T}} (1 - e^{-\lambda_2 T t}) \tag{8.30c}$$

This is again a first-order differential equation, which may be solved through the use of an integrating factor.

$$p = e^{\int \lambda_{3T} dt} = e^{\lambda_{3T} t} \tag{8.31}$$

The solution then becomes

$$
\begin{aligned}
N_3 &= \frac{1}{e^{\lambda_{3T} t}} \int e^{\lambda_{3T} t} \phi N_1 \sigma_{a_1} \frac{\lambda_2}{\lambda_{2T}} (1 - e^{-\lambda_2 T t}) - \frac{C}{e^{\lambda_{3T} t}} \\
&= \frac{\phi N_1 \sigma_{a_1}}{e^{\lambda_{3T} t}} \frac{\lambda_2}{\lambda_{2T}} \int [e^{\lambda_{3T} t} - e^{(\lambda_{3T} - \lambda_{2T}) t}] \, dt - C e^{-\lambda_{3T} t} \\
&= \phi N_1 \sigma_{a_1} \frac{\lambda_2}{\lambda_{2T}} \left(\frac{1}{\lambda_{3T}} - \frac{e^{-\lambda_2 T t}}{\lambda_{3T} - \lambda_{2T}} \right) - C e^{-\lambda_{3T} t} \tag{8.32}
\end{aligned}
$$

The constant of integration may be evaluated for the initial conditions that there is unirradiated material, bismuth in this case, at $t = 0$, and there will be no radioactive product, polonium in this case; so $N_3 = 0$.

$$0 = \phi N_1 \sigma_{a_1} \frac{\lambda_2}{\lambda_{2T}} \left(\frac{1}{\lambda_{3T}} - \frac{1}{\lambda_{3T} - \lambda_{2T}} \right) - C$$

$$C = \phi N_1 \sigma_{a_1} \frac{\lambda_2}{\lambda_{3T}(\lambda_{2T} - \lambda_{3T})} \tag{8.33}$$

Substitute Eq. (8.33) back into Eq. (8.32) and note that $N_1 \sigma_{a_1} = V\Sigma a_1$; that is, in this case N_1 is total number, not number density.

$$N_3 = \phi V \Sigma a_1 \frac{\lambda_2}{\lambda_{2T}} \frac{\lambda_{2T}(1 - e^{-\lambda_{3T}t}) - \lambda_{3T}(1 - e^{-\lambda_{2T}t})}{\lambda_{3T}(\lambda_{2T} - \lambda_{3T})} \qquad (8.34a)$$

If the probabilities of removal by neutron capture are small, as is commonly the case ($\sigma\phi \ll \lambda$), then $\lambda_{2T} \approx \lambda_2$ and $\lambda_{3T} \approx \lambda_3$; so

$$N_3 = \frac{\phi V \Sigma a_1 [\lambda_2(1 - e^{-\lambda_3 t}) - \lambda_3(1 - e^{-\lambda_2 t})]}{\lambda_3(\lambda_2 - \lambda_3)} \qquad (8.34b)$$

Figure 8.13 shows the grams of polonium present per kilogram of bismuth irradiated as a function of time. The smooth upper curve shows the buildup of polonium during irradiation as decreed by Eq. (8.34b). After various irradiation times the irradiated material is removed from the neutron flux and decay will occur. After the shorter irradiation periods (t_{rad}) the decay of activated ^{210}Bi will offset the ^{210}Po decay for a period of time and the polonium content

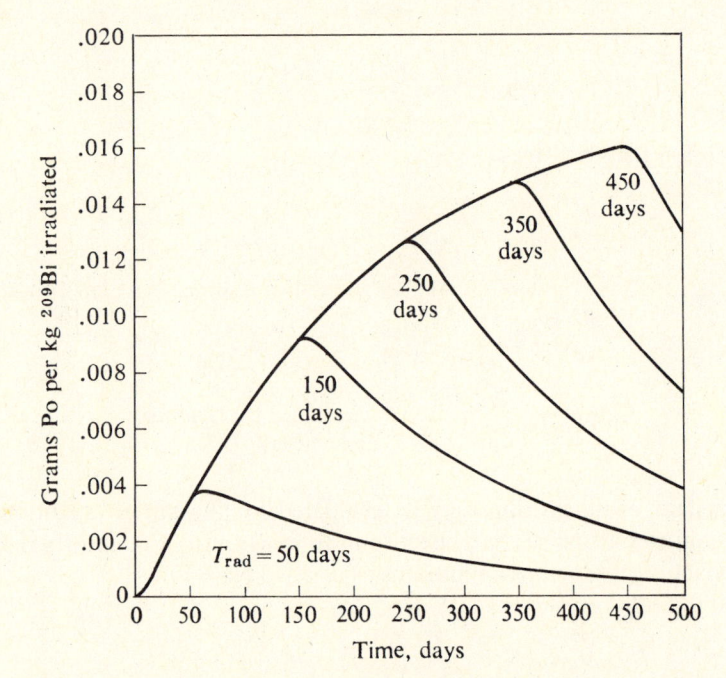

Fig. 8.13 Polonium production by irradiation of ^{209}Bi for various times (T_{rad}) followed by removal from the reactor. During irradiation the thermal flux is 10^{14} neutrons/cm^2 s at a temperature of 200°C.

will rise to a maximum before starting to fall. After longer irradiation, say 350 days, the decay rate of the polonium on removal is larger than its formation from the decay of activated bismuth. Thus, the amount of polonium will start to fall immediately on removal from the thermal neutron flux.

After removal from the flux the dual decay is described by Eq. (3.31), but the initial conditions differ from those of a freshly separated sample due to the fact that polonium is already present.

$$N_3 = \frac{\lambda_2}{\lambda_3 - \lambda_2} N_2' e^{-\lambda_2(t - t_{rad})} + C e^{-\lambda_3(t - t_{rad})} \tag{8.35}$$

When $t - t_{rad} = 0$, $N_2' = $ the number of ^{210}Bi atoms at time of removal, and $N_3' = $ the number of ^{210}Po atoms at the time of removal.

$$N_3' = \frac{\lambda_2 N_2'}{\lambda_3 - \lambda_2} + C$$

$$C = N_3' - \frac{\lambda_2 N_2'}{\lambda_3 - \lambda_2} \tag{8.36}$$

Thus, after removal,

$$N_3 = \frac{\lambda_2}{\lambda_3 - \lambda_2} N_2'(e^{-\lambda_3(t - t_{rad})} - e^{-\lambda_2(t - t_{rad})}) + N_3' e^{-\lambda_3(t - t_{rad})} \tag{8.37a}$$

If Eq. (8.37a) is differentiated and set equal to zero, the time at which a maximum amount of polonium is present can be determined.

$$(t - t_{rad})_{max} = \frac{\ln\left[\frac{(\lambda_2 - \lambda_3)\lambda_3 N_3'}{\lambda_2^2 N_2'} + \frac{\lambda_3}{\lambda_2}\right]}{\lambda_3 - \lambda_2} \tag{8.37b}$$

Example 5 If 1 kg of bismuth is irradiated for 100 days in a thermal flux of 10^{14} neutrons/cm^2 s at a temperature of 200°C, determine the mass of ^{210}Po which will be present on removal. How long after removal will the ^{210}Po peak? How many grams of ^{210}Po will be present at this time?
The number of ^{210}Bi target nuclei will be

$$N_1 = \frac{10^3 \text{ g} \times 6.023 \times 10^{23} \text{ (atoms/g atom)}}{208.98 \text{ g/g atom}}$$

$$= 2.882 \times 10^{24} \text{ atoms } ^{209}\text{Bi}$$

The macroscopic cross section for these atoms will be

$$N_1 \sigma_{a_1} = N_1 \frac{\sqrt{\pi}}{2} \sqrt{\frac{293}{T}} \sigma_a^{209}$$

$$= 2.882 \times 10^{24} \times \frac{\sqrt{\pi}}{2} \times \sqrt{\frac{293}{473}} \times 0.015 \times 10^{-24}$$

$$= 0.030\,04 \text{ cm}^2$$

The decay constants for the 5-day ^{210}Bi and 138-day ^{210}Po are

$$\lambda_2 = (0.693/5.0) \times 24 \times 3600 = 1.6041 \times 10^{-6}\,\text{s}^{-1}\,(0.1386\,\text{day}^{-1})$$

$$\lambda_3 = (0.693/138) \times 24 \times 3600 = 0.581 \times 10^{-7}\,\text{s}^{-1}\,(0.005\,02\,\text{day}^{-1})$$

Since the probability of neutron absorption is very small for both ^{210}Bi and ^{210}Po, λ_{2T} may be replaced by λ_2 and λ_{3T} may be replaced by λ_3.

Thus, at the end of the 100-day irradiation of the original natural bismuth, the number of polonium atoms present will be

$$N_3 = N_3' = (\phi V \Sigma_{a_1}/(\lambda_2 - \lambda_3)) \left[\frac{\lambda_2}{\lambda_3} (1 - e^{-\lambda_3 t}) - (1 - e^{-\lambda_2 t}) \right]$$

$$= \frac{10^{14}\,(\text{neutrons/cm}^2\,\text{s}) \times (0.03\,004\,\text{cm}^2) \times 1 \text{ atom Po/neutron}}{(16.041 - 0.581) \times 10^{-7}\,\text{s}^{-1}}$$

$$\times \left[\frac{16.041 \times 10^{-7}}{0.581 \times 10^{-7}} (1 - e^{-0.502}) - (1 - e^{-13.86}) \right]$$

$$= 1.9194 \times 10^{19} \text{ atoms } ^{210}\text{Po}$$

The corresponding mass of polonium produced in the 1 kg of bismuth will be

$$m_3 = \frac{1.9194 \times 10^{19} \text{ atoms Po} \times 210 \text{ g } ^{210}\text{Po/g atom}}{6.023 \times 10^{23} \dfrac{\text{atoms } ^{210}\text{Po}}{\text{g atom}}}$$

$$= 0.0067 \text{ g Po}$$

Before determining the time at which the maximum amount of polonium will be present, the number of ^{210}Bi atoms present at the end of irradiation must be determined. For the ^{210}Bi the 100-day irradiation period is 20 half-lives and thus saturation exists.

$$N_2 = N_2' = \frac{\phi N_1 \sigma_{a_1}}{\lambda_2} = \frac{10^{14} \times 0.030\,04}{16.041 \times 10^{-7}}$$

$$= 1.8704 \times 10^8 \text{ atoms}$$

The time after removal at which the maximum amount of Po will be present will be

$$(t - t_{rad})_{max} = \frac{\ln\left[\frac{(\lambda_2 - \lambda_3)\lambda_3 N_3'}{\lambda_2^2 N_2} + \frac{\lambda_3}{\lambda_2}\right]}{\lambda_3 - \lambda_2}$$

$$= \frac{\ln\left[\frac{(16.041 - 0.581) \times 0.581 \times 1.9194}{16.041^2 \times 0.18\,704} + \frac{0.581}{16.041}\right]}{0.00\,502 - 0.1386}$$

$$= 6.955 \text{ days}$$

When the irradiated material is held for approximately a week after irradiation, the maximum number of polonium atoms will be present.

$$N_{3\,max} = \frac{\lambda_2 N_2'}{\lambda_2 - \lambda_3}\left[e^{-\lambda_3(t - t_{rad})max} - e^{-\lambda_2(t - t_{rad})max}\right] + N_3' e^{-\lambda_3(t - t_{rad})max}$$

$$= \frac{16.041 \times 10^{-7} \times 1.8074 \times 10^{18}}{(16.041 - 0.581)10^{-7}}\left(e^{-0.005\,02 \times 6.955} - e^{-0.1386 \times 6.955}\right)$$

$$+ 1.9194 \times e^{-0.005\,02 \times 6.955}$$

$$= 2.044 \times 10^{19}\ ^{210}\text{Po atoms}$$

$$m_{3\,max} = \frac{2.044 \times 10^{19} \times 210}{6.023 \times 10^{23}} = 0.007\,127 \text{ g Po}$$

By waiting 6.995 days after removal before chemical processing there would be an increase of 6.8 percent in the polonium recovery. Again, it should be noted that an increase of this sort will occur only at the shorter irradiation periods where the decay rate of the bismuth exceeds that of the polonium.

Tritium Activation in Borated Water

In pressurized water reactors the addition of boric acid acts as a chemical shim. That is, the boron acts as a poison which, as it is burned out, offsets the reduction in reactivity due to fuel burnup. Both thermal and fast-neutron reactions are involved in the activation of significant amounts of tritium. The following reactions predominate:

(1) ^{10}B(n, 2α)T which has a threshold of 1 MeV with a cross section increasing from 15 mbarn at 1 MeV and increasing to 75 mbarn at 5 MeV and remaining constant from 5 to 10 MeV. A cross section of 50 mbarn may be used for the fast flux ($E > 1$ MeV).

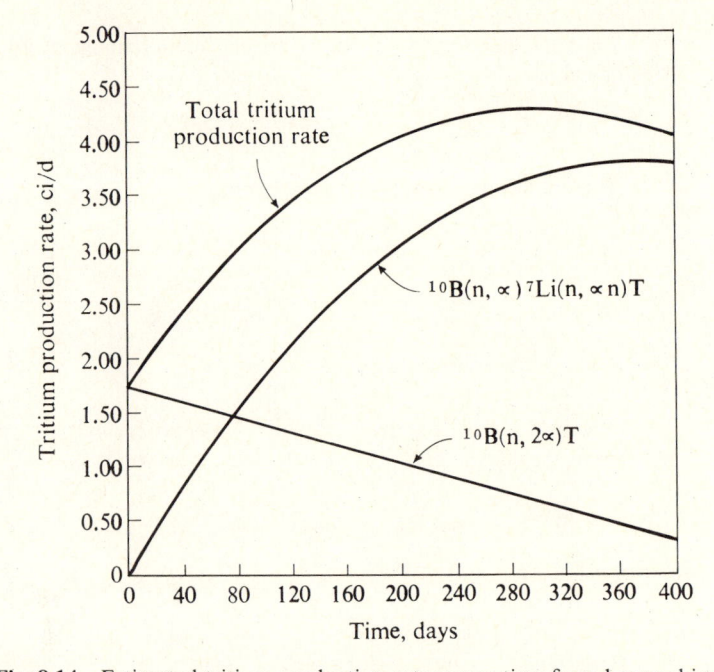

Fig. 8.14   Estimated tritium production rate versus time from boron shim in a 1000-Mw(e) PWR.

(2) $^{10}B(n, \alpha)^{7}Li(n, n\alpha)T$, a duplex reaction where the first reaction is a thermal neutron reaction and the second has a 3 MeV threshold whose cross section increases linearly from 0 at 3 MeV to 400 mbarn at 6 MeV and remains constant from 6 to 10 MeV.   The cross section for the $^{7}Li(n, n\alpha)T$ reaction can also be taken as approximately 50 mbarn for the fast flux ($E > 1$ MeV).

Figure 8.14 shows the tritium production rate from boron shim in the coolant of a typical 1000-Mw(e) PWR.   The accumulated tritium activity must be considered as contributing to the hazard of any leakage from the primary loop and one of the contributors to the seriousness of a loss of coolant accident.

The rate from the first reaction falls off as the chemical shim is reduced from an initial level of 1500 to 400 ppm after 350 days of reactor full-power operation.   In the duplex reaction the rate builds up as the ^{7}Li concentration builds up due to ^{7}Li activation by thermal neutrons.

Slowing Down of Neutrons

Neutrons slow from an average energy of 2 MeV at birth, as the result of fission, to 0.025 eV for 2200 m/s thermal neutrons.   This energy reduction is accom-

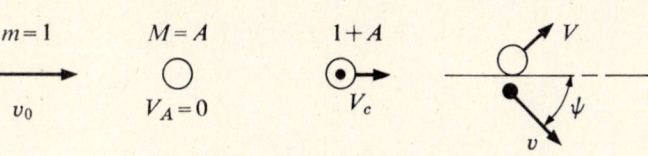

Fig. 8.15 Laboratory system scatter.

plished essentially by elastic scattering in which both kinetic energy and momentum are conserved. More energy is lost in scattering with light nuclei; therefore, materials of this sort (light water, heavy water, beryllia, graphite, etc.) make effective moderators.

For elastic scattering one must take measurements in the laboratory system where the observer and the target nucleus are both stationary. The rest mass of the neutron is 1 u and the struck nucleus has a rest mass of A u. Figure 8.15 shows the scattering process in the laboratory system.

v_0 = original neutron velocity in laboratory system, cm/s

$V_A = 0$ = velocity of struck nucleus in laboratory system

V_c = velocity of compound nucleus in laboratory system, cm/s

V = velocity of scattered nucleus in laboratory system, cm/s

v = velocity of scattered neutron in laboratory system, cm/s

ψ = scattering angle in lab system of emergent neutron with respect to original direction of motion

v_1 = incident neutron velocity in COM system, cm/s

V_1 = velocity of struck nucleus in COM system, cm/s

V_2 = velocity of scattered nucleus in COM system, cm/s

v_2 = velocity of scattered neutron in COM system, cm/s

θ = neutron scattering angle in COM system

Since the struck nucleus has little or no velocity, the momentum of the compound nucleus is equal to the momentum of the incident neutron.

$$1 \times v_0 = (1 + A)V_c \tag{8.38}$$

Thus, the velocity of the compound nucleus is

$$V_c = \frac{v_0}{1 + A} \tag{8.39}$$

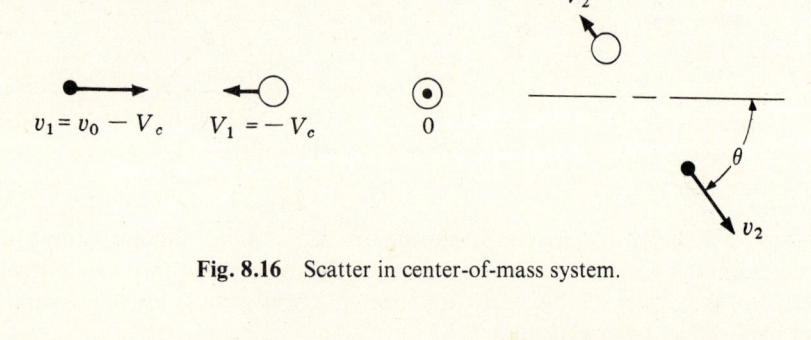

Fig. 8.16   Scatter in center-of-mass system.

To examine the breakup of the compound nucleus it is convenient to transfer to the center-of-mass (COM) system (see Fig. 8.16). To do this, the observer must imagine that he or she is traveling at the speed and in the direction that the compound nucleus will have after the collision. This amounts to subtracting V_c from both the velocity of the incident neutron and the target nucleus. Note that in this system the compound nucleus will appear to hang stationary after the collision.

The velocity of the incoming neutron in the COM system is

$$v_1 = v_0 - V_c = v_0 - \frac{v_0}{1 + A} = \frac{Av_0}{1 + A} \tag{8.40}$$

In the COM system the kinetic energy before the collision must equal the kinetic energy of the particles as they fly apart. The binding energy to form and break up the compound nucleus cancels and, therefore, only KE must be considered. The KE available for the compound nucleus is the sum of the KE's of the incoming neutron with a velocity $(v_0 - V_c)$ and the nucleus with a velocity $- V_c$.

$$\text{KE}_{\text{COM}} = \tfrac{1}{2}(v_0 - V_c)^2 + \tfrac{1}{2}A(-V_c)^2$$

Substituting from Eqs. (8.40) and (8.39), we obtain

$$\text{KE}_{\text{COM}} = \frac{1}{2}\left(\frac{A}{1 + A}\right)^2 v_0^2 + \frac{1}{2}A\left(\frac{v_0}{1 + A}\right)^2$$

$$= \frac{A}{1 + A}\,\text{KE}_0 \tag{8.41}$$

where KE_0 is the original KE of the neutron in the laboratory system. The same result may be obtained in the laboratory system by subtracting the KE of the compound nucleus from that of the incident neutron.

Note that the KE_{COM} will be 235/236 that of the incident neutron, KE_0, when a ^{235}U nucleus is struck, whereas it is only $\frac{1}{2}KE_0$ when a light hydrogen nucleus is hit. Thus, the difference between the laboratory and the center of mass systems is more pronounced for lighter elements.

In Fig. 8.16, KE_{COM} is shared by the scattered particles which fly apart in opposite directions in the COM system. The KE imparted to the compound nucleus is equal to that of the emergent particles.

$$\frac{A}{1+A}\frac{v_0^2}{2} = \frac{AV_2^2}{2} + \frac{v_2^2}{2} \tag{8.42}$$

The momentum of the compound nucleus is zero in the COM system, and thus the momenta of the particles as they fly apart are equal and opposite.

$$1 \times v_2 + AV_2 = 0$$
$$v_2 = -AV_2 \tag{8.43}$$

When Eqs. (8.42) and (8.43) are combined,

$$\frac{AV_2^2}{2} + \frac{A^2V_2^2}{2} = \frac{[A/(1+A)]v_0^2}{2} = \frac{Av_2^2}{2A^2} + \frac{v_2^2}{2}$$

scattered neutron

$$v_2 = \frac{Av_0}{1+A} \tag{8.44}$$

compound nucleus

$$V_2 = \frac{v_0}{1+A} \tag{8.45}$$

It is useful to convert back to the laboratory system to compare the KE of the scattered neutron in this system with its original value, KE_0. As shown in Fig. 8.17, this is accomplished by adding the velocity of the compound nucleus, V_c, to the COM velocity of the scattered neutron, v_2. By using the Pythagorean theorem, we get

$$v^2 = (v_2 \sin \theta)^2 + (v_2 \cos \theta + V_c)^2 \tag{8.46}$$

Substituting from Eqs. (8.45) and (8.39) yields

$$v^2 = \left(\frac{A}{1+A}v_0 \sin \theta\right)^2 + \left(\frac{A}{1+A}v_0 \cos \theta + \frac{v_0}{1+A}\right)^2$$

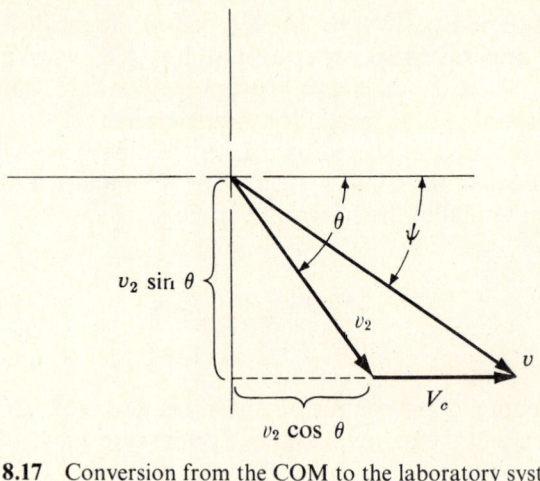

Fig. 8.17   Conversion from the COM to the laboratory system.

which reduces to

$$v^2 = \frac{(A^2 + 2A \cos \theta + 1)v_0^2}{(1 + A)^2} \tag{8.47}$$

The ratio of the new neutron kinetic energy, KE, to its original value, KE_0, is

$$\frac{KE}{KE_0} = \frac{v^2/2}{v_0^2/2} = \frac{A^2 + 2A \cos \theta + 1}{(1 + A)^2} \tag{8.48}$$

Note that this energy ratio is maximum when $\theta = 0$.

$$\left(\frac{KE}{KE_0}\right)_{max} = \frac{A^2 + 2A + 1}{(1 + A)^2} = 1 \tag{8.49}$$

This indicates that with forward scatter the neutron energy is unchanged.

The minimum value of the energy ratio occurs when $\theta = \pi$.

$$\alpha = \left(\frac{KE}{KE_0}\right)_{min} = \frac{A^2 - 2A + 1}{(1 + A)^2} = \frac{(A - 1)^2}{(A + 1)^2} \tag{8.50}$$

For hydrogen ($A = 1$) the value of α is

$$\alpha_H = \left(\frac{1 - 1}{1 + 1}\right)^2 = 0$$

This indicates that backscatter of a neutron by a hydrogen atom can cause the neutron to lose all its energy in a single collision. For Be ($A = 9$),

$$\alpha_{Be} = \frac{(9 - 1)^2}{(9 + 1)^2} = 0.64$$

This relatively light moderator material can cause a neutron to lose a maximum of 36 percent of its energy in a single collision. For ^{235}U ($A = 235$),

$$\alpha_{235} = \frac{(235 - 1)^2}{(235 + 1)^2} = 0.983$$

This heavy fuel atom can reduce a neutron's energy by a maximum of 1.7 percent on a single collision.

These few examples have illustrated the advantage of using light nuclei as moderating materials.

Scatter in the COM System

In the COM system the breakup of the compound nucleus is independent of its mode of formation. Neutrons scatter in a random or isotropic manner in *all* directions.

A number of neutrons, n, are considered to scatter in an isotropic manner about a point. The sphere shown in Fig. 8.18 is considered to have a unit radius. The probability of scatter into the angle between θ and $\theta + d\theta$ is the ratio of the area of the elemental ring subtended by the differential angular element to the total area of the unit sphere, or it is the ratio of the solid angle subtended by the elemental ring to the total solid angle.* It may be expressed as

$$\frac{2\pi \sin \theta \, d\theta}{4\pi} = \frac{\sin \theta \, d\theta}{2} \tag{8.51a}$$

When the probability of scatter into the differential angle is multiplied by the total number of neutrons scattering, one has the differential number of neutrons scattering into the angle between θ and $\theta + d\theta$.

$$dn = \tfrac{1}{2}n \sin \theta \, d\theta \tag{8.51b}$$

---

* A *solid angle* is defined as the angle intercepted by a cone on a surface of a unit sphere, which has a total solid angle of 4π steradians.

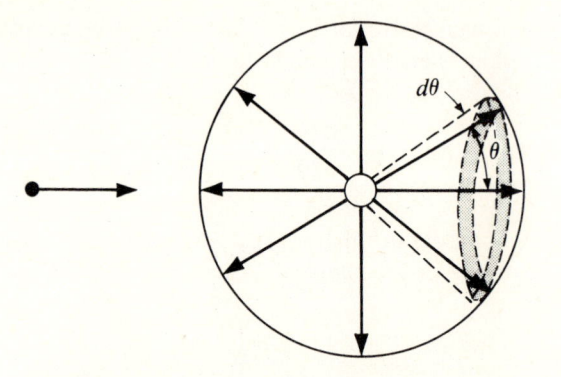

Fig. 8.18   Isotropic scatter in the COM system.

Logarithmic Energy Decrement

The energy loss of a neutron on collision is dependent on the scattering angle, as shown in Eq. (8.48). A convenient measure of energy loss is logarithmic energy decrement, ξ. It is defined as the average drop in the logarithm of neutron energy per collision. The result will turn out to be independent of energy level.

$$\xi = \overline{\ln E_0 - \ln E} = -\ln \overline{(E/E_0)} \tag{8.52}$$

The product of the number of neutrons scattering into the angle between θ and $\theta + d\theta$ is multiplied by the log decrement at the angle θ.

$$-\ln \frac{E}{E_0} \, dn = \frac{n}{2} \sin \theta \, d\theta \left[-\ln \frac{A^2 + 2A \cos \theta + 1}{(1 + A)^2} \right] \tag{8.53}$$

Equation (8.53) is then integrated between 0 and π to give the sum of the log decrements of all n neutrons. Therefore, the average value can be found by dividing by n.

$$\xi = -\ln (\overline{E/E_0}) = \left(\frac{1}{n}\right) \int_0^\pi -\ln \left[\frac{A^2 + 2A \cos \theta + 1}{(1 + A)^2}\right] \frac{n}{2} \sin \theta \, d\theta \tag{8.54}$$

Let

$$x = \frac{A^2 + 2A \cos \theta + 1}{(1 + A)^2}$$

$$dx = \frac{-2A \sin \theta \, d\theta}{(1 + A)^2}$$

The limits for the integration must be changed to the proper values of x. When

$$\theta = 0, \qquad x = 1$$
$$\theta = \pi, \qquad x = \alpha$$

Therefore,

$$\xi = \int_1^\alpha \frac{\ln x}{2} \frac{(A+1)^2}{2A}\, dk = \frac{(A+1)^2}{4A} \int_1^\alpha \ln x\, dx$$

Note that

$$\frac{(A+1)^2}{4A} = \frac{(A+1)^2}{(A+1)^2 - (A-1)^2} = \frac{1}{1 - [(A-1)^2/(A+1)^2]} = \frac{1}{1-\alpha}$$

so that

$$\xi = \frac{1}{1-\alpha} \int_1^\alpha \ln x\, dx = \frac{1}{1-\alpha} (x \ln x - x)_1^\alpha$$

$$= 1 + \frac{\alpha}{1-\alpha} \ln \alpha \tag{8.55}$$

or combining with Eq. (8.50), we get

$$\xi = 1 + \frac{(A-1)^2}{2A} \ln \frac{A-1}{A+1} \tag{8.56}$$

Thus, the average loss in the log of the energy per collision is a function of the mass of the struck nucleus and is independent of energy level. Equation (8.56) can be approximated rather closely by

$$\xi = \frac{2}{A + \frac{2}{3}} \tag{8.57}$$

For $A = 2$ Eq. (8.57) is in error by only 3.3 percent. The accuracy improves with larger values of A, as shown by Fig. 8.19.

Let $E_1, E_2, \ldots, E_n$ represent average neutron energies after each of n collisions during the slowing-down process from some original energy, E_0, to reach some lower energy, E_n.

$$\ln \frac{E_0}{E_n} = \ln \left(\frac{E_0}{E_1} \cdot \frac{E_1}{E_2} \cdot \frac{E_2}{E_3}, \ldots, \frac{E_{n-1}}{E_n} \right)$$

$$= \ln \left(\frac{E_0}{E_1} \right)^n = n \ln \frac{E_0}{E_1}$$

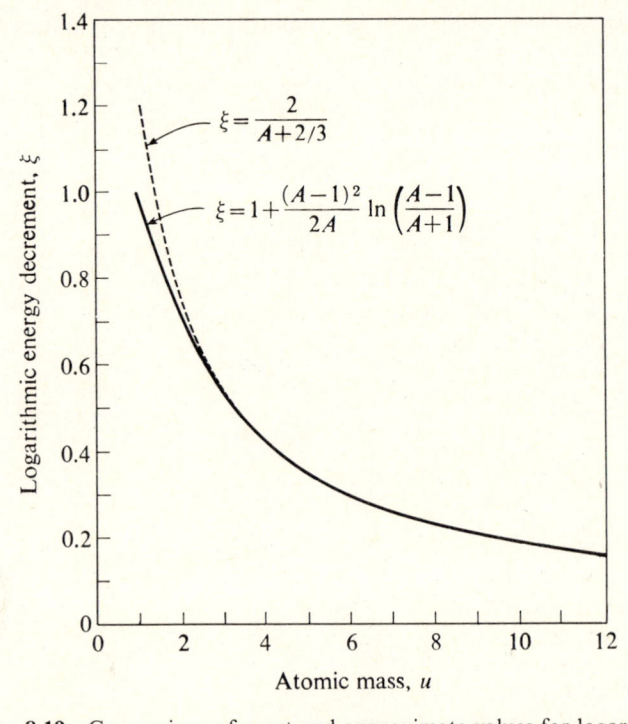

Fig. 8.19   Comparison of exact and approximate values for logarithmic energy decrement.

If E_n represents the average thermal neutron energy, the number of collisions to thermalize is

$$n = \frac{\ln (E_0/E_n)}{\xi} \qquad (8.58)$$

Example 6   Compute the number of collisions in beryllium required to reduce 2.0-MeV neutrons to 0.025 eV.

$$\xi = \frac{2}{A + \frac{2}{3}} = \frac{2}{9 + \frac{2}{3}} = 0.207$$

$$n = \frac{\ln (E_0/E_n)}{\xi} = \frac{\ln (2.0 \times 10^6/0.025)}{0.207} = 86 \text{ collisions}$$

Note that for fast neutrons energy loss will occur on subsequent collisions until thermal energies are reached.   Further collisions are as apt to have a gain in energy as a loss in energy.   The larger ξ, the smaller the number of collisions to thermalize and hence the more effective is the material as a moderator.

However, the log energy decrement (or the number of collisions to thermalize) does not completely describe the excellence of a material as a moderator. As neutrons interact with moderator nuclei there must be a high probability for scatter and a low probability of being absorbed.

Macroscopic Slowing-Down Power

Macroscopic slowing-down power (MSDP) is the product of the log decrement times the macroscopic scattering cross section for epithermal neutrons.

$$\text{MSDP} = \xi \Sigma_s^{\text{epi}} \tag{8.59}$$

This indicates how rapidly slowing down will occur in material. It represents the slowing-down power of all the nuclei in a cubic centimeter of material. For a light gas, such as helium, there would be a good log decrement but a poor slowing-down power because of the small probability of scatter. Not only is σ_s small, but the atom density of the gas is much too low to be attractive.

For a compound or a mixture the log decrement can be found as the sum of the macroscopic slowing-down powers divided by the total macroscopic scattering cross section of all the constituents.

$$\xi = \frac{\xi_1 \Sigma_{s1}^{\text{epi}} + \xi_2 \Sigma_{s2}^{\text{epi}} + \xi_3 \Sigma_{s3}^{\text{epi}}}{\Sigma_{s1}^{\text{epi}} + \Sigma_{s2}^{\text{epi}} + \Sigma_{s3}^{\text{epi}}} \tag{8.60}$$

Moderating Ratio

The macroscopic slowing-down power still does not tell the complete story about the effectiveness of a moderator. An element such as boron has a high log decrement and a good slowing-down power, but it is a poor moderator because of its high probability of absorbing neutrons. This can be accounted for by dividing the macroscopic slowing-down power by the macroscopic thermal absorption cross section. This is called the *moderating ratio*.

$$\text{MR} = \frac{\xi \Sigma_s^{\text{epi}}}{\Sigma_a^{\text{th}}} \tag{8.61}$$

For a single element this reduces to

$$\text{MR}_1 = \frac{\xi \sigma_s^{\text{epi}}}{\sigma_a^{\text{th}}} \tag{8.62}$$

Table 8.1 Comparison of the Moderating Characteristics of Materials

Material	ξ	No. Collisions to Thermalize	MSDP	Moderating Ratio
H_2O	0.927	19	1.425	62
D_2O	0.510	35	0.177	4830
He	0.427	42	8.87×10^{-6}*	51
Be	0.207	86	0.1538	126
B	0.171	105	0.092	0.00086
C	0.158	114	0.083	216

* 1 atm and 20°C.

and for a two-component mixture or a compound of two elements it is

$$MR_2 = \frac{\xi_1 \Sigma_{s1}^{epi} + \xi_2 \Sigma_{s2}^{epi}}{\Sigma_{a1}^{th} + \Sigma_{a2}^{th}} \tag{8.63}$$

A good moderator, then, has a high moderating ratio and a large macroscopic slowing-down power.

Table 8.1 indicates the relative merits of several materials as moderators. Light water has a high log decrement and a good macroscopic slowing-down power, but because of a 0.332-barn absorption of the $_1^1H$ cross section it has the lowest moderating ratio of any of the commonly used moderators. Its availability and low cost justify its use, even though the use of enriched fuel is required to achieve criticality in a reactor core.

The vastly superior moderating ratio of heavy water is offset by its extremely high cost. Beryllium and carbon (graphite) have similar moderating ratios. The difficulty of fabrication of beryllium, its cost, and its toxicity are not conducive to its common use. Graphite has been commonly used as a moderator because of its availability in quantities, its low cost, and ease of fabrication.

Helium is ruled out on the basis of its low density, which results in an extremely poor MSDP. Boron is also a poor moderator, but in this case it is because of the high cross section for absorption. It does find use, however, as a neutron absorber in control rods.

Beryllium hydride (BeH_2) is a moderator that has recently been shown to reduce the critical size of a spherical core, even below that for polyethylene (CH_2). Its number density of H and Be atoms is high and it has a positive effect on the reactivity of the core because of an $(n, 2n)$ reaction for the beryllium. Amorphous BeH_2 has a density of only $620 \ kg/m^3$, but a crystalline form with a density of $780 \ kg/m^3$ has been produced.

Average Value of the Cosine of the Scattering Angle

Figure 8.20 shows that when the velocity of the compound nucleus is added to the velocity of scattered monoenergetic neutrons in the COM system, the scatter in the laboratory system is anisotropic. There is preferential forward scatter.

The average value of the cosine of the scattering angle in the COM system is found by multiplying the number of neutrons scattering into the angle between θ and $\theta + d\theta$ by $\cos \theta$. When this product is integrated from 0 to π and divided by the number of scattered neutrons, the average value of $\cos \theta$ results.

$$\overline{\cos \theta} = \frac{1}{n} \int_0^\pi \cos \theta \, \frac{n}{2} \sin \theta \, d\theta = \frac{1}{4} \int_0^\pi \sin 2\theta \, d\theta$$

$$\left(= \frac{1}{8} \cos 2\theta \Big|_0^\pi = 0 \right) \tag{8.64}$$

For this value of the cosine, θ is 90°, which indicates that just as many neutrons scatter forward as scatter backward. The scatter is isotropic.

This will not be so in the laboratory system. Figure 8.17 shows how the cosine of the scattering angle in the laboratory system can be expressed in terms of the COM scattering angle, θ.

$$\cos \psi = \frac{v_2 \cos \theta + V_c}{\sqrt{(v_2 \cos \theta + V_c)^2 + (v_2 \sin \theta)^2}}$$

$$= \frac{\dfrac{Av_0}{A+1} \cos \theta + \dfrac{v_0}{A+1}}{\sqrt{\left(\dfrac{Av_0}{A+1} \cos \theta + \dfrac{v_0}{A+1}\right)^2 + \left(\dfrac{Av_0 \sin \theta}{A+1}\right)^2}}$$

$$\left(= \frac{A \cos \theta + 1}{\sqrt{A^2 + 2A \cos \theta + 1}} \right) \tag{8.65}$$

The average value of the scattering angle in the laboratory system can now be determined.

$$\cos \psi = \frac{v_2 \cos \theta + V_c}{\sqrt{(v_2 \cos \theta + V_c)^2 + (v_2 \sin \theta)^2}}$$

$$\left(\overline{\cos \psi} = \int_0^\pi \frac{A \cos \theta + 1}{\sqrt{A^2 + 2A \cos \theta + 1}} \, \frac{\sin \theta}{2} \, d\theta = \frac{2}{3A} \right) \tag{8.66}$$

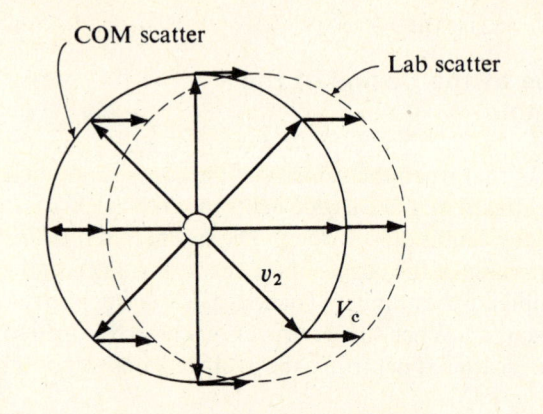

Fig. 8.20   Isotropic scatter in the COM system and an isotropic scatter in the laboratory system.

For graphite,

$$\overline{\cos \psi} = \frac{2}{3 \times 12} = 0.056 \qquad (\psi = 86.8°)$$

which indicates nearly isotropic scatter in the laboratory system.
    For hydrogen,

$$\overline{\cos \psi} = \tfrac{2}{3} \qquad (\psi = 48.2°)$$

which indicates quite strong forward scatter.

Transport Mean Free Path

The *transport mean free path* is a scattering mean free path that has been corrected for the somewhat greater distance traveled in the laboratory system due to the preferential forward scatter.   It can be determined by dividing the scattering mean free path by 1, minus the average value of cos ψ.   Its reciprocal is the macroscopic transport cross section.

$$\lambda_{tr} = \frac{\lambda_s}{1 - \overline{\cos \psi}} = \frac{1}{\Sigma_s(1 - \overline{\cos \psi})} = \frac{1}{\Sigma_{tr}} \qquad (8.67)$$

Example 7 Compute the transport mean free path for thermal neutrons in beryllia.

$$\rho_{BeO} = 2.70 \text{ g/cm}^3$$

$$N_{Be} = N_0 = \frac{\rho \times 6.023 \times 10^{23}}{A}$$

$$= \frac{2.7 \times 6.023 \times 10^{23}}{25.01} = 6.51 \times 10^{22} \text{ atoms/cm}^2$$

$$\overline{\cos \psi_{Be}} = \frac{2}{9 \times 3} = 0.0741$$

$$\overline{\cos \psi_0} = \frac{2}{16 \times 3} = 0.0417$$

$$\lambda_{tr} = \frac{1}{N_{Be}\,\sigma_{sBe}(1 - \overline{\cos \psi_{Be}}) + N_0\,\sigma_{s0}(1 - \overline{\cos \psi_0})}$$

$$= \frac{1}{6.51 \times 10^{22} \times 7 \times 10^{-24}(1 - 0.0741)}$$

$$+ \frac{1}{6.51 \times 10^{22} \times 4.2 \times 10^{-24}(1 - 0.0417)}$$

$$= \frac{1}{0.422 + 0.262} = 1.46 \text{ cm}$$

Problems

1. A collimated beam of 2200 m/s neutrons has an intensity of 100 000 neutrons/cm^2 s.
   (a) What would be its intensity after passing through a sheet of 2.5-mm-thick boral? Boral has a density of 2.56 g/cm^3 and its composition is 50 percent Al, 40 percent B, and 10 percent C by weight.
   (b) What would be the beam intensity after passing through a sheet of pure aluminum of the same thickness?

2. If $n(E) = dn/dE$, show that by starting with Eq. (8.10),

$$n(E) = \frac{2\pi n_0 E^{1/2}}{(\pi k T)^{3/2}} e^{-E/kT}$$

and the most probable kinetic energy is $kT/2$.

The temperature in a reactor core is 500°C. What will be the most probable thermal neutron velocity? What will be the average thermal neutron velocity? What will be the most probable kinetic energy?

3. For the neutron chopper it would be convenient to have an expression for the effective neutron temperature for a well-thermalized flux that would give the effective neutron temperature in terms of the time t at which $n(t)$ peaks. Start with Eq. (8.11) and transform $n(v)$ to $n(t)$, differentiate, and set this expression equal to zero. Suppose that a particular chopper setup has a flight distance of 2.5 m between spinning disk (chopper) and the BF_3 counter. The multichannel analyzer uses a time interval of 20 μs for each channel and the fortieth channel has accumulated the maximum number of counts. What is the effective neutron temperature?

4. At 500°C, 3 percent enriched UO_2 has a density of 10.5 g/cm.[3] For thermal neutrons compute the following:
   (a) The macroscopic fission cross section
   (b) The absorption mean free path
   (c) The mean free path for the interaction of thermal neutrons by either scatter or absorption

5. Three gold foils are irradiated in a thermal flux of 10^{10} neutrons/cm^2 s. The foil is 2.5 cm in diameter and 0.25 mm thick. The effective neutron temperature during irradiation is 37°C.
   (a) How many curies will each foil emit upon removal after irradiation for periods of 2 h, 2 days, and 2 months, respectively?
   (b) What will be the activity of the foil that was irradiated for 2 days at the end of the 24-h period following removal?
   (c) The flux is raised to 5×10^{13} neutrons/cm^2 s. For a foil irradiated for 2 days, what will be its activity on removal? What will it be 24 h after removal?

6. An iron Charpy specimen 10 mm $\times$ 10 mm $\times$ 55 mm is irradiated in a thermal flux of 10^{12} neutrons/cm^2 s for 30 days at 100°C. The resultant activity is due to the 0.31 percent ^{58}Fe atom which is present. Determine:
   (a) The activity (Bq) on removal
   (b) The activity 60 days after removal

7. A cobalt wire 0.305 m in length and 1.6 mm in diameter is left in a subcritical reactor core for 2 yr. The flux is 10^6 neutrons/cm^2 s.
   (a) What will be the activity (μCi) of the wire on removal?
   (b) What will be its activity after another 2 yr out of the reactor core? The effective neutron temperature during the irradiation is 100°C.

8. Ten grams of tantalum is exposed to a thermal flux of 5×10^{13} neutrons/cm^2 s for 115 days at an effective neutron temperature of 500°C.
   (a) On removal, what will be its activity?
   (b) After 25 days out of the core, what will be the activity?
   (c) Repeat with the flux reduced to 10^9 neutrons/cm^2 s.

9. The indium and cadmium covered indium foils of Example 8.4 are placed at new locations where the thermal flux is 10^5 neutrons/cm^2 s and are irradiated for 35 min. Assume that the ratio of thermal neutron absorptions to epithermal resonance absorptions is independent of position. Calculate the count rate of the cadmium covered foil 20 min after removal from the core and the count rate of the bare foil counted 25 min after removal.

10. One kilogram of thorium is irradiated for 1 yr in a thermal flux of 10^{14} neutrons/cm² s at an effective neutron temperature of 650°C. Note that the short half-life of ^{233}Th (22.1 min) allows us to ignore the ^{233}Th and assume that neutron absorption by ^{232}Th leads directly to ^{233}Pa. Also, for the ^{233}U its long half-life $(1.62 \times 10^5 \text{ yr})$ means that its probability of decay is much smaller than its probability of removal by neutron absorption. Hence, $\phi\sigma_a^U$ replaces λ_U during irradiation. How many grams of ^{233}Pa and ^{233}U will be present at the end of the irradiation period? By what percent will the uranium yield increase if the Pa is allowed to decay before separation?

11. A 1000-MW(e) pressurized water reactor (PWR) is to have a boron shim which decreases from 1500 ppm to 400 ppm over a 350-day period. Tritium is produced due to the ^{10}B(n, 2α)T and ^{10}B(n, α)^{7}Li(n, αn)T reactions. The fast flux $(E > 1 \text{ MeV})$ is 6×10^{13} neutrons/cm² s and the cross section for the fast reactions can be taken as 50 mbarn for the ^{10}B(n, 2α)T and 75 mbarn for the ^{7}Li(n, αn)T reactions. The ^{10}B(n, α)^{7}Li reaction occurs due to thermal neutron absorption in a thermal flux of 4×10^{13} neutrons/cm² s. The water is at an average temperature of 288°C and a pressure of 13.79 MPa. It has a volume of 18.14 m³. There is a primary coolant turnover rate of 0.12 percent/day due to leakage from the system. Estimate the rate of tritium production (Ci/day) when the reactor has been on line for 200 days at the given power (flux) level.

12. Show that $\overline{\cos \psi} = \int_0^\pi \cos \psi \sin \theta \, d\theta/2 = \frac{2}{3}A$.

13. For a Be moderator at 427°C, determine:
    (a) The number of collisions to thermalize
    (b) Macroscopic slowing-down power , 1539
    (c) Moderating ratio
    (d) Transport mean free path

14. 2-MeV neutrons are being thermalized in beryllia (BeO) at 537°C having a density of 2.69 g/cm³. Compute the following:
    (a) The log decrement
    (b) The macroscopic slowing-down power
    (c) The number of collisions to thermalize
    (d) The moderating ratio
    (e) For elastic scattering with a Be atom, the value of α and also E/E_0 when $\theta = 90°$

15. For beryllium hydride having a density of 780 kg/m³, repeat parts (a) through (d) of Prob. 13. Compare the effectiveness of BeO and BeH₂ as moderators.

References

1. Glasstone, S., and M. C. Edlund, *The Elements of Nuclear Reactor Theory*. Princeton, N.J.: D. Van Nostrand Company, 1952.
2. Glasstone, S., and A. Sesonske, *Nuclear Reactor Engineering*. Princeton, N.J.: D. Van Nostrand Company, 1963.

3. Murray, R. L., *Introduction to Nuclear Engineering*. Englewood Cliffs, N.J.: Prentice-Hall, Inc., 1961.

4. El-Wakil, M. M., *Nuclear Power Engineering*. New York: McGraw-Hill Book Company, 1962.

5. Lamarsh, J. R., *Introduction to Nuclear Reactor Theory*. Reading, Mass.: Addison-Wesley Publishing Co., Inc., 1966.

6. King, C. D. G., *Nuclear Power Systems*. New York: Macmillan Publishing Co., Inc., 1964.

7. Guinn, V. P., "Activation Analysis," *Industrial Research* **6**, No. 10 (October 1964), pp. 30–36.

8. *Reactor Physics Constants*, 2d ed., ANL 5800, July 1962.

9. John, J., H. R. Lukens, and H. L. Schlesinger, "Trace Analysis—the Nuclear Way," *Industrial Research* **13**, No. 9 (September 1971), pp. 49–51.

10. Ray, J. W., "Tritium in Power Reactors," *Reactor and Fuel-Reprocessing Technology* **12**, No. 1 (Winter 1968–1969), pp. 19–26.

11. Briggs, R. B., "Tritium in Molten Salt Reactors," *Reactor Technology* **14**, No. 4 (Winter 1971–1972), pp. 335–342.

12. Onega, R. J., *An Introduction to Fission Reactor Theory*. Blacksburg, Va.: University Publications, 1975.

13. Lamarsh, J. R., *Introduction to Nuclear Engineering*. Reading, Mass.: Addison-Wesley Publishing Co., Inc., 1975.

14. Duffey, D., J. P. Balogna, and P. F. Wiggins, "Analysis of Geothermal Power Plant Water Using Gamma Rays from Capture of Californium-252 Neutrons," *Nuclear Technology* **27**, No. 2 (November 1975), pp. 488–499.

15. Rao, K. S., and M. Srinivasan, "BeH$_2$ as a Moderator in Minimum Critical Mass Systems," *Nuclear Technology* **49**, No. 2 (July 1980), pp. 315–320.

The Steady-State Reactor Core

In a thermal reactor core fast neutrons are born of fission. They slow to thermal energies by collisions with moderator nuclei. Some are then absorbed by fissionable nuclei with the subsequent fissions producing a new generation of neutrons. The ratio of neutrons in the new generation to the number in the previous generation is called the *multiplication factor.* In a core of finite size there is a probability of neutron leakage and a probability that absorption will occur before leakage. The sum of these probabilities is unity; the neutrons either leak out or they do not leak out.

Infinite Multiplication Factor

In a core of infinite extent there can be no leakage. For such a core the infinite multiplication factor, k_∞, is the ratio of the number of neutrons, n', in the current generation to the number in the previous generation, n.

$$k_\infty = \frac{n'}{n} \tag{9.1}$$

In an actual core it is necessary to study the diffusion of neutrons from the center toward the physical boundaries, where they may leak out and be lost for subsequent fissions. The effective multiplication factor is the product of the nonleakage probability, P_{NL}, and the infinite multiplication factor.

$$k_{\text{eff}} = k_\infty P_{NL} \tag{9.2}$$

The nonleakage probabilities can be determined only after the diffusion of neutrons and the critical size of the core are examined later in this chapter.

For a reactor to be critical the effective multiplication factor must be unity. Thus, there is a constant number of neutrons in each generation and the fission

energy is released at a constant rate. When k_{eff} is greater than unity, the reactor is said to be supercritical and the power level will rise exponentially. Great care must be exercised that the rate of increase be kept within reasonable limits. When k_{eff} is less than 1, the reactor is subcritical and there will be a decrease in neutron population and power.

For a real core to be critical ($k_{eff} = 1$), k_∞ must be larger than 1 at the ambient temperature prior to startup to allow for:

(1) Leakage of neutrons.
(2) Buildup of fission fragments, some of which have very significant absorption cross sections.
(3) Consumption of fissionable nuclei. This may be partially offset by conversion of ^{232}Th to ^{233}U, ^{238}U to ^{239}Pu, and so on. (In a true breeder reactor there will be a net gain in fissionable nuclei. This is a goal toward which the nuclear industry is striving).
(4) Changes in temperature and pressure in the core.

Four-Factor Equation

The infinite multiplication factor can be evaluated as the product of:

(1) The fast fission factor, ε
(2) The resonance escape probability, p
(3) The thermal utilization factor, f
(4) The reproduction factor, η

$$k_\infty = \varepsilon p f \eta \qquad (9.3)$$

Taking an infinite core with n fast neutrons that have been produced by thermal neutrons absorbed in fissionable fuel, let us examine the life history of these neutrons.

First, their number can be increased slightly, due to fast fission. Not only can the high-energy neutrons cause fission in the fissionable isotopes, but the fertile nuclei, ^{232}Th and ^{238}U, have a small cross section for fission above a threshold energy of about 1 MeV. In a homogeneous core, there is little probability of much fast fission as there are apt to be several collisions with the more numerous moderator nuclei before a neutron will collide with a fuel nucleus. In a heterogeneous core where the fuel is in sizable chunks (rods, pins, plates, pellets, etc.), the fast neutrons must travel through fuel for some distance before contacting the moderator. Here there may be enough fast collisions to increase the neutron population by several percent. The fast fission factor, ε, is defined as the ratio of the total number of fast neutrons to the number of fast neutrons induced by thermal fission. The total number of fast neutrons is then

$$n\varepsilon = \text{total fast neutrons}$$

These fast neutrons scatter and slow down. Absorption cross sections are small in the fast region and absorption can be ignored until neutrons reach epithermal energies where large resonances exist for both ^{238}U and ^{232}Th. These fertile isotopes, if present, will prevent many neutrons from ever reaching the thermal region. The *resonance escape probability* is the ratio of the number of neutrons thermalized to the total fast neutrons. It expresses the probability that a neutron will escape resonant capture and will reach thermal. The number of thermal neutrons available for fission is now

$$n\varepsilon p = \text{number of thermal neutrons}$$

A method of evaluating the resonance escape probability will be considered after the discussion of the four-factor equation is completed.

Not all thermal neutrons are absorbed in fuel. The *thermal utilization factor* is the ratio of the number of neutrons absorbed in the fuel to the total number of absorptions in fuel, moderator, cladding, and so on.

$$f = \frac{\Sigma_{abs}^{fuel} \phi_{fuel}}{\Sigma_{abs}^{fuel} \phi_{fuel} + \Sigma_{abs}^{mod} \phi_{mod} + \Sigma_{abs}^{clad} \phi_{clad}} \tag{9.4}$$

In a heterogeneous core there can be considerable difference in the flux in the fuel, the flux with the moderator, and the flux in the cladding. If the difference is small or nonexistent, as in a homogeneous core, the fluxes cancel and

$$f = \frac{\Sigma_{abs}^{f}}{\Sigma_{abs}^{f} + \Sigma_{abs}^{m} + \Sigma_{abs}^{cl}} = \left(\frac{\sigma_a^f}{\sigma_a^f + \dfrac{N_m \sigma_a^m}{N_f} + \dfrac{N_{cl} \sigma_a^{cl}}{N_f}} \right) \tag{9.5}$$

Thus, the number of thermal neutrons absorbed in the fuel becomes

$$n\varepsilon pf = \text{thermal neutrons absorbed in fuel}$$

The *reproduction factor*, η, is the number of fast neutrons produced per thermal neutron absorbed in the fuel. It is determined by multiplying the ratio of fission absorptions in the fuel to the total absorptions in the fuel by v, the number of fast neutrons emitted per fission. Originally, η was known as the thermal fission factor, when the primary interest was on thermal reactors. However, with the development of fast reactors η is of great interest for energies from 0.1 to 2 MeV. Thus, the reproduction factor (Reference 15) seems more appropriate, as it is a satisfactory term regardless of the energy level involved.

$$\eta = \frac{\Sigma_{fis}^{f}}{\Sigma_{abs}^{f}} v \tag{9.6}$$

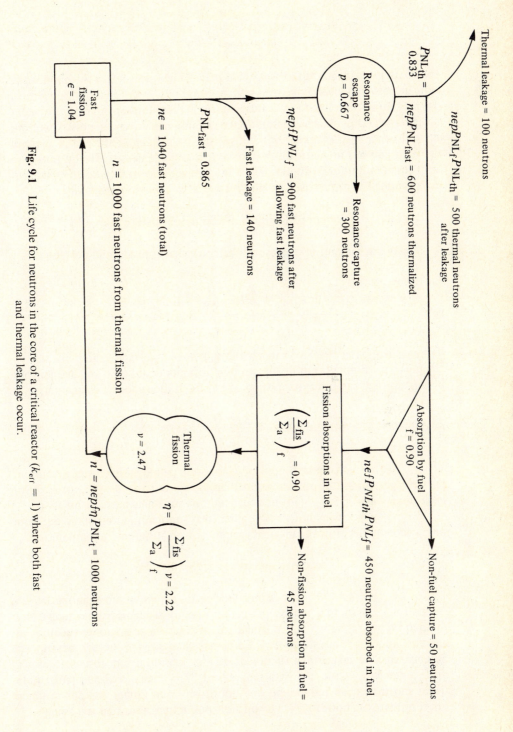

Fig. 9.1 Life cycle for neutrons in the core of a critical reactor ($k_{eff} = 1$) where both fast and thermal leakage occur.

The total number of fast neutrons in the next generation is then

$$n' = n\varepsilon p f \eta = n k_\infty \qquad \text{(9.7a)}$$

so that

$$k_\infty = \frac{n'}{n} = \varepsilon p f \eta \qquad \text{(9.7b)}$$

Checking the units on k_∞:

$$k_\infty = \qquad \varepsilon \qquad \times \qquad p \qquad \times \qquad f \qquad \times \qquad \eta$$

| $\dfrac{\text{fast } n}{\substack{\text{fast } n \text{ from} \\ \text{thermal fission}}}$ | $\dfrac{\text{thermal } n}{\text{fast } n}$ | $\dfrac{\text{thermal } n \text{ abs.}}{\substack{\text{in fuel} \\ \text{thermal } n}}$ | $\dfrac{\substack{\text{fast } n \text{ from} \\ \text{thermal fission}}}{\substack{\text{thermal } n \text{ absorbed} \\ \text{in fuel}}}$ |

$$\text{(9.7c)}$$

Figure 9.1 shows the life cycle for neutrons in a finite critical core. Leakage and absorption just balance the production of neutrons by both fast and thermal fission with the result that each successive generation has the same number of neutrons.

There is a question as to what should be considered as the fuel. We could say it is either just the thermally fissile atoms, the heavy metal atoms, or the fuel material (the heavy metal atoms plus any alloy content: oxygen, carbon, zirconium, etc.). The first could really be only applied to thermal reactors, but the latter two are independent of reactor type (i.e., fast or thermal). It is most common to consider the fuel as the heavy-metal atoms. However, if one is consistent in what is chosen for the numerator of f and the denominator of η, the value of Σ_a^f cancels when they are multiplied together to obtain k_∞.

Calculation of Resonance Escape Probability

In considering the resonance escape probability observe Fig. 9.2, which shows the absorption cross section for ^{238}U. Radiative capture cross sections reach about 7×10^3 barns at 6.7 eV, 5.4×10^3 barns at 21 eV, and 4.3×10^3 barns at 37 eV. Also, fertile ^{232}Th shows two close peaks of 500 and 700 barns between 20 and 25 eV and two lesser peaks between 50 and 75 eV. Therefore, for both of these resonance absorbers the absorption cross sections are strongly energy dependent. One must consider the neutron slowing-down density, $q(E)$, which is the number of neutrons per cm^3 s slowing down past a given energy level, E.

In Fig. 9.3 it can be seen that the slowing-down density at the lower end of the energy interval, $q(E)$, is less than the value at the upper end, $q(E + \Delta E)$, by the number of neutrons which have been absorbed (leakage being neglected).

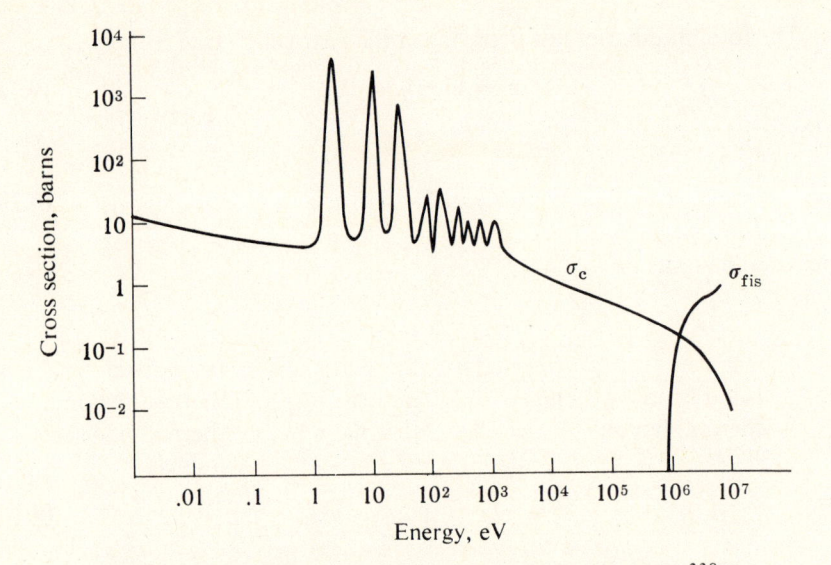

Fig. 9.2   Absorption (capture and fission) cross sections for ^{238}U.

The probability of absorption, given a collision, is

$$P_{abs} = \frac{\Sigma_a}{\Sigma_a + \Sigma_s} \tag{9.8}$$

The number of scattering collisions per neutron in the energy interval ΔE is

$$n = \frac{\Delta \ln E}{\xi} = \frac{\Delta E}{\xi E} \tag{9.9}$$

The total number of neutrons scattering into the energy interval will be

$$\frac{\Delta E}{\xi E} q(E)$$

Neutrons will leave the energy interval by scattering and absorption equivalent to

$$\Sigma_a \phi(E) \Delta E + \Sigma_s \phi(E) \Delta E$$

Under steady-state conditions the number of neutrons entering equals the number leaving:

$$q(E) \frac{\Delta E}{\xi E} = (\Sigma_a + \Sigma_s)\phi(E) \Delta E \tag{9.10}$$

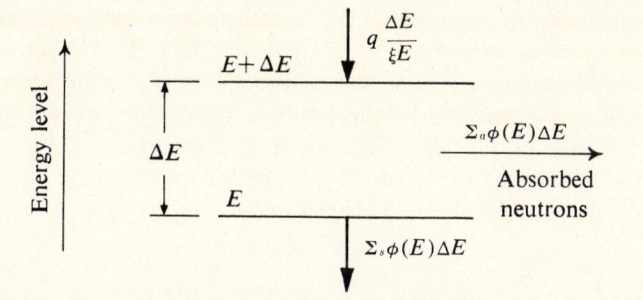

Fig. 9.3 Neutron slowing-down density with absorption as energy drops from $E + \Delta E$ to E.

The neutron flux in an interval is

$$\phi(E) = \frac{q(E)}{\xi E(\Sigma_a + \Sigma_s)} \tag{9.10a}$$

Since some neutrons are absorbed while slowing down in ΔE, the slowing-down density, q, will change by an amount equal to the number of neutrons absorbed

$$\Delta q = \Sigma_a \phi(E) \, \Delta E \tag{9.10b}$$

Substituting Eq. (9.10a) into Eq. (9.10b) and expressing the result in differential form,

$$\frac{dq}{q} = \frac{\Sigma_a}{\Sigma_a + \Sigma_s} \frac{dE}{\xi E} \tag{9.11}$$

Equation (9.11) is then integrated from the average fast-neutron slowing-down density at birth (q_0) and average energy at birth (E_0) to the values after thermalization, q_{th} and E_{th}.

$$\int_{q_0}^{q_{th}} \frac{dq}{q} = \int_{E_0}^{E_{th}} \frac{\Sigma_a}{\Sigma_a + \Sigma_s} \frac{dE}{\xi E} \tag{9.12}$$

$$\ln \frac{q_{th}}{q_0} = \int_{E_0}^{E_{th}} \frac{\Sigma_a}{\Sigma_a + \Sigma_s} \frac{dE}{\xi E} = -\int_{E_{th}}^{E_0} \frac{\Sigma_a}{\Sigma_a + \Sigma_s} \frac{dE}{\xi E} \tag{9.13}$$

Taking antilogs of both sides and realizing that the ratio of thermal slowing-down density to the initial fast slowing-down density is essentially the definition of resonance escape probability:

$$p = q_{th}/q_0 = \exp\left(-\int_{E_{th}}^{E_0} \frac{\Sigma_a}{\Sigma_a + \Sigma_s} \frac{dE}{\xi E}\right) \tag{9.14}$$

Through the epithermal region the scattering cross sections are essentially constant and are primarily contributed by the moderator. On the other hand, absorption cross sections vary wildly in the epithermal region, the principal resonances being contributed by the fertile resonance absorber atoms (^{232}Th or ^{238}U) that may be present. For the present discussion assume that only ^{238}U is present and that σ_a^{238} and N_{238} represent the absorption cross section and atom density for this resonance absorber.

$$p = \exp\left(-\int_{E_{th}}^{E_0} \frac{N_{238}\sigma_a^{238}}{N_{238}\sigma_a^{238} + \Sigma_s} \frac{dE}{\xi E}\right) = \exp\left(-\frac{N_{238}}{\xi\Sigma_s} \int_{E_{th}}^{E_0} \frac{\sigma_a^{238}\Sigma_s}{N_{238}\sigma_a^{238} + \Sigma_s} \frac{dE}{E}\right)$$

$$= \exp\left[-\frac{N_{238}}{\xi\Sigma_a} \int_{E_{th}}^{E_0} (\sigma_a^{238})_{\text{eff}} \frac{dE}{E}\right] \tag{9.15}$$

where $(\sigma_a^{238})_{\text{eff}}$, the effective resonance absorber cross section, is defined as

$$(\sigma_a^{238})_{\text{eff}} = \frac{\sigma_a^{238}\Sigma_s}{N_{238}\sigma_a^{238} + \Sigma_s} = \frac{\sigma_a^{238}}{\dfrac{N_{238}\sigma_a^{238}}{\Sigma_s} + 1} \tag{9.16}$$

The effective resonance integral is given as

$$I_{\text{eff}} = \int_{E_{th}}^{E_0} (\sigma_a^{238})_{\text{eff}} \frac{dE}{E} \tag{9.17}$$

making

$$p = \exp\left(-\frac{N_{238}}{\xi\Sigma_s} I_{\text{eff}}\right) \tag{9.18}$$

When the effective resonance integral for a homogeneous reactor is laboriously evaluated by mechanical integration through the resonance region, it can be represented for either ^{232}Th or ^{238}U by the equation

$$I_{\text{eff}} = 3.9\left(\frac{\Sigma_s}{N_{238}}\right)^{0.415} \tag{9.19}$$

Note that Σ_s/N_{238} must be expressed in barns and the effective resonance integral has the units of barns. Observe that I_{eff} is a function only of the scattering cross section per resonance absorber atom (Σ_s/N_{238}).

Equation (9.19) is satisfactory for ratios of Σ_s/N_{238} less than 1000 barns. The effective resonance integral for ^{238}U runs from 9.25 barns for pure metal to an upper limit of 240 barns for an infinitely dilute mixture of ^{238}U in moderator. For thorium the values run between 11.1 barns for pure metal and 69.8 barns for the infinitely dilute mixture.

Example 1 Compute the infinite multiplication factor for a homogeneous mixture of 200 moles of graphite per mole of 5 percent enriched uranium. From the mole ratio and the given enrichment

$$\frac{N_{238}}{N_{235}} = \frac{0.95}{0.05} = 19$$

$$\frac{N_C}{N_{235}} = \frac{200 \text{ moles C/mole U}}{0.05 \text{ mole } ^{235}\text{U/mole U}} = 4000 \frac{\text{mole C}}{\text{mole } ^{235}\text{U}}$$

For a homogeneous mixture $\varepsilon = 1$,

$$I_{\text{eff}} = 3.9\left(\frac{\Sigma_s}{N_{238}}\right)^{0.415} = 3.9\left(\frac{\dfrac{N_U}{N_{235}}\sigma_s^u + \dfrac{N_C}{N_{235}}\sigma_s^C}{N_{238}/N^{235}}\right)^{0.415}$$

$$= 3.9\left(\frac{20 \times 8.3 + 4000 \times 4.66}{19}\right)^{0.415}$$

$$= 3.9(978)^{0.415} = 68 \text{ barns}$$

$$\xi_C = \frac{2}{A + \frac{2}{3}} = \frac{2}{12.67} = 0.158 \qquad \xi_U = \frac{2}{238.6} = 0.0084$$

$$\xi_{\text{av}} = \frac{N_U \sigma_s^U \xi_U + N_C \sigma_s^C \xi_C}{N_U \sigma_s^U + N_C \sigma_s^C}$$

$$= \frac{20 \times 8.3 \times 0.0084 + 4000 \times 4.66 \times 0.158}{20 \times 8.3 + 4000 \times 4.66} \approx 0.158$$

$$p = \exp\left(-\frac{N_{238}}{\xi\Sigma_s} I_{\text{eff}}\right) = \exp\left[-\frac{1}{0.158(978)} \times 68\right]$$

$$= e^{-0.442} = 0.641$$

$$f = \frac{\sigma_a^{235} + (N_{238}/N_{235})\sigma_a^{238}}{\sigma_a^{235} + \dfrac{N_{238}}{N_{235}}\sigma_a^{238} + \dfrac{N_C}{N_{235}}\sigma_a^C}$$

$$= \frac{694 + 19(2.71)}{694 + 19(2.71) + 4000(0.0034)} = 0.985$$

$$\eta = \frac{\Sigma_f^{235}}{\Sigma_a^U} v = \frac{\sigma_f^{235} v}{\sigma_a^{235} + \dfrac{N_{238}}{N_{235}}\sigma_a^{238}}$$

$$= \frac{582 \times 2.43}{694 + 19(2.71)} = 1.898$$

$$k_\infty = \varepsilon p f \eta = 1.0 \times 0.641 \times 0.985 \times 1.898 = 1.198$$

Note that in calculating f and η the fuel was considered as being the fissionable ^{235}U and the fertile ^{238}U. One could use only the fissionable isotope as the fuel without affecting k_∞; however, the values of f and η would be different. Care must be taken that a similar basis is used for evaluating Σ_a^f in these two items.

Heterogeneous Cores

In Example 1 the value of k_∞, is 1.198; allowing for a modest neutron leakage the mixture of graphite and 5 percent enriched uranium should experience little difficulty in achieving criticality. However, in Problem 3 the same calculation with natural uranium as a fuel will result in $k_\infty = 0.778$. The homogeneous mixture with 200 moles of graphite per mole of natural uranium cannot achieve criticality even in an infinite core. Why then could CP-1, the original reactor constructed by Enrico Fermi and his associates at the University of Chicago, go critical? Its fuel was natural uranium and its moderator was graphite.

The answer is that it was a heterogeneous assembly of graphite blocks and lumps of natural uranium. In a heterogeneous reactor where the neutrons must travel through a significant amount of fuel before entering the moderator there may be a gain in the total fast neutron population of several percent, as indicated by the fast fission factor, ε. An even more important improvement occurs in the resonance escape probability, p. Fast neutrons are born in the fuel but are mainly slowed down in the moderator. After thermalization by the moderator they must diffuse back into the fuel elements. Any epithermal neutrons diffusing into the fuel are quickly absorbed by the large ^{238}U resonances in the outer layer of fuel. This effect is called *self-shielding* and it permits the interior of the fuel to see few, if any, epithermal neutrons. The result is a marked improvement in the resonance escape probability since only a small fraction of the fuel volume is involved in resonance capture.

The thermal utilization factor is not as good in a heterogeneous core as it is in a homogeneous one. If Eq. (9.4) is rearranged

$$f = \frac{\Sigma_a^f}{\Sigma_a^f + \dfrac{\phi_m}{\phi_f}\Sigma_a^m + \dfrac{\phi_{cl}}{\phi_f}\Sigma_a^{cl}} \tag{9.20}$$

it can be seen that if the flux in the moderator and clad are larger than the average value in the fuel the thermal utilization factor will suffer a reduction. The ratios ϕ_m/ϕ_f and ϕ_{cl}/ϕ_f are known as the *thermal disadvantage factors* for the moderator and clad, respectively. Figure 9.4 shows the variation of fast and thermal flux in a heterogeneous core. Fast neutrons are born in the fuel

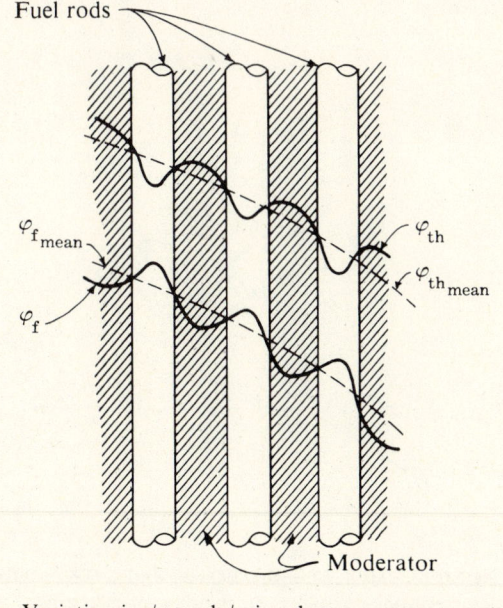

Fig. 9.4 Variation in ϕ_{th} and ϕ_f in a heterogeneous reactor lattice.

and leak into the moderator, where they are lost to the thermal group by slowing-down collisions. The reverse is true of the thermal flux, where the slow neutrons are born in the moderator as their fast ancestors lose energy. The fuel elements act as a strong sink or absorber of the thermal neutrons. Thus, the thermal flux dips in the fuel and peaks in the moderator.

As power levels have increased in reactors, it has become necessary to use smaller-diameter rods or pins or thinner-plate-type elements to prevent excessive temperatures in the fuel. The smaller each individual element, the closer a heterogeneous core approaches the homogeneous core. The result is that the small dips in flux can be averaged out and the core can be treated as though it were homogeneous without serious error in many instances.

Neutron Current Density

As neutrons diffuse through matter, if the scattering is isotropic, "simple" diffusion theory will describe their travels. For a heavier moderator such as graphite, the deviation from isotropic scatter is not serious, as shown in Chapter 8, where for graphite $\overline{\cos \psi} = 0.056$. Later in this chapter a correction will be introduced to allow for the anisotropy of the scatter, which is especially necessary with lighter moderators. Further, the "simple" diffusion theory

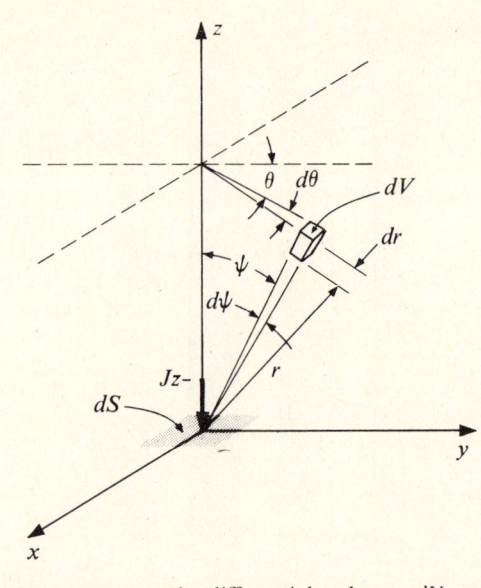

Fig. 9.5 Neutron scatter in differential volume, dV, contributes a differential current density, dJ_{z-}, to the flow of neutrons through the area, dS.

assumes that a monoenergetic group of neutrons be considered where the velocity is the average value for the group. Since scattering cross sections do not vary strongly with neutron energy, an average cross section for scatter can be used.

Neutron current density is the number of neutrons per second crossing a unit area normal to the direction of neutron flow.

Figure 9.5 shows a differential volume, dV, from which neutrons will scatter through the area dS downward, contributing a differential current flow in the z direction, dJ_{z-}. The net current flow in the z direction is found by subtracting the downward current flow in the z direction, J_{z-}, from the upward flow from the lower hemisphere, J_{z+}. Thus,

$$J_z = J_{z+} - J_{z-} \tag{9.21}$$

The elemental volume in terms of spherical coordinates is

$$dV = r \, d\psi \, r \sin \psi \, d\theta \, dr = r^2 \sin \psi \, d\theta \, d\psi \, dr \tag{9.22}$$

The neutron flux is position dependent and at any position of the differential volume, dV, the number of scatterings per unit time within the elemental volume

will be $\Sigma_s \phi\, dV$. The effective surface area as seen from dV is $\cos \psi\, dS$. The fraction of neutrons scattered through the effective area, dS, is $\cos \psi\, dS/4\pi r^2$, assuming that there are no interactions as the neutrons travel from dV to dS. Interactions between these two locations result in an attenuation by a factor of $e^{-\Sigma r}$. If the medium is assumed to be only weakly absorbing, then $\Sigma \approx \Sigma_s$ and the neutron flow through dS in the z direction becomes

$$dJ_{z-}\, dS = \Sigma_s \phi\, dV\, \frac{\cos \psi\, dS}{4\pi r^2}\, e^{-\Sigma_s r} \tag{9.23}$$

Rearranging, substituting Eq. (9.22) for dV, and integrating over the entire upper hemisphere gives

$$J_{z-} = \frac{\Sigma_s}{4\pi} \int_0^{2\pi} \int_0^{\pi/2} \int_0^{\infty} \phi\, e^{-\Sigma_s r}\,(\cos \psi \sin \psi\, d\theta\, d\psi\, dr) \tag{9.24}$$

The flux is a function of position:

$$\phi = f(x, y, z) = f'(r, \theta, \psi) \tag{9.25}$$

ϕ may be expanded in a Maclaurin series in terms of the flux and its derivatives at the origin:

$$\phi(x, y, z) = \phi_0 + x\left(\frac{\partial \phi}{\partial x}\right)_0 + y\left(\frac{\partial \phi}{\partial y}\right)_0 + z\left(\frac{\partial \phi}{\partial z}\right)_0$$

$$+ \frac{x^2}{2}\left(\frac{\partial^2 \phi}{\partial x^2}\right)_0 + \frac{y^2}{2}\left(\frac{\partial^2 \phi}{\partial y^2}\right)_0 + \frac{z^2}{2}\left(\frac{\partial^2 \phi}{\partial z^2}\right)_0 + \cdots \tag{9.26}$$

Fortunately, any terms resulting from the second-order partial derivatives cancel and the results would be exactly the same if only the first-order terms were considered in Eq. (9.26). Since the flux gradient in the x or y directions makes no contribution to the neutron flow in the z direction, Eq. (9.24) becomes

$$J_{z-} = \frac{\Sigma_s}{4\pi} \phi_0 \int_0^{2\pi} \int_0^{\pi/2} \int_0^{\infty} e^{-\Sigma_s r} \cos \psi \sin \psi\, d\theta\, d\psi\, dr$$

$$+ \frac{\Sigma_s}{4\pi}\left(\frac{\partial \phi}{\partial z}\right)_0 \int_0^{2\pi} \int_0^{\pi/2} \int_0^{\infty} \underbrace{r \cos \psi}_{z}\, e^{-\Sigma_s r} \cos \psi \sin \psi\, d\theta\, d\psi\, dr \tag{9.27}$$

Integrating, we get

$$J_{z-} = \frac{\Sigma_s \phi_0}{4\pi} \left[\frac{-e^{-\Sigma_s r}}{\Sigma_s} \right]_0^\infty \frac{1}{2} \sin^2 \psi \Big]_0^{\pi/2} \theta \Big]_0^{2\pi}$$

$$+ \frac{\Sigma_s}{4\pi} \left(\frac{\partial \phi}{\partial z} \right)_0 \left[\frac{e^{-\Sigma_s r}}{\Sigma_s^2} (-\Sigma_s r - 1) \right]_0^\infty - \frac{\cos^3 \psi}{3} \Big]_0^{\pi/2} \theta \Big]_0^{2\pi} \quad \text{(9.28)}$$

Substituting limits yields

$$J_{z-} = \frac{\Sigma_s \phi_0}{4\pi} \left(\frac{1}{\Sigma_s} \right) \frac{1}{2} (2\pi) + \frac{\Sigma_s}{4\pi} \left(\frac{\partial \phi}{\partial z} \right)_0 \frac{1}{\Sigma_s^2} \left(\frac{1}{3} \right) (2\pi)$$

$$= \frac{\phi_0}{4} + \frac{1}{6\Sigma_s} \left(\frac{\partial \phi}{\partial z} \right)_0 \quad \text{(9.29)}$$

The upward current flow through ds from the lower hemisphere is found by a similar integration with the limits ψ going from π to $\pi/2$. This results in

$$J_{z+} = \frac{\phi_0}{4} - \frac{1}{6\Sigma_s} \left(\frac{\partial \phi}{\partial z} \right)_0 \quad \text{(9.30)}$$

Taking the difference in (9.30) and (9.29) results in the net current flow in the z direction,

$$J_z = -\frac{1}{3\Sigma_s} \left(\frac{\partial \phi}{\partial z} \right)_0 \quad \text{(9.31)}$$

In a similar manner the component of current in the x and y directions can be obtained. When the three are added vectorially,

$$J = -\left(\frac{1}{3\Sigma_s} \right) \left[\frac{\partial \phi}{\partial x} i + \frac{\partial \phi}{\partial y} j + \frac{\partial \phi}{\partial z} k \right] = -D \, (\text{grad } \phi) \quad \text{(9.32)}$$

This is known as *Fick's law of diffusion.* It indicates that the neutron current will flow in a direction opposite to a positive gradient of the flux. It will be in proportion to the diffusion constant, D, and current will be zero when flux is maximum.

The diffusion constant is corrected for anisotropic scatter by using the transport mean free path or the macroscopic transport cross section in place of the scattering values.

$$D = \frac{1}{3\Sigma_{tr}} = \frac{1}{3\Sigma_s(1 - \overline{\cos \psi})} = \frac{\lambda_{tr}}{3} \quad \text{(9.33)}$$

Development of Diffusion Equation

As neutrons diffuse through a reactor core they may (a) be absorbed by fuel, moderator, coolant, cladding, structure, and so on; (b) leak out at the core boundaries; or (c) act as a source for new fission neutrons. If one considers the neutrons in a differential volume, dV, an expression can be developed with the aid of Fig. 9.6 for the net neutron leakage from this elemental volume. First consider only neutrons leaking into the front face in the y direction.

$$L_y = J_y \, dx \, dz = -D\left(\frac{\partial \phi}{\partial y}\right) dx \, dz \tag{9.34}$$

From the rear face the leakage is

$$L_{y+dy} = J_{(y+dy)} \, dx \, dz = -D\left[\frac{\partial \phi}{\partial y} + \frac{\partial(\partial \phi/\partial y)}{\partial y} \, dy\right] dx \, dz \tag{9.35}$$

The net leakage in the y direction is the difference between Eqs. (9.35) and (9.34).

$$L_{y_{net}} = L_{y+dy} - L_y = -D\left(\frac{\partial \phi}{\partial y} + \frac{\partial^2 \phi}{\partial y^2} \, dy\right) dx \, dz$$

$$+ D\frac{d\phi}{\partial y} \, dx \, dz = -D\frac{\partial^2 \phi}{\partial y^2} \, dx \, dy \, dz \tag{9.36}$$

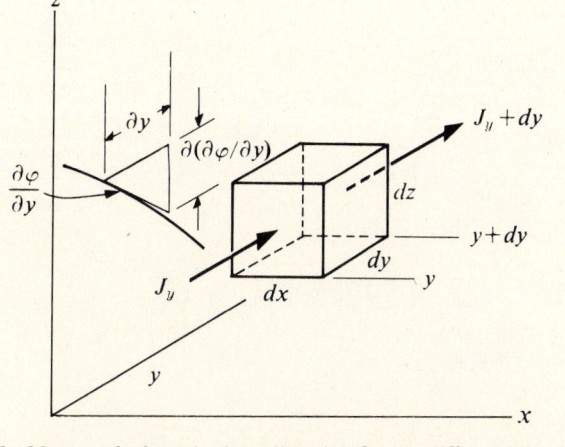

Fig. 9.6. Neutron leakage in the y direction from a differential volume.

Similarly,

$$L_{x_{net}} = -D \frac{\partial^2 \phi}{\partial x^2} \, dx \, dy \, dz \qquad (9.37)$$

$$L_{z_{net}} = -D \frac{\partial^2 \phi}{\partial z^2} \, dx \, dy \, dz \qquad (9.38)$$

The total leakage for a unit volume (neutrons/cm^3 s) is

$$L_t = -D\left(\frac{\partial^2 \phi}{\partial x^2} + \frac{\partial^2 \phi}{\partial y^2} + \frac{\partial^2 \phi}{\partial z^2}\right) = -D\nabla^2 \phi \qquad (9.39)$$

The expression for the neutrons absorbed in a unit volume per second is

$$\text{no. abs.} = \phi \Sigma_a \qquad (9.40)$$

If fuel is present in the volume being considered, there will be a source of neutrons due to fission. For thermal neutrons, k_∞ new thermal neutrons will appear for each neutron absorbed and thus the source strength is

$$S_{th} = k_\infty \phi \Sigma_a \qquad (9.41)$$

The rate of change of neutron population density is equal to the rate of production, less the leakage, less the absorption of neutrons:

$$\frac{\partial n}{\partial t} = k_\infty \Sigma_a \phi - (-D\nabla^2 \phi) - \Sigma_a \phi \qquad (9.42)$$

For a steady-state core,

$$\frac{\partial n}{\partial t} = 0 \qquad (9.43)$$

and, therefore, under these conditions

$$0 = D\nabla^2 \phi + (k_\infty - 1)\Sigma_a \phi \qquad (9.44)$$

Dividing by Σ_a, we get

$$0 = \frac{D}{\Sigma_a} \nabla^2 \phi + (k_\infty - 1)\phi \qquad (9.45)$$

The ratio D/Σ_a is called the *square of the thermal diffusion length, L^2*. Therefore,

$$L = \sqrt{D/\Sigma_a} = \sqrt{\lambda_{tr}\lambda_a/3} \tag{9.46}$$

The diffusion length can be thought of as more or less a "representative average" distance for interaction as a neutron diffuses.

Substituting the diffusion length into the steady-state diffusion equation and rearranging yields

$$0 = \nabla^2\phi + \left(\frac{k_\infty - 1}{L^2}\right)\phi \tag{9.47a}$$

$$0 = \nabla^2\phi + B^2\phi \tag{9.47b}$$

B is called the *buckling* of a reactor since this second-order partial differential equation is analogous to the one that describes the buckling of a column. In one-group theory only the leakage of thermal neutrons from the core will be considered. Later two-group theory will be developed where the leakage of fast neutrons must also be taken into account.

$$B^2 = \frac{k_\infty - 1}{L^2} \tag{9.48a}$$

This is sometimes called the *material buckling* for a core lattice. Its magnitude bears an inverse relation to the size that a core must have to be critical. That is, the overall neutron production must just balance the absorption and leakage during steady-state operation. After the next example the diffusion equation will be solved for several simple geometries to illustrate this point.

When Eq. (9.48a) is rearranged,

$$1 = k_\infty\left(\frac{1}{B^2L^2 + 1}\right) \tag{9.48b}$$

and Eq. (9.48b) is compared with Eq. (9.2), it is seen that for the steady-state critical reactor ($k_{eff} = 1$) the bracket $[1/(B^2L^2 + 1)]$ represents the nonleakage probability for the thermal group of neutrons.

$$P_{NL_{th}} = \frac{1}{B^2L^2 + 1} \tag{9.48c}$$

Example 2   Determine the material buckling for the mixture of 200 moles of graphite per mole of 5 percent enriched uranium.*   The core temperature is 20°C. For a critical core of these materials, what will be the thermal nonleakage probability?

$$\rho_G = 1.6 \text{ g C/cm}^3 \qquad \rho_U = 18.9 \text{ g U/cm}^3$$

$$\text{vol. U} = \frac{238 \text{ g/mole}}{18.9 \text{ g/cm}^3} = 12.6 \text{ cm}^3/\text{mole U}$$

$$\text{vol. graphite} = \frac{200 \text{ moles C/mole U} \times 12 \text{ g C/mole C}}{1.60 \text{ g C/cm}^3}$$

$$= 1500 \text{ cm}^3/\text{mole U}$$

$$\text{total vol.} = 1500 + 12.6 = 1512.6 \text{ cm}^3 \text{ mixture/mole U}$$

$$N_U = \frac{6.024 \times 10^{23} \text{ atoms U/g mole U}}{1512.6 \text{ cm}^3/\text{g mole U}}$$

$$= 3.98 \times 10^{20} \text{ atoms U/cm}^3 \text{ mixture}$$

$$N_{235} = 0.05 \times 3.98 \times 10^{20} = 0.199 \times 10^{20} \text{ atoms } ^{235}\text{U/cm}^3 \text{ mixture}$$

$$N_{238} = 0.95 \times 3.98 \times 10^{20} = 3.78 \times 10^{20} \text{ atoms } ^{238}\text{U/cm}^3 \text{ mixture}$$

$$N_C = 200 \text{ atoms C/atom U} \times 3.98 \times 10^{20} \text{ atoms U/cm}^3$$

$$= 7.96 \times 10^{22} \text{ atoms C/cm}^3 \text{ mixture}$$

$$(\Sigma_{\text{tr}})_{\text{th}} = N_C \sigma_{s_{\text{th}}}^C (1 - \overline{\cos \psi_C}) + N_U \sigma_{s_{\text{th}}}^U (1 - \overline{\cos \psi_U})$$

$$= 7.96 \times 10^{22} \times 4.8 \times 10^{-24} \left[1 - \frac{2}{3(12)} \right]$$

$$+ 3.98 \times 10^{20} \times 8.3 \times 10^{-24} \left[1 - \frac{2}{3(238)} \right]$$

$$= 0.361 + 0.0033 = 0.364 \text{ cm}^2/\text{cm}^3 \text{ mixture}$$

$$\Sigma_a = [N_{235} \sigma_a^{235} + N_{238} \sigma_a^{238} + N_C \sigma_a^C] \frac{\sqrt{\pi}}{2} \sqrt{\frac{293}{T}}$$

$$\Sigma_a = [0.199 \times 10^{20} \times 694 \times 10^{-24} + 3.78 \times 10^{20} \times 2.73 \times 10^{-24}$$

$$+ 7.96 \times 10^{22} \times 0.0034 \times 10^{-24}] \frac{\sqrt{\pi}}{2} \times 1$$

$$= 0.0151 \times \frac{\sqrt{\pi}}{2} = 0.0134 \text{ cm}^2/\text{cm}^3 \text{ mixture}$$

---

* *Enrichment* is the atom percent of ^{235}U in uranium when it is increased above the abundance as found in nature. Thus, 5 percent enriched uranium will have 5 percent of its atoms ^{235}U and 95 percent of them ^{238}U.

Since the core temperature is 20°C, the only correction to Σ_a is for the average velocity being larger than the most probable ($\sqrt{\pi/2}$) × 20°C.

$$L^2 = \frac{1}{3\Sigma_a\Sigma_{tr}} = \frac{1}{3(0.364)(0.0134)} = 68.2 \text{ cm}^2$$

$$B^2 = \frac{k_\infty - 1}{L^2} = \frac{1.198 - 1}{68.2} = 0.002\,90 \text{ cm}^{-2}$$

$$B = 0.0538 \text{ cm}^{-1}$$

$$P_{NL_{th}} = \frac{1}{B^2L^2 + 1} = \frac{1}{(0.002\,90)(68.2) + 1} = 0.835$$

Infinite Slab Reactor

The diffusion equation, (9.47), will first be solved for an infinite slab reactor in order to determine the slab thickness for criticality. Figure 9.7 shows the slab with the origin taken at its center. Since the slab is considered infinite there is no leakage of neutrons in the y or z direction. Leakage occurs only in the x direction through the slab faces. There will therefore be a flux gradient (and

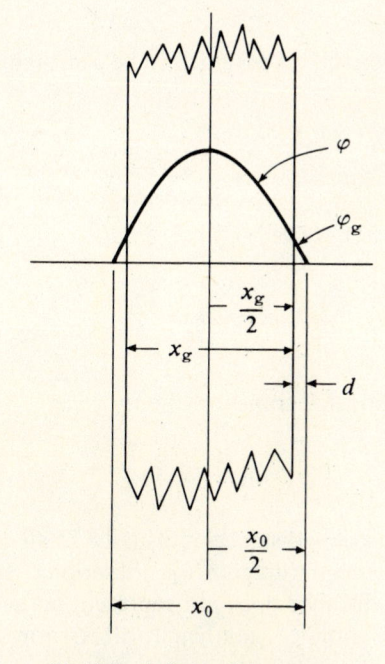

Fig. 9.7 Infinite slab reactor.

hence neutron flow) only in the x direction. The flux must fall as one approaches either face from the center in order to have a current flow toward the outside. The flux, however, is still finite at the slab faces ($x = \pm x_g/2$) and does not fall to zero until the extrapolated half-thickness is reached ($x = \pm x_0/2$).

$$\frac{x_0}{2} = \frac{x_g}{2} + d \tag{9.49}$$

d is the extrapolation distance by which the geometrical boundary of the core must be extended for the flux to drop to zero.

At the boundary of a bare core neutrons stream out into space and few, if any, are scattered back (the atmosphere is for all intents like a vacuum to the escaping neutrons). Thus, the return current J_{x-} may be set equal to zero in Eq. (9.29).

$$J_{x-} = 0 = \frac{\phi}{4} + \frac{D}{2}\frac{d\phi}{dx} \tag{9.50}$$

From this it follows that

$$\frac{d\phi}{dx} = -\frac{\phi}{2D} \tag{9.51}$$

If the extrapolation of the flux is assumed to be a straight line, the slope is also

$$\frac{d\phi}{dx} = \frac{-\phi_g}{x_0/2 - x_g/2} = -\frac{\phi_g}{d} \tag{9.52}$$

Equating (9.51) and (9.52), we get

$$d = 2D = \frac{2\lambda_{tr}}{3} \tag{9.53}$$

The use of more sophisticated transport theory gives

$$d = 0.71\lambda_{tr} \tag{9.54}$$

Since the extrapolation distance is usually small in comparison to the critical dimensions of a core, the difference is not serious. It must be remembered that the extrapolation distance does not represent an actual finite distance in the reactor, only a simplified mathematical treatment of the boundary conditions.

The flux in the infinite slab varies only in the x direction; the diffusion equation reduces to an ordinary second-order linear differential equation.

$$0 = \nabla^2 \phi + B^2 \phi = \frac{d^2\phi}{dx^2} + B^2 \phi \tag{9.55}$$

Using operator notation, we have

$$D^2 + B^2 = 0$$

$$D = \pm Bi$$

The imaginary roots for the operator give a sine and cosine solution of the form

$$\phi = A_1 \cos Bx + A_2 \sin Bx \tag{9.56}$$

The boundary conditions demand that:

(1) The flux drops to zero at the extrapolated boundaries; it is finite at the geometrical boundaries. When

$$x = \pm \frac{x_0}{2}, \qquad \phi = 0$$

(2) The flux be symmetrical and finite about the origin. When

$$x = 0, \qquad \frac{d\phi}{dx} = 0$$

Taking the second requirement first, determine the expression for the flux gradient.

$$\frac{d\phi}{dx} = -A_1 B \sin Bx + A_2 B \cos Bx \tag{9.57}$$

At the origin the slope of the flux is zero, $\sin Bx = 0$, and $\cos Bx = 1$. Therefore,

$$0 = 0 + A_2 B$$

and since B is real and positive,

$$\phi = A_1 \cos Bx \tag{9.58}$$

Applying the first condition, that at the extrapolated boundary the flux must be zero, we obtain

$$0 = A_1 \cos B \frac{x_0}{2}$$

This can occur when $Bx_0/2$ is equal to odd multiples of $\pi/2$.

$$\frac{Bx_0}{2} = \frac{\pi}{2}, \frac{3\pi}{2}, \frac{5\pi}{2}, \cdots \tag{9.59}$$

From this, various values of x_0, called *eigenvalues*, will satisfy the differential equation. Fortunately, only the first value or fundamental eigenvalue is needed to describe the flux in a minimum-size critical reactor.

$$x_0 = \frac{\pi}{B}, \frac{3\pi}{B}, \frac{5\pi}{B}, \cdots \tag{9.60}$$

$$\downarrow \quad \downarrow \quad \downarrow$$

fundamental eigenvalue $\hookrightarrow$ harmonic eigenvalues

$$\phi = A_1 \cos \frac{\pi x}{x_0}, \underbrace{A_{12} \cos \frac{3\pi x}{x_0}, A_{13} \cos \frac{5\pi x}{x_0}, \cdots}_{\text{drop for critical reactor}} \tag{9.61}$$

Retaining only the fundamental value, the flux for a steady-state critical infinite slab reactor can be written as

$$\boxed{\phi = A \cos \frac{\pi x}{x_0}} \tag{9.62}$$

The value π/x_0 is known as the geometric buckling for this reactor configuration. It will be different for other core geometries, but in each case must equal the value of the material buckling for the critical core.

Note that A is an arbitrary constant, which in this case is equal to the maximum flux at the center of the core. To raise the flux in an operating reactor, the control rods are removed sufficiently to allow the reactor to be slightly supercritical and for the flux to increase in a controlled manner. At the desired flux level the control rods are inserted just enough to stabilize the flux and return the value of k_{eff} to 1. In theory, then, a reactor can be critical at any flux level.

Example 3 For the homogeneous graphite–uranium mixture considered in Examples 1 and 2, what will be the thickness of a critical infinite slab?

$$x_0 = \frac{\pi}{B} = \frac{\pi}{0.0538} = 58.4 \text{ cm}$$

$$d = 0.71\lambda_{\text{tr}} = \frac{0.71}{\Sigma_{\text{tr}}} = \frac{0.71}{0.364} = 1.93 \text{ cm}$$

$$x_g = x_0 - 2d = 58.4 - 2(1.93) = 54.5 \text{ cm}$$

Flux Distribution in a Rectangular Parallelepiped

To extend the solution for the infinite slab to a real core where the flux varies in each of the principal directions, consider the rectangular parallelepiped, as shown in Fig. 9.8, with the origin at the center of the core. The diffusion equation is written as

$$\left(\frac{\partial^2 \phi}{\partial x^2} + \frac{\partial^2 \phi}{\partial y^2} + \frac{\partial^2 \phi}{\partial z^2} + B^2 \phi = 0 \right) \tag{9.63}$$

The equation can be solved by the separation of variables. Assume a solution which is the product of three functions, X, Y, and Z, each of which is a function of only a single variable, or

$$\phi = X(x)Y(y)Z(z) = XYZ \tag{9.64}$$

This is to say that the flux variation in the x direction is independent of that in the y and z directions. The flux variation in the y or z direction is similarly independent of variation in the other two directions. Differentiating Eq. (9.64) twice with respect to each variable yields

$$\frac{\partial \phi}{\partial x} = YZ \frac{\partial X}{\partial x} \qquad \frac{\partial^2 \phi}{\partial x^2} = YZ \frac{\partial^2 X}{\partial x^2} \tag{9.65a,b}$$

$$\frac{\partial \phi}{\partial y} = XZ \frac{\partial Y}{\partial y} \qquad \frac{\partial^2 \phi}{\partial v^2} = XZ \frac{\partial^2 Y}{\partial y^2} \tag{9.66a,b}$$

$$\frac{\partial \phi}{\partial z} = XY \frac{\partial Z}{\partial z} \qquad \frac{\partial^2 \phi}{\partial z^2} = XY \frac{\partial^2 Z}{\partial z^2} \tag{9.67a,b}$$

Substituting these second-order partial derivatives into Eq. (9.63), we get

$$YZ \frac{\partial^2 X}{\partial x^2} + XZ \frac{\partial^2 Y}{\partial y^2} + XY \frac{\partial^2 Z}{\partial z^2} + B^2 XYZ = 0 \tag{9.68}$$

Dividing by XYZ gives us

$$\frac{1}{X} \frac{\partial^2 X}{\partial x^2} + \frac{1}{Y} \frac{\partial^2 Y}{\partial y^2} + \frac{1}{Z} \frac{\partial^2 Z}{\partial z^2} + B^2 = 0 \tag{9.69}$$

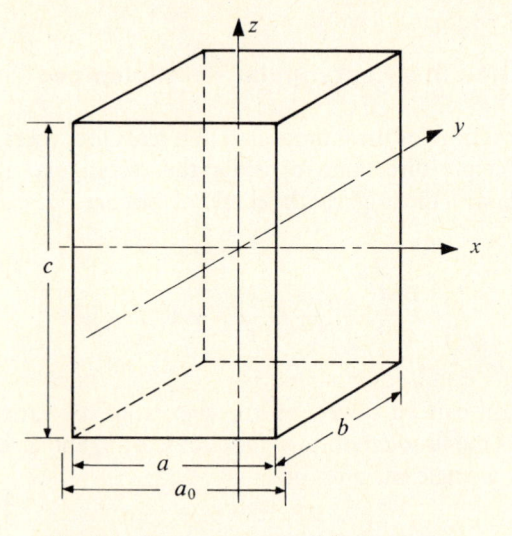

Fig. 9.8   Rectangular parallelepiped core.

Each of the first three terms in Eq. (9.69) is a function of a single variable. Therefore, each of the three must be equal to a constant.

$$\frac{1}{X}\frac{\partial^2 X}{\partial x^2} = -\alpha^2 \tag{9.70a}$$

$$\frac{1}{Y}\frac{\partial^2 Y}{\partial y^2} = -\beta^2 \tag{9.70b}$$

$$\frac{1}{Z}\frac{\partial^2 Z}{\partial z^2} = -\gamma^2 \tag{9.70c}$$

From this the square of the buckling becomes

$$B^2 = \alpha^2 + \beta^2 + \gamma^2 \tag{9.71}$$

Taking Eq. (9.70a), and using a total second derivative since it is a function of x only, the form is similar to Eq. (9.55).

$$\frac{d^2 X}{dx^2} + \alpha^2 X = 0 \tag{9.72a}$$

The solution of this equation is

$$X = A'_1 \cos \alpha x + A'_2 \sin \alpha x \qquad \text{(9.72b)}$$

When this is multiplied by YZ for particular values of y and z an expression for the flux variation in the x direction results.

The boundary conditions are:

(1) When $x = a_0/2$, $X = 0$.
(2) When $x = 0$, $dX/dx = 0$.

Applying these conditions, in a manner similar to that for the infinite slab,

$$\alpha = \frac{\pi}{a_0} \qquad \text{(9.73a)}$$

Similarly,

$$\beta = \frac{\pi}{b_0} \qquad \text{(9.73b)}$$

and

$$\gamma = \frac{\pi}{c_0} \qquad \text{(9.73c)}$$

The three functions of a single variable that are multiplied to give the flux are

$$X = A' \cos \frac{\pi}{a_0} x \qquad \text{(9.74a)}$$

$$Y = A'' \cos \frac{\pi}{b_0} y \qquad \text{(9.74b)}$$

$$Z = A''' \cos \frac{\pi}{c_0} z \qquad \text{(9.74c)}$$

The flux can be written

$$\phi = A \cos \frac{\pi x}{a_0} \cos \frac{\pi y}{b_0} \cos \frac{\pi z}{c_0} \qquad \text{(9.75)}$$

if cubic, $a_0 = b_0 = c_0$

If the core is a cube, the buckling then becomes

$$B^2 = 3\frac{\pi^2}{a_0^2}$$ (9.76)

or

$$\left(a_0 = \sqrt{3}\,\frac{\pi}{B} \right)$$ (9.77)

The extrapolated length of the side of a cubic core is greater than the extrapolated thickness of an infinite slab of the same materials by a factor of $\sqrt{3}$.

The geometric dimensions set the critical mass of fuel required to achieve criticality. However, it should be noted that the minimum critical mass is not an end in itself. This is because the core must be large enough to dissipate the thermal energy being generated without damage to the fuel, the fuel cladding, or the structural elements of the core.

Spherical Reactor Core

The spherical core is attractive because its low surface area-to-volume ratio cuts neutron leakage to a minimum. This results in the minimum amount of fuel to achieve criticality.

Since the flux varies only in the radial direction, the diffusion equation can be written in spherical coordinates as (Fig. 9.9)

$$\left(\nabla^2\phi = \frac{d^2\phi}{dr^2} + \frac{2}{r}\frac{d\phi}{dr} \right)$$ (9.78)

Substituting Eq. (9.78) into the diffusion equation yields

$$\left(\frac{d^2\phi}{dr^2} + \frac{2}{r}\frac{d\phi}{dr} + B^2\phi = 0 \right)$$ (9.79)

Let

$$u = \phi r$$ (9.80a)

Taking the derivative of u with respect to r gives

$$\frac{du}{dr} = \phi + r\frac{d\phi}{dr}$$ (9.80b)

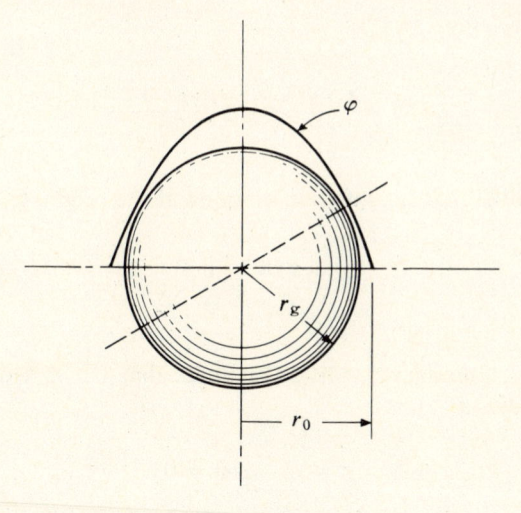

Fig. 9.9 Spherical reactor core and flux variation along a diameter.

and the second derivative is

$$\frac{d^2 u}{dr^2} = r \frac{d^2 \phi}{dr^2} + 2 \frac{d\phi}{dr} \tag{9.80c}$$

Thus,

$$\frac{1}{r} \frac{d^2 u}{dr^2} = \frac{d^2 \phi}{dr^2} + \frac{2}{r} \frac{d\phi}{dr} \tag{9.80d}$$

Substituting Eqs. (9.80a) and (9.80d) into Eq. (9.79) gives

$$\frac{1}{r} \frac{d^2 u}{dr^2} + B^2 \frac{u}{r} = 0 \tag{9.81}$$

which reduces to

$$\frac{d^2 u}{dr^2} + B^2 u = 0 \tag{9.82}$$

Therefore,

$$u = C_1 \cos Br + C_2 \sin Br = \phi r \tag{9.83}$$

and

$$\phi = \frac{C_1 \cos Br}{r} + \frac{C_2 \sin Br}{r} \qquad (9.84)$$

If it is required that the flux remain finite at the origin, $C_1 = 0$, since

$$\lim_{r \to 0} \frac{\cos Br}{r} = \frac{1}{0} = \infty$$

The second boundary condition requires that the flux drop to 0 at the extrapolated radius. When

$$r = r_0 \qquad \phi = 0$$
$$0 = C_2 \sin Br_0$$

This can occur only when

$$
\begin{array}{ccc}
Br_0 = 0 & \pi & 2\pi,\ 3\pi,\ \text{etc.} \\
\downarrow & \downarrow & \downarrow \\
\text{trivial} & \text{fundamental} & \text{harmonic eigenvalues} \\
 & \text{eigenvalue} &
\end{array} \qquad (9.85)
$$

The first value gives a trivial solution and the fundamental eigenvalue is

$$Br_0 = \pi \qquad (9.86a)$$

Thus,

$$r_0 = \frac{\pi}{B} \qquad (9.86b)$$

and

$$r_g = \frac{\pi}{B} - 0.71\lambda_{tr} \qquad (9.87)$$

The flux in the spherical core is

$$\phi = \frac{C_2 \sin \pi r/r_0}{r} \qquad (9.88)$$

Power Developed by a Spherical Core

The flux level at any point in a core will determine the number of interactions by fission. The differential power developed, dP, in a unit volume of core is

$$dP = G\Sigma_{fis} \phi \, dV \tag{9.89}$$

where G is energy released per fission (8.9×10^{-18} kWh/fission).

For the spherical geometry

$$dV = 4\pi r^2 \, dr \tag{9.90}$$

Substituting Eq. (9.90) into Eq. (9.89) and integrating from the origin to the geometrical radius will give the power developed by the core.

$$P = 4\pi G\Sigma_{fis} \int_0^{r_g} r^2 \frac{C_2 \sin Br}{r} \, dr$$

$$= 4\pi G\Sigma_{fis} C_2 \int_0^{r_g} r \sin Br \, dr$$

$$= 4\pi G\Sigma_{fis} C_2 \left(\frac{1}{B^2} \sin Br_g - \frac{r_g}{B} \cos Br_g \right) \tag{9.91}$$

It may be convenient to replace C_2 by an expression in terms of the maximum flux, ϕ_{max}.

$$\lim_{r \to 0} \phi = \frac{C_2 \sin Br}{r} = \frac{0}{0}$$

Since this is indeterminate, differentiate both numerator and denominator (L'Hôpital's rule) and check this limit.

$$\lim_{r \to 0} \phi = \lim_{r \to 0} \frac{C_2 B \cos Br}{1} = C_2 B = \phi_{max} \tag{9.92a}$$

$$C_2 = \frac{\phi_{max}}{B} \tag{9.92b}$$

Substituting (9.92b) into (9.91), the power developed by the core becomes

$$P = \frac{4\pi G\Sigma_{fis} \phi_{max}}{B^2} \left(\frac{1}{B} \sin Br_g - r_g \cos Br_g \right) \tag{9.93}$$

Example 4   For the homogeneous graphite–uranium mixture used in Examples 1, 2, and 3, what will be the critical radius of a bare critical sphere? What must be the maximum flux if the core is to produce 1 kW(th)? Compare the flux at the center of the core to that at the geometrical radius.

$$r_g = \frac{\pi}{B} - 0.71\lambda_{tr} = \frac{\pi}{0.0538} - 1.93 = 56.5 \text{ cm}$$

$$\Sigma_{fis} = N_{235}\sigma_{fis}^{235}\frac{\sqrt{\pi}}{2}$$

$$= 0.199 \times 10^{20} \times 582 \times 10^{-24} \times \frac{\sqrt{\pi}}{2}$$

$$= 0.0103 \text{ cm}^2/\text{cm}^3_{mix}$$

$$\phi_{max} = \frac{B^2 P}{4\pi G \Sigma_{fis}[(1/B)\sin Br_g - r_g\cos Br_g]}$$

$$= \frac{0.0538^2 \times 1}{4\pi \times 8.9 \times 10^{-18} \times 0.0103 \times 3600[(1/0.0538)\sin(0.0538 \times 56.5) - 56.5\cos(0.0538 \times 56.5)]}$$

$$= 1.200 \times 10^{10}\frac{\text{neutrons}}{\text{cm}^2\text{ s}}$$

$$\frac{\phi_{max}}{\phi_{r_g}} = \frac{\phi_{max}}{(\phi_{max}/Br_g)\sin Br_g} = \frac{Br_g}{\sin Br_g}$$

$$= \frac{0.0538 \times 56.5}{\sin[0.0538(56.5)]} = 29.9$$

Cylindrical Core

To determine the buckling of a cylindrical core the diffusion equation must have the Laplacian operator expressed in cylindrical coordinates. It becomes

$$\frac{\partial^2\phi}{\partial r^2} + \frac{1}{r}\frac{\partial\phi}{\partial r} + \frac{\partial^2\phi}{\partial z^2} + B^2\phi = 0 \tag{9.94}$$

Using the method of separation of variables, as was done for the rectangular parallelepiped, the flux becomes

$$\phi = AJ_0\left(\frac{2.405r}{r_0}\right)\cos\frac{\pi z}{z_0} \tag{9.95}$$

where r_0 is the extrapolated core radius, z_0 is the extrapolated core height, and $J_0(2.405r/r_0)$ is a Bessel function of the first kind of zero order.

The square of the buckling for the cylinder is

$$B^2 = \left(\frac{2.405}{r_0}\right)^2 + \left(\frac{\pi}{z_0}\right)^2 \qquad (9.96)$$

Reflected or Blanketed Reactor Cores

A reactor core is frequently surrounded by a reflecting material to reduce the ratio of peak flux to the flux at the edge of the core fuel and to reduce the amount of fissionable material required to achieve criticality. Essentially, good moderator materials are effective reflectors. Neutrons leaking into material having a large macroscopic scattering (or transport) cross section will have some scattering collisions return neutrons to the fuel lattice, thus producing additional fissions. Since there are few absorptions in the reflector, the thermal flux may actually increase slightly beyond the outermost fuel, as shown in Fig. 9.10. The one-group theory to be developed here will not show this rise in flux as one moves into the reflector; however, it will show the higher relative flux at the core lattice boundary and the savings in fuel.

If the reflector surrounding a core contains a subcritical amount of a multiplying medium, such as ^{233}U, in the thermal Molten Salt Breeder Reactor (MSBR), it is called a *blanket*. Neutron capture in the uranium produces fission, while capture by the fertile thorium contributes to the possibility of breeding. In a fast reactor core significant fission takes place in the blanket even if only fertile ^{238}U is present, since many fast neutrons exceed the threshold energy for fast fission.

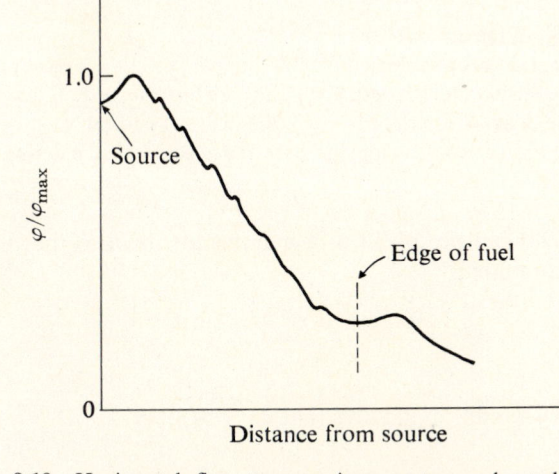

Fig. 9.10 Horizontal flux traverse in a water-moderated, natural uranium-fueled subcritical assembly.

Spherical Core with Finite Blanket (or Reflector)

Two diffusion equations are required to describe the flux in this two-region system. In the core itself there is a source term due to fission.

$$k_\infty^c \Sigma_a^c \phi_c - (-D_c \nabla^2 \phi_c) - \Sigma_a^c \phi_c = 0 \tag{9.97}$$

When the constants are combined into the material buckling for the core, this becomes

$$\nabla^2 \phi_c + B_c^2 \phi_c = 0 \tag{9.98}$$

Equation (9.98) describes the core flux, ϕ_c, from the origin to the core radius, r_c.

A blanket contains a source term $k_\infty^b \Sigma_a^b \phi_b$, which would be missing in a reflector, where k_∞^b would be 0. The diffusion equation for a blanket is

$$k_\infty^b \Sigma_a^b \phi_b - (-D_b \nabla^2 \phi_b) - \Sigma_a^b \phi_b = 0 \tag{9.99}$$

Rearranging, and letting $(1 - k_\infty^b)/L_b^2 = B_b^2$, the square of the buckling for the blanket material,

$$\nabla^2 \phi_b - B_b^2 \phi_b = 0 \qquad K = \frac{1}{L} \tag{9.100}$$

This equation describes the blanket flux, ϕ_b, from the core boundary at $r = r_c$ to the extrapolated boundary of the reflector at $r = r_c + T$. The following boundary conditions will determine the four constants that will result from the solutions of the two diffusion equations (9.98) and (9.100):

(1) The flux, ϕ_c, at the center of the core must remain finite.
(2) The flux, ϕ_b, at the extrapolated blanket boundary $(r_c + T)$ must be zero.
(3) At the core–blanket interface $(r = r_c)$, the core flux must equal the blanket flux.
(4) At the core–blanket interface $(r = r_c)$, the neutron current must be continuous. Therefore, the current leaving the core must equal that entering the blanket $(J_c = J_b)$.

Equation (9.98) can be solved in a manner identical to that used for the bare core. The core flux is

$$\phi_c = \frac{C_1 \cos B_c r}{r} + \frac{C_2 \sin B_c r}{r} \tag{9.101}$$

The first boundary condition dictates that $C_1 = 0$ since

$$\lim_{r \to 0} \frac{C_1 \cos B_c r}{r} = \frac{C_1}{0} = \infty$$

Thus, the core flux is described as

$$\phi_c = \frac{C_2 \sin B_c r}{r} \qquad (9.102)$$

Returning to the blanket equation and expressing it in terms of spherical coordinates where the flux varies only in a radial direction (Fig. 9.11) will give

$$\frac{d^2\phi_b}{dr^2} + \frac{2}{r}\frac{d\phi_b}{dr} - B_b^2\phi_b = 0 \qquad (9.103)$$

Letting

$$u = \phi_b r \qquad (9.104a)$$

$$\frac{du}{dr} = r\frac{d\phi_b}{dr} + \phi_b \qquad (9.104b)$$

$$\frac{d^2u}{dr^2} = r\frac{d^2\phi_b}{dr^2} + 2\frac{d\phi_b}{dr} \qquad (9.104c)$$

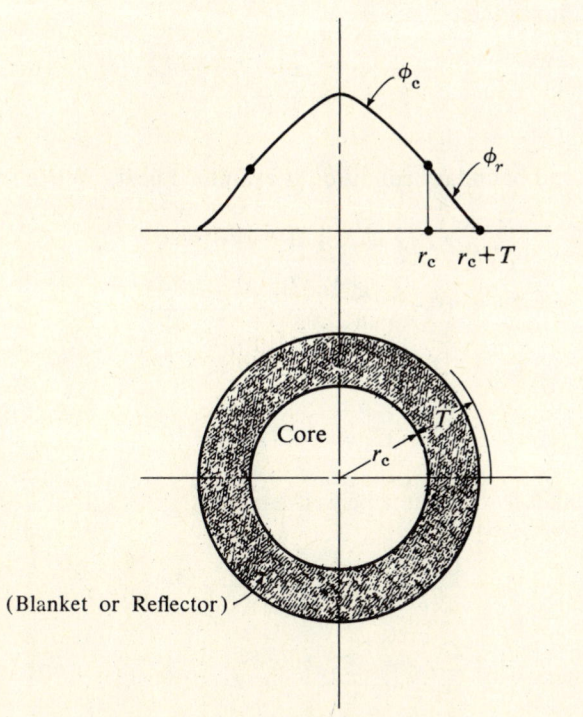

Fig. 9.11 Spherical core with finite blanket or reflector and flux variation along a diameter.

Dividing Eq. (9.104c) by r yields

$$\frac{1}{r}\frac{d^2u}{dr^2} = \frac{d^2\phi_b}{dr^2} + \frac{2}{r}\frac{d\phi_b}{dr} \tag{9.105}$$

Hence,

$$\frac{d^2u}{dr^2} - B_b^2 u = 0 \tag{9.106}$$

Using operator notation, we get

$$D^2 - B_b^2 = 0$$

$$D = \pm B_b$$

giving as a solution

$$u = C_3 e^{B_b r} + C_4 e^{-B_b r} = \phi_b r \tag{9.107}$$

The blanket flux is then

$$\phi_b = \frac{C_3 e^{B_b r}}{r} + \frac{C_4 e^{-B_b r}}{r} \tag{9.108}$$

The second boundary condition is examined next. When

$$r = r_c + T \qquad \phi_r = 0$$

$$0 = \frac{C_3 e^{B_b(r_c + T)}}{r_c + T} + \frac{C_4 e^{-B_b(r_c + T)}}{r_c + T}$$

Therefore,

$$C_4 = -C_3 e^{2B_b(r_c + T)}$$

The blanket flux can now be written.

$$\phi_b = \frac{C_3 e^{B_b r}}{r} - \frac{C_3 e^{2B_b(r_c + T)}}{r} e^{-B_b r} \tag{9.109}$$

$$= \frac{C_3}{r} e^{B_b(r_c + T)} \left[e^{B_b(r - r_c - T)} - e^{-B_b(r - r_c - T)} \right]$$

$$= \frac{C_3 e^{B_b(r_c + T)}}{r} 2 \sinh \left[B_b(r - r_c - T) \right] \tag{9.110}$$

Let $C_3' = 2C_3$ and then

$$\phi_r = \frac{C_3' e^{B_b(r_c + T)}}{r} \sinh \left[B_b(r - r_c - T) \right] \tag{9.111}$$

Equating fluxes at the interface, we obtain

$$\frac{C_2 \sin B_c r_c}{r_c} = \frac{C_3' e^{B_b(r_c + T)}}{r_c} \sinh (-B_b T)$$

$$C_2 = -\frac{C_3' e^{B_b(r_c + T)}}{\sin B_c r_c} \sinh B_b T \tag{9.112}$$

Turning to the final boundary condition, the neutron currents at the interfaces must be evaluated.

$$J_c \big|_{r=r_c} = -D_c \frac{d\phi_c}{dr} = \frac{-D_c C_2 (r_c B_c \cos B_c r_c - \sin B_c r_c)}{r_c^2} \tag{9.113}$$

$$J_b \big|_{r=r_c} = -D_b \frac{d\phi_r}{dr} = \frac{-D_b C_3' e^{B_b(r_c + T)} [B_b r_c \cosh (-B_b T) - \sinh (-B_b T)]}{r_c^2} \tag{9.114}$$

Setting Eqs. (9.113) and (9.114) equal and substituting Eq. (9.112), we have

$$D_c \frac{\sinh B_b T}{\sin B_c r_c} (r_c B_c \cos B_c r_c - \sin B_c r_c)$$

$$= -D_b (B_b r_c \cosh B_b T + \sinh B_b T) \tag{9.115}$$

Rearranging yields

$$\frac{D_c}{D_b} (1 - r_c B_c \cot B_c r_c) = r_c B_b \coth B_b T + 1 \tag{9.116}$$

Notice that this rather cumbersome expression for the core radius, r_c, involves core and blanket properties (D_c, B_c, D_b, and B_b) plus the blanket thickness, T. In its present form Eq. (9.116) may be solved for r_c only by trial and error if the other parameters are known. However, if the core and blanket diffusion constants are approximately equal ($D_c \approx D_b$),

$$\cot B_c r_c = -\frac{B_b}{B_c} \coth B_b T \tag{9.117}$$

Cotangents in the second quadrant are negative and will have the same magnitude but opposite sign from the cotangent of the supplement of the angle.

$$\cot(\pi - B_c r_c) = -\cot B_c r_c \tag{9.118}$$

Therefore,

$$\cot(\pi - B_c r_c) = \frac{B_b}{B_c} \coth B_b T \tag{9.119}$$

Now, solving for r_c explicitly, we get

$$r_c = \frac{\pi - \cot^{-1}[(B_b/B_c)\coth B_b T]}{B_c} \tag{9.120}$$

When the blanket thickness increases, the coth B_b decreases toward 1 as a limit. When the blanket thickness is $2.65/B_b$,

$$\coth 2.65 = 1.01$$

and the coth is only 1 percent greater than if the blanket were infinitely thick. For practical purposes if $B_b \geq 2.65$, the reflector can be considered infinitely thick and the expression for the core radius becomes

$$r_c = \frac{\pi - \cot^{-1}(B_b/B_c)}{B_c} \tag{9.121}$$

Reflector savings, δ, is the difference between the radius of an unreflected core and the radius of a core with reflector.

$$\delta = \frac{\pi}{B_c} - d - \frac{\pi - \cot^{-1}[(B_b/B_c)\coth B_b T]}{B_c}$$

$$= \frac{\cot^{-1}[(B_b/B_c)\coth B_b T]}{B_c} - d \tag{9.122}$$

For a reflector that produces no neutrons, the blanket buckling, B_b, becomes the reciprocal of the reflector's diffusion length $(1/L_b)$. This is true in the following example, where graphite is the reflector.

Example 5   For the homogeneous graphite–uranium mixture used previously, determine the critical radius of a spherical core reflected with an infinite thickness of graphite. What will be the reflector savings, the ratio of the peak

flux to the flux at the core–reflector interface, and the maximum flux when the reflected core develops 1 kW(th)?

$$N_c = \frac{1.6 \times 6.03 \times 10^{23}}{12} = 8.03 \times 10^{22} \frac{\text{graphite nuclei}}{\text{cm}^3}$$

$$\Sigma_a^b = 8.03 \times 10^{22} \times 0.0034 \times 10^{-24} \times \frac{\sqrt{\pi}}{2} = 2.42 \times 10^{-4} \text{ cm}^{-1}$$

$$D_b = \frac{\lambda_{tr}}{3} = \frac{1}{3\Sigma_{tr}} = \frac{1}{3 \times 8.03 \times 10^{22} \times 4.8 \times 10^{-24}[1 - (\frac{2}{3} \times 12)]}$$

$$= 0.915 \text{ cm}$$

$$L_b^2 = \frac{D_b}{\Sigma_a} = \frac{0.915}{2.42 \times 10^{-4}} = 3780 \text{ cm}^2$$

$$L_b = 61.5 \text{ cm}$$

$$\cot(\pi - B_c r_c) = \frac{1}{B_c L_b} = \frac{1}{0.0538 \times 61.5} = 0.302$$

$$\pi - B_c r_c = 1.277$$

$$r_c = \frac{\pi - 1.277}{0.0538} = 34.7 \text{ cm}$$

$$\delta = \frac{\pi}{B_c} - d - r_c = \frac{\pi}{0.0538} - 1.93 - 34.7$$

$$= 21.9 \text{ cm}$$

$$\frac{\phi_{max}}{\phi_{r_c}} = \frac{B_c r_c}{\sin B_c r_c} = \frac{0.0538(34.7)}{\sin(0.0538 \times 34.7)} = 1.95$$

$$\phi_{max} = \frac{B_c^2 P}{4\pi G \Sigma_{fis}[(1/B_c) \sin B_c r_c - r_c \cos B_c r_c]}$$

$$= \frac{(0.0538)^2 \times 1}{4 \times \pi \times 8.9 \times 10^{-18} \times 0.0103 \times 3600[(1/0.0538) \sin (0.0538}$$

$$\times 34.7) - 34.7 \cos (0.0538 \times 34.7)]$$

$$= 2.51 \times 10^{11} \text{ neutrons/cm}^2 \text{ s}$$

Comparing the results of Examples 4 and 5, it can be seen that the reflected core has a volume which is only 23.2 percent that of the bare core. This means that only 23.2 percent as much fissionable ^{235}U will be required for criticality. Although the peak flux for a given power is an order of magnitude greater, the ratio of ϕ_{max}/ϕ_{r_c} is only 1.95 as compared to a ratio of 29.0 in the bare core. The flux is considerably flattened by the addition of a reflector, thus making the consumption of fissionable nuclei much more uniform across the core.

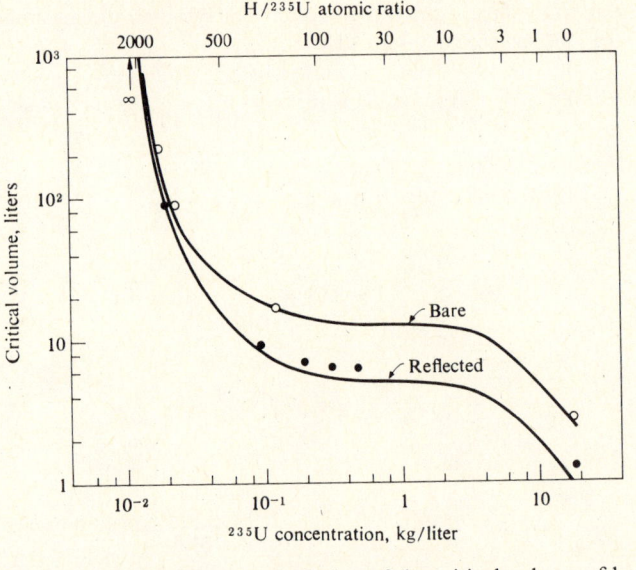

Fig. 9.12 Transport theory calculation of the critical volume of bare and water reflected (reflector thickness = 20 cm) spheres as a function of ^{235}U concentration (or $H/^{235}U$ atomic ratio) of homogeneous $^{235}U-H_2O$ systems. The open and filled circles represent experimental results. [From H. Soodak, *Reactor Handbook*, Vol. 3: *Physics*. New York: Interscience Publishers, 1962.]

Figure 9.12 shows the advantage of using a reflector for a spherical core for a $^{235}U-H_2O$ system. When the $H/^{235}U$ atomic ratio is 100, the bare core requires a critical volume of about 14 liters, as compared to 7 liters (experimental) for the reflected core. The saving of 50 percent in the fuel inventory for the core is very worthwhile.

In Chapter 8, beryllium hydride (BeH_2) was mentioned as an effective moderator, whose high H and Be atom densities and an (n, $2n$) reaction in the beryllium contribute to a minimum critical volume for a BeO reflected spherical core. Figure 9.13 shows how the critical mass is reduced by increasing the BeH_2 density for the three principal fissile isotopes, ^{233}U, ^{235}U, and ^{239}Pu. For densities greater than 680 kg/m^3, the critical mass required is less than that for polyethylene. Polycrystalline BeH_2 has been produced with a density of 780 kg/m^3 and this results in critical masses approximately 15 percent less than when using polyethylene as the moderator. The hydrogen atom density is only 3.5 percent greater than for the polyethylene, so most of the difference can be attributed to the superior nuclear qualities of the Be atoms, as compared to carbon.

The importance of the reflector in achieving a critical configuration was emphasized with the successful orbiting of the SNAP-10A space nuclear

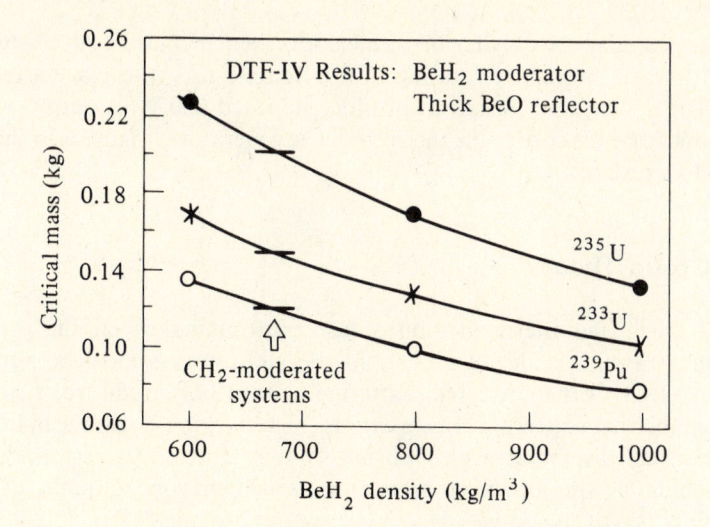

Fig. 9.13 Critical mass variation with BeH_2 density for 0.40-m-thick BeO-reflected spherical assemblies. [From K. S. Rao and M. Srinivasan, *Nuclear Technology* **49**, No. 2 (July 1980), p. 318.]

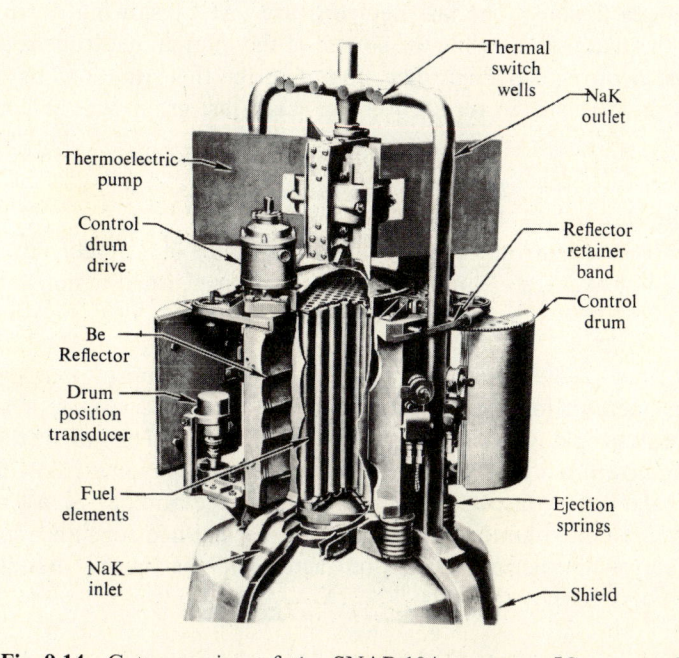

Fig. 9.14 Cutaway view of the SNAP-10A reactor. [Courtesy of Atomics International, a division of North American Aviation, Inc.]

auxiliary power system. After the vehicle was launched into a successful orbit, criticality was achieved by rotation of the Be reflectors into place around the core. The system then began to produce its rated 600 W(e) output with the heat supplied by the core to the thermoelectric generator. Figure 9.14 shows the SNAP-10A reactor.

Two-Group Theory

Between birth and thermalization a fast neutron makes on the average n collisions, traveling a distance $(\lambda_{tr})_f$ in the laboratory system between each collision. $(\lambda_{tr})_f$ is the corrected scattering or transport mean free path based on the epithermal scattering cross sections. The slowing-down mean free path, λ_{sl} is the total distance traveled during slowing down. Its reciprocal is the macroscopic slowing-down cross section, which may be thought of as the probability of slowing down.

$$\lambda_{sl} = n(\lambda_{tr})_f = \frac{1}{\Sigma_{sl}} \tag{9.123}$$

In two-group theory the neutrons are considered as either fast or thermal. The thermal neutrons diffuse, having leakage and absorption. The absorption further causes fission. The fast neutrons are lost by slowing down. Their source is thermal fission, while the source of the thermal neutrons is provided by the fast neutrons slowing down. To describe this situation the diffusion equation must be written for each group. The fast group will be considered first.

$$-(-D_f)\nabla^2\phi_f - \Sigma_{sl}\phi_f + f\eta\Sigma_{ath}\phi_{th} = 0$$

| ↑ | ↑ | ↑ |
| fast neutron leakage | slowing down (removal by thermalization) | source of fast neutrons from thermal fission |

$$\tag{9.124}$$

Note that the leakage involves a fast diffusion constant, D_f, and the Laplacian operator for the fast flux, $\nabla^2\phi_f$. The second term represents the number of slowing-down interactions in a unit volume. This can be thought of as slowing down, more or less, in one collision. Leakage and slowing down must be balanced by fast-neutron production. The number of fast neutrons produced in a unit volume is the number of thermal neutrons absorbed times the product of the thermal utilization factor and the thermal fission factor. Dividing Eq. (9.124) by D_f gives

$$\nabla^2\phi_f - \frac{\Sigma_{sl}}{D_f}\phi_f + \frac{f\eta\Sigma_{ath}\phi_{th}}{D_f} = 0 \tag{9.125}$$

The ratio of D_f/Σ_{sl} in the fast neutron diffusion equation is analogous to L^2 in the one-group diffusion equation, (9.47a). It can be designated as L_s^2, the square of the slowing-down length.

$$L_s^2 = \frac{D_f}{\Sigma_{sl}} = \frac{n_f \lambda_{tr}^2}{3} \tag{9.126}$$

Equations (8.48) and (8.57) can be combined with Eq. (9.126) to give

$$L_s^2 = \frac{\ln (E_0/E_{th})}{3\xi(\Sigma_s)^2(1 - \overline{\cos \psi})^2} \tag{9.127}$$

$$L_s = \sqrt{\frac{\lambda_{sl}\lambda_{tr}}{3}} \tag{9.128}$$

Thus, the slowing-down length is seen to be more or less an average of the mean free path for slowing down and the mean free path for scatter. The fast diffusion equation has now become

$$\nabla^2\phi_f - \frac{\phi_f}{L_s^2} + \frac{f\eta\Sigma_{a_{th}}\phi_{th}}{D_f} = 0 \tag{9.129}$$

In a similar manner a thermal diffusion equation can be written as

$$-(-D_{th})\nabla^2\phi_{th} - \Sigma_{a_{th}}\phi_{th} + \varepsilon p\Sigma_{sl}\phi_f = 0$$

$$\uparrow \qquad\qquad \uparrow \qquad\qquad \uparrow$$

leakage of thermal neutrons	absorption of thermal neutrons	source thermal neutrons from the slowing down of fast neutrons

(9.130)

This equation is similar to Eq. (9.43) for one-group theory, except that the source of thermal neutrons is from the slowing down of fast neutrons. The number of slowing-down interactions can be increased slightly by the fast fission factor, ε, and is reduced by the resonance escape probability, p. Rewriting, we obtain

$$\nabla^2\phi_{th} - \frac{\phi_{th}}{L^2} + \frac{\varepsilon p\Sigma_{sl}\phi_f}{D_{th}} = 0 \tag{9.131}$$

Experiment has shown that fast and thermal fluxes in a large bare core will have the same shape, differing primarily in magnitude (see Fig. 9.15). It is assumed that in both instances, for steady-state operation, the same buckling can be used in the diffusion equation.

$$\nabla^2\phi_{th} + B^2\phi_{th} = 0 \tag{9.132}$$

$$\nabla^2\phi_f + B^2\phi_f = 0 \tag{9.133}$$

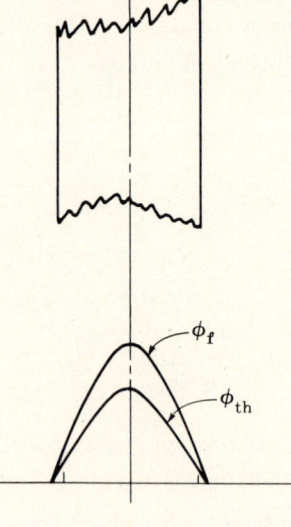

Fig. 9.15　Fast and thermal neutron flux distributions in a bare slab reactor.

Comparing Eq. (9.132) with (9.131), and Eq. (9.133) with (9.129), it is seen that

$$B^2\phi_{\text{th}} = -\frac{\phi_{\text{th}}}{L^2} + \frac{\varepsilon p \Sigma_{s1} \phi_f}{D_{\text{th}}} \tag{9.134}$$

and

$$B^2\phi_f = -\frac{\phi_f}{L_s^2} + \frac{f\eta \Sigma_a \phi_{\text{th}}}{D_f} \tag{9.135}$$

Solving Eq. (9.135) for ϕ_f, we get

$$\phi_f = \frac{f\eta \Sigma_a \phi_{\text{th}} L_s^2}{D_f(B^2 L_s^2 + 1)} \tag{9.136}$$

Substituting Eq. (9.136) into (9.134) yields

$$(B^2 L^2 + 1)\phi_{\text{th}} = \frac{\varepsilon p \Sigma_{s1} L^2 f\eta \Sigma_a \phi_{\text{th}} L_s^2}{D_{\text{th}} D_f(B^2 L_s^2 + 1)} \tag{9.137a}$$

Simplifying, we obtain

$$(B^2 L^2 + 1)(B^2 L_s^2 + 1) = k_\infty \tag{9.137b}$$

Rearranging yields

$$1 = k_\infty \left(\frac{1}{B^2 L^2 + 1} \right) \left(\frac{1}{B^2 L_s^2 + 1} \right) \tag{9.137c}$$

At the critical condition the effective multiplication factor is unity. For two-group theory the nonleakage probability is the product of the thermal nonleakage probability and the nonleakage probability for fast neutrons. Since Eq. (9.48c) defined the thermal nonleakage probability, the fast non-leakage probability is

$$P_{NL_f} = \frac{1}{B^2 L_s^2 + 1} \tag{9.138}$$

The total nonleakage probability is

$$P_{NL} = \frac{1}{B^4 L_s^2 L^2 + B^2 L_s^2 + B^2 L^2 + 1} \tag{9.139}$$

For large cores with small values of buckling, the $B^4 L_s^2 L^2$ term can be omitted and

$$P_{NL} = \frac{1}{B^2 (L_s^2 + L^2) + 1} \tag{9.140}$$

The sum of the squares of the diffusion length and the slowing-down length is known as the square of the migration length, M^2:

$$M^2 = L_s^2 + L^2 \tag{9.141}$$

When M^2 replaces L^2 in Eq. (9.48a), this is often called *modified one-group theory*. Buckling is handled as with one-group theory and the results are those for *two-group theory* when the $B^4 L_s^2 L^2$ term can be ignored.

$$B^2 = \frac{k_\infty - 1}{M^2} \tag{9.142}$$

Example 6   Compare the diameter of the critical sphere in Example 4 with that for the same material using both modified one-group theory and full two-group theory.
   First the slowing-down length must be determined.

$$E_{av} = \tfrac{3}{2} kT = 1.5 \times 0.025 = 0.0375 \text{ eV}$$

This is the average neutron energy for the thermal group.

The number of collisions to thermalize is

$$n = \frac{\ln (E_0/E_{th})}{\cdot \quad \xi} = \frac{\ln [(2 \times 10^6)/0.0375]}{0.158} = 113$$

$$(\Sigma_{tr})_f = N_C \sigma_{s_f}^C (1 - \overline{\cos \psi_C}) + N_U \sigma_{s_f}^U (1 - \overline{\cos \psi_U})$$

$$= 7.96 \times 10^{22} \times 4.66 \times 10^{-24}\left(1 - \frac{2}{3 \times 12}\right)$$

$$+ 3.98 \times 10^{20} \times 8.3 \times 10^{-24}\left[1 - \frac{2}{3(238)}\right]$$

$$= 0.351 + 0.0033 = 0.354 \text{ cm}^2/\text{cm}^3$$

$$L_s^2 = \frac{n_f}{3\Sigma_{tr}^2} = \frac{113}{3 \times (0.354)^2} = 302 \text{ cm}^2$$

$$M^2 = L_s^2 + L^2 = 302 + 68.2 = 370 \text{ cm}^2$$

$$B^2 = \frac{k_\infty - 1}{M^2} = \frac{1.198 - 1}{370} = 0.000\,536$$

$$B = 0.0232$$

$$r_g = \frac{\pi}{B} - d = \frac{\pi}{0.0232} - 1.93 = 135.2 - 1.93$$

$$= 133.3 \text{ (modified one-group theory)}$$

$$B^4 L_s^2 L^2 + B^2(L_s^2 + L^2) + (1 - k_\infty) = 0$$

$$B^4 + \frac{L_s^2 + L^2}{L_s^2 L^2} B^2 - \frac{k_\infty - 1}{L_s^2 L^2} = 0$$

$$B^4 + \frac{370}{302(68.2)} B^2 - \frac{0.198}{302(68.2)} = 0$$

$$B^4 + 0.017\,963 B^2 - 0.000\,009\,61 = 0$$

$$B^2 = \frac{-0.017\,963 \pm \sqrt{0.017\,963^2 + 4(0.000\,009\,61)}}{2}$$

$$= \frac{-0.017\,96 + 0.019\,00}{2} = 0.000\,52$$

$$B = 0.0228$$

$$r_g = \frac{\pi}{0.0228} - 1.93 = 137.6 - 1.93 = 135.7 \text{ cm (two-group theory)}$$

Fermi age theory assumes that neutrons lose energy in a smooth manner, rather than in discrete steps. For a heavier moderator like graphite, where 113 collisions are required for thermalization, this model is quite adequate. It is

less satisfactory for a light-water moderator, where, on the average, only 19 collisions are required.

Consideration of the variation in the slowing-down density during thermalization shows that the fast nonleakage probability is

$$P_{NL_f} = e^{-B^2 \tau} \qquad (9.143)$$

where $\tau = L_s^2$ = Fermi age. The Fermi age and the square of the slowing-down length are identical. When the fast nonleakage probability is expanded in a series

$$P_{NL_f} = \frac{1}{1 + L_s^2 B^2 + L_s^4 B^4 + \cdots} \qquad (9.144)$$

For larger cores where B^2 is small, the terms beyond the second one in the denominator contribute very little. Then the results of Fermi age theory and the two-group theory used here are comparable.

Comparison of One-Group, Modified One-Group, and Two-Group Theory for a Molten Salt Core

The molten salt reactor concept is being studied because of its possible development as a thermal breeder (see Chapter 13). A typical salt with mole fractions of 0.72 LiF, 0.16 BeF_2, 0.115 ThF_4, and 0.005 $^{233}UF_4$ will have an infinite multiplication factor of only 0.1435. The value is low because of insufficient moderation. As graphite is added to a core assembly the value of k_∞ will increase until it surpasses 1 at 84.5 percent graphite and 15.5 percent salt content. As the graphite content is further increased, k_∞ increases until there is about 97.5 percent graphite and 2.5 percent salt. Then it falls rapidly to 0 for pure graphite. The reasons for this can be inferred from Fig. 9.16. As the salt fraction increases, the thermal utilization factor, f, increases because a larger fraction of the neutron captures take place in the fissile ^{233}U. At the same time the resonance escape probability, p, decreases as the probability of resonance capture by the fertile ^{232}Th nuclei increases with the larger salt fraction. The other two factors in the four-factor equation remain unchanged, and thus a peak occurs for k_∞.

The buckling shown in Fig. 9.16 is for modified one-group theory. It indicates that a volume fraction of approximately 5 percent salt will result in the minimum core size. The fact that a 13 percent salt fraction has been proposed for a molten salt breeder core is an indication that minimum core size is not always a primary consideration.

Figure 9.17 shows the effect of varying the mole fraction of $^{233}UF_4$ on the critical size and power developed by a cylindrical core (diameter = height)

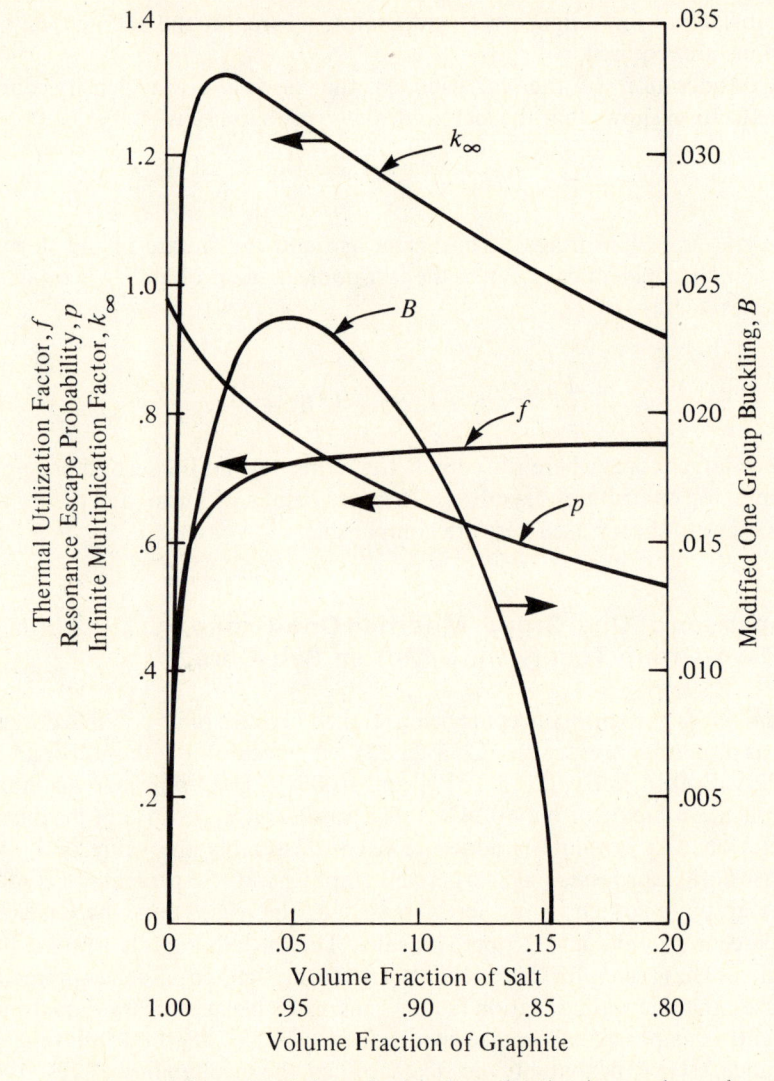

Fig. 9.16 Effect of variation of salt volume fraction in a molten salt core with graphite as the moderator. Salt composition (mole fractions): 0.72 ^{7}LiF, 0.16 BeF$_2$, 0.115 ThF$_4$, and 0.005 ^{233}UF$_4$. Average core temperature 650°C.

using one-group, modified one-group, and two-group theory. Note that the one-group theory, which ignores the leakage of fast neutrons, results in smaller critical sizes and much less power for a given peak flux. The closeness of the curves for the modified one-group and two-group theories indicates that the two models are in close agreement for this combination of core materials. It also emphasizes the importance of considering the fast leakage. With smaller

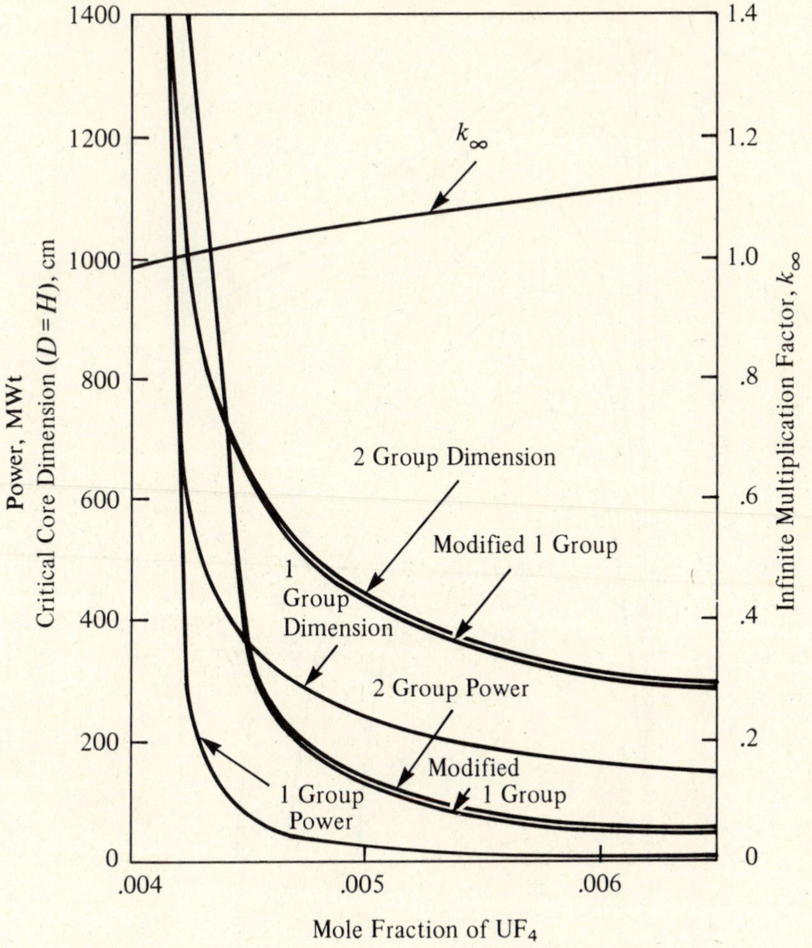

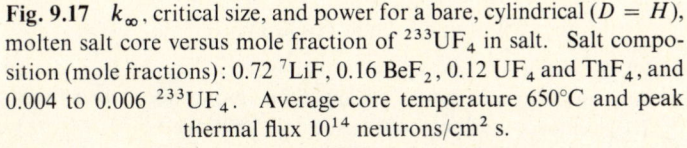

Fig. 9.17 k_∞, critical size, and power for a bare, cylindrical ($D = H$), molten salt core versus mole fraction of $^{233}UF_4$ in salt. Salt composition (mole fractions): 0.72 ^7LiF, 0.16 BeF_2, 0.12 UF_4 and ThF_4, and 0.004 to 0.006 $^{233}UF_4$. Average core temperature 650°C and peak thermal flux 10^{14} neutrons/cm² s.

mole fractions of $^{233}UF_4$ in the salt ($k_\infty - 1$) is smaller, reducing the buckling and increasing the core size. For large power reactors this is desirable, as the specific core power (kW/liter) may be limited by heat transfer and materials considerations.

Figure 9.18 illustrates the effect of three different mole fractions of $^{233}UF_4$ on k_∞ and B (modified one-group) as the salt fraction is increased. For a given salt fraction larger amounts of fissile material will increase k_∞ and B, resulting in smaller critical sizes.

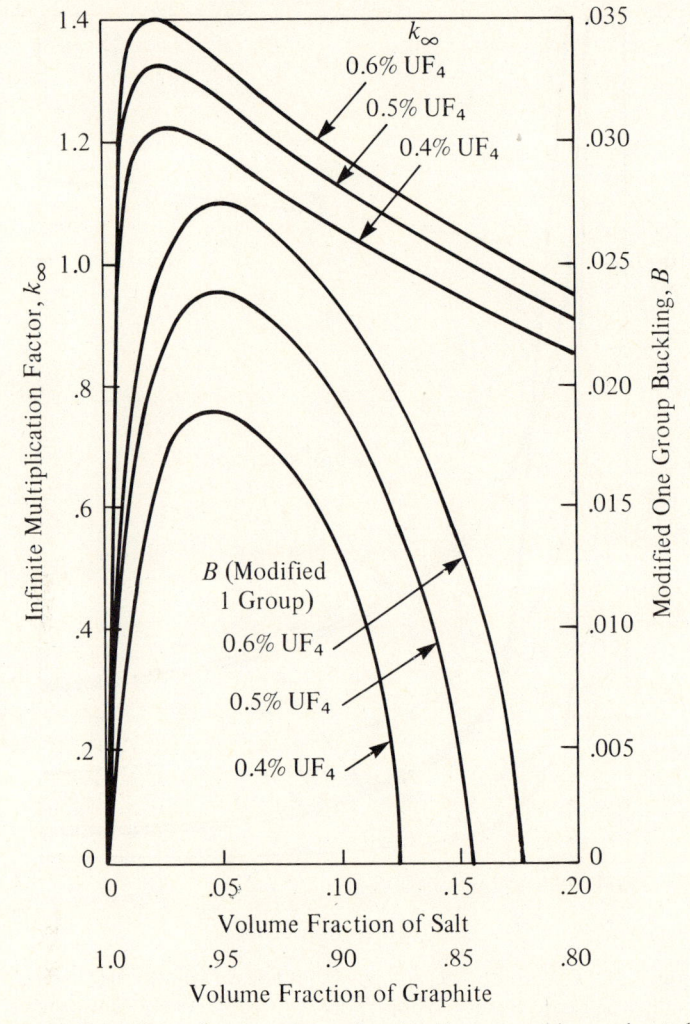

Fig. 9.18 Effect of salt volume fraction in a graphite-moderated molten salt core on k_∞ and B (modified one-group) for three different $^{233}UF_4$ contents. Salt composition (mole fractions): 0.72 ^7LiF, 0.16 BeF_2, (0.12 UF_4) ThF_4, and $^{233}UF_4$, as labeled on curves. Core temperature 650°C.

Subcritical Blanket for Molten Salt Core

It may be noted in Fig. 9.16 that as the salt fraction increases beyond 15.5 percent the value of k_∞ will drop below unity. If the core proper is surrounded by a graphite–salt blanket region containing, say, 35 to 40 volume percent salt, this blanket will act as a reflector. It also will have some conversion by

neutron capture in the fertile thorium and a subcritical amount of ^{233}U fission. The blanket will make a significant contribution to the core's ability to breed, and will develop a modest amount of power—as will be illustrated by Prob. 9.15.

Fast Reactors

In a fast-reactor system the average neutron energy lies between 0.1 and 0.5 MeV. There is no moderator material to cause scattering collisions and produce a large degradation of energy. Steam, gas, and liquid metals are being considered as coolants; liquid sodium is the most popular current choice. A principal requirement of fast-reactor coolants is that they should not appreciably moderate neutrons. Thus, light nuclei should be kept to a minimum in a fast core.

At high neutron energies the ratio of fission cross section to capture cross section is significantly greater, so that more neutrons are available for absorption by fertile nuclei. ^{239}Pu is most attractive for a fast-reactor system since η, the number of fast neutrons emitted per neutron absorbed by the fissile nucleus, is 2.92 at 1 MeV and 2.4 for the neutron spectrum in an oxide-fueled fast reactor, as compared to 2.11 at thermal energies. Thus, the oxide-fueled breeder will have 0.4 neutron per fission to divide among breeding, leakage, and parasitic capture in nonfuel material.

The fission mean free path and the diffusion length for a fast spectrum of neutrons are larger than in a thermal system, producing a tendency for the fast reactor to leak more neutrons. This requires that the enrichment of the fuel be increased by a factor of about 5. To get the same output per kilogram of fissile material in the more highly enriched fuel, the power density of the core must increase similarly. The excellence of the heat transfer characteristics of sodium is some help. Also, the fuel rod diameters are smaller than in water-cooled reactors, the diameters being about 0.5 cm rather than the order of 1.25 cm for thermal systems.

The small fuel rod size coupled with the longer diffusion length tends to allow the core to be treated as homogeneous with even less error than is involved in a thermal reactor.

One-Group Theory for Fast Cores

One-group theory can be applied fairly effectively to the determination of the critical size of a fast reactor, provided that properly averaged cross-section values for the neutron spectrum are used. For example, a metallic fuel produces a harder (higher average energy) spectrum than does an oxide-fueled reactor, and carbide-fueled reactors are intermediate between the other two types. The presence of oxygen or carbon in the fuel provides some moderation due to elastic collisions with the light nuclei, resulting in a softer neutron spectrum. Metallic fuels with

their harder spectrum would give a higher breeding ratio but would be unable to stand the temperatures resulting from the high power density. Carbide fuels are being developed, but at the present time it appears that at least the first generation of fast breeders will be designed for oxide fuel. Appendix E lists values of cross sections and η appropriate for an oxide-fueled fast-neutron spectrum.

For a fast core the expression for the infinite multiplication factor reduces to

$$k_\infty = f\eta \tag{9.145}$$

Since few neutrons reach thermal energies the resonance escape probability, p, is no longer a problem of significance and the fast fission factor, ε, is inappropriate due to the lack of a significant number of thermal fissions. If one expands Eq. (9.145),

$$k_\infty = \frac{\Sigma_a^{\text{fuel}}\eta}{\Sigma_a^{\text{fuel}} + \Sigma_a^{\text{clad}} + \Sigma_a^{\text{Na}}} \tag{9.146}$$

If more than one fissionable species is present, the $\Sigma_a^{\text{fuel}}\eta$ terms are added to account for the contribution of each to the next generation of neutrons (see Example 7). Note that the fertile isotopes ^{232}Th, ^{238}U, ^{240}Pu, and ^{242}Pu can provide significant numbers of neutrons in a fast spectrum.

Example 7   An oxide-fueled fast reactor is to have a core containing 3.5 volume percent ^{239}PuO$_2$, 26.5 volume percent ^{238}UO$_2$, 25 volume percent stainless steel clad and structure, and 45 volume percent sodium as coolant.
(a) Compute the infinite multiplication factor for the core.

Material	Σ_a	× vol. frac. =	net Σ_a
^{239}PuO$_2$	0.060	0.035	0.002 1
^{238}UO$_2$	0.008	0.265	0.002 12
Na	4×10^{-5}	0.45	0.000 018
SS	0.0015	0.25	0.000 375
		1.00	0.004 613 cm^{-1}

$$k_\infty = \frac{\Sigma_a^{\text{fuel}}\eta}{\Sigma_a^{\text{fuel}} + \Sigma_a^{\text{SS}} + \Sigma_a^{\text{Na}}}$$

$$= \frac{0.0021 \times 2.4 + 0.002\ 12 \times 0.4}{0.004\ 61}$$

$$= 1.092 + 0.184 = 1.276$$

Note the contribution of the ^{238}U to k_∞.

(b) Compute the diffusion length, material buckling, and the nonleakage probability for the core.

Material	Σ_{tr}	× vol. frac.	= net Σ_{tr}
$PuO_2 + UO_2$	0.18	0.30	0.054
Na	0.08	0.45	0.036
SS	0.25	0.25	0.0625
		1.00	0.1525 cm^{-1}

$$L^2 = D_f/\Sigma_a = \frac{1}{3\Sigma_{tr}\Sigma_a} = \frac{1}{3 \times 0.1525 \times 0.004\,613}$$

$$= 475 \text{ cm}^2$$

$$L = 21.8 \text{ cm}$$

$$B^2 = \frac{k_\infty - 1}{L^2} = \frac{1.276 - 1}{475} = 0.000\,580 \text{ cm}^{-2}$$

$$B = 0.0241 \text{ cm}^{-1}$$

$$P_{NL} = \frac{1}{1 + B^2 L^2} = \frac{1}{1 + 0.000\,58 \times 475}$$

$$= 0.785$$

(c) What would be the radius of a bare critical spherical core?

$$r = \frac{\pi}{B} - \frac{2}{3}\lambda_{tr} = 3.1416/0.0241 - \frac{2}{3} \times \frac{1}{0.1525}$$

$$= 131.2 - 4.4 = 126.8 \text{ cm}$$

Fast Core Blanket

The tendency for neutrons to leak from a fast reactor core makes it desirable to use a reflector, which is termed a *blanket*. This blanket contains fertile material so that, in addition to scattering neutrons back into the core proper, there will be conversion due to neutron absorption by the fertile nuclei. Also, there will be some neutron production because of fast fission in the fertile material. The effective use of a blanket contributes significantly to the ability of the reactor as a whole to breed more new fissile nuclei than are consumed. If this happens, the reactor is then a *breeder*.

The blanket can contain a higher fraction of fertile material than the sum of the fertile plus fissile material in the core proper, since there is less fission and less resultant heat generated. Because of the lower heat generation, less coolant is required and the fertile "fuel" elements can have a larger diameter, reducing the amount of cladding, as illustrated in Examples 7 and 8.

Example 8   A 0.61-m-thick blanket is added to the spherical core of Example 7. The blanket is to contain 50 volume percent $^{238}UO_2$, 35 volume percent sodium, and 15 percent stainless steel cladding and structure.
(a) Determine the reflector buckling.

Material	Σ_a	$\times$ vol. frac. =	net Σ_a
$^{238}UO_2$	0.008 15	0.50	0.004 08
Na	4×10^{-5}	0.35	0.000 014
SS	0.0015	0.15	0.000 225
		1.00	0.004 319 cm^{-1}

$$k_\infty = \frac{\eta_{238}\Sigma_a^{238}}{\Sigma_a^{total}} = \frac{0.4 \times 0.004\,08}{0.004\,319} = 0.378$$

Material	Σ_{tr}	$\times$ vol. frac. =	net Σ_{tr}
$^{238}UO_2$	0.18	0.50	0.090
Na	0.08	0.35	0.028
SS	0.25	0.15	0.0375
		1.00	0.1555 cm^{-1}

$$L_b^2 = \frac{1}{3\Sigma_{tr}\Sigma_a} = \frac{1}{3 \times 0.1555 \times 0.004\,32} = 496 \text{ cm}^2$$

$$L_b = 22.3 \text{ cm}$$

$$B_b^2 = \frac{1 - k_{\infty b}}{L_b^2} = \frac{1 - 0.378}{496} = 0.001\,253 \text{ cm}^{-2}$$

$$B_b = 0.0354 \text{ cm}^{-1}$$

(b) What will be the critical radius of the core when surrounded by the blanket?

$$d = \frac{2}{3}\lambda_{tr} = \frac{2}{3\Sigma_{tr}} = \frac{2}{3 \times 0.1555} = 4.3 \text{ cm}$$

$$D_b = \frac{1}{3\Sigma_{tr}^b} = \frac{1}{3 \times 0.1555} = 2.14 \text{ cm}$$

$$D_c = \frac{1}{3\Sigma_{tr}^c} = \frac{1}{3 \times 0.1525} = 2.19 \text{ cm}$$

Therefore, D_b and D_c are approximately equal.

$$r_c = \frac{\pi - \cot^{-1}\left[(B_b/B_c)\coth TB_b\right]}{B_c}$$

$$= \frac{\pi - \cot^{-1}\{(0.0354/0.0241)\coth\left[(2 \times 12 \times 2.54 + 4.3(0.0354)\right]\}}{0.0241}$$

$$= 105.9 \text{ cm}$$

The addition of the 0.61-m blanket results in a reduction of 41.8 percent in the volume of core materials required to achieve criticality. In addition, a large fraction of the neutrons leaking from the core are captured by fertile nuclei in the blanket and are converted to fissile nuclei making a substantial contribution to the ability of the fast core to breed.

Multigroup Calculations

To approach the actual situation where flux and cross sections are energy dependent, it is necessary to break the neutron population into numerous energy groups. Each group may gain neutrons by slowing down from higher-energy groups and the higher-energy groups will receive neutrons directly from fission. Neutrons are lost from a group by slowing down to the next lower group, by absorption (sometimes producing fission), or by leakage out of the core. For some of the fast groups absorption can be ignored, but in the epithermal and thermal regions it must be considered. For the thermal group there is no loss by slowing down. Proper cross sections must be used for each energy group.

The complexity of multigroup calculations demands the use of a digital computer to solve the complex array of simultaneous equations resulting from writing the diffusion equation for each group. The use of reflectors, blankets, and fuel loadings where either enrichment or lattice geometry varies will mean that separate equations must be written for each of the various regions for each group.

Computer codes are available for various combinations of fuel and moderator. Caution must be exercised that the code being used is correct for the calculation at hand. A code for a water-moderated, UO_2-fueled core will be of little use if applied to a graphite-moderated, sodium-cooled core fueled with UC.

Problems

1. Compute the thermal fission factor for 4 percent enriched UC (uranium monocarbide).

2. Compute the thermal utilization factor for a core where there are 25 moles of heavy water per mole of 4 percent enriched UC. What would be the value if light water replaced the heavy water?

3. Repeat Example 1, using natural uranium as the fuel.

4. Compute the infinite multiplication factor for a homogeneous mixture containing 10 moles of ^{232}Th and 500 moles of Be per mole of ^{233}U.

5. A reactor lattice consists of 0.635-cm-diameter UO_2 elements with 0.025-cm-thick Zr clad spaced 2.5 cm on centers in a square lattice, as shown in Fig. 9P.1. The moderator

$I_{eff} = 42.7 \times 10^{-24}$ $\varepsilon_{av} = .199$ $P = .51$ $f = .992$ $\eta = 2.033$ $\varepsilon = 1.0$

$K_{\infty} = 1.030$

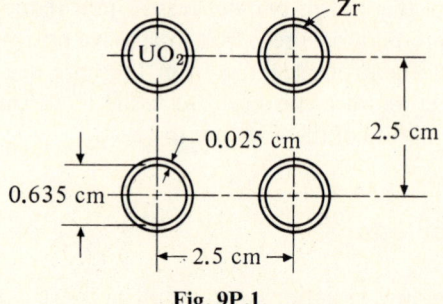

Fig. 9P.1.

is D_2O. Compute the infinite multiplication factor, k_∞. The fuel enrichment is 2 percent and the density of the D_2O is 0.77 g/cm³ at the average core temperature of 300°C. Consider the UO_2 density to be 10.5 g/cm³.

6. Using one-group theory for the lattice of Prob. 5, determine:
   (a) The material buckling
   (b) The thickness of an infinite critical slab
   (c) The diameter of a bare critical sphere
   (d) The dimensions of a bare critical cube
   (e) The height of a critical core with a square base 1.2 m on a side

7. For the bare spherical core of Prob. 6 the maximum flux is to be 10^{12} neutrons/cm² s. What power [kW(th)] will be developed?

8. Develop an expression similar to Eq. (9.93) for a bare critical cubic core. How many kW(th) will be developed in the core of Prob. 6(d) if the peak flux is 10^{12} neutrons/cm² s?

9. A spherical core having its lattice defined as in Prob. 5 is to be reflected with heavy water. The peak flux is to be 10^{12} neutrons/cm² s.
   (a) Find the critical radius of the core when the extrapolated reflector thickness is equal to the thermal diffusion length for the D_2O. What power will it develop?
   (b) Determine the critical radius and the power developed if the reflector were infinitely thick.

10. For a critical infinite slab with an infinitely thick reflector develop an expression similar to Eq. (9.120) for the critical slab thickness.

11. A spherical shell reactor has been suggested to serve as an intense source of radiant thermal energy. Such a core fueled by a uniform dispersion of uranium carbide particles in a graphite matrix would operate as a thermal reactor. Use one-group theory to develop an expression for the critical extrapolated outer-shell radius, assuming the inner radius is given. Also, determine an expression for the flux distribution in the shell.

   For a core whose material buckling is equal to 0.0269 cm^{-1} and whose interior radius is 100 cm, determine the outer extrapolated critical radius.

12. Write a computer program to solve Eq. (9.116) for a reflected spherical core using the lattice of Prob. 5 and a graphite reflector. Plot the reflector savings versus reflector thickness.

13. An infinite slab is bare on one face and has an infinite reflector on the other face. Develop an expression for the critical thickness of the slab. Assume that $D_r \approx D_c$.

14. For the semireflected slab of Prob. 13, determine the core thickness if the core contains 200 moles of graphite moderator per mole of 5 percent enriched uranium. The reflector is graphite; how far from the bare face will the flux be maximum?

15. A molten salt core is to operate at 650°C with 13 percent salt and 87 percent graphite by volume in the core, and 65 percent graphite and 35 percent salt in the blanket. The salt is made up of the following mole fractions:

$$LiF = 0.72 \qquad BeF_2 = 0.16 \qquad UF_4 = 0.005 \qquad ThF_4 = 0.115$$

The peak core flux is to be 10^{14} neutrons/cm^2 s. Shown below are some characteristics of the core.

	Core	Blanket
Thermal utilization factor	0.746 7	0.756 2
Resonance escape probability	0.617 3	0.383 6
Thermal fission factor	2.300 4	2.300 4
Fast fission factor	1.00	1.00
Macroscopic cross sections (cm^{-1})		
Thermal scatter	0.425 2	0.393 3
Epithermal scatter	0.410 0	0.374 3
Thermal transport	0.402 2	0.392 9
Epithermal transport	0.387 9	0.355 1
Absorption	0.007 848	0.020 862
Fission	0.005 392	0.014 52
Collisions to thermal	108.3	114.2

Utilizing one-group theory for a spherical core and blanket, what will be the reflector savings and the thermal power developed when the blanket thickness is 20 cm?

16. Compute the slowing-down length and the migration length for the lattice described in Prob. 5.

17. Using modified one-group theory and two-group theory, compute the radius of a bare critical sphere constructed with the lattice described in Prob. 5.

18. A molten salt core having 13 percent salt and 87 percent graphite by volume was described in Prob. 15. Use that information with one-group, modified one-group, and two-group theories to determine the radius and the height for a bare, critical, cylindrical core whose diameter equals its height.

19. Develop an expression for the power developed in a cylindrical core. Calculate the power developed by the bare, critical, cylindrical core of Prob. 18 using one-group, modified one-group, and two-group theories. The peak thermal flux is to be 10^{14} neutrons/cm^2 s.

20. A liquid metal–cooled fast reactor is fueled with 20 percent enriched uranium dioxide which occupies 25 percent of the core volume, the remainder being 25 percent stainless steel and 50 percent sodium. Determine the infinite multiplication factor for the core. What would be the critical radius of a bare cylindrical core whose height is equal to its diameter?

21. Determine the peak flux in the previous core if it is to develop 1000 MW(th).

22. A spherical fast reactor core containing the same percentages of fuel, coolant, and stainless steel as the core for Prob. 20 is to have a 20-cm thick blanket. The blanket is to contain 50 percent ThO_2, 20 percent stainless steel, and 30 percent sodium. Determine the critical core radius and the power developed in the core and blanket if the peak fast flux is 5×10^{14} neutrons/cm^2 s.

References

1. Glasstone, S., and M. C. Edlund, *The Elements of Nuclear Reactor Theory*. Princeton, N.J.: D. Van Nostrand Company, 1952.
2. Glasstone, S., and A. Sesonske, *Nuclear Reactor Engineering*. Princeton, N.J.: D. Van Nostrand Company, 1963.
3. Murray, R. L., *Nuclear Reactor Physics*. Englewood Cliffs, N.J.: Prentice-Hall, Inc., 1957.
4. El-Wakil, M. M., *Nuclear Power Engineering*. New York: McGraw-Hill Book Company, 1962.
5. Stephenson, R., *Introduction to Nuclear Engineering*. New York: McGraw-Hill Book Company, 1954.
6. Hawley, J. P., *SNAP 10A Reactor Operation*, ASME Paper No. 65–WA/NE–6, 1965.
7. Soodak, H., ed., *Reactor Handbook*, Vol. 3: *Physics*. New York: Interscience Publishers, 1962.
8. Lamarsh, J. R., *Introduction to Nuclear Reactor Theory*. Reading, Mass.: Addison-Wesley Publishing Co., Inc., 1966.
9. Murray, R. L., *Introduction to Nuclear Engineering*. Englewood Cliffs, N.J.: Prentice-Hall, Inc., 1961.
10. King, C. D. G., *Nuclear Power System*. New York: Macmillan Publishing Co., Inc., 1964.
11. Almenas, K., "A Proposal for Using Nuclear Reactors as Thermal Radiation Sources," *Nuclear Technology* 10, No. 1 (January 1971), pp. 22–32.
12. Onega, R. J., *An Introduction to Fission Reactor Theory*. Blacksburg, Va.: University Publications, 1975.
13. Lamarsh, J. R., *Introduction to Nuclear Engineering*. Reading, Mass.: Addison-Wesley Publishing, Co., Inc., 1975.
14. Murray, R. L., *Nuclear Energy*. Elmsford, N.Y.: Pergamon Press, Inc., 1980.
15. Henry, A., *Nuclear Reactor Analysis*. Cambridge, Mass.: The MIT Press, 1975.
16. Ouderstadt, J. J., and L. J. Hamilton, *Nuclear Reactor Analysis*. New York: McGraw-Hill Book Company, 1976.

10

Transient Reactor
Behavior
and Control

Previous chapters have dealt with reactors in which the flux (or neutron population) varies only with position   Perhaps as important as the steady-state behavior is the behavior of the neutron population when it varies with time as well as position.

The change from steady state in reactor neutron population is referred to as *reactivity change*.   Strictly speaking, the percent change in effective multiplication factor is the reactivity of a reactor.   Any change in power level (up or down) is accompanied by a reactivity change.   Reactivities are of two kinds:

(1) Temporary deliberate change in power level to a new steady value
(2) Accidental uncontrolled increase in power

These reactivities can be compensated for by:

(1) Control devices
(2) The reactor itself (self-regulating)
(3) Safety devices

Neutron Lifetime

The total neutron lifetime, l, is the sum of the slowing-down time, l_s, and the thermal lifetime, l_{th}.   The *slowing-down time* can be described as the time a neutron spends in slowing from fission to thermal energy, and the *thermal lifetime* as the time a neutron spends diffusing at thermal energies before absorption.   Compared with the thermal lifetime in a thermal reactor, the slowing-down time is short.   Thus, the neutron lifetime is frequently taken

equal to the thermal lifetime. This neutron lifetime is also called the *generation time* or the total cycle time. In equation form,

$$l = l_{th} + \ell_s \approx \ell_{th} \tag{10.1}$$

For a core of infinite extent, neutrons must diffuse until they are absorbed. The lifetime in an infinite core, $l_{th\infty}$, is found by dividing the absorption mean free path, λ_a, by the average neutron velocity, $\bar{v}$.

$$l_{th\infty} = \frac{\lambda_a}{\bar{v}} \tag{10.2}$$

If the reactor is finite, only $(N \times P_{NL_{th}})$ neutrons remain in the core to contribute to the effective lifetime of the generation. $N(1 - P_{NL_{th}})$ neutrons leak out and make no contribution to the thermal lifetime. Thus, the effective lifetime will be shorter for a finite reactor.

$$Nl_{th} = Nl_{th\infty}P_{th} + N(1 - P_{th})0$$

$$l_{th} = l_{th\infty}P_{th} = \frac{l_{th\infty}}{1 + B^2L^2} \tag{10.3}$$

Substituting Eqs. (10.3) and (10.2) into (10.1), we get

$$l \approx \frac{1}{\Sigma_a \bar{v}(1 + B^2L^2)} \tag{10.4}$$

Table 10.1 compares thermal lifetimes and slowing-down lifetimes for some common moderators. Notice that in each case the thermal lifetime is greater than the slowing-down time. In a fast reactor, of course, there is no true slowing-down time; the resulting neutron lifetime is about 10^{-7} s.

Table 10.1 Neutron Lifetimes for Pure Moderators

	Slowing-Down Time (s)	Thermal Lifetime (s)
C	1.5×10^{-4}	1.8×10^{-2}
H_2O	5.6×10^{-6}	2.1×10^{-4}
D_2O	4.3×10^{-5}	1.4×10^{-1}
Be	5.7×10^{-5}	3.7×10^{-3}

Example 1   Calculate the effective lifetime of a neutron in the reactor of Example 2 in Chapter 9, where

$$\frac{N_C}{N_U} = 200 \qquad\qquad L^2 = 68.2 \text{ cm}^2$$

$$B^2 = 0.00290 \text{ cm}^{-2} \qquad\qquad \Sigma_a = 0.0134 \text{ cm}^{-1}$$

$$\bar{v} = 2.482 \times 10^5 \text{ cm/s}$$

Neglecting the effect of neutron slowing-down time gives us

$$l = \frac{1}{\Sigma_a \bar{v}(1 + B^2 L^2)}$$

$$l = \frac{1}{0.0134(2.482 \times 10^5)(1 + 0.002\,90 \times 68.2)}$$

$$= \frac{1}{0.0134(2.482 \times 10^5)(1.216)}$$

$$= 2.50 \times 10^{-4} \text{ s}$$

This is much less than that for pure carbon. Comparison with Table 10.1 shows the effect on the lifetime of the increased neutron absorption in uranium.

Reactivity

When a reactor is operating at steady state, as discussed in Chapter 9, the effective multiplication factor, k_{eff}, is unity. While the power is increasing or decreasing (a reactivity change), the multiplication factor differs from unity. This difference is referred to as the *excess multiplication factor*:

$$\Delta k_{\text{eff}} = k_{\text{eff}} - 1 \qquad\qquad (10.5)$$

The *reactivity*, ρ, is the ratio of excess multiplication factor to effective multiplication factor:

$$\rho = \frac{\Delta k_{\text{eff}}}{k_{\text{eff}}} = \frac{k_{\text{eff}} - 1}{k_{\text{eff}}} \qquad\qquad (10.6)$$

Under steady-state conditions $\rho = 0$, and under normal power level changes near steady state, $\rho \approx \Delta k_{\text{eff}}$. Reactivity has been given units of dollars ($) and cents (¢) and inhours. More discussion of units will be encountered later. Since Δk_{eff} represents the fractional increase in the number of neutrons from one generation to the next, then the increase in neutron flux per generation

equals $\phi \, \Delta k_{eff}$. The increase in neutron flux per unit of time is found by dividing the flux increase per generation by the neutron lifetime (generation time), l.

$$\frac{d\phi}{dt} = \frac{\phi \, \Delta k_{eff}}{l} \tag{10.7}$$

It is plain that the right side of Eq. (10.7) is the neutron flux gained per generation divided by the total time per generation, or the number of neutrons gained per unit time.

Assuming that Δk_{eff} is time independent and integrating, we get

$$\phi = \phi_0 e^{(\Delta k_{eff}/l)t} \tag{10.8}$$

where ϕ_0 represents the initial or steady-state neutron flux. Define the reactor period (or e folding time), T, as the time for the flux to change by a factor of e.

$$T = \frac{l}{\Delta k_{eff}} \tag{10.9}$$

$$\phi = \phi_0 e^{t/T} \tag{10.10}$$

For the reactor to be at steady state $\phi = \phi_0$, and therefore $T = \infty$. Reactor period is a very important concept and is one of the most responsive indicators of reactor conditions; all reactors employ an automatic safety system that will *scram* the reactor if the period gets too small. During power changes the period must be kept as large as possible to prevent dangerous power excursions. A period of less than 7 to 10 s represents a definitely dangerous condition.

Example 2 The neutron lifetime for a neutron produced by fission in a thermal reactor is about 10^{-3} s. Using this value and a reactivity of $\frac{1}{2}$ of 1 percent, determine the power increase of a reactor in 2 s.

$$T = \frac{l}{\Delta k_{eff}}$$

We can use a value of $\Delta k_{eff} = 0.005$ since ρ will be very nearly equal to Δk_{eff}.

$$T = \frac{0.001}{0.005} = \tfrac{1}{5} \text{ s}$$

$$\frac{\phi}{\phi_i} = e^{5(2)} = 22\,026$$

Since power is proportional to neutron flux the power would also increase to 20 000 times its original value. Any human-made structure would have a difficult time withstanding this power excursion! Of course, if the Δk_{eff} is decreased or the effective neutron lifetime increased, or both, the reactor period will be lengthened. This is the direction in which to strive for reactor control.

Delayed Neutrons

Fortunately, not all neutrons occurring as a result of fission appear simultaneously. From observation, the neutrons appear in seven distinct groupings (see Table 10.2) according to their mean lifetimes from fission to absorption. The overwhelming majority (over 99 percent) are prompt neutrons. These are emitted directly from fission within about 10^{-13} s after fission begins. The remainder are the delayed neutrons, the result of neutron decay of some of the fission fragments. The decay of 55.72-s half-life ^{78}Br along one branch of its decay diagram (Fig. 10.1) involves β^- emission followed by the instantaneous emission of a neutron by the excited state of ^{87}Kr. Thus, the half-life of the first decay group is characterized by that of the β^- decay of ^{87}Br. Similarly, for the second group, 22.72 s, ^{137}I decays by β^- emission followed by the ejection of a neutron.

$$^{137}\text{I} \xrightarrow{\ \beta^-\ } {}^{137}\text{Xe} \xrightarrow{\ n\ } {}^{136}\text{Xe}$$

Even though the delayed neutrons are a very small percentage of the total neutrons, they can have a significant effect on the reactivity because their mean lifetimes are long. Also, their energies are smaller than those of prompt neutrons (i.e., less fast leakage), and therefore they are more effective in absorption than an equal number of prompt neutrons. Table 10.2 presents the fraction of prompt $(1 - \beta)$ and delayed (β) neutrons for each of the three fissionable fuels. The groupings are not exact, but they represent the best available observations grouped into the most representative lifetimes.

Table 10.2 Prompt and Delayed Neutron Groups from Thermal Fission

Group i	Energy for ^{235}U Fission (*MeV*)	Half-Life for ^{235}U Fission (*s*)	Percentage of Total Neutrons from Fission ^{235}U (β_i)	^{233}U (β_i)	^{239}Pu (β_i)
0	~2	~10^{-8}	99.359	99.736	99.790
1	0.250	55.72	0.021	0.023	0.007
2	0.560	22.72	0.140	0.079	0.063
3	0.430	6.22	0.126	0.066	0.044
4	0.620	2.30	0.253	0.073	0.069
5	0.430	0.61	0.074	0.014	0.018
6		0.23	0.027	0.009	0.009
		$\sum_{1}^{6} \beta_i =$	0.641	0.264	0.210

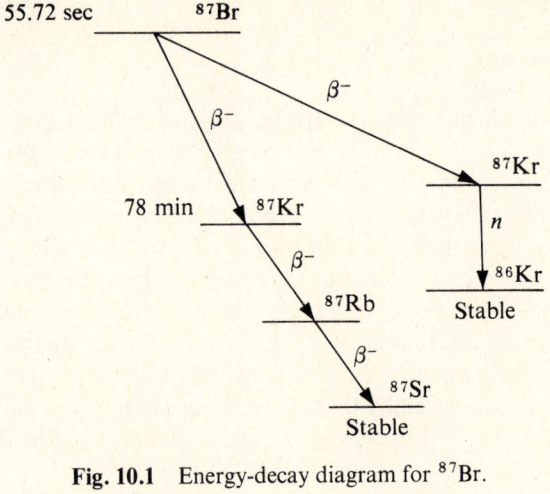

Fig. 10.1   Energy-decay diagram for ^{87}Br.

Average Neutron Lifetime

We can now find an average lifetime for a neutron in a thermal reactor. This is a straight averaging process and is best done by means of an example.

> **Example 3**   Find the average neutron lifetime, $\bar{l}$, of all neutrons in a ^{235}U reactor. Note that the mean lifetime of a delay group is the reciprocal of its decay constant.

$$\bar{l} = \frac{\sum\limits_{i=0}^{6} \beta_i l_{mi}}{\sum\limits_{i=0}^{6} \beta_i} \qquad \bar{l}_{mi} = \frac{1}{\lambda_i} = \frac{t_{1/2}}{0.693} \qquad \textbf{(10.11)}$$

i	β_i	l_{mi}	$\beta_i l_{mi}$
0	99.359	$\sim 10^{-3}$	0.099 36
1	0.021	80.40	1.688
2	0.140	32.79	4.591
3	0.126	9.01	1.135
4	0.253	3.32	0.840
5	0.074	0.88	0.065
6	0.027	0.33	0.009

$$\sum \beta_i = 100 \qquad \sum_{i=0}^{6} \beta_i l_{mi} = 8.428$$

$$\bar{l} = \frac{8.43}{100} = 0.0843 \text{ s}$$

This shows that only 0.641 percent of the total neutrons have increased the effective generation time by a factor of approximately 84. This, of course, can effectively increase the reactor period by a like amount. Without the delayed neutrons control would indeed be difficult. Returning to Example 2, with an average lifetime of 0.0843 s for a ^{235}U-fueled reactor, the period becomes

$$T = \frac{l}{\Delta k_{\text{eff}}} = \frac{0.0843}{0.005} = 16.86 \text{ s}$$

The resulting period, using l, becomes the limiting value for very, very small reactivity increases. After 2 s the new flux level will be

$$\phi = \phi_0 e^{t/T} = \phi_0 e^{2/16.8} = 1.126\phi_0$$

Effect of Delayed Neutrons

Figure 10.2 shows the effect of reactivity changes on the neutron flux. The dashed curve shows the relative neutron flux for a Δk_{eff} if there were no delayed neutrons present. It is merely a plot of Eq. (10.8). The curve labeled $\Delta k_{\text{eff}} > 0$ (with delayed neutrons), however, shows what actually happens when excess reactivity is introduced. At first the reactor behaves as if all neutrons are prompt because the delayed neutrons from the increased flux are not yet effective.

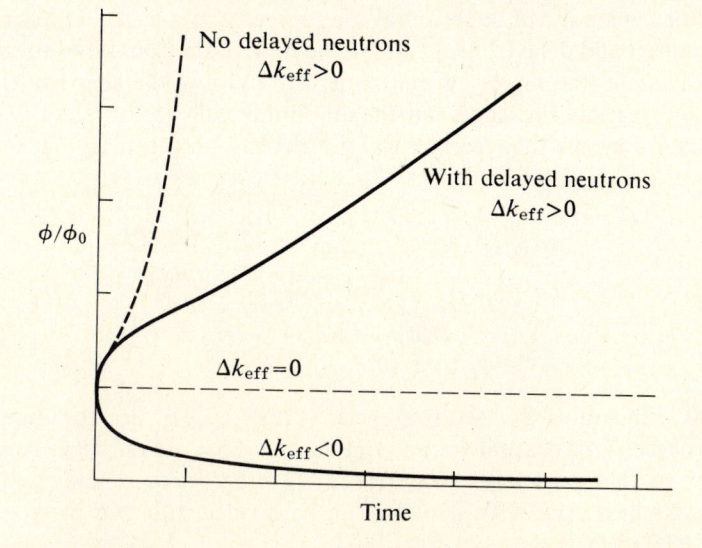

Fig. 10.2 Effect of reactivity change on neutron flux in a critical reactor.

After a few seconds the delayed neutrons appear and the rate of flux increase begins to level off. The flux is still increasing, but the number of delayed neutrons entering the reactor is still smaller than the number of neutrons being delayed. Therefore, the rate of increase of flux gradually decreases and approaches a constant value. The rate of flux increase finally approaches a value determined by what is called the *stable reactor period*. More will be said later concerning the stable period. When $\Delta k_{\text{eff}} < 0$, the rate of decrease of flux is very sharp at first. When the delayed neutrons appear they now tend to flatten out the curve. Since the flux is dying out, the short-lived delayed neutrons disappear completely and the curve approaches a slope whose value is determined by the longest-lived neutron group.

Diffusion Equation for a Transient Reactor

In Chapter 9 the diffusion equation was developed and solved for various geometries. In a transient reactor we have exactly the same conditions and materials as in a critical reactor, except for the simplification that the neutron flux does not change with time. The equation governing the neutron flux will be (9.42), which is repeated here:

$$\frac{\partial n}{\partial t} = D\nabla^2\phi - \phi\Sigma_a + S \tag{10.12}$$

where S represents some source of neutrons for the next fission. From the previous discussion it can be seen that the source, S, in a transient reactor will involve prompt and delayed neutrons. If we now restrict ourselves to reactors near steady state (the region of most interest anyway), the shape of the flux distribution will stay the same, but the amplitude will rise or decrease. This approximation means that the flux will not depend on geometry but on time. We can say, therefore, that

$$\nabla^2\phi \approx -B^2\phi \qquad \text{and} \qquad \frac{\partial n}{\partial t} = \frac{dn}{dt}$$

$$\frac{dn}{dt} = -DB^2\phi - \phi\Sigma_a + S \tag{10.13}$$

Figure 10.3 illustrates the neutron cycle. $c_1, c_2, \ldots, c_6$ are the number of neutrons of each type per unit volume and $\lambda_1, \lambda_2, \ldots, \lambda_6$ are the decay constants. The neutrons that are left are available for slowing, leakage, absorption, and fission in the next cycle. This source can be divided into two parts, prompt (S_p) and delayed (S_d):

$$S = S_p + S_d \tag{10.14}$$

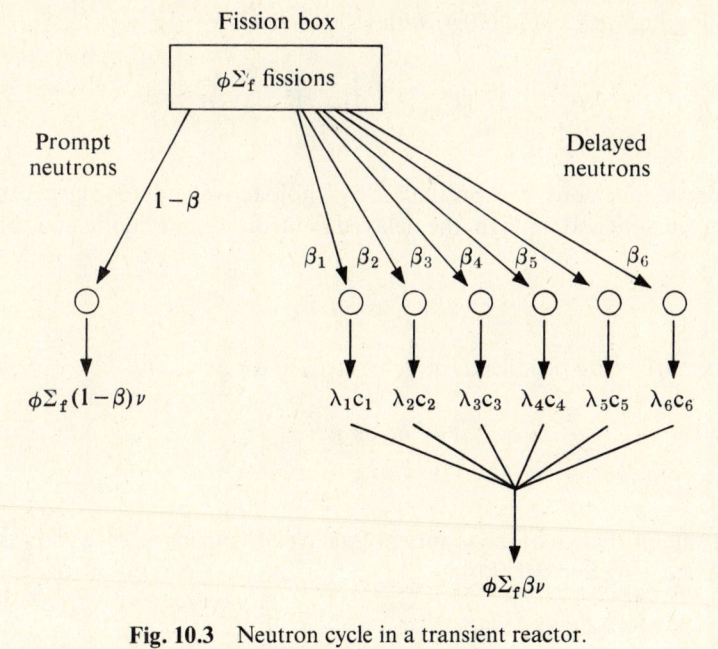

Fission box

$\phi\Sigma_f$ fissions

Prompt
neutrons

Delayed
neutrons

$1-\beta$

β_1 β_2 β_3 β_4 β_5 β_6

$\phi\Sigma_f(1-\beta)\nu$

$\lambda_1 c_1$ $\lambda_2 c_2$ $\lambda_3 c_3$ $\lambda_4 c_4$ $\lambda_5 c_5$ $\lambda_6 c_6$

$\phi\Sigma_f\beta\nu$

Fig. 10.3 Neutron cycle in a transient reactor.

The prompt source reaching thermal is that from the $1-\beta$ fraction of prompt neutrons. In a manner similar to Eq. (9.41),

$$S_p = \phi\Sigma_a k_\infty P_f(1-\beta) \tag{10.15}$$

The delayed source is made up of the remainder of neutrons, which we have already divided into their six groups.

$$S_d = \sum_{i=1}^{6} C_i \lambda_i \tag{10.16}$$

where C_i is the number of delayed neutrons in the ith group actually reaching thermal energies $= pP_f c_i$. Substituting Eqs. (10.14), (10.15), and (10.16) into (10.13) gives

$$\frac{dn}{dt} = -\phi\Sigma_a\left[B^2\frac{D}{\Sigma_a} + 1\right] + \phi\Sigma_a k_\infty P_f(1-\beta) + \sum_{1}^{6} C_i\lambda_i$$

$$= -\frac{\phi\Sigma_a}{P_{\text{th}}} + \phi\Sigma_a k_\infty P_f(1-\beta) + \sum_{1}^{6} C_i\lambda_i$$

$$= \frac{\phi\Sigma_a}{P_{\text{th}}}\left[k_{\text{eff}}(1-\beta) - 1\right] + \sum_{1}^{6} C_i\lambda_i \tag{10.17}$$

Combining Eqs. (8.22) and (10.4) with (10.17) yields

$$\frac{dn}{dt} = \frac{n}{l_{th}} [k_{eff}(1 - \beta) - 1] + \sum_{1}^{6} C_i \lambda_i \qquad (10.18)$$

The delayed neutrons are produced by radioactively decaying precursors. These precursors and, in turn, the delayed neutrons are decaying at a rate

$$-\lambda_i C_i$$

while they are being produced (or generated) at a rate

$$\frac{k_{eff} \beta_i n}{l_{th}}$$

The net rate of change of the number of delayed neutrons is, therefore, the sum of those decaying and generating:

$$\frac{dC_i}{dt} = -\lambda_i C_i + \frac{k_{eff} \beta_i n}{l_{th}} \qquad (10.19)$$

The solution of the seven simultaneous Eqs. (10.18) and (10.19) yields an expression for reactivity in terms of reactor period and neutron lifetime. This equation is called the *inhour equation*:

$$\rho = \frac{l}{k_{eff} T} + \sum_{1}^{6} \frac{\beta_i}{1 + \lambda_i T} \qquad (10.20)$$

In Eq. (10.20) the thermal neutron lifetime has been replaced by the total neutron lifetime. A more accurate development and solution of Eqs. (10.18) and (10.19) would yield this result. The first term of Eq. (10.20) refers to the prompt neutrons; the six succeeding terms apply to the delayed neutrons.

If Eqs. (10.18) and (10.19) are solved for n, the number of neutrons per unit volume, the parameter T will also appear. For positive values of reactivity this T has six negative values and one positive value. The negative values die out quickly, leaving the positive value to dominate. It is this positive number that is called the *stable reactor period* and is a result of the solutions to Eqs. (10.18) and (10.19), which were plotted in Fig. 10.2. The slope of the curve is determined by the stable reactor period. For negative values of reactivity ($k_{eff} < 1$), all values of T are negative. Each term then decays exponentially to zero, and the stable period is determined by the longest-lived delayed neutron group.

Units of Reactivity

The *inhour* is the amount of reactivity that will make the stable reactor period 1 h:

$$\rho(\text{inhours}) = \frac{l}{3600k_{\text{eff}}} + \sum_{1}^{6} \frac{\beta_i}{1 + 3600\lambda_i} \qquad \textbf{(10.21)}$$

A more common unit of reactivity is the dollar ($). It is an amount of reactivity equivalent to β. It follows that a cent is one-hundredth of a dollar.

Example 4   Find the amount of reactivity corresponding to 1 inhour for ^{235}U. What is this in units of dollars?

$$\rho = \frac{l}{3600k_{\text{eff}}} + \sum_{1}^{6} \frac{\beta_i}{1 + 3600\lambda_i}$$

Instead of using all six delayed neutron groups, combine the six groups into one group of average lifetime:

$$\rho = \frac{l}{3600k_{\text{eff}}} + \frac{\beta l_d}{l_d + 3600} \qquad \textbf{(10.22)}$$

where l_d is the average lifetime of delayed neutrons:

$$l_d = \frac{\sum_{1}^{6} \beta_i l_{mi}}{\beta} = \frac{0.08335}{0.00641}$$

$$= 13.0 \text{ s}$$

$$\rho = \frac{10^{-3}}{3600} + \frac{0.006\,41(13)}{13 + 3600}$$

$$= 0.0278 \times 10^{-5} + 2.306 \times 10^{-5}$$

$$= 2.33 \times 10^{-5}$$

Therefore, 1 inhour is equivalent to 2.33×10^{-5} units of reactivity. This corresponds to

$$\$ = \frac{2.33 \times 10^{-5}}{0.006\,41} = \$0.003\,63$$

The inhour equation is frequently plotted graphically so that for a given reactivity the period can be determined. With the graph available, various

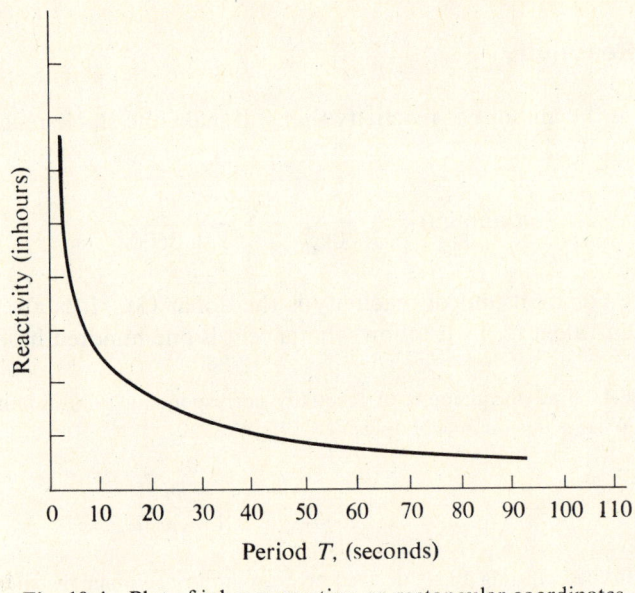

Fig. 10.4 Plot of inhour equation on rectangular coordinates.

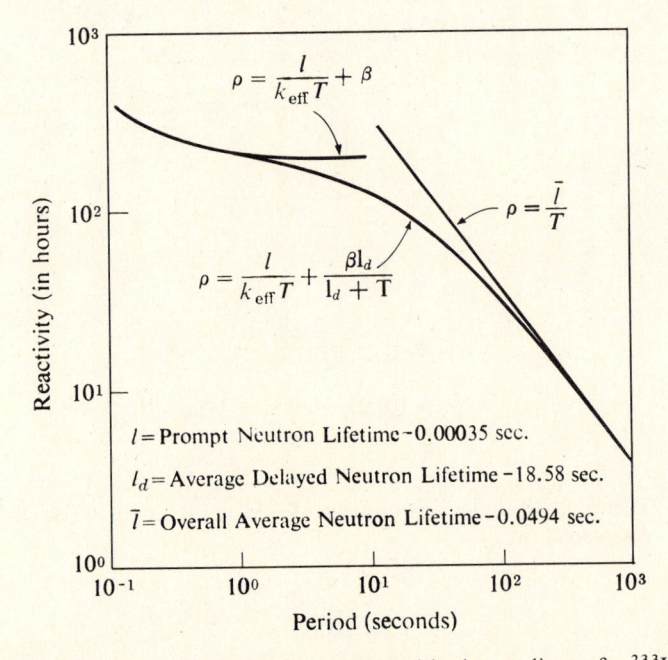

$$\rho = \frac{l}{k_{\text{eff}}T} + \beta$$

$$\rho = \frac{\bar{l}}{T}$$

$$\rho = \frac{l}{k_{\text{eff}}T} + \frac{\beta l_d}{l_d + T}$$

l = Prompt Neutron Lifetime – 0.00035 sec.

l_d = Average Delayed Neutron Lifetime – 18.58 sec.

$\bar{l}$ = Overall Average Neutron Lifetime – 0.0494 sec.

Fig. 10.5 Plot of inhour equation on logarithmic coordinates for ^{233}U-fueled reactor which is beryllium moderated with 1 mole ^{233}U per 10 moles Th and 500 moles Be.

parameters can be determined in terms of the reactivity added or taken away from the reactor. The resulting period is then available from the graph. Figures 10.4 and 10.5 show a plot of the inhour equation. The reactivity used with these graphs would be evaluated in terms of such physical factors as control rod worth, temperature change, fuel depletion or addition, reflector adjustment or removal, moderator or coolant changes, and core size or shape changes.

Limiting Cases of Δk_{eff}

Suppose that Δk_{eff} and ρ are very very small ($\ll \beta$). The inhour equation (using one group of delayed neutrons) becomes

$$\frac{\Delta k_{eff}}{k_{eff}} = \frac{l + \beta \bar{l}_d}{T} \qquad (10.23)$$

because T will be large compared with $\bar{l}_d$ in the denominator:

$$\Delta k_{eff} = \frac{\bar{l}}{T} \qquad (10.24)$$

This shows that for very small reactivity all the delayed neutron emitters contribute to the increase of flux, and the period may be computed using the average lifetime of all the neutrons.

On the other hand, if Δk_{eff} and ρ are large, the period will be small, and the inhour equation becomes

$$\frac{k_{eff} - 1}{k_{eff}} \approx \frac{l}{k_{eff} T} + \beta \qquad (10.25)$$

$$T = \frac{l}{k_{eff} - 1 - \beta k_{eff}} = \frac{l}{k_{eff}(1 - \beta) - 1} \qquad (10.26)$$

The prompt excess multiplication factor is sometimes defined as

$$\Delta k_{exp} = k_{eff}(1 - \beta) - 1 \qquad (10.27)$$

Substituting, Eq. (10.26) becomes

$$T = \frac{l}{\Delta k_{exp}} \qquad (10.28)$$

This shows that for large reactivity the $(1 - \beta)$ prompt neutrons govern the period, since the delayed neutrons cannot follow the reaction.

An interesting transition takes place when $\Delta k_{exp} = 0$. At this point

$$\Delta k_{exp} = k_{eff}(1 - \beta) - 1 = 0$$

$$\Delta k_{eff} = \beta k_{eff}$$

$$\rho = \beta$$

If this should be the case the reactor would be prompt-critical. This is a dangerous condition since the reactor is now critical on prompt neutrons alone. When $\rho < \beta$ the reactor is said to be delayed-critical; thus the delayed neutrons control the reactor. Needless to say, a reactor should always be delayed-critical.

> **Example 5** A control rod is suddenly withdrawn from a water-moderated ^{235}U reactor. The period is seen to be 30 min. What is the reactivity in dollars? From Table 10.1 the thermal neutron lifetime of the water is seen to be 2.1×10^{-4} s. It can be assumed that a period of 30 min is long and that Eq. (10.24) applies.
>
> $$\Delta k_{eff} = \frac{\bar{l}}{T}$$
>
> For convenience, let us use one group of delayed neutrons and the thermal neutron lifetime in the pure moderator.
>
> $$\Delta k_{eff} = \frac{2.1 \times 10^{-4} + 0.006\,41(13.0)}{30(60)}$$
>
> $$\rho \approx \Delta k_{eff} = 4.65 \times 10^{-5}$$
>
> $$\$ = \frac{\rho}{\beta} = \frac{4.65 \times 10^{-5}}{0.006\,41}$$
>
> $$= 0.007\,24$$

Natural Reactivity Changes

In an operating reactor many reactivity changes occur as a result of physical or nuclear changes in the system. They take place over a relatively long period of time and result directly from reactor operation. These effects are due mainly to temperature changes, pressure changes, buildup of poisons, fuel depletion, and fuel buildup. To compensate for decreases in reactivity, excess reactivity $(\Delta \rho)$ is built into reactors initially. Control rods are inserted far into the core

Table 10.3 Factors Affecting Reactivity

Factor	Effect
Temperature increase	Decrease
Pressure increase (large ^{10}B)	Decrease
Steam formation	Decrease
Fission product accumulation	Decrease
Fuel depletion	Decrease
Fuel breeding	Increase

and, as reactivity changes, their equilibrium position can be raised to balance the natural reactivity change. To simplify the solution to these reactivity changes it is usually assumed that either the reactivity changes almost instantaneously or that the reactivity changes so slowly that flux remains constant. Table 10.3 shows the expected change in reactivity for some material factors in reactor operation. Succeeding paragraphs discuss these in more detail.

Temperature Effects on Reactivity

The temperature coefficient of reactivity (α) is made up of three parts:

(1) A nuclear temperature coefficient arising from a change in cross section with changing neutron temperature $(\partial\rho/\partial T)_{N,B^2}$

(2) A density temperature coefficient arising from a change in material density N with changing temperature $(\partial\rho/\partial T)_{\sigma,B^2}$

(3) A volume temperature coefficient arising from a change in geometric buckling when temperature changes $(\partial\rho/\partial T)_{\sigma,N}$

The temperature coefficient of reactivity is then given by

$$\alpha = \frac{d\rho}{dT} = \left(\frac{\partial\rho}{\partial T}\right)_{N,B^2} + \left(\frac{\partial\rho}{\partial T}\right)_{\sigma,B^2} + \left(\frac{\partial\rho}{\partial T}\right)_{\sigma,N} \qquad \textbf{(10.29)}$$

The temperature coefficient (α) should be (and is) negative for stable operation. A negative α means that as temperature increases, reactivity decreases, the fissioning rate ($\phi\sigma_f$) decreases, and the heat transfer decreases causing temperature to fall back toward its original value. In this case the reactor is said to be self-regulating.

Example 6   Let us develop an expression for the change in reactivity due to temperature variation at very nearly steady state. For a large reactor, Eqs. (9.2) and (9.140) give the effective multiplication factor,

$$k_{\text{eff}} = \frac{k_\infty}{1 + (L_s^2 + L^2)B^2}$$

and Eq. (10.6) gives the reactivity

$$\rho = \frac{k_{\text{eff}} - 1}{k_{\text{eff}}}$$

Therefore,

$$\rho = 1 - \frac{1 + (L_s^2 + L^2)B^2}{k_\infty} = 1 - \frac{1 + M^2 B^2}{k_\infty} \tag{10.30}$$

In Eq. (10.30) L, L_s, and k_∞ are temperature dependent while B^2 is dependent on core volume, which is, in turn, dependent on temperature.

The nuclear temperature coefficient $(\partial \rho / \partial T)_{N, B^2}$ will be dependent almost entirely on thermal diffusion length. If we assume that neutron cross sections follow the $1/v$ law, then the absorption cross section is inversely proportional to the square root of the absolute temperature. Since the square of the thermal diffusion length, L^2, is also inversely proportional to the absorption cross section it will increase as the temperature increases. This increase in L also increases the neutron leakage. The result of this fairly large increase in L causes the reactivity, ρ, to decrease [Eq. (10.30)], thereby making the nuclear temperature coefficient negative.

The buckling, B^2, is a geometrical quantity and has little effect on the nuclear temperature coefficient. The slowing-down length, L_s, is inversely proportional to the scattering cross section. The scattering cross section is nearly independent of neutron energy, and therefore temperature has practically no effect on L_s.

Because of the characteristics of ε, p, f, and η temperature has relatively little effect on k_∞. Because of the definitions of ε and η they will not be affected by changes in temperature. Also, f will not be affected in a homogeneous reactor. However, the thermal disadvantage factor will decrease somewhat with temperature increase, causing a slight increase in f in a heterogenous reactor. The resonance escape probability, p, will decrease slightly with increase in temperature because of the increase in resonance absorption. This increase in resonance absorption arises chiefly through the *Doppler effect*. The Doppler effect means that the resonance peaks increase in width with increasing material temperatures. The phenomenon is analogous to frequency changes due to the relative motion of sources of sound. The decrease in p, however, is small and

will tend to cancel any increase in f. The net effect on k_∞ due to increased temperature is, therefore, practically zero.

In determining $(\partial\rho/\partial T)_{N,B^2}$, the nuclear temperature coefficient, the partial reactivity with respect to the square of the thermal diffusion length is multiplied by the partial derivative of the square of the thermal diffusion length with respect to temperature.

$$\left(\frac{\partial\rho}{\partial T}\right)_{N,B^2} = \frac{\partial\rho}{\partial L^2} \times \frac{\partial L^2}{\partial T}$$

Letting σ_a' be a reference cross section at datum temperature T', the cross section may be expressed as a function of temperature, T.

$$\sigma_a = \sigma_a' \frac{T'^{1/2}}{T^{1/2}}$$

The square of the thermal diffusion length then becomes

$$L^2 = \frac{\lambda_{\text{tr}} T^{1/2}}{3N\sigma_a' T'^{1/2}}$$

and

$$\frac{\partial L^2}{\partial T} = \frac{1}{2}\left(\frac{\lambda_{\text{tr}} T^{-1/2}}{3N\sigma_a' T'^{1/2}}\right)\frac{T}{T}$$

$$= \frac{1}{2}\left(\frac{\lambda_{\text{tr}} T^{1/2}}{3N\sigma_a' T'^{1/2}}\right)\frac{1}{T} = \frac{1}{2}\frac{L^2}{T}$$

Differentiating the reactivity with respect to the square of the thermal diffusion length gives

$$\frac{\partial\rho}{\partial L^2} = -\frac{B^2}{k_\infty}$$

and

$$\left(\frac{\partial\rho}{\partial T}\right)_{N,B^2} = -\frac{B^2 L^2}{2k_\infty T} \qquad \textbf{(10.31a)}$$

The density temperature coefficient, $(\partial\rho/\partial T)_{\sigma,B^2}$, will be dependent on the change in nuclear density of reactor materials. An increase in reactor temperature will cause nuclear densities to decrease and also reduce the macroscopic cross sections. The mean free path, thermal diffusion length, and slowing-down length will increase because they are inversely proportional to the density. Equation (10.30) then reveals that reactivity, ρ, will decrease, neutron leakage

will increase, and density temperature coefficient will be negative. The effect of density changes on k_∞ will cancel one another, leaving k_∞ practically independent of temperature-caused density changes.

To establish the density temperature coefficient, $(\partial\rho/\partial T)_{\sigma,B^2}$, the partial derivative of reactivity with respect to the square of the migration length is multiplied by the partial of the square of the migration length with respect to temperature.

$$\left(\frac{\partial\rho}{\partial T}\right)_{\sigma,B^2} = \frac{\partial\rho}{\partial M^2}\frac{\partial M^2}{\partial T}$$

Differentiating Eq. (10.30) with respect to the square of the migration length, we get

$$\frac{\partial\rho}{\partial M^2} = -\frac{B^2}{k_\infty}$$

Both L^2 and L_s^2 vary inversely as N^2:

$$L^2 + L_s^2 = M^2 = \frac{1}{N^2}\left(\frac{1}{3\sigma_a\sigma_{tr_{th}}} + \frac{n}{3\sigma_{tr_f}^2}\right)$$

The number density of atoms in the core is inversely proportional to the specific volume of the core material. The specific volume is in turn proportional to the cube of the increase in linear dimension due to thermal expansion. Let N', v', and M' represent fixed values at the reference temperature T'. $\bar\alpha$ is the temperature coefficient of linear expansion of the core material.

$$\frac{v}{v'} = \left(\frac{N'}{N}\right) = [1 + \bar\alpha(T - T')]^3$$

$$M^2 = \left(\frac{N'}{N}\right)^2 M'^2 = [1 + \bar\alpha(T - T')]^6 M'^2$$

Differentiating with respect to temperature yields

$$\frac{\partial M^2}{\partial T} = 6[1 + \bar\alpha(T - T')]^5 M'^2\bar\alpha$$

Then

$$\left(\frac{\partial\rho}{\partial T}\right)_{\sigma,B^2} = -\frac{B^2}{k_\infty}6[1 + \bar\alpha(T - T')]^5\bar\alpha\frac{M^2}{[1 + \bar\alpha(T - T')]^6}$$

$$= \frac{-6B^2M^2\bar\alpha}{k_\infty[1 + \bar\alpha(T - T')]}$$

When the deviation from criticality is small,

$$k_{\text{eff}} \simeq 1 = \frac{k_\infty}{1 + B^2 M^2}$$

$$B^2 M^2 = k_\infty - 1$$

and for small temperature changes

$$1 + \bar{\alpha}(T - T') \cong 1$$

Then

$$\left(\frac{\partial \rho}{\partial T}\right)_{\sigma, B^2} = \frac{-6\bar{\alpha}(k_\infty - 1)}{k_\infty} \qquad \textbf{(10.31b)}$$

The volume temperature coefficient, $(\partial \rho / \partial T)_{\sigma, N}$, depends only on the geometrical buckling. Buckling is inversely proportional to change in radius, and a volume increase due to temperature increase will decrease buckling. Equation (10.30) shows that reactivity will then increase. This will make the temperature coefficient positive.

To develop the volume temperature coefficient, use the buckling of a spherical core. The resultant expression will be approximately correct for other compact core configurations. The radius of the sphere is a function of the extrapolated radius of the core,

$$\frac{r_e}{r_e'} = 1 + \bar{\alpha}(T - T')$$

where r_e' is a reference radius at the base temperature T'. By developing $(\partial \rho / \partial T)_{\sigma, N}$ in a manner similar to that used for Eq. (10.31b), the volume temperature coefficient may be shown to be

$$\left(\frac{\partial \rho}{\partial T}\right)_{\sigma, N} = 2\bar{\alpha}\frac{k_\infty - 1}{k_\infty} \qquad \textbf{(10.31c)}$$

Fortunately, the two negative coefficients usually more than overcome the one positive coefficient and leave the overall temperature coefficient of reactivity, α, negative. Table 10.4 summarizes the various contributions to temperature coefficient of reactivity.

The total temperature coefficient can be evaluated by summing the three partial coefficients

$$\frac{d\rho}{dT} = -\frac{B^2 L^2}{2k_\infty T} - 6\bar{\alpha}\frac{k_\infty - 1}{k_\infty} + 2\bar{\alpha}\frac{k_\infty - 1}{k_\infty} \qquad \textbf{(10.32)}$$

where $\bar{\alpha}$ is a temperature coefficient of linear expansion.

Table 10.4 Temperature Effects on Reactivity for Increasing Temperature

Parameter	Effect on Coefficient		
	Nuclear	Density	Volume
σ_s	$\longleftrightarrow$	—	—
σ_a	$\downarrow$	—	—
N	—	$\downarrow$	—
Σ_s	$\longleftrightarrow$	$\downarrow$	—
Σ_a	$\downarrow$	$\downarrow$	—
L	$\uparrow$	$\uparrow$	—
L_s	$\longleftrightarrow$	$\uparrow$	—
λ	$\uparrow$	$\uparrow$	—
ϵ	$\longleftrightarrow$	$\longleftrightarrow$	—
η	$\longleftrightarrow$	$\longleftrightarrow$	—
p	$\downarrow$	$\longleftrightarrow$	—
f	$\uparrow$ possibly	$\longleftrightarrow$	—
k_∞	$\longleftrightarrow$	$\longleftrightarrow$	—
B	—	—	$\downarrow$
Net effect	Negative	Negative	Positive

$\longleftrightarrow$ Unchanged
$\downarrow$ Decrease
$\uparrow$ Increase

A positive temperature coefficient would be very dangerous and care is taken to ensure that it is negative. In homogeneous reactors the negative temperature coefficient is the largest, but heterogeneous reactors with liquid coolants may also have large density variations providing large negative coefficients. Solid moderated gas-cooled reactors are not subject to density variations and have the smallest negative temperature coefficients. In a ^{235}U-fueled fast reactor the temperature coefficient may be positive because the Doppler effect actually increases fission. To prevent this condition, the ^{235}U content is limited so that the ^{238}U contribution will maintain the temperature coefficient negative at operating temperature.

Fission Product Accumulation

Some of the fission products or their decay products are strong neutron absorbers. These poisons accumulate during reactor operation, causing a decrease in reactivity. Two fission decay products, ^{135}Xe and ^{149}Sm, exhibit

extremely high absorption cross sections for thermal neutrons: 2.72×10^6 barns and 4.08×10^4 barns, respectively. As a decay product ^{135}Xe has a 5.6 percent yield and a 0.3 percent yield as a direct fission fragment. ^{149}Sm has a 1.4 percent yield as a fission decay product. These absorbers, therefore, build up to significant amounts, and the neutrons they tend to absorb must be recognized. ^{135}Xe is a product in the decay chain

$$^{135}\text{Te} \xrightarrow[\text{2 min}]{\beta} {}^{135}\text{I} \xrightarrow[\text{6.7 h}]{\beta} {}^{135}\text{Xe} \xrightarrow[\text{9.2 h}]{\beta} {}^{135}\text{Cs} \xrightarrow[\text{3} \times 10^6 \text{ yr}]{\beta} {}^{135}\text{Ba}$$

and ^{149}Sm is the stable product of the decay chain

$$^{149}\text{Nd} \xrightarrow[\text{1.8 h}]{\beta} {}^{149}\text{Pm} \xrightarrow[\text{55 h}]{\beta} {}^{149}\text{Sm}$$

In a steady-state reactor secular equilibrium exists and the poisons will decay at the same rate they are produced. The primary effect of the poisons is reduction in thermal utilization. Three processes can exist which affect the rate of change of the poison nuclei. They are production, radioactive decay, and loss by absorption of neutrons. Table 10.5 indicates the rates at which these changes take place.

When secular equilibrium exists, the production of an isotope must equal the loss of that isotope. From Table 10.5, for the chain including ^{135}Xe,

$$\phi \Sigma_f y_{\text{Te}} + \phi \Sigma_f y_{\text{Xe}} + \lambda_\text{I} N_\text{I} = \lambda_\text{I} N_\text{I} + \lambda_{\text{Xe}} N_{\text{Xe}} + \phi \Sigma_{\text{Xe}} \sigma_{a\text{Xe}}$$

$$N_{\text{Xe}} = \frac{\phi \Sigma_f (y_{\text{Te}} + y_{\text{Xe}})}{\lambda_{\text{Xe}} + \phi \sigma_{a\text{Xe}}} \tag{10.33}$$

Table 10.5 Rates of Change of Poison Nuclei in a Reactor

Isotope	Production	Decay	Removal by Neutron Absorption
^{135}I	$\phi \Sigma_f y_{\text{Te}}$	$\lambda_\text{I} N_\text{I}$	$\phi N_\text{I} \sigma_{al} \approx 0$
^{135}Xe	$\phi \Sigma_f y_{\text{Xe}}$ $\lambda_\text{I} N_\text{I}$	$\lambda_{\text{Xe}} N_{\text{Xe}}$	$\phi N_{\text{Xe}} \sigma_{a\text{Xe}}$
^{135}Pm	$\phi \Sigma_f y_{\text{Nd}}$	$\lambda_{\text{Pm}} N_{\text{Pm}}$	
^{149}Sm	$\lambda_{\text{Pm}} N_{\text{Pm}}$		$\phi N_{\text{Sm}} \sigma_{a\text{Sm}}$

For the decay chain including ^{149}Sm,

$$\phi \Sigma_f y_{Nd} + \lambda_{Pm} N_{Pm} = \lambda_{Pm} N_{Pm} + \phi N_{Sm} \sigma_{aSm} \qquad (10.34)$$

$$N_{Sm} = \frac{\Sigma_f y_{Nd}}{\sigma_{aSm}} \qquad (10.35)$$

Notice that the equilibrium concentration of ^{149}Sm is independent of neutron flux. To calculate the change in reactivity, use Eq. (10.6):

$$\rho = \frac{k_{eff} - 1}{k_{eff}} = 1 - \frac{1}{k_{eff}}$$

$$= 1 - \frac{1 + B^2 M^2}{k'_\infty} \qquad (10.36)$$

where k'_∞ refers to the point where secular equilibrium exists. Considering that ρ varies mostly with k_∞ and differentiating without taking the limit, we get

$$\frac{\Delta \rho}{\Delta k_\infty} = \frac{1 + M^2 B^2}{k'^2_\infty} \qquad (10.37)$$

$$\Delta \rho = \frac{1 + M^2 B^2}{k'^2_\infty} \Delta k_\infty \qquad (10.38)$$

Assuming that the change in k_∞ is due to a change in f, we have

$$\Delta k_\infty = \eta \varepsilon p \, \Delta f \qquad (10.39)$$

$$\frac{\Delta k_\infty}{k'_\infty} = \frac{\Delta f}{f'} \qquad (10.40)$$

where k'_∞ and f' are evaluated at the point of secular equilibrium. Translating this to the original critical state yields

$$\frac{\Delta k_\infty}{k'_\infty} = \frac{\Delta f}{f'} = \frac{f' - f}{f'} \frac{f}{f} = -f \, \Delta \left(\frac{1}{f}\right) \qquad (10.41)$$

where f refers to the original critical condition.

$$f = \frac{\Sigma_{au}}{\Sigma_{au} + \Sigma_{anf}}$$

$$\frac{1}{f} = 1 + \frac{\Sigma_{anf}}{\Sigma_{au}}$$

$$\Delta\left(\frac{1}{f}\right) = \frac{1}{\Sigma_{au}} \Delta\Sigma_{anf} \tag{10.42}$$

The only change in the nonfuel absorption cross section, Σ_{anf}, is that due to poison; therefore,

$$\Delta\Sigma_{anf} = \Sigma_{ap} \tag{10.43}$$

$$\Delta\left(\frac{1}{f}\right) = \frac{\Sigma_{ap}}{\Sigma_{au}} \tag{10.44}$$

Substituting, we have

$$\frac{\Delta k_{\infty}}{k'_{\infty}} = -f \frac{\Sigma_{ap}}{\Sigma_{au}} \tag{10.45}$$

Substituting Eq. (10.45) into Eq. (10.38) and dropping the prime notation yields

$$\Delta\rho = -\frac{1 + M^2 B^2}{k_{\infty}} f \frac{\Sigma_{ap}}{\Sigma_{au}}$$

$$= -\frac{f}{k_{\text{eff}}} \frac{\Sigma_{ap}}{\Sigma_{au}} \tag{10.46}$$

Equation (10.46) gives the change in ρ due to poison, which produces a change in f. For ^{135}Xe,

$$\Sigma_{ap} = \Sigma_{a\text{Xe}} = N_{\text{Xe}} \sigma_{a\text{Xe}} \tag{10.47}$$

$$\Delta\rho_{\text{Xe}} = -\frac{f}{k_{\text{eff}}} \frac{(y_{\text{Te}} + y_{\text{Xe}})\phi\Sigma_{fu}\sigma_{a\text{Xe}}}{(\lambda_{\text{Xe}} + \phi\sigma_{a\text{Xe}})\Sigma_{au}} \tag{10.48}$$

For ^{149}Sm,

$$\Sigma_{ap} = N_{\text{Sm}} \sigma_{a\text{Sm}}$$

$$\Delta\rho_{\text{Sm}} = -\frac{f}{k_{\text{eff}}} \frac{y_{\text{Nd}}\Sigma_{fu}}{\Sigma_{au}} \tag{10.49}$$

Equations (10.48) and (10.49) give the change in reactivity at equilibrium concentration of poison.

Example 7   Find the equilibrium change in reactivity due to xenon and samarium buildup for a reactor fueled with 20 percent enriched uranium. The average neutron flux is 10^{13} neutrons/cm^2 s and the thermal utilization factor is 0.87. At thermal energies

$$\Sigma_{au} = N_U \sigma_{au} = \sigma_{a238} N_{238} + \sigma_{a235} N_{235}$$

$$= 2.71(0.80 N_U) + 694(0.20 N_U) = 140.97 N_U$$

$$\Sigma_{fu} = N_U \sigma_f = \sigma_{f235} N_{235} = 582(0.20 N_U) = 116.4 N_U$$

$$\sigma_{aXe} = 2.72 \times 10^6 \text{ barns} \qquad \sigma_{aSm} = 4.08 \times 10^4 \text{ barns}$$

$$\lambda_{Xe} = 2.1 \times 10^{-5} \, \text{s}^{-1} \qquad k_{eff} = 1$$

$$\Delta\rho_{Xe} = -\frac{f(y_{Te} + y_{Xe})\phi\sigma_{aXe}\Sigma_{fu}}{(\lambda_{Xe} + \phi\sigma_{aXe})\Sigma_{au}}$$

$$= -\frac{0.87(0.059)10^{13}(2.72 \times 10^6)(10^{-24})(116.6 N_U)}{[2.1 \times 10^{-5} + 10^{13}(2.72 \times 10^6)(10^{-24})]140.97 N_U}$$

$$= -23.8 \text{ percent} \quad \text{or} \quad \frac{0.0238}{0.006\,41} = \$3.72$$

$$\Delta\rho_{Sm} = -\frac{0.87(0.014)(116.4 N_U)}{140.97 N_U}$$

$$= -1.003 \text{ percent or} \quad \frac{0.010\,03}{0.006\,41} = \$1.56$$

Notice that the change in reactivity due to samarium buildup is independent of flux.

Fission Product Poisoning After Shutdown

During steady-state operation (particularly for fluxes below 10^{15} neutrons/cm^2 s), radioactive decay of xenon helps limit its equilibrium concentration. After reactor shutdown, however, the process is reversed. Now there are no neutrons to be absorbed by the xenon and because of its radioactive nature the xenon concentration builds up as iodine decays. This situation proceeds in a manner similar to transient equilibrium (Chapter 3) since the decay constant of the parent, ^{135}I, is slightly greater than that of the daughter, ^{135}Xe. The increased negative reactivity to be overcome is referred to as *xenon override*, and it can rise to large values. Figure 10.6 shows the effect of xenon poisoning on reactivity as a function of time after shutdown. If it is desired to start up a reactor shortly after shutdown a large excess reactivity must be built into the reactor to "override" the poisoning; otherwise, a day or more must elapse before the reactor can be restarted. This becomes particularly important in

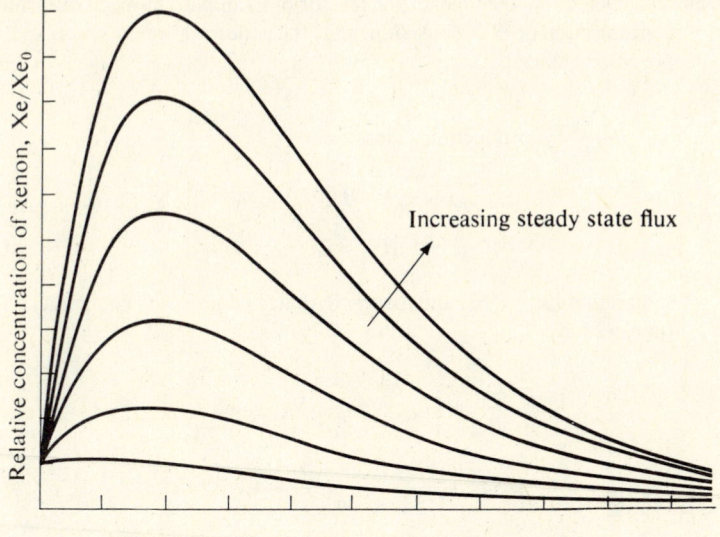

Fig. 10.6   Effect of xenon poisoning on reactivity.

propulsion reactors. Of course, the reactor can be gradually shut down to reduce the equilibrium concentration before complete shutdown. The transient concentration after shutdown can be evaluated with the aid of Table 10.5.

$$\frac{dN_{Xe}}{dt} = \text{production} - \text{loss} \tag{10.50}$$

$$\frac{dN_{Xe}}{dt} = \lambda_I N_I - \lambda_{Xe} N_{Xe}$$

$$= \lambda_I (N_I)_0 e^{-\lambda_I t} - \lambda_{Xe} N_{Xe} \tag{10.51}$$

where the subscript 0 refers to the equilibrium concentration. Solving Eq. (10.51), we get

$$N_{Xe} = (N_{Xe})_0 e^{-\lambda_{Xe} t} + \frac{(N_I)_0 \lambda_I}{\lambda_{Xe} - \lambda_I} (e^{-\lambda_I t} - e^{-\lambda_{Xe} t}) \tag{10.52}$$

By combining Eqs. (10.33), (10.46), and (10.52), an expression for the reactivity change as a function of time after shutdown can be obtained. Alternatively, Eq. (10.50) could be evaluated for the time required to build up a specific concentration of xenon or some other poison.

Example 8   Determine the time for the reactor of Example 6 to reach its equilibrium concentration of ^{135}Xe. Assume that the concentration of xenon and iodine are zero at startup.

$$\frac{dN_{Xe}}{dt} = \text{production} - \text{loss}$$

$$= \phi \Sigma_f (y_{Te} + y_{Xe}) + \lambda_I N_I - \lambda_I N_I - \lambda_{Xe} N_{Xe} - \phi N_{Xe} \sigma_{aXe}$$

$$= \phi \Sigma_f (y_{Te} + y_{Xe}) - N_{Xe}(\lambda_{Xe} + \phi \sigma_{aXe}) \tag{10.53}$$

Integrating Eq. (10.53) and assuming that Σ_f does not vary significantly with time, we have

$$\frac{dN_{Xe}}{dt} + (\lambda_{Xe} + \phi \sigma_{aXe}) N_{Xe} = \phi \Sigma_f (y_{Te} + y_{Xe})$$

$$N_{Xe} = \frac{\phi \Sigma_f (y_{Te} + y_{Xe})}{\lambda_{Xe} + \phi \sigma_{aXe}} + C e^{-(\lambda_{Xe} + \phi \sigma_{aXe})t}$$

At $t = 0$, $N_{Xe} = 0$:

$$0 = \frac{\phi \Sigma_f (y_{Te} + y_{Xe})}{\lambda_{Xe} + \phi \sigma_{aXe}} + C$$

$$C = -\frac{\phi \Sigma_f (y_{Te} + y_{Xe})}{\lambda_{Xe} + \phi \sigma_{aXe}}$$

$$N_{Xe} = \frac{\phi \Sigma_f (y_{Te} + y_{Xe})}{\lambda_{Xe} + \phi \sigma_{aXe}} [1 - e^{-(\lambda_{Xe} + \phi \sigma_a Xe)t}]$$

If we assume that the time to reach equilibrium is essentially equal to the time to reach 99 percent of the equilibrium value,

$$0.99 = 1 - e^{-(\lambda_{Xe} + \phi \sigma_a Xe)t}$$

$$0.01 = e^{-(21 \times 10^{-6} + 10^{13} \times 2.72 \times 10^{-18})}$$

$$= e^{-48.2 \times 10^{-6}t}$$

$$48.2 \times 10^{-6}t = 4.605$$

$$t = 0.956 \times 10^5 \text{ s or } 1.108 \text{ days}$$

Thus, the xenon builds up fairly slowly and concentration after shutdown will not become important if the reactor is operated for only short periods of time, say an hour or two.

Chemical Shim and Volume Control

The control requirements of the initial reactor core are in excess of the equilibrium core requirements because all of the fuel is fresh in the initial core. The initial core requirements are met in pressurized water reactors by the combined effects of control rods and a chemical shim. In boiling water reactors, the control requirements are met by movable control rods and burnable poison in the form of gadolinia (Gd_2O_3) mixed in with fuel in several fuel rods.

The chemical shim that offsets the initial high reactivity in a pressurized water reactor is in the form of boric acid. Boric acid has the advantage of being depleted as the fuel is depleted, and since it is uniformly distributed, local flux depressions are avoided.

At the beginning of core life, boron concentration in the coolant is about 1500 ppm and is reduced at the rate of about 100 ppm per month throughout core life. A small amount of water (80 to 120 gpm) is continuously drawn from the primary coolant and processed through the chemical and volume control system. In addition to the long-term control of reactivity due to fuel burnup, this system also maintains chemicals for corrosion control, reduces fission products, provides seal water for the reactor coolant pumps, and provides makeup water.

A portion of the water withdrawn in the system is automatically diluted with unborated water and reinjected into the coolant system to reduce the boron concentration. The borated water that is removed is eventually recycled through an evaporator to concentrate the boron for reuse.

The same system can be used to reduce reactivity when shutting the reactor down and in controlling xenon concentrations.

Fuel Depletion

If in addition to its own fuel a reactor contains fertile material, some new fuel will be produced as the original fuel is burned up. Reactors designed specifically to produce fissionable fuel are frequently called *breeder reactors*. In a reactor such as this, the amount of fissionable fuel produced is greater than the amount of fuel burnup, and the reactivity increases unless it is properly controlled. Even in a thermal ^{235}U reactor that contains ^{238}U there will be some conversion into ^{239}Pu according to the reaction

$$^{238}U + {}_0n^1 \longrightarrow {}^{239}U \xrightarrow{\beta-} {}^{239}Np \xrightarrow{\beta-} {}^{239}Pu$$

This production of ^{239}Pu occurs for intermediate neutron energies as well as for thermal energies.

The *conversion ratio*, CR, is the number of fissionable nuclei produced per fissionable nucleus removed from the reactor by all processes. The neutrons that are absorbed by ^{238}U during a cycle are absorbed either by resonance absorption or by thermal absorption. The conversion ratio, CR, is therefore the sum of these two factors.

Resonance absorption by ^{238}U

The total number of fast neutrons produced per number of ^{235}U nuclei undergoing reactions is $\eta_{235}\varepsilon$, where η_{235} is the number of neutrons produced per thermal neutron absorbed by ^{235}U.

A fraction, p, of neutrons will escape resonance absorption while slowing down, and a fraction, P_{NL_f}, of neutrons will not leak while slowing. Therefore, the number of neutrons absorbed by ^{238}U per number of thermal neutrons absorbed by ^{235}U nuclei is

$$\eta_{235}\varepsilon(1 - p)P_{NL_f}$$

Since this is the number of neutrons absorbed by ^{238}U, it is also the number of ^{239}Pu nuclei produced in the resonance region per ^{235}U nucleus removed.

Thermal absorption by ^{238}U

The number of thermal neutrons absorbed by ^{238}U nuclei per thermal neutron absorbed by ^{235}U is

$$\frac{\Sigma_{a238}}{\Sigma_{a235}}$$

This is also the number of ^{239}Pu nuclei produced at thermal per ^{235}U nucleus removed. Therefore, the conversion ratio is

$$\text{CR} = \eta_{235}\varepsilon(1 - p)P_f + \frac{\Sigma_{a238}}{\Sigma_{a235}} \tag{10.45}$$

Reactivity Change in a Uranium Reactor

Even though the change in reactivity due to a fuel change is a complicated matter, it can be evaluated from the change in effective multiplication factors. Considering only *one absorption* process:

$$\frac{\text{number neutrons produced by new nuclei}}{\text{number neutrons produced by old nuclei}} = \frac{\sigma_{a\,\text{new}}\eta_{\text{new}}}{\sigma_{a\,\text{old}}\eta_{\text{old}}} \tag{10.55}$$

Equation (10.55) is equivalent to the ratio of new multiplication factor to original multiplication factor for *one absorption*. Since we must consider many absorption processes,

$$\frac{\text{neutrons produced}}{\text{neutron used up}} = \frac{\text{nuclei produced}}{\text{neutron used up}}$$

$$\times \frac{\text{number neutrons produced by new nucleus}}{\text{number neutrons produced by old nucleus}}$$

$$\frac{\text{neutrons produced}}{\text{neutron used up}} = \text{CR} \times \frac{\sigma_{a\,\text{new}}\eta_{\text{new}}}{\sigma_{a\,\text{old}}\eta_{\text{old}}} \tag{10.56}$$

Equation (10.56) is equivalent to the ratio of new multiplication factor to ^{235}U nuclei removed by all processes. Considering now the change in multiplication factor due to depletion of ^{235}U:

$$\frac{\Delta\rho}{^{235}\text{U removed}} = \text{CR}\,\frac{\sigma_{a\text{Pu}}\eta_{\text{Pu}}}{\sigma_{a235}\eta_{235}} - 1$$

$$\Delta\rho = \left(\text{CR}\,\frac{\sigma_{a\text{Pu}}\eta_{\text{Pu}}}{\sigma_{a235}\eta_{235}} - 1\right)R \tag{10.57}$$

where R is the fraction of ^{235}U nuclei removed after a period of time t. It must be remembered that plutonium is not only produced by the conversion process but also is removed by absorption of neutrons. Equation (10.57) applies only to the reactivity change due to depletion of original fissionable fuel and does not include the reactivity change due to absorption of neutrons by plutonium.

The reactivity change, therefore, can be positive, negative, or zero. Except in the case of breeder reactors, the reactivity change is negative. This means that excess reactivity must be built into the reactor initially to compensate for the loss during the core lifetime.

The excess reactivity built into a reactor has been controlled with the Spectral Shift Control Reactor (SSCR). In the SSCR the reactivity was controlled by varying the concentration of D_2O–H_2O moderator. Since the slowing-down power of the D_2O is much less than that of H_2O, a large concentration of D_2O effectively increases the critical mass of the reactor. As fuel is burned up and poisons decrease the reactivity, the concentration of D_2O was decreased by adding ordinary water. This increased the slowing-down power of the moderator and reduced the critical mass. The D_2O concentration is periodically reduced throughout the life of the core. It is claimed that an SSCR has a longer core life than normal and that the reactor can be operated at full power all the time. Other advantages claimed were higher average power, since the reactor is operated with the control rods fully out; higher thermal efficiency; uniform fuel burnup; lower capital and fuel costs; and a larger conversion ratio.

Example 9   What is the theoretical conversion ratio for a natural uranium fueled reactor if the reactivity is to remain constant? Neglect any neutron losses such as poison or absorption by reactor structural materials.

$$\Delta\rho = 0 = \left(\mathrm{CR}\, \frac{\sigma_{a\mathrm{Pu}}\eta_{\mathrm{Pu}}}{\sigma_{a235}\eta_{235}} - 1 \right)R$$

$$\mathrm{CR}_{\Delta\rho=0} = \frac{\sigma_{a235}\eta_{235}}{\sigma_{a\mathrm{Pu}}\eta_{\mathrm{Pu}}}$$

$$\eta_{\mathrm{Pu}} = \frac{\Sigma_{f\mathrm{Pu}}}{\Sigma_{a\mathrm{Pu}}}\nu_{\mathrm{Pu}}$$

$$= \frac{746}{1026}(2.90) = 2.11$$

$$\eta_{235} = \frac{\Sigma_{f235}}{\Sigma_{a235}}\nu_{235} = \frac{582}{694}(2.42)$$

$$= 2.07$$

$$\mathrm{CR}_{\Delta\rho=0} = \frac{694(2.07)}{1026(2.11)} = 0.664 \text{ or about } \tfrac{2}{3}$$

Control of Power Distribution

Power density and power distribution within the reactor core must be controlled to keep the reactor within operating limits. The design limit power densities, which are the maximum allowable in normal operation, are of concern in two general areas:

(1) Fuel safety limits
(2) Accident limits

Fuel safety limits are dictated by fuel rod mechanical and metallurgical design. They generally relate to 1 percent fuel-clad strain, departure from nucleate boiling, and fuel centerline melting. Accident safety limits, on the other hand, relate specifically to each type of accident. Fuel cladding temperature limit during loss of coolant accident is generally the limiting accident condition, although cladding oxidation, zirconium–water reaction, uncontrolled rod movement, and other accident considerations could provide limiting conditions. Whichever factor leads to lowest power density, of course, determines the maximum allowable operating limit.

Power density distributions cover a wide range of shapes and magnitudes; maximum characteristics of these distributions are of primary interest in controlling power distribution. To simplify specifying limits, the concepts of

peaking factors for PWRs and *maximum average planar linear heat generation rate* (MAPLHGR) for BWRs is used. Peaking factors are an expression of maximum to average power density. They may be in several forms, indicating total or partial components of the distributions, and they may refer to measured or calculated values. The concept of MAPLHGR was developed as a limit for BWRs because the peak cladding temperature during a LOCA is determined primarily by fuel bundle power.

The most commonly used expression for power density is the thermal power produced per unit of length of fuel rod. It is called the *linear power density* or *linear heat generation rate*, and is expressed in kW/ft. Fuel safety limits would limit the linear power density to 20 to 30 kW/ft; however, accident limits dictate maximum linear power densities on the order of 12 to 15 kW/ft. These design power densities are the limit of allowed operations. During almost all operating periods, reactors operate at power densities much less than the maximum allowable.

Power distributions in PWRs are quantified in terms of hot channel factors mentioned previously. The factors of major interest are:

(1) *Total Peaking Factor*, F_Q, defined as the ratio of maximum local power density to the average power density
(2) *Engineering Factor*, F_Q^E, the allowance on heat flux required by manufacturing tolerances
(3) *Radial Peaking Factor*, F_{xy}^N, the ratio of maximum power density to average power density in the horizontal plane of peak local power
(4) *Axial Peaking Factor*, F_S^N, the ratio of the maximum power density to the average power density in the vertical plane of peak power

An equation for the total peaking factor is

$$F_Q = F_{xy}^N \times F_z^N \times F_Q^E \qquad (10.58)$$

Other allowances such as for measurement uncertainty, fuel densification, and so on, can be added if desired.

The horizontal power shape, F_{xy}^N, is determined by the fuel enrichment, burnable poison, and the presence or absence of a bank of control rods. Xenon, samarium, and moderator density effects on the radial power distribution are small. Radial power distributions are, therefore, relatively fixed. The axial power shape, F_z^N, is under control of the operator through the operation of control rods and operation of the chemical and volume control system. Effects that cause axial power variations include rod motion, fuel burnup, moderator density, xenon oscillation, and Doppler effect. An indication of axial power profile is usually given by signals from long ion chambers outside the reactor vessel placed parallel to the axis of the core. These signals are generally used to maintain the peaking factor within limits during normal operation. The

specific control performed varies from reactor vendor to vendor. In a few cases reactors are provided with "incore" neutron detectors for flux mapping and for control of power distribution.

For pressurized water reactors typical limiting F_Qs at full power range from 2.15 to 2.2 BWRs approach power distribution a little differently than PWRs. They generally are designed to maintain a constant radial and axial shape distribution through combinations of void control, rod placement, and axial placement of burnable poisons. Power-level change is accomplished by core coolant control, which will maintain approximately the same power distribution at both part and full load. Monitoring of the power distribution is accomplished by fission chambers placed in the core.

Operating limits in BWRs are met by specifying a peak allowable linear heat rate. The peak cladding temperature that would be reached during a loss of coolant accident in a BWR is determined primarily by the fuel bundle power rather than the rod power. Therefore, limiting power distribution to limit cladding temperature during a LOCA is quantified in terms of MAPLHGR, which is defined as the highest heat generation rate in a 6-in. segment of a bundle anywhere in the reactor, and its value determines the peak cladding temperature during a LOCA. The axial location of the MAPLHGR has a small effect on peak cladding temperature. Peak cladding temperature is insensitive to local peaking within a bundle, but MAPLHGR limits vary with fuel burnup.

At full power, the MAPLHGR limit can also be expressed as a limit on total peaking factor because of the insensitivity of MAPLHGR to axial location and local peaking. For boiling water reactors typical MAPLHGRs range from 11 to 14 kW/ft. Corresponding F_Qs are from 2.2 to 3.0.

Reactivity Changes in Fast Reactors

Small changes in reactivity can be introduced into a fast reactor in much the same fashion as into a thermal reactor. These allow power increases with long periods. The fast fission in ^{232}Th or ^{238}U causes the fraction of delayed neutrons to be slightly larger in a fast core than a thermal core. Table 10.6 shows delayed neutron fraction for various fissile and fertile species. Since the fraction of delayed neutrons from ^{232}Th and ^{238}U is so much greater than that from thermal reactor fuels, a smaller reactivity change will give the same period (see Fig. 10.5).

The prompt neutron lifetime is the order of 10^{-7} s for a fast reactor compared to 10^{-3} s for a thermal core. This results in even a faster rate of increase in flux and power for a prompt critical excursion, making prompt criticality all the less desirable.

Some of the reactivity changes discussed earlier in this chapter do not occur in fast reactors. Since there are no thermal neutrons, changes in temperature alone do not change the neutron spectrum. The ^{135}Xe and ^{149}Sm cross sections are fairly small for fast neutron energies, so their effect can be neglected.

Table 10.6 Delayed Neutron Fraction (β) and Delayed Neutron Lifetimes (l_d) for Fast and Thermal Fission

Species	β_f	$\bar{l}_{d_f}$	β_{th}	$\bar{l}_{d_{th}}$
^{232}Th	0.022	10.02	—	—
^{233}U	0.0027	17.89	0.002 64	18.58
^{235}U	0.0065	12.75	0.0064	13.14
^{238}U	0.0157	7.68	—	—
^{239}Pu	0.0021	14.63	0.0021	15.71
^{240}Pu	0.0026	13.12	—	—

Three significant factors affecting reactivity changes of fast reactors are:

(1) The Doppler effect
(2) The sodium void coefficient
(3) Fuel rod expansion

Doppler Effect

During a prompt critical power excursion in a fast reactor it is important to have an inherent mechanism for the rapid insertion of negative reactivity. There is no time for a mechanical scram system to operate. The Doppler effect may provide significant negative reactivity. As the temperature increases in the material through which neutrons are diffusing, there tends to be a broadening of any resonance peaks. This is because the velocities of the target atoms can add to or subtract from the neutron velocity near the resonance region, broadening the resonance peak but keeping the area under the curve constant.

The net reactivity temperature coefficient depends to a large extent on opposite Doppler broadening of resonance peaks in the fissile and fertile fuel. Increasing temperature results in increased absorption in the fissile material (*positive coefficient*) and decreased absorption in the fertile material (*negative coefficient*). Therefore, there must be a limit to the ratio of fissile to fertile material in order to maintain a negative coefficient of reactivity in the fuel.

The important resonances involving the Doppler effect in fast reactors in the range of 0.5 to 20 keV are on the lower tail of a fast-reactor spectrum. The softer the spectrum, the more neutrons there are in this low energy region and the larger the Doppler effect. A metal-fueled reactor has the hardest spectrum, followed by carbide fuels and then oxide fuels.

In a small metal-fueled reactor the Doppler effect is not significant, but in reactors using a lot of sodium a significant number of neutrons reach the 0.5 to 20 keV range. This results in the negative temperature coefficient arising from the Doppler effect. Furthermore, it requires extremely accurate data to

determine the lower energy neutron flux and the reactivity changes. The addition of BeO to large fast reactors has been studied in some detail (Reference 8). The purpose of the BeO'is to degrade the energy spectrum and thus enhance the negative Doppler coefficient. Table 10.7 shows two cores, where case B has added 8.3 volume percent BeO to soften the spectrum. Note that the mean neutron energy is 0.23 MeV for the hard spectrum of case A and 0.10 MeV for the soft spectrum of case B. The Doppler coefficient is twice as big for case B. Also, Δk for sodium loss from core and blanket is -0.0051 compared to a positive value of 0.0070 for the hard spectrum. Figure 10.7 shows the dynamic response to the addition of $1.50 reactivity in 10 ms to the two cores. For case A fuel failure would occur probably followed by boiling and expulsion of sodium from the core before the scram could act at 400 ms. The fuel in case B will survive this reactivity insertion, the major difference being in the values of the Doppler coefficients. Note the magnitude and short duration of the power pulse in both cases.

The introduction of BeO into a reactor may prove difficult, practically, because the potential power generation of the reactor will be reduced due to the lower fuel volume.

The Doppler coefficient can cause a reverse effect in a cold sodium accident when the sodium temperature drops. This could occur with a large power reduction, not accompanied by a reduction in flow rate, or if a slug of cool sodium is pumped into a hot core unintentionally.

The Doppler coefficient is linked to the sodium void coefficient because of the hardening of the flux due to a void. This can reduce the Doppler coefficient by as much as a factor of 2.

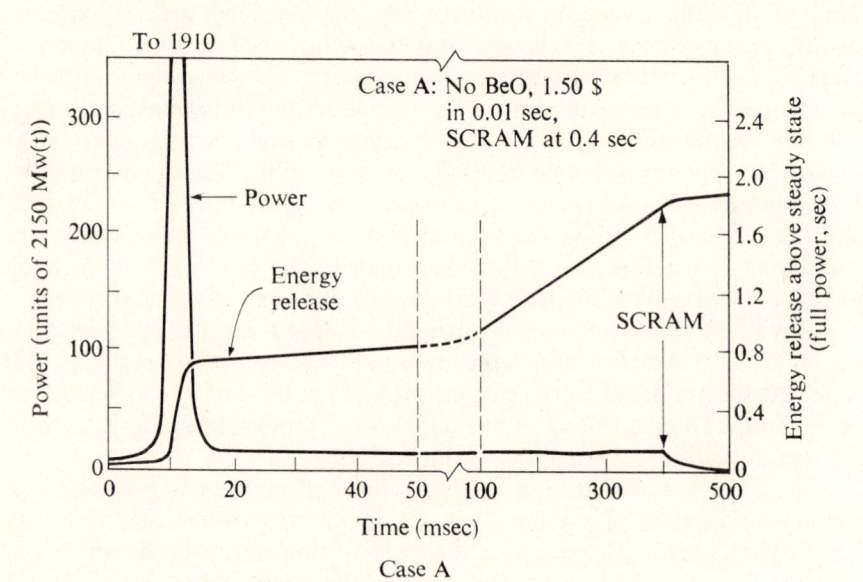

Case A

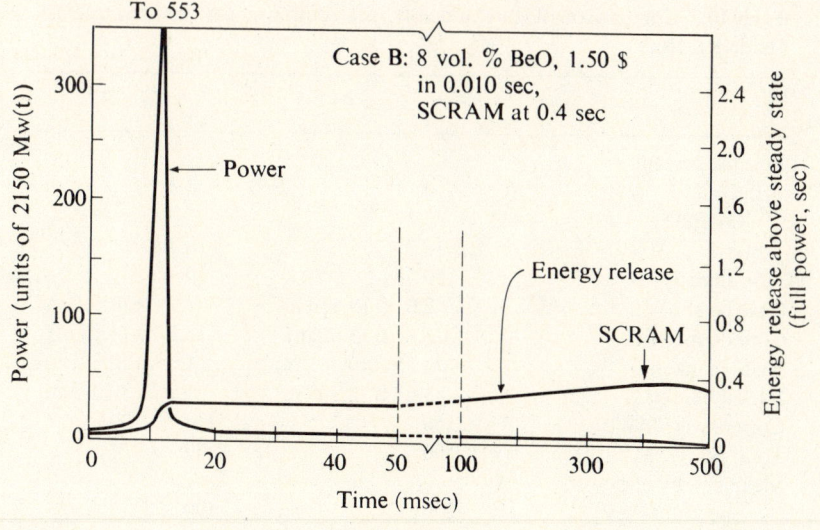

Case B

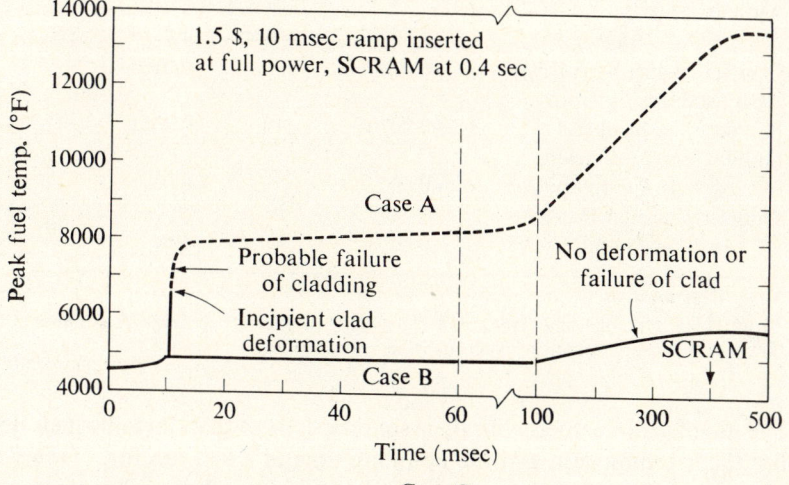

Case C

Fig. 10.7 Comparison of responses hard (case A) and soft (case B) spectrum cores for 1000-MW(e) fast oxide reactors. $1.5 10-ms ramp inserted at full power. (A) Transient power and energy release (hard). (B) Transient power and energy release (soft). (C) Transient fuel temperatures for both cases. [From H. H. Hummel and D. Okrent, *Reactivity Coefficients in Large Fast Power Reactors.* Chicago: American Nuclear Society, Inc., 1970.]

Table 10.7 Comparison of Hard and Soft Spectrum Cores for 1000-Mw(e) Fast Oxide Reactors

	Case A	Case B
Volume percent		
BeO	0	8.3
UO$_2$–PuO$_2$	33.3	25
Steel	16.7	16.7
Sodium	50	50
Core height	2.0 ft (0.610 m)	2.0 ft (0.610 m)
Core diameter	11.8 ft (3.65 m)	13.7 ft (4.18 m)
Axial blankets	1.25 ft (0.381 m)	1.25 ft (0.381 m)
Radial blankets	1.0 ft (0.305 m)	1.0 ft (0.305 m)
Core power, MW(th)	2150	2150
Total power, MW(th)	2500	2500
Mean energy of power spectrum (keV)	230	100
Fraction of fissions below 9 keV	0.12	0.28
Total breeding ratio	1.34	1.12
Atom percent (^{239}Pu + ^{241}Pu)	11.5	14.7
Doppler coefficient $T(dk/dt)$	−0.006	−0.012
Δk for sodium loss from core and blankets*	+0.0070	−0.0051
Sodium power coefficient $\Delta k/1\%\Delta P$ at full power	+0.04¢	−0.03¢
Doppler power coefficient $\Delta k/1\%\Delta P$ at full power	−0.8¢	−1.8¢

* $\beta \approx 0.004$ for both case A and case B.
From H. H. Hummel, and D. Okrent, *Reactivity Coefficients in Large Fast Power Reactors.* Chicago: American Nuclear Society, Inc., 1970.

The use of ceramic fuels with their softer spectrum than metallic fuels does penalize the breeding gain and the doubling time in a fast reactor. However, the long burnup times make ceramic fuels desirable. Lower breeding gain along with the Doppler effect must share the economic penalty.

Sodium Void Coefficient

Sodium coolant in the reactor expands as the reactor heats up or the number of sodium atoms decreases. This results in a decrease in the number of neutrons

absorbed. While this is small in power reactors, changes resulting from other considerations are important.

One of the primary safety considerations in the design of a fast-breeder system is that loss of part or all of the sodium will not result in a net positive increase in reactivity. Such a condition might result from a rupture in the sodium system, blockage of flow in coolant channels causing sodium vapor formation, or the introduction of a significant amount of entrained gas. Regardless of the cause, the effect is to produce a general hardening of the neutron spectrum because of the reduction of the number of scattering collisions in the void region due to the removal of coolant nuclei. Also, with the removal of the sodium the log energy decrement, ξ, is reduced, contributing to the harder flux. Figure 10.8 shows how η increases with energy, indicating that there will be more neutrons released per fission at higher average energies. It can be noted that the increase for ^{233}U is less than that for ^{235}U or ^{239}Pu. This effect tends to produce a negative reactivity change in fissile material. On the other hand, when fertile material is present, the harder spectrum causes more fissions and consequently a positive reactivity change.

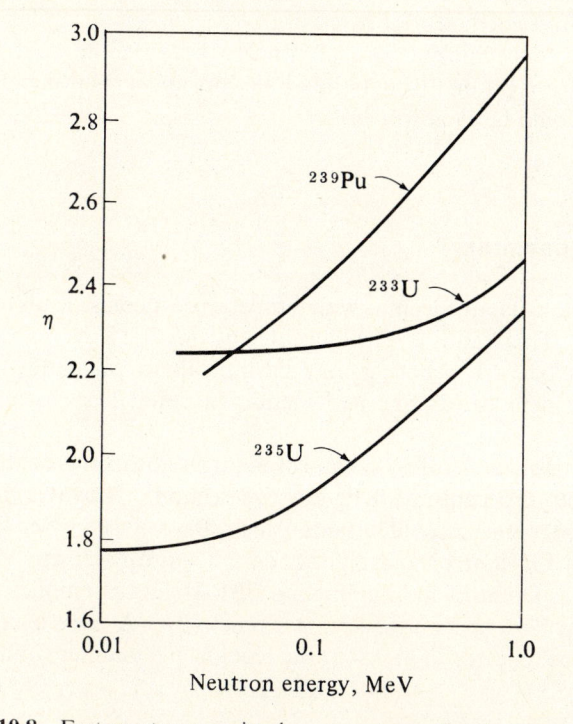

Fig. 10.8 Fast neutrons emitted per neutron absorbed, η, versus neutron energy for ^{233}U, ^{235}U, and ^{239}Pu.

Leakage (L^2B^2) is increased because Σ_s is decreased by removal of the sodium atoms from the voided volume. Thus, the transport mean free path is increased by the reduced scattering cross section. Leakage is related to neutron current flow, which is, in turn, proportional to the flux gradient. A volume near the center of the core is much less affected since

$$\left.\frac{\partial\phi}{\partial\bar{y}}\right|_{r=0} = 0$$

while the slope increases toward the outer boundaries, increasing leakage from peripheral volumes.

In small cores B^2 is large and the leakage overrides the positive reactivity effect of the hardened spectrum. In large power reactors this may not happen because B^2 is too small to allow the change in L^2 to have sufficient effect to override the positive effect of the harder spectrum. A local void at the center of a core tends to produce a positive contribution to reactivity due to the small neutron current flow at this location, but near the outer boundary of the core the effect would be negative. Overall, then, the result in large reactors is to produce a positive coefficient of reactivity.

Positive void coefficients are of concern because the void might cause autocatalytic propagation through the whole core and a resulting loss of coolant flow accident could be even worse.

Fuel Rod Expansion

Lateral expansion of long fuel pins with temperature increase leads to a reduction in fuel density. This results in a leakage of neutrons and a decrease in reactivity. In reactors fueled with pellets it may be difficult to calculate this reactivity because in accident conditions fuel meltdown could lead to an increase in reactivity.

In fast reactors control by means of neutron absorbing control rods may not be satisfactory because of their low rod section. However, in the Enrico Fermi fast breeder reactor, control rods using ^{10}B were used. The Fermi reactor used two rods for control and eight rods for shutting down. In other fast reactors, such as Dounray in England and EBR-II, fuel assemblies are moved in to increase reactivity and out to decrease reactivity. A third means of control is by moving the reflector in or out of the reactor in a manner similar to the fuel movement.

Problems

1. Compute the average neutron lifetime for ^{233}U and ^{239}Pu. Half-lives for delayed neutron groups are:

-.049 s   -.033 s

	^{233}U	^{239}Pu
1	55.00	54.28
2	20.57	23.04
3	5.00	5.60
4	2.13	2.13
5	0.615	0.618
6	0.277	0.257

2. In Prob. 1, what will be the reactivity for a reactor period of 1 min?

3. What is the reactivity in cents for a reactor period of 10 s and a prompt neutron lifetime of 2×10^{-4} s? The fuel is ^{235}U.

4. For a reactor period of 8 s, calculate the reactivity of a U-fueled reactor. The prompt neutron lifetime is approximately 3.5×10^{-4} s.

5. At what value of ρ would the reactor of Prob. 3 become prompt-critical?

6. In shutting down a reactor, $k_{eff} < 1$ and the delayed neutron contribution becomes effective some time after the prompt neutrons have disappeared. Eventually, the flux decreases with a period determined by the longest-lived delayed neutron group. What will be the time necessary to shut down a ^{235}U reactor if the average neutron flux goes from 1.75×10^{13} to 10^3 neutrons/cm^2 s?

7. What is the reactivity (cents and inhours) of a carbon-moderated ^{235}U-fueled reactor when the power is suddenly increased, resulting in a reactor period of 12 min?

8. Derive Eq. (10.31c).

9. A 75 percent enriched ^{235}U–H$_2$O-moderated reactor has been operating at equilibrium for some time. The average neutron flux is 10^{14} neutrons/cm^2 s, and the moderator fuel ratio is 500 moles H$_2$O per mole U. Determine the reactivity change due to xenon and samarium. The mean core temperature is 300°C.

10. Write a Fortran program to calculate the xenon concentration at any time after reactor shutdown.

11. Develop an electronic analog circuit to calculate the relative xenon buildup after reactor shutdown for various neutron fluxes. Is there a limiting value of flux for which no xenon buildup occurs? Neglect the absorption cross section of ^{135}I.

12. A reactor is loaded with 25 kg of 3 percent enriched fuel. It operates continuously for 1 yr at a power output of 10 000 kW. Determine the change in reactivity due to fuel consumption. The resonance escape probability is 0.677 and fast neutron nonleakage probability is 0.973.

13. In starting up a reactor it is desired to reach full power in 1 min. Full power is an increase by a factor of 10^4. What period is required?

14. Discuss the containment problem of a sodium-cooled fast reactor, especially the prevention of a sodium–air reaction.

15. A large PWR has $\Sigma_f = 0.005$ cm^2/cm^3 and is operating with a mean thermal flux of 3.5×10^{13} neutrons/cm^2 s. Assume a thermal utilization factor (based on ^{235}U) equal to 0.87. The mean core temperature is 550°F. How much excess reactivity must the reactor have at shutdown to be able to override the maximum xenon buildup? How long after shutdown does the xenon reach its peak?

   At the time of an unscheduled shutdown the reactor has $2 worth of excess reactivity. It takes 8 h to correct the difficulty. Can the reactor be restarted immediately? If not, how long after the original shutdown must the operator wait to be able to go critical again?

16. A 3 percent enriched $^{235}_{92}$U-fueled reactor has a resonance escape probability of 0.75 and a fast nonleakage probability of 0.95. What is its conversion ratio?

17. A thermal reactor uses ^{239}Pu as fuel and ^{232}Th as the fertile material. If the resonance escape probability is 70 percent and the fast nonleakage probability is 95 percent, determine the conversion ratio. Assume that there are three neutrons absorbed by Pu per neutron absorbed in thorium.

   If the reactor produces 1500 MW(th) for 10 months and is loaded with 2000 kg of Pu, determine the reactivity change due to plutonium consumption.

18. A PWR rated at 2700 MW(th) is composed of 157 fuel assemblies, each containing 264 active fuel rods. Each fuel rod is 12 ft long. If the limiting power density is 13.4 kW/ft, what is the corresponding total peaking factor? Assuming an engineering factor of 1.05, what is the limiting axial peaking factor for a radial peaking factor of 1.435?

19. Recent practice in designing PWRs has been to allow total peaking factors to be somewhat lower than earlier reactors, thereby allowing higher power densities. For the reactor in Prob. 18, what linear power density would result if the axial peaking factor was 1.5? Practically, how would this be accomplished in the reactor?

References

1. *Proceedings of the Second United Nations International Conference on the Peaceful Uses of Atomic Energy*, Vol. 11, Geneva, 1958.
2. Bonilla, C. F., ed., *Nuclear Engineering*. New York: McGraw-Hill Book Company, 1957.
3. Liverhant, S. E., *Elementary Introduction to Nuclear Reactor Physics*. New York: John Wiley & Sons, Inc., 1960.
4. Deuster, R. W., and Z. Levine, *Economics of Spectral Shift Control*, ASME Paper No. 60-WA-335, 1960.
5. Glasstone, S., and A. Sesonske, *Nuclear Reactor Engineering*. New York: D. Van Nostrand Company, 1963.

6. El-Wakil, M. M., *Nuclear Power Engineering.* New York: McGraw-Hill Book Company, 1962.
7. *Reactor Handbook*, Vol. 2, U.S. Atomic Energy Commission, 1960.
8. Henley, J., and J. Lewins, eds., *Advances in Nuclear Science and Technology*, Vol. 5. New York: Academic Press, Inc., 1969.
9. Hummel, H. H., and D. Okrent, *Reactivity Coefficients in Large Fast Power Reactors.* American Nuclear Society, Inc., 1970.
10. Ash, M., *Nuclear Reactor Kinetics.* New York: McGraw-Hill Book Company, 1979.

Radiation Damage and Reactor Materials Problems

The interactions between radiation and matter can produce profound changes in the properties of a material. These may prove to be beneficial, as in the case of the strengthening in a thermal core of an austenitic stainless steel without serious loss in ductility, or they may court disaster, as illustrated by the sharp increase in the ductile–brittle transition temperature with low-carbon structural steels.

To appreciate these changes fully, one must take the basic approach of the solid-state physicist who studies the interaction of a single type of radiation under carefully controlled conditions. This approach is gradually improving our understanding of the behavior of materials in a radiation field. However, many situations in a reactor are too complex to yield an answer by this fundamental approach. Thus, the engineering or phenomenological approach is used to give a specific answer to a particular problem, although it may shed little light on the understanding of the entire process.

Radiation Damage to Crystalline Solids

Crystalline materials have their atoms arranged in a well-defined lattice where each atom has a fixed rest position. Ideally, there should be no imperfections in the atomic arrangement; as a practical matter, however, imperfections exist due to the presence of impurities and alloying elements. Further irregularities can be introduced by plastic deformation. Any alteration in the regularity of the lattice will alter the properties of a material. Nuclear radiation tends to destroy order and alter properties by producing several types of defects.

(1) Vacancy-interstitial pairs (Frenkel defects) are formed when energetic particles collide with atoms ejecting them from stable lattice sites. There may be a cascading effect because of a knocked-on atom having received sufficient energy to eject

more atoms. These displaced atoms may finally lose their energy and occupy positions other than normal lattice sites, thus becoming interstitials. The presence of interstitials and vacancies makes it more difficult for dislocations to move through the lattice, usually increasing the strength and reducing the ductility of a material.

These atom displacements which are produced in a reactor core are due primarily to the scattering of fast neutrons. The neutrons of 1 MeV energy and above have 90 percent of the total neutron energy and are responsible for the major amount of lattice disruption.

(2) Impurity atoms are produced by nuclear transmutations. Neutron capture in a reactor produces an isotope which, in turn, may be unstable and produce an entirely new atom as it decays. For example, ^{210}Po is produced in a reactor by irradiation of ^{209}Bi.

$$^{209}_{83}\text{Bi} + ^{1}_{0}\text{n} \longrightarrow \quad ^{210}_{83}\text{Bi*} \longrightarrow \quad ^{210}_{84}\text{Po*} + _{-1}^{0}\text{e}$$
$$\quad\quad\quad\quad\quad\quad\quad\quad\quad\quad\quad\quad\quad\quad\quad\quad \longmapsto \, ^{206}_{82}\text{Pb} + ^{4}_{2}\alpha \quad \textbf{(11.1)}$$

The 138-day half-life ^{210}Po will produce 138 W of heat energy per gram when freshly separated. This material was used as the heat source in the SNAP-3 project.

For most metallic materials long irradiations at high flux levels are necessary to produce significant property changes due to impurity buildup. However, a semiconductor such as germanium may have large changes in conductivity due to the gallium and arsenic atoms that are introduced as the activated Ge isotopes decay.

In stainless steels, trace amounts of boron undergo an n, α reaction. The helium thus generated forms bubbles and leads to the deterioration of mechanical properties, as discussed more fully later in this chapter.

It should be noted that the production of impurities is largely due to the absorption of thermal neutrons, which have much larger absorption cross sections than do fast neutrons.

Thermal neutron capture produces many gammas in the range 5 to 10 MeV. The recoil energy given to the nucleus emitting the gamma is often high enough to cause several Frenkel defects with resultant property changes.

(3) Replacement collisions occur when a moving interstitial ejects an atom from a lattice site and then lacks sufficient energy to leave itself. This is of particular interest in the study of the disordering of ordered solid solutions. It has little meaning if only a single type of atom is present in a lattice or if a solid solution has a random arrangement.

The β' phase in the Cu–Zn alloy system is an example of a body-centered cubic structure with an ordered solid solution. AgCu_3 is a face-centered material with an ordered arrangement. Both these materials show increased amounts of order with low-temperature irradiation.

(4) Spikes are caused by the intense local heating as knocked-on atoms and fission fragments energize particles along their track. This may occur as a high degree of excitation of the atoms without their leaving a stable lattice position (thermal spike) or as a shower of secondary displacements which drive interstitials into the surrounding lattice (displacement spike). In either case there is intense local

Table 11.1 Effect of Radiation on the Mechanical Properties of Plastics Irradiated in Air at 25°C

Material	Property	Initial Value	Dose Rate (kGy/h)	Thickness (cm)	Effect at Gy Dose* 5 × 10⁴	10⁵	10⁶	10⁷	10⁸
High-density (linear) polyethylene—Marlex 50	Tensile strength	29.5 MPa	10	0.005	A	D			B
	Elongation	600%			D				
Low-density (branched) polyethylene—Alathon	Tensile strength	9.65 MPa	20	0.48		A	A	A	D
	Elongation	250%					B	A	C
	Notch impact str.	38.6 J/cm				A	B	D	A
Nitrogen-containing thermoplastic—nylon	Tensile strength	52.4 MPa	20	0.25	A	B	C	C	D
	Elongation	62%			A	B	C	C	
	Notch impact str.	9.64 J/cm			B	C	C	D	D
Halogen-containing thermoplastic—Teflon	Tensile strength	23.4 MPa	10	0.36	A	A	A	C	D
	Elongation	250%			C	D	D	D	D
Thermoset—cast phenolic	Tensile strength	75.8 MPa	20	0.46	C	D	A	B	D
	Elongation	2%					A	A	D
	Notch impact str.	1.83 J/cm					A	A	D
Thermoset—cellulose pulp-filled urea–formaldehyde	Tensile strength	53.8 MPa	20	0.32			A	A	D
	Elongation	0.5%					A	C	C
	Notch impact str.	1.07 J/cm					A	A	D

* A, 80% of initial value retained; B, 50–80% of initial value retained; C, 10–50% of initial value retained; D, <10% of initial value retained.

Data abstracted from R. O. Bolt and J. G. Carroll, *Radiation Effects on Organic Materials*. New York: Academic Press, Inc., 1963.

heating with temperatures, sometimes rising well above the melting point. If melting occurs, the new lattice may form on the old lattice with new vacancies and interstitials replacing the original ones. Recrystallization and phase change may occur. The sudden cooling in a spike area for a steel will result in the local formation of hard, brittle martensite.

A knock-on atom with an energy of 300 eV will have a range of 10 to 100 Å. Its energy is transferred to lattice vibrations. Considering this as a point source of heating, a region 30Å in diameter would be heated to 1086°C in 5×10^{-12} s. After 20×10^{-12} s the heat-affected zone has expanded to 60 Å, but the mean temperature has dropped to 150°C. In like manner a 100-MeV fission fragment in uranium might produce a spike 4×10^4 Å long and 100 Å in diameter with temperatures reaching 4000°C. Localized melting, diffusion, and phase changes are all possible with such heating in the spike area.

(5) Ionization effects are caused by the passage through a material of gamma rays, or charged particles. This is particularly important with materials which have either ionic or covalent bonding. Materials such as insulators, dielectrics, plastics, lubricants, hydraulic fluids, and rubber are among those which are sensitive to ionization. Plastics with long-chain-type molecules having varying amounts of cross-linking may have sharp changes in properties due to irradiation. Table 11.1 indicates that, in general, plastics suffer varying degrees of loss in unirradiated properties after exposure to high radiation fields. It is interesting to note that nylon begins to suffer degradation of its toughness at relatively low doses but suffers little loss in strength. The high-density (linear) polyethylene Marlex 50 loses both strength and ductility at relatively low doses.

In general, rubber will harden on irradiation. However, butyl or Thiokol rubbers will soften or even become liquid with high radiation doses. Natural rubber will retain considerable flexibility on exposure to a gamma dose of 8.7×10^3 kJ/kg. The exposure limit is usually considered to be 4×10^3 kJ/kg for static applications and 5×10^2 kJ/kg for dynamic conditions. Radiation resistance of these elastomers is quite dependent on the conditions for curing and processing, curing agents, fillers, and antioxidants. Carbon black is often used as a filler and provides a degree of radiation resistance.

Those liquids to be used as lubricants or hydraulic fluids which have the aromatic ring type of structure show an inherent radiation resistance. It is important that oils and greases be evaluated for their radiation resistance if they are to be applied in such an environment.

Metals with shared electrons which are relatively free to wander through the lattice are affected very little by ionization.

Amorphous Materials

An ideal amorphous material has no ordered structure and might be considered to have the maximum possible deviation from a crystalline state. There are areas of limited order in all real amorphous substances. The major difference between amorphous solids and liquids is the difference in the time for ordered areas and disordered areas to interchange. For many liquids this time is 10^{-10}

to 10^{-11} s, for glycerine it is 10^{-7} s, and for glass 10^{-1} s near the melting point but 10^{8} s at room temperature.

In a disordered array atomic displacements have little effect, but ionization can be very important.

Number of Atom Displacements per Neutron Scattering Collision

When an energetic neutron strikes an atom it is necessary to transfer a minimum displacement energy, E_d, of approximately 25 eV to eject the struck atom from its lattice site. The ejected atom is called a *primary knock-on* and it will travel through the lattice causing ionization, heating, and secondary displacements. The maximum energy transfer, T_m, to the primary knock-on of mass A occurs when the neutron of energy E is backscattered ($\theta = 180°$). It can be shown that

$$T_m = \frac{4AE}{(A + 1)^2} \tag{11.2}$$

We are particularly interested in determining the number of displacements which can be produced by each primary knock-on atom. The rate at which displacements occur, $\dot{n}_d$, can be given as

$$\dot{n}_d = N\phi(E)\sigma_d(E) \tag{11.3}$$

where N is the number density of target nuclei (atoms/cm^3), $\phi(E)$ is the fast flux (neutrons/cm^2 s), and $\sigma_d(E)$ is the cross section for both primary and secondary displacements. The latter is defined as

$$\sigma_d(E) = \int_{E_d}^{T_m} P(T)v(T)K(E, T) \, dT \tag{11.4}$$

where $P(T)$ is the probability that an atom receiving energy T is displaced, $K(E, T)$ is the differential cross section for the transfer of kinetic energy T from a neutron of energy E, and $v(T)$ is the total number of displacements in a cascade originating from a primary recoil whose energy is T. The integration occurs between E_d, the displacement threshold energy, and T_m, the maximum energy that can be transferred to a target atom by a neutron of energy E. At high energies the primary knock-on (ion) will lose energy primarily by ionization and excitation interactions as it passes through the lattice. As the knock-on loses energy it tends to pick up free electrons, effectively reducing its charge. As a result, the principal mechanism for energy losses progressively changes from one of ionization and excitation at high energies to one of elastic collisions with the lattice atoms at low energies. These elastic collisions produce secondary

displacements. For simplification we will assume an ionization threshold energy ($E_i \approx 1000\,A$) in eV above which the primary knock-on causes only ionization and below which it produces only displacements. We will further assume that:

(1) There is a sharp displacement threshold

$$P(T) = 0 \qquad \text{if } T < E_d$$

$$P(T) = 1 \qquad \text{if } T \geq E_d$$

where P is the probability of displacement.

(2) There is hard sphere scattering such that

$$K(E, T) = \sigma_s / T_m$$

(3) The Kinchin–Pease model may be used to define $v(T)$:

$$v(T) = T/2E_d \qquad \text{for } T \leq E_i$$

$$v(T) = E_i/2E_d \qquad \text{for } T > E_i$$

Note the $2E_d$ in the denominator. Since it takes E_d to displace the atom originally in the site, another E_d will be required if the striking atom is to leave also.

Using the assumptions above in Eq. (11.4), we obtain

$$\dot{n}_d = N\phi\sigma_s \left(\int_{E_d}^{E_i} \frac{T}{2E_d T_m}\, dT + \int_{E_i}^{T_m} \frac{E_i}{2E_d T_m}\, dT \right) \tag{11.5}$$

Dividing through by the rate of scattering intersections gives the number of displacements per scattering collision. Carrying out the integration yields

$$\frac{\dot{n}_d}{N\phi\sigma_s} = \frac{E_i^2 - E_d^2}{4E_d T_m} + \frac{E_i T_m - E_i^2}{2T_m E_d} \tag{11.6}$$

If $T_m < E_i$, the second integral is dropped and the upper limit on the first becomes T_m, so that

$$\frac{\dot{n}_d}{N\phi\sigma_s} = \frac{T_m^2 - E_d^2}{4E_d T_m} \tag{11.7}$$

but, $E_d \ll T_m$, so

$$\frac{\dot{n}_d}{N\phi\sigma_s} = \frac{T_m}{4E_d} \tag{11.8}$$

Now if $T_m > E_i$, we must retain both terms and

$$\frac{\dot{n}_d}{N\phi\sigma_s} = \frac{E_i^2}{4E_d T_m} + \frac{2E_i T_m - 2E_i^2}{4T_m E_d}$$

$$= \frac{E_i}{4E_d}(2 - E_i/T_m) \qquad (11.9)$$

Example 1   **(a)**   Compare the maximum number of displacements per neutron scattering collision for 1-MeV neutrons in aluminum and zirconium. For aluminum:

$$E_i = 1000A = 1000 \times 27 = 2.7 \times 10^4 \text{ eV}$$

$$T_m = \frac{4AE}{(A+1)^2} \times \frac{4 \times 27 \times 10^6}{28 \times 28} = 13.8 \times 10^4 \text{ eV}$$

In this case the maximum energy of the primary knock-on, T_m, is greater than the ionization energy, E_i.

$$\frac{\dot{n}_d}{N\phi\sigma_s} = \frac{E_i}{4E_d}(2 - E_i/T_m)$$

$$= \frac{2.7 \times 10^4}{4 \times 25}(2 - 2.7 \times 10^4/13.8 \times 10^4)$$

$$= 488 \text{ displacements per scattering collision}$$

For zirconium:

$$E_i = 1000 \times 91 = 9.1 \times 10^4 \text{ eV}$$

$$T_m = \frac{4 \times 91 \times 10^6}{92 \times 92} = 4.32 \times 10^4 \text{ eV} < E_i$$

$$\frac{\dot{n}_d}{N\phi\sigma_s} = \frac{T_m}{4E_d} = \frac{4.34 \times 10^4}{4 \times 25} = 435 \text{ displacements per scattering collision}$$

(b)   Determine the number of displacements per zirconium atom per day predicted by the number of displacements per scattering collision. Assume a fast flux of 5×10^{15} neutrons/cm^2-s.

$$\frac{\dot{n}_d}{N} = \phi\sigma_s \text{ (no. displ./scattering collision)}$$

$$= 5 \times 10^{15} \text{ neutrons/cm}^2 \text{ s} \times 6.2 \times 10^{-24} \text{ cm}^2/\text{Zr atom}$$
$$\times 435 \text{ displ./scat. col.} \times (3600 \times 24) \text{ s/day}$$
$$= 1.17 \text{ displ./day}$$

Note that in part (a) of this example, the number of displacements per scattering collision is the same order of magnitude for both zirconium and aluminum even though the atomic weights are quite different. This occurs because in aluminum most of the energy dissipation in the cascade is by ionization

and we have not, in the model used, taken any of this as producing displacements. On the other hand, in zirconium we take no account of energy dissipation by ionization. On this basis the model probably underestimates the number of displacements in aluminum and overestimates the number in zirconium.

As to whether the calculations are more appropriate for Al or Zr, one can argue either way. In Al the error in calculated displacements is greater because of the low ionization threshold discussed above. On the other hand, because ranges of knock-ons are greater in light elements (larger T_m) the cascade will be more spread out, displacements further apart, and there will be less tendency for mechanical relaxation of the displacement cascade. The calculation above, estimating 1.17 displacements per Zr atom per day, indicates the tremendous amount of disruption of the lattice structure caused by fast neutrons.

The model used tends to overestimate the number of displacements because of several significant omissions: replacement collisions as well as focusing and channeling effects have not been considered. Because of the directional nature of lattices, displaced atoms may travel preferentially in certain directions having only glancing collisions without removing struck atoms from stable lattices sites. Other limitations on using results of the calculations are that no consideration has been given to a thermal relaxation (mechanical relaxation and radiation annealing) and thermal recovery effects, which will be discussed shortly.

Temperature and Mobility Effects

As temperatures increase, the vibration of atoms in the lattice increases, improving the probability that an interstitial atom might migrate to a vacancy and erase both defects. This ability of displaced atoms to diffuse in the lattice is known as *mobility*. The damage due to atom displacements may be healed either during or after irradiation if temperatures are high enough to allow the mobility necessary for atoms to migrate.

Figure 11.1 shows in a schematic manner the way that fast neutrons in a reactor core introduce changes in mechanical properties of materials. Atom displacements produce vacancies, interstitials, and spikes.

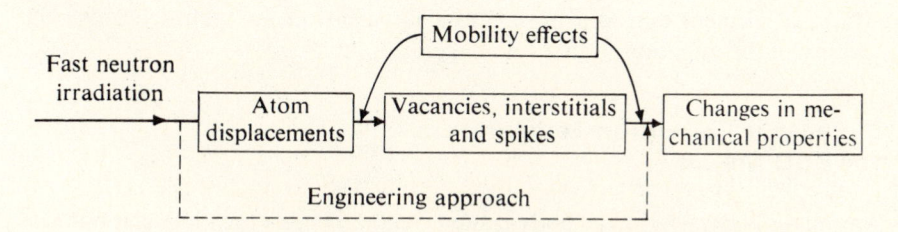

Fig. 11.1   Radiation damage in solids due to fast neutron irradiation.

Table 11.2 Effect of Fast-Neutron Irradiation on the Mechanical Properties of Metals

Material	Integrated Fast Flux (NVT)	Radiation Temperature (°C)	Tensile Strength (MPa)	Yield Strength (MPa)	Elongation (%)
Austenitic SS	0	—	576	235	65
Type 304	1.2×10^{21}	100	720	663	42
Low-carbon	0	—	517	276	25
steel	2×10^{19}	80	676	634	6
A-212 (0.2% C)	1×10^{20}	80	800	752	4
	2×10^{19}	293	703	524	9
	2×10^{19}	404	579	293	14
Aluminum	0	—	124	65	28.8
6061-0	1×10^{20}	66	257	177	22.4
Aluminum	0	—	310	265	17.5
6061-T6	1×10^{20}	66	349	306	16.2
Zircaloy-2	0	—	276	155	13
	1×10^{20}	138	310	279	4

The effect these displacements will have on properties will be determined by the amount of mobility introduced either during or after irradiation.

The data in Table 11.2 illustrate the effect of fast-neutron irradiation on the properties of several structural alloys used in reactors. Aluminum is satisfactory at low temperatures in research reactors. As temperatures rise in power reactors, zirconium becomes attractive to temperatures of 540°C. For higher cladding temperatures stainless steel and other high-temperature alloys can be used despite cross sections which are much higher than those for aluminum and zirconium. In general, irradiation increases tensile strength and yield strength at the expense of ductility. The effects are much the same as those produced by cold work. Particularly with the face-centered cubic materials, such as 304 SS and 6061 aluminum, the marked increase in yield strength with only modest loss in ductility can be considered as a distinct improvement in properties.

The data for 0.2 percent C steel show that at temperatures over 370°C there is a marked self-annealing effect during irradiation. The temperature of the structural element during exposure to a fast-neutron flux will determine the severity of the radiation-induced damage.

Increase in Transition Temperature for BCC Metals

An unhappy characteristic of body-centered cubic metals is their loss of toughness at reduced temperatures. For ferritic steels this transition occurs between room

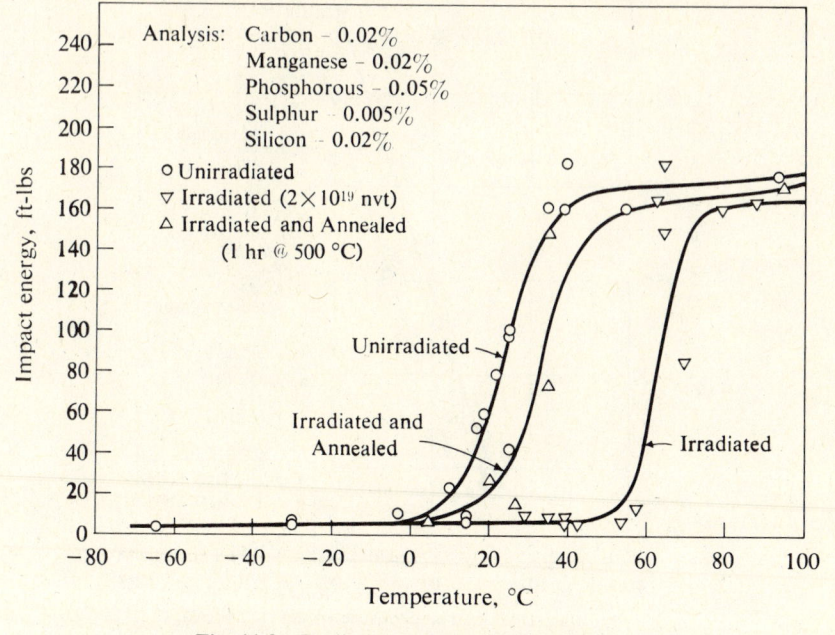

Fig. 11.2   Radiation damage to Swedish iron.

temperature and −73°C, depending on the alloy content and melting practice used.   Figure 11.2 shows results on hot-rolled Swedish iron, which illustrate this loss in impact absorption ability at reduced temperatures.

Irradiation to 2×10^{19} NVT raised the transition temperature by 60°C. Post-irradiation annealing at 500°C for 1 h returned the transition temperature to within 10°C of its value in the unirradiated state.   These results indicate that the transition temperature can be raised in a ferritic iron alloy quite significantly (several hundred degrees in some cases) on exposure to a large integrated fast-neutron flux.   It is also indicated that upon annealing the increased mobility allows much of the radiation damage to heal itself.   Figure 11.3 shows the increase in the nil ductility transition temperature for a representative group of low-carbon-steel alloys irradiated at temperatures below 232°C.   Many current reactors have core vessel wall temperatures in the range of 200 to 290°C, so that increase in the nil ductility transition temperature is of very real concern.

Figure 11.4 shows mechanical twinning in a ferrite crystal of hot-rolled Swedish iron, indicating that twinning was a mode of deformation as well as slip in this material.

For reasons of economics, the core vessels of large power reactors have been constructed of low-carbon steels.   The loss of ductility and increase in the transition temperature of these alloys is a serious concern to reactor designers.   The use of steels that are vacuum melted or vacuum poured will reduce the original transition temperature of the steel.   Also, in many instances the core operating

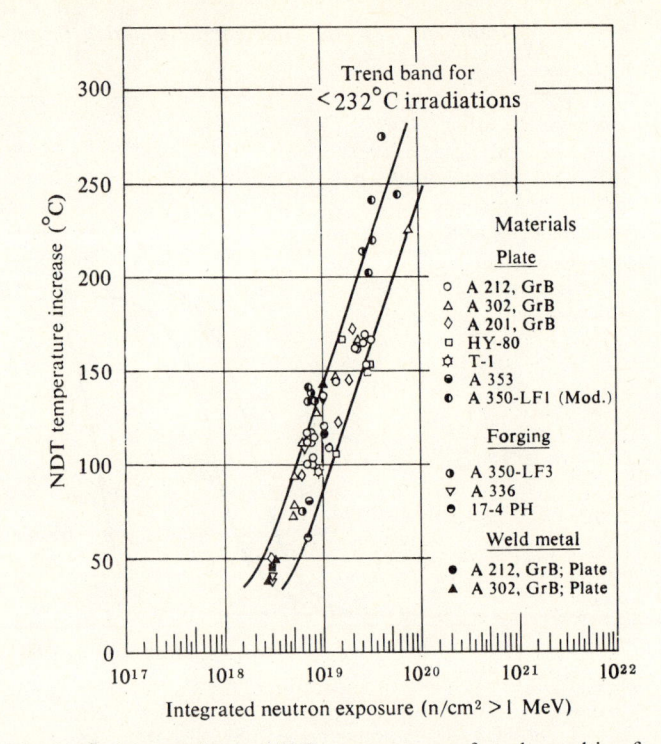

Fig. 11.3 Increase in the NDT temperatures of steels resulting from irradiation at temperatures below 232°C. [From M. T. Simnad and L. P. Zumwalt, *Materials and Fuels for High-Temperature Nuclear Energy Applications.* Cambridge, Mass.: The MIT Press, 1964.]

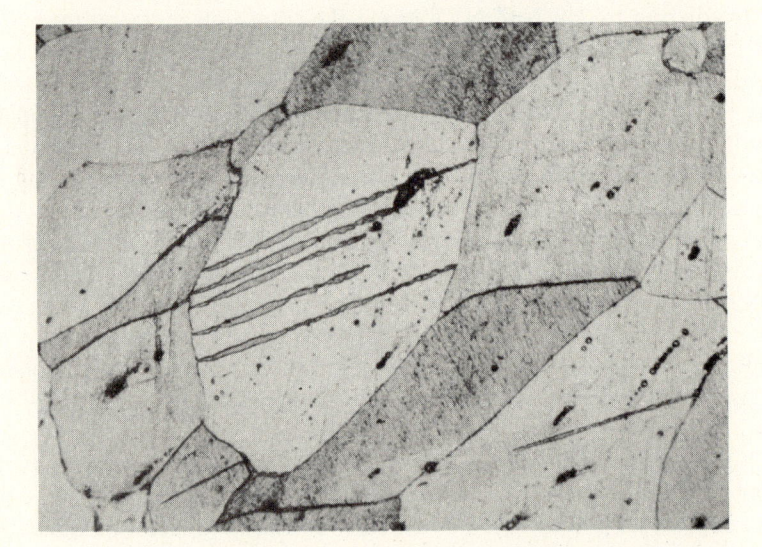

Fig. 11.4 Mechanical twins near fracture in Charpy specimen of hot-rolled Swedish iron. 200X magnification.

temperature is high enough to slow the rate at which radiation damage accumulates. Research is continuing in this area to improve our understanding of the changes that take place and to improve our ability to design safe pressure vessels for reactor cores.

Stainless Steels in Fast Reactors

For stainless steel (see Table 11.2) exposed to a thermal reactor fluence of 10^{21} neutrons/cm^2, the tensile properties show some increase in ultimate strength, an almost threefold gain in the yield strength, and a drop of about one-third in ductility. High-temperature tests show a dramatic loss in strength and ductility. The damage is thought to be due largely to helium-filled bubbles where the gas has been produced by the n, α reaction of thermal neutrons with ^{10}B. Boron contents of the order of a few ten-thousandths of a percent are sufficient to produce the effects noted. The boron will burn out, however, after an exposure of about 10^{21} neutrons/cm^2 and little further damage will be noted. Figure 11.5 shows how for boron contents of 2 and 10 ppm the number of helium atoms formed reaches a plateau in less than 100 days at power.

Operation of fast reactors, such as EBR-II and Dounray, to fluences in excess of 2×10^{22} neutrons/cm^2 indicates that significant amounts of swelling occur in austenitic stainless steel cladding and structure. These fast fluxes contain a significant number of neutrons at energies in excess of 3 MeV. Above this

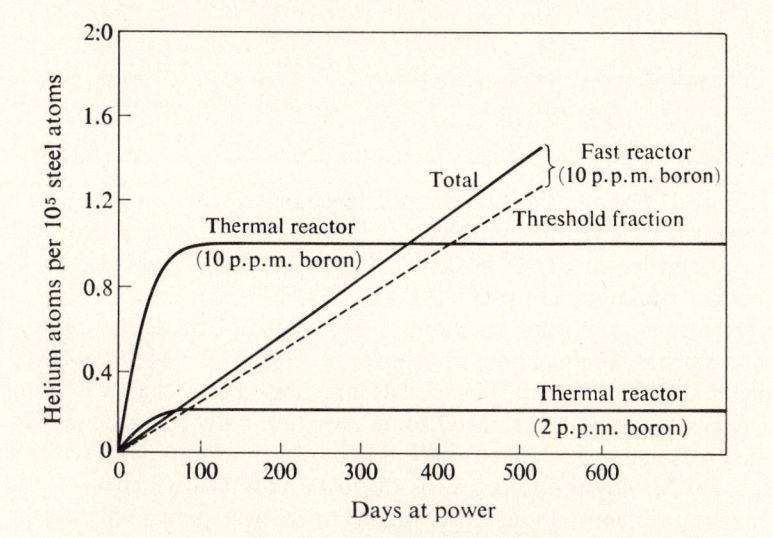

Fig. 11.5 Helium production in fast thermal fluxes. Flux in the fast reactor 2.5×10^{15} neutrons/cm^2 s; flux in thermal reactor 1.6×10^{14} neutrons/cm^2 s. [From A. S. Fraser et al., *Nature* **211** (July 16, 1966), p. 291.]

energy level most materials will undergo an (n, α) reaction. Thus, void formation is no longer dependent on boron. The (n, α) reaction in boron will only
occur at a rate approximately 1 percent of that in a thermal reactor. Here, as
shown in Figure 11.5, the helium formation does not saturate, but is linear with
time. Further, the void formation is uniform throughout the steel, where in
the thermal reactor it is limited to zones about precipitates where the boron
may be concentrated. The voids are larger than would be predicted by the
amount of helium present. This effect may be due to condensation of vacancies
at sites where a few helium atoms have nucleated a bubble.

These voids can cause gross swelling of stainless steels where the volumetric
change may amount to several percent for fluences in excess of 10^{22} neutrons/
cm^2. As exposures at high burnups are ultimately expected to reach 10^{24}
neutrons/cm^2, this swelling must be accommodated in the design of a fast reactor
core. Correlation of available data has led to the development of an empirical
equation (Reference 28) to predict the percent volume change for 304 stainless
steel.

$$\Delta V/V = A(\phi t)^m + B \qquad (11.10)$$

where 	$T =$ operating temperature, °C

$\phi t =$ fluence $\times 10^{-22}$, (neutrons/cm^2) $\times 10^{-22}$

$A = 2.65 \times 10^{-8}(T - 348)^3 - 1.54 \times 10^{-5}(T - 348)^2$
$+ 2.24 \times 10^{-3}(T - 348)$
$B = 0.2(1 - e^{-1.12\phi t})/(1 + e^{0.1(T - 480)})$

$m = 0.872 + 2.98 \times 10^{-3} T$

Figure 11.6 shows the temperature dependence of this volumetric change,
which peaks in the vicinity of 500°C. A fluence of 10^{23} neutrons/cm^2 will produce a volume increase of 18 percent at 490°C. At 600°C and the same fluence,
the increase has dropped to 4.8 percent.

The results from burst tests run on segments of EBR-II fuel-pin cladding
taken at various core locations are shown in Table 11.3. Material taken from
the plenum region where the fluence was low showed little change in properties.
That below the midplane showed some radiation-induced hardening but still
retained reasonable ductility, as indicated by the transgranular fracture shown
in Figure 11.7a. For those specimens above the midplane, which were irradiated
at temperatures only about 50°C higher, there was a marked loss in both
strength and ductility. The fractures became transgranular as indicated by
Figure 11.7b. Figure 11.8 shows how radiation-induced phase change has occurred in this cladding with the intergranular fracture. It contributes to the
deterioration in the properties of the material.

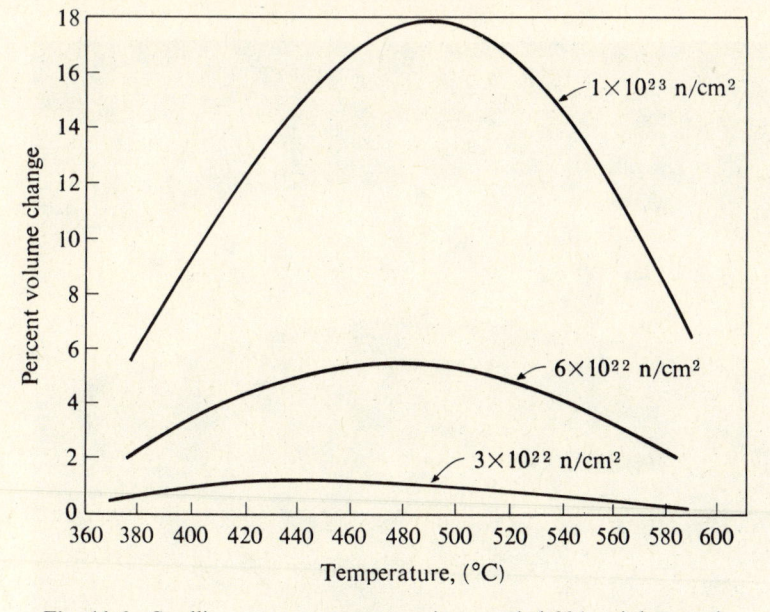

Fig. 11.6 Swelling versus temperature in annealed 304 stainless steel which has been irradiated to fast fluences of 3×10^{22}, 6×10^{22}, and 1×10^{23} neutrons/cm². [Courtesy of Hanford Engineering Development Laboratory, USAEC, operated by the Westinghouse Hanford Company.]

Table 11.3 Fuel-Pin Cladding Burst Test Results

Operating Temperature (°C)	Fluence (neutrons/cm²) ($E > 0.1$ MeV)	Failure Stress (MPa)	Strain ($\Delta D/D\%$)	Specimen Location
475	1.2×10^{21}	407.4	18.4	Plenum
445	5.5×10^{21}	497.5	9.6	Below midplane
475	5.8×10^{21}	151.8	1.0	Above midplane
480	4.0×10^{21}	187.3	0.3	Above midplane
495	5.0×10^{21}	268.2	0.2	Above midplane
—	Control	392.7	19.2	—

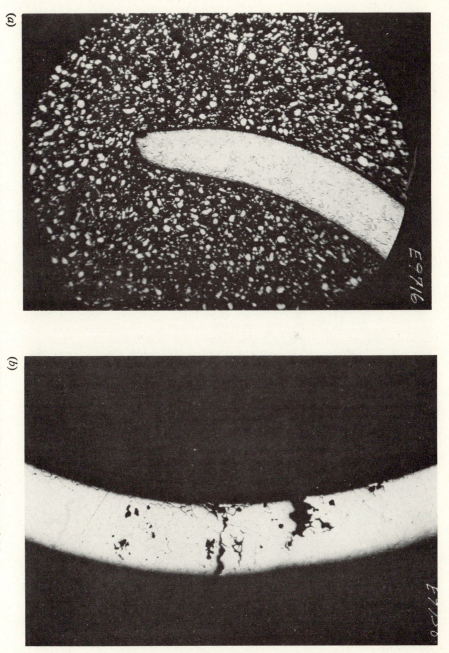

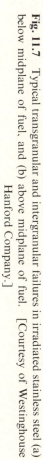

Fig. 11.7 Typical transgranular and intergranular failures in irradiated stainless steel (a) below midplane of fuel, and (b) above midplane of fuel. [Courtesy of Westinghouse Hanford Company.]

(a)

(b)

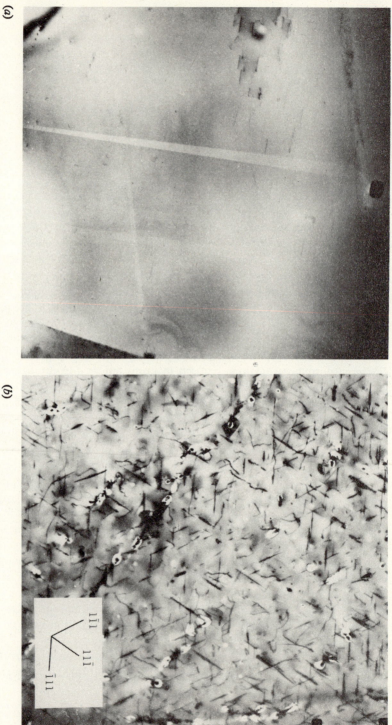

Fig. 11.8 Fast-neutron irradiation-induced precipitation of $M_{23}C_6$ carbides in Type 316 stainless steel. (a) Unirradiated control sample, annealed 3050 h at 480°C. (b) Sample irradiated in EBR-II at 480°C to 0.8×10^{22} neutrons/cm^2 (>0.1 MeV). [Courtesy of Westinghouse Hanford Company.]

Comparison Between Thermal and Fast-Neutron Damage

It has been noticed that there is significantly less damage due to irradiation of metals at cryogenic temperatures in a fast flux (see Reference 32) than is observed for the same fluence in various reactor spectra at similar temperatures. Further, the recovery of radiation damage during post-irradiation isochronal annealing after irradiation in a fast flux is different from the recovery after irradiation in the various reactor spectra. Isochronal annealing involves heating from the cryogenic temperature to successively higher annealing temperatures. After holding at each annealing temperature for perhaps 5 min, the specimens are cooled back to the cryogenic temperature, and their electrical resistance is measured before heating to the next higher temperature.

Since much of the radiation behavior of materials for the liquid-metal fast breeder reactor (LMFBR) has been in fluxes containing a large thermal neutron component, it seemed desirable to try comparing the damage in thermal and fast fluxes. The study of Horak and Blewitt (see Reference 32) was performed in the cryogenic facilities of the CP-5 reactor at the Argonne National Laboratory with irradiation in liquid helium at 4.5K.

Defect production as the result of thermal neutron irradiation is largely due to the recoil of nuclei that have absorbed a thermal neutron and emitted one or more gammas—with a total energy of 5 to 10 MeV. The recoil energy of the nucleus may produce a number of Frenkel defects by ejection of an atom that has absorbed a neutron from its own lattice site and subsequent collisions. Table 11.4 indicates the gamma emission spectra for Cu, Ni, Fe, Ti, and Pd. However, since 2 to 3 MeV is required to provide the $2 E_d = 50$ eV necessary to

Table 11.4 Gamma Ray Emission Spectra After Thermal Neutron Capture for Various Metals

Metal	Number of Gamma Rays per Neutron Capture						
	0–1 *MeV*	1–2 *MeV*	2–3 *MeV*	3–5 *MeV*	5–7 *MeV*	7–9 *MeV*	>9 *MeV*
Cu	>0.68	0.47	0.26	0.30	0.27	0.32	0
Ni	>0.84	0.40	0.23	0.23	0.34	0.62	0.008
Fe	>0.75	0.60	0.27	0.23	0.25	0.38	0.021
Ti	>0.54	1.60	0.16	0.24	0.78	0.01	0.002
Pd	>0.53	0.33	0.75	0.62	0.19	0.03	0.0001

From J. A. Horak and T. H. Blewitt, *Nuclear Technology* **27**, No. 3 (November 1975), p. 433.

create a Frenkel pair, Horak and Blewitt suggest using only the gammas above 3 MeV from Table 11.4 to evaluate $\bar{E}_t$. Summing the average recoil energy, $\bar{E}_{Ri}$, each of the four energy groups multiplied by the probability of a gamma ray emission per neutron capture, p_i, in each energy interval will result in the average energy transferred to the lattice by each (n, γ) capture, $\bar{E}_t$.

$$\bar{E}_T = \sum_{i=4}^{i=7} E_{Ri} p_i \tag{11.11}$$

The defect production by thermal neutrons is proportional to the product of the (n, γ) cross section, σ_α, multiplied by $\bar{E}_t$. Table 11.5 shows that the values of $\sigma_\alpha \bar{E}_t$ range from 1042 for Fe to 22 440 for Pd; thus, there is a large variation in the thermal neutron defect production rates. There are three fcc metals, one bcc, and one hcp in the group used for the study. Also, iron and nickel are the principal constituents of the austenitic stainless steels used in the LMFBR program.

The rate of defect production by the fast neutrons is proportional to the average elastic scattering in the energy range 0.1 to 2.0 MeV.

The fluence for the thermal irradiation was 5.7×10^{17} neutrons/cm^2, and for the fast flux it was 5.8×10^{17} neutrons/cm^2. Thus, the fluences are essentially equal; the total irradiation-induced defect concentration and the isochronal recovery differences are due to the energy-dependent behavior of the two energy ranges. The measured defect concentrations are determined from the increase in resistivity per the Frenkel pair after (in the case of the thermal irradiation) these are found by subtracting out the effect of the increase in resistivity due to transmutations. Observe that for thermal irradiation the ratios of calculated to measured defect concentrations vary from 1.0 to 2.9, while for fast-flux irradiation the ratios vary from 2.3 to 6.5. For thermal irradiation since recoil energies are relatively low, they produce only a few displacements per gamma emitted. Thus, the displacements are quite uniformly distributed, and the Kinchin–Pease model gives a fairly good prediction of the number of displacements. Fast neutron scattering collisions transmit much higher energies to the struck nuclei; effects such as focusing collisions, channeling events, annihilation of point defects during a cascade, and so on, tend to make the number of defects smaller than predicted by the simple displacement theory. Also, the defects tend to be more clustered.

Figure 11.9 shows the isochronal recovery for nickel after both thermal and fast irradiation. Not only is the recovery somewhat faster for the material that has been thermally irradiated, but its resistivity drops below the initial value— this effect being termed *super recovery*. This is attributed to the fact that with thermal irradiation the point defects are homogeneously distributed and thus can contribute to long-range diffusion. Precipitation of interstitial impurity atoms such as oxygen, hydrogen, nitrogen, or carbon, on formation of additional phases, reduces the resistivity below its value for the unirradiated material.

Table 11.5 Comparison of Radiation Damage Produced by Fast and Thermal Neutrons

Element	Lattice Type	σ_a (barns)	$\bar{E}_T$ (eV)	$\sigma_a \bar{E}_T$ (eV–barns)	σ_s (barns)	Defect Concentration ($af \times 10^{-5}$)				Ratio Calculated to Measured Defect Concentration	
						Measured		Calculated			
						C_{th}	C_F	C_{th}	C_F	R_{th}	R_F
Cu	fcc	3.77	384	1474	3.50	0.9	16.2	1.0	106	1.7	6.5
Ni	fcc	4.8	336	2611	3.25	1.1	18.0	2.7	106	2.5	5.9
Fe	bcc	2.5	411	1042	2.50	0.9	25.4	1.1	86	1.2	3.4
Ti	hcp	5.8	429	2552	2.50	2.4	43.3	2.3	100	1.0	2.3
Pd	fcc	170.0	127	22 440	4.50	6.5	16.1	19.6	81.4	2.9	5.1

Adapted from J. A. Horak and T. H. Blewitt, *Nuclear Technology.* **27**, No. 3 (November 1975), pp. 416–438.

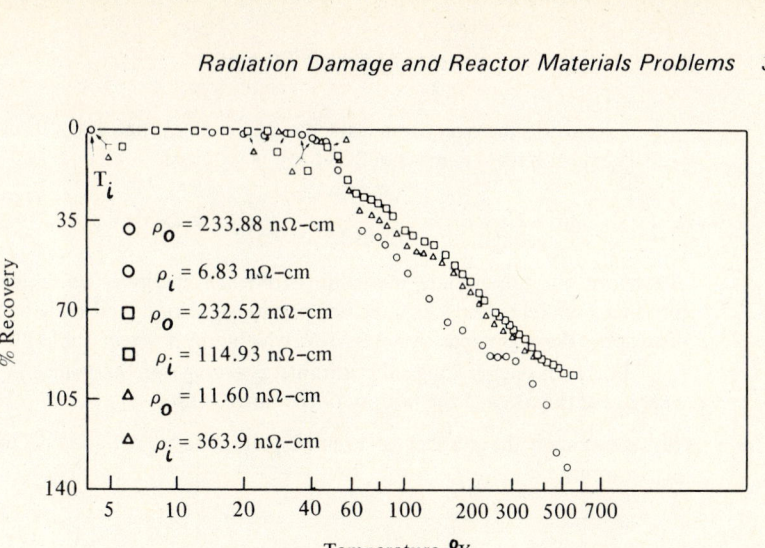

FLUENCE

O = 5.7 x 10^{17} thermal neutron/cm^2
□ = 5.8 x 10^{17} fast neutron/cm^2 E > 0.1 MeV
△ = 2.0 x 10^{17} fast neutron/cm^2 E > 0.1 MeV

Fig. 11.9 Isochronal recovery for nickel versus logarithm of absolute temperature after thermal- and two fast-neutron irradiations at 4.5 K. [From J. A. Horak and T. H. Blewitt, *Nuclear Technology* **27**, No. 3 (November 1975), p. 433.]

With fast-neutron irradiation, there is a much greater tendency for clustering of the defects, which prevents their participation in long-range diffusion, and thus no super recovery is observed after annealing.

Example 2 A sample of iron is irradiated in a thermal flux of 5×10^{13} neutrons/cm^2 s at a temperature of 300°C.

(a) What is the average energy transferred to the lattice by gamma radiation upon capture of a thermal neutron? A momentum balance between the gamma and the recoiling nonrelativistic iron atom gives

$$V = \frac{E_\gamma}{mc^2}$$

The recoil energy of the iron atom is then

$$E_{\text{Fe}} = \frac{mV^2}{2} = \frac{E_\gamma^2}{2mc^2} = \frac{E_\gamma^2 \times 10^6}{2 \times 55.847 \times 931.478}$$

$$= 9.612 E_\gamma^2 \qquad \text{eV}$$

To provide the iron atom with the $2E_d = 50$ eV necessary to create a Frenkel defect, a gamma of 2.28 MeV is required.

$$E_\gamma = \sqrt{\frac{E_{Fe}}{9.612}} = \sqrt{\frac{50}{9.612}} = 2.28 \text{ MeV}$$

Therefore, one can exclude any gammas of lesser energy as being unable to produce Frenkel defect. As indicated previously, Horak and Blewitt suggest using only those gammas above 3 MeV, which are shown in Table 11.4.

For each energy range the arithmetic average of the values of recoil energies at the top and the bottom of the range is used.

(b) Determine the number of Frenkel pairs created per thermal neutron capture.

Energy Group (*MeV*)	E_{Rn} (eV)	E_{Ri} (eV)	E_z (eV)	P_i	$\bar{E}_R p_i$
3–5	86.5	240.3	163.4	0.23	37.6
5–7	240.3	471.0	355.7	0.25	88.9
7–9	471.0	778.6	624.8	0.38	237.4
9	(use 10 MeV)		961.2	0.021	20.2
					$\bar{E}_T = 384.1$ eV

$$\bar{\nu}_{th} = \frac{\bar{E}_T}{2E_d} = \frac{384.1}{50} = 7.68$$

or seven Frenkel pairs are created by each thermal neutron absorption.

(c) Determine the rate at which Frenkel pairs are generated (F.P./cm^{3}s).

$$N = \frac{\rho N_{AV}}{A} = \frac{7.87 \times 6.023 \times 10^{23}}{55.847} = 8.488 \times 10^{22} \text{ atoms Fe/cm}^3$$

$$\sigma_a = 2.62 \frac{\sqrt{\pi}}{2} \sqrt{\frac{293}{573}} = 1.66 \text{ barns}$$

$$\dot{n}_{F.P.} = N\phi_{th}\sigma_a\bar{\nu}_{th} = 8.488 \times 10^{22} \times 5 \times 10^{13} \times 1.66 \times 10^{-24} \times 7$$
$$= 4.932 \times 10^{13} \text{ F.P./cm}^3 \text{ s}$$

Graphite

Graphite may be used effectively as a moderator and/or reflector in sodium- or gas-cooled reactors. Not only are its nuclear characteristics (i.e., log decrement, slowing-down power, and moderating ratio) very satisfactory, but it has unusually good mechanical and thermal properties at temperatures in excess of 2000°C.

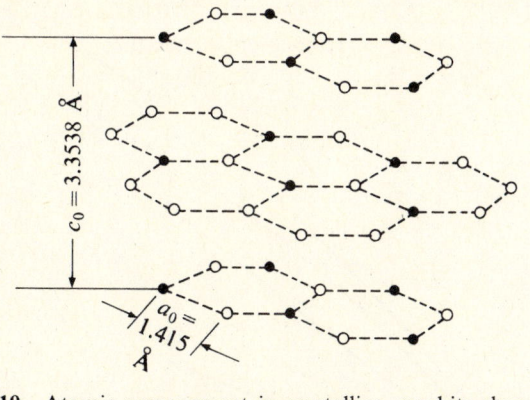

Fig. 11.10 Atomic arrangement in crystalline graphite showing the ABABA type of stacking.

The structure of graphite is hexagonal, wherein the atoms in alternate planes are offset so that there is an **ABABA** stacking sequence. This is shown in Figure 11.10. Only one-half of the atoms in the B plane lie directly over the atoms in the A plane. Bonding in the hexagonal planes is covalent, but in the *c* direction between planes weak van der Waals bonding exists. Because some of the electrons are free to move in the layers, ionization has little effect. Because of the weak forces between layers, displaced atoms lodge between the planes and tend to force the layers apart, increasing the *c* distance.

A minimum energy of 25 eV is required for a neutron to dislodge a carbon atom from a stable lattice site. A 2-MeV fast neutron will displace an average of 60 carbon atoms from stable lattice sites before falling below the 25-eV energy level. These primary carbon atoms will displace a total of about 20 000 secondary atoms for each fast neutron slowed down. As irradiation temperatures rise, more of these displaced atoms will be able to self-anneal by returning to stable lattice sites. Figure 11.11 indicates the increase in *c* spacing as a function of radiation exposure and irradiation temperature. Irradiation exposure above 300°C produces relatively little increase compared to irradiation at lower temperatures.

Graphite is usually produced by extrusion of a mixture of ground petroleum coke and a pitch binder, followed by baking at 800°C and graphitization at 2500 to 3000°C. During extrusion the coke particles are aligned in the direction of extrusion. This results in anisotropy in the final graphite. The thermal conductivity parallel to the direction of extrusion of a graphite which was graphitized at 2800°C is 2.3 W/cm °C, as compared to 1.4 W/cm °C in the perpendicular direction. Similarly, the electrical resistivity is 7×10^{-4} Ω-cm in the parallel direction and 9×10^{-4} Ω-cm in the *c* direction. The thermal conductivities are 4.0 W/cm °C in the layer direction and 0.8 W/cm °C perpendicular to the layer direction.

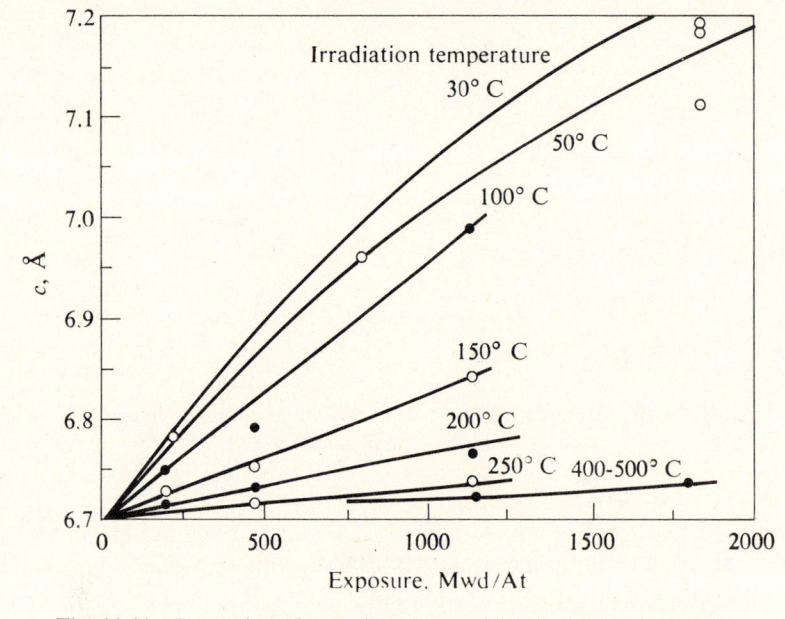

Fig. 11.11   Expansion of *c* spacing at several irradiation temperatures. [From R. E. Nightingale, H. H. Yoshikawa, and E. M. Woodruff, *Nuclear Graphite*. New York: Academic Press, Inc., 1962, Chap. 9.]

Irradiation produces gross dimensional changes in graphite. Below 300°C there will be significant expansion in a direction perpendicular to the extrusion direction and contraction in the parallel direction. Figure 11.12 shows the magnitude of these changes for three grades of graphite. Differences in the graphitization temperature and the character of the raw materials produce different amounts of anistropy and crystallinity, causing differing amounts of dimensional instability. Above 300°C irradiation can produce contraction both parallel and perpendicular to the extrusion axis, as is shown by Figure 11.13. For long exposures over the lifetime of a reactor these dimensional changes can cause serious difficulty if they are not accounted for in reactor design.

Another problem with graphite irradiated at low temperatures is the accumulation of stored energy. On heating to a temperature high enough to allow the atoms sufficient mobility to return to a stable lattice site, this energy may be released, producing a further rise in temperature; this, if not carefully controlled, may lead to combusion of the carbon. In the early 1950s the operator at the English Windscale Reactor tried to anneal the radiation damage to the moderator graphite by raising the flux (and hence the core temperature) to a high enough level to release the stored energy in a controlled manner. Because of a faulty thermocouple, the flux level was raised

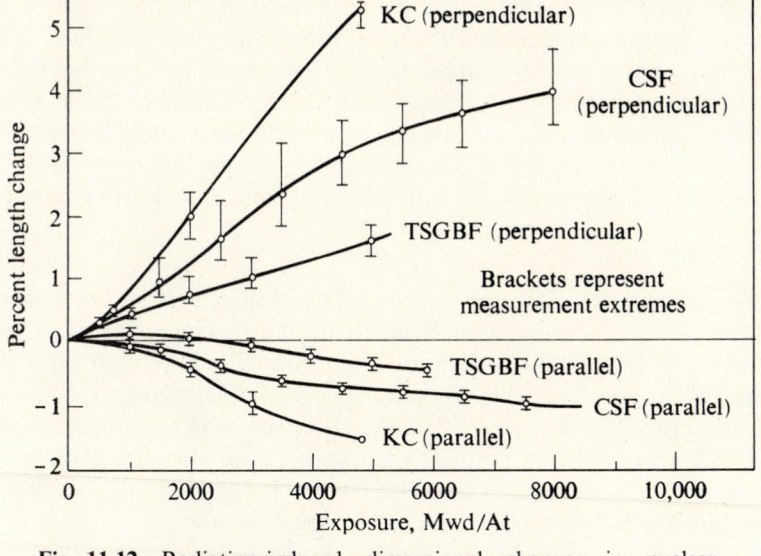

Fig. 11.12 Radiation-induced dimensional changes in nuclear graphites at approximately 30°C. Note that 1 MWd/At $= 0.87 \times 10^{17}$ NVT, $E > 0.18$ MeV. [From J. M. Davidson, E. M. Woodruff, and H. H. Yoshikawa, "High Temperature Radiation Induced Contraction in Graphite," *Proceedings of the Fourth Conference on Carbon.* Elmsford, N.Y. Pergamon Press, Inc., 1960, p. 600.]

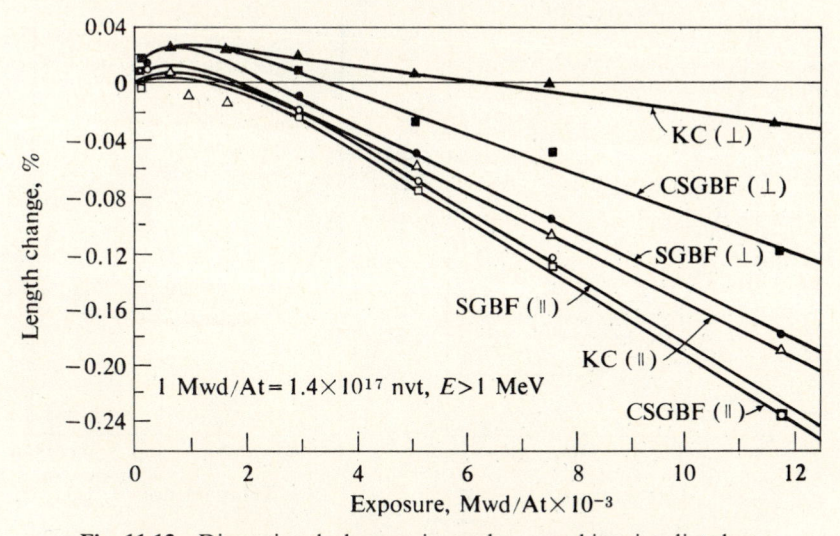

Fig. 11.13 Dimensional changes in nuclear graphites irradiated at 400 to 500°C. [From R. E. Nightingale, H. H. Yoshikawa, and E. M. Woodruff, *Nuclear Graphite.* New York: Academic Press, Inc., 1962, Chap. 9.]

too high.   This unfortunate event taught the nuclear engineer what conventional power engineers had long known: carbon is a fine conventional fuel.   The resultant fire released some fission products to the atmosphere, but the contamination of surrounding countryside was less severe than had been anticipated.   If irradiation occurs at high enough temperatures, displaced atoms have sufficient mobility to seek lattice sites and stored energy presents little problem.   This is the case in high-temperature gas-cooled reactors moderated with graphite.

Pyrolytic graphite

Pyrolytic graphite is deposited on a substrate directly from a carbonaceous vapor phase.   Methane and natural gas are the most common media to provide the carbon.   The crystallites of pyrolytic carbon (PC) have their layer planes parallel to the surface on which they are deposited.   The good crystallite alignment can produce a density of 2.22 g/cm^3, which is close to the theoretical density.   Most commercial graphite has a density around 1.8 g/cm^3 or less.   The high density of PC makes it a desirable coating for carbide fuels because of its ability to contain fission fragments.   One of the remarkable characteristics of this material is the anisotropy of its properties due to the nature of its structure. Electrical conductivity, thermal conductivity, and tensile strength are significantly greater in directions parallel to the *ab* basal planes than in the *c* direction, which is perpendicular to the basal planes.   The tensile strength is about 10

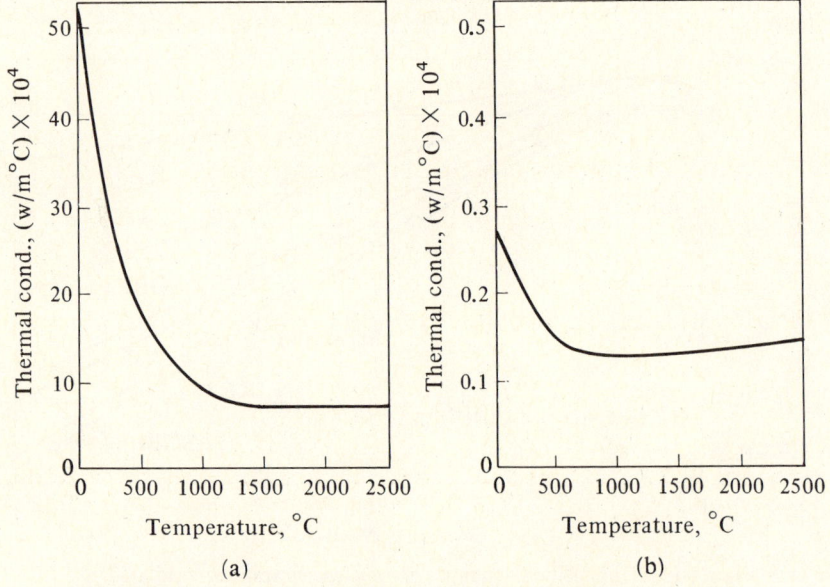

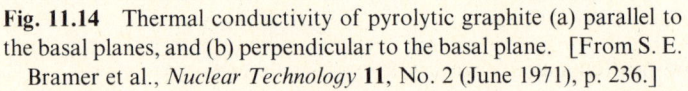

Fig. 11.14   Thermal conductivity of pyrolytic graphite (a) parallel to the basal planes, and (b) perpendicular to the basal plane.   [From S. E. Bramer et al., *Nuclear Technology* **11**, No. 2 (June 1971), p. 236.]

times as great parallel to the basal planes as it is in the perpendicular c direction. The electrical conductivity is three orders of magnitude greater in the parallel direction than it is in the perpendicular direction. The temperature dependence of the thermal conductivity of the pyrolytic graphite both parallel and perpendicular to the basal planes is shown in Figures 11.14a and b. When this material is used as a reentry heat shield for a radioisotopic heat source (Reference 26) this anistropy of the thermal conductivity is very useful. The low conductivity in the c direction discourages heat flow through the shield. The large lateral conductivity transfers heat away from the stagnation point to reduce the maximum temperature due to aerodynamic heating.

Nuclear Fuels

Fissionable metals suffer from radiation damage in a manner similar to that encountered by structural alloys, with the additional problems introduced by the high-energy fission fragments and the heavy gases xenon and krypton, which appear among the fission products. Each fission produces two fragments which share 167 MeV of kinetic energy in inverse proportion to their atomic masses. These will have a range on the order of several hundred angstroms as they produce their spikes. The gas formation produces eventual swelling of the fuel and may place the cladding under considerable pressure as well. There are several methods that may be used to cope with these problems, but first a consideration of the metallurgy of uranium, plutonium, and fertile thorium is in order.

Uranium

Domestic carnotite ore contains as little as 0.6 to 1 kg of U_3O_8 per ton of ore. Leaching in H_2SO_4 and an ion exchange process result in relatively pure U_3O_8, which is reacted with hydrogen to produce UO_2.

$$U_3O_8 + 2H_2 \rightarrow 3UO_2 + 2H_2O$$

The uranium dioxide is reacted with hydrofluoric acid to give the green salt of uranium

$$UO_2 + 4HF \rightarrow UF_4 + 2H_2O$$

If the uranium is to be converted to the metallic state directly, it can be reduced with magnesium.

$$UF_4 + 2Mg \rightarrow 2MgF_2 + U$$

The reaction is exothermic; because of the difference in densities of the liquid metal and molten slag separation takes place effectively.

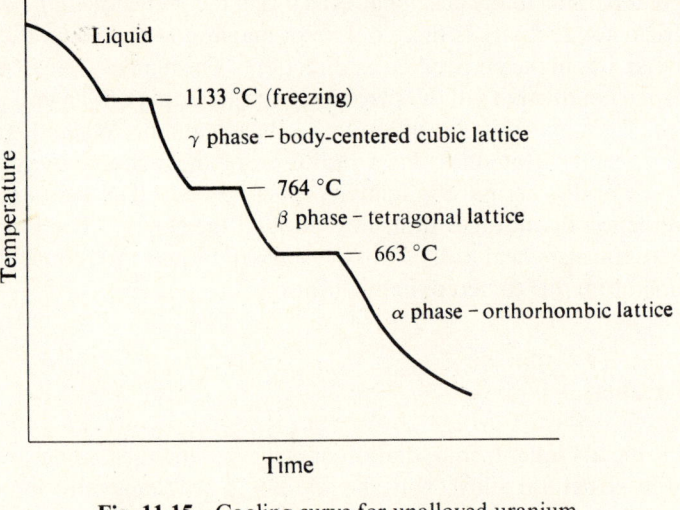

Fig. 11.15   Cooling curve for unalloyed uranium.

When the uranium is to be enriched by the gaseous diffusion process in current plants or by gaseous centrifuge plants being constructed, the green salt must be first converted to uranium hexafluoride, a gas.

$$UF_4 + F_2 \rightarrow UF_6$$

Isotopic separation is accomplished by the slightly different masses of ^{235}U and ^{238}U and their different diffusion rates to the top and bottom of a diffusion column. A multitude of stages in cascade is required for significant enrichment. Enrichments as high as 93 percent are available. The hexafluoride is reconverted to green salt by hydrogen.

$$UF_6 + H_2 \rightarrow UF_4 + 2HF$$

Often the enriched uranium is reduced with calcium.

$$UF_4 + 2Ca \rightarrow 2CaF_2 + U$$

Uranium exists with three separate crystal structures in the solid state (see Figure 11.15). On solidification, the material exists as the γ phase with a conventional body-centered cubic lattice. It is relatively soft at these temperatures and can be formed with relative ease, particularly by extrusion.

At 764°C it shifts to the β phase which has a complex tetragonal lattice with 30 atoms per unit cell. Figure 11.16 shows the way in which alternate planes in the lattice have the atoms placed.

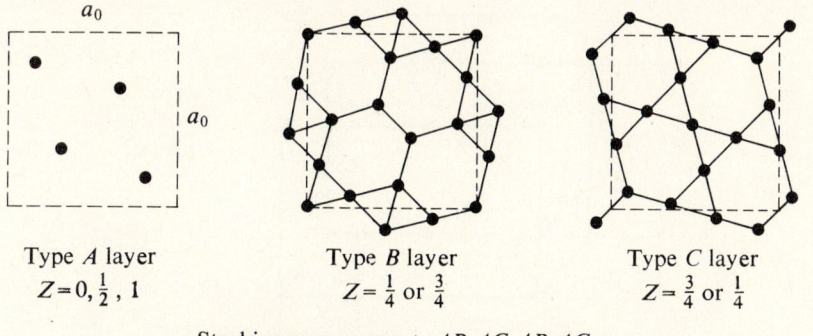

Type *A* layer
$Z = 0, \frac{1}{2}, 1$

Type *B* layer
$Z = \frac{1}{4}$ or $\frac{3}{4}$

Type *C* layer
$Z = \frac{3}{4}$ or $\frac{1}{4}$

Stacking arrangement, *AB AC AB AC*

Fig. 11.16 (010) projections of the β-uranium structure. This phase with its complicated lattice tends to be stiffer than the adjacent α and γ phases. Little effort has been made to work β-phase uranium. [From A. N. Holden, *Physical Metallurgy of Uranium*. Reading, Mass.: Addison-Wesley Publishing Co., Inc., 1958, assigned to General Manager, U.S. Atomic Energy Commission.]

Alpha-phase uranium, with its unusual orthorhombic lattice, will form when the metal cools below 663°C. The structure of α-uranium is shown in Figure 11.17.

This lattice structure has properties that are strongly anisotropic, as evidenced by the coefficient of thermal expansion. It is strongly positive in the *a* and *c* directions, whereas it is actually negative in the *b* direction. One can imagine the extension in the *c* direction straightening out the accordion pleating to reduce the *b* dimension.

When α uranium undergoes plastic deformation during a working process certain planes in each crystal will tend to rotate into the rolling direction, setting up preferred orientation. The character of the preferred orientation is dependent not only on the method of working, but also on the temperature at which the work is accomplished. Figure 11.18 illustrates the difference in growth between two different types of α-deformed uranium with different working conditions and randomly oriented α-uranium.

Growth is accounted for by the mechanism called *thermal ratcheting*. On heating above about 350°C, as adjacent crystals try to grow in different directions, the resultant stresses are relieved by grain boundary flow. On cooling, when the direction of motion would like to reverse, the lattice is too rigid to allow the accommodation. The net result is growth in the (010) direction.

Cold-worked α-uranium retains its preferred orientation when annealed to allow recrystallization in the α region. Heating into the β region, followed by quenching back to the α region, will produce random orientation of the α uranium. The treatment may be repeated for even better results.

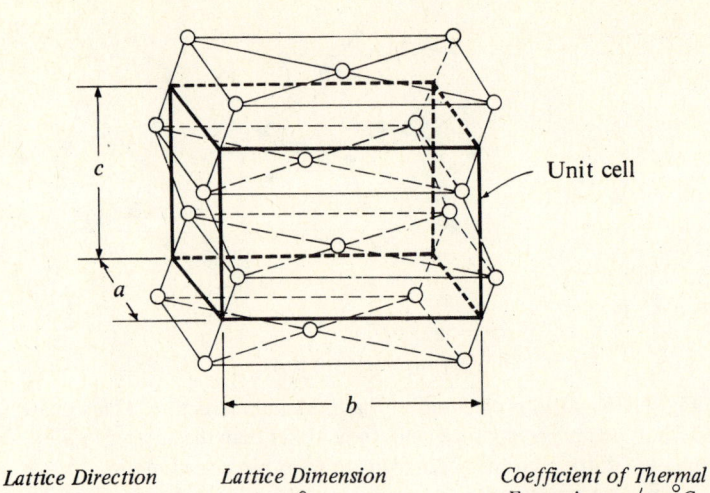

Lattice Direction		Lattice Dimension $\overset{\circ}{A}$	Coefficient of Thermal Expansion cm/cm °C (25-325°C)
a	100	2.852	47.7
b	010	5.865	−4.3
c	001	4.945	43.0

Fig. 11.17 Structure of orthorhombic α-phase uranium, its lattice dimensions, and its coefficient of thermal expansion.

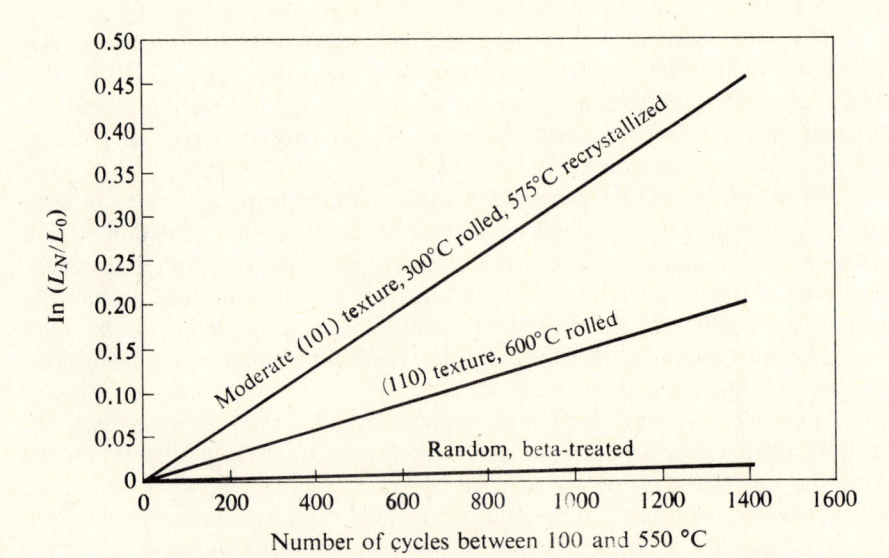

Fig. 11.18 Effect of preferred orientation on growth of uranium during thermal cycling. [From A. N. Holden, *Physical Metallurgy of Uranium*. Reading, Mass.: Addison-Wesley Publishing Co., Inc.; assigned to General Manager, U.S. Atomic Energy Commission.]

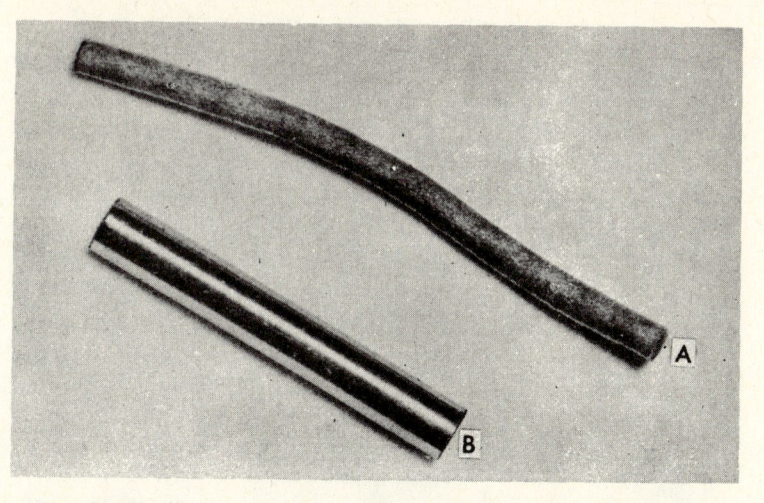

Fig. 11.19 (a) Growth of a highly textured uranium rod due to irradiation compared with a dummy; (b) the size of the original uranium rod. [From A. N. Holden, *Physical Metallurgy of Uranium.* Reading, Mass.: Addison-Wesley Publishing Co., Inc.; assigned to General Manager, U.S. Atomic Energy Commission.]

In randomly oriented α-uranium this growth due to thermal cycling will manifest itself by a roughening of the surface as some grains grow outward and others contract at the surface. This pimpling may be minimized by fine grain size. One of the disadvantages of γ-worked uranium is its tendency toward coarse grain size, with an attendant tendency to suffer from surface roughening.

When uranium is involved in the fission process the radiation damage produces growth in a manner somewhat like that due to thermal cycling if preferred orientation exists. One difference in irradiation growth is that individual crystals will undergo large changes in dimension where they will not do so on thermal cycling. It is now felt that the generation of dislocation loops in α-uranium accounts for its growth on irradiation.

As a heavy fission fragment is brought to rest in a displacement spike, a core of vacancies is surrounded by a shell of interstitials. The interstitials form on (010) planes and the vacancies on (110) planes, producing growth in the *b* direction and shrinkage in the *a* direction (see References 6 and 7). Figure 11.19 shows an extreme amount of growth that can occur under irradiation.

As the burnup of uranium increases, the generation of xenon and krypton tends to cause swelling and gross distortion of the metal. Not only are xenon and krypton isotopes stable end products of the decaying fission fragments, but they also provide some unstable isotopes which are merely an intermediate product in the decay chain. One of the major challenges in alloying metallic uranium is the attainment of better dimensional stability under irradiation.

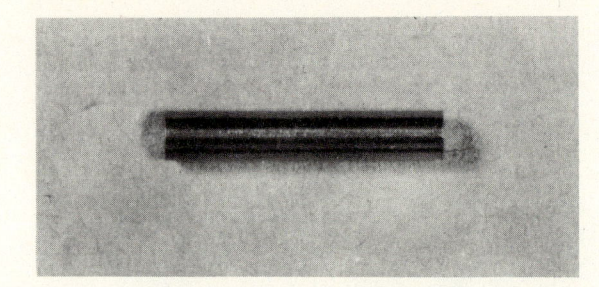

Before Irradiation

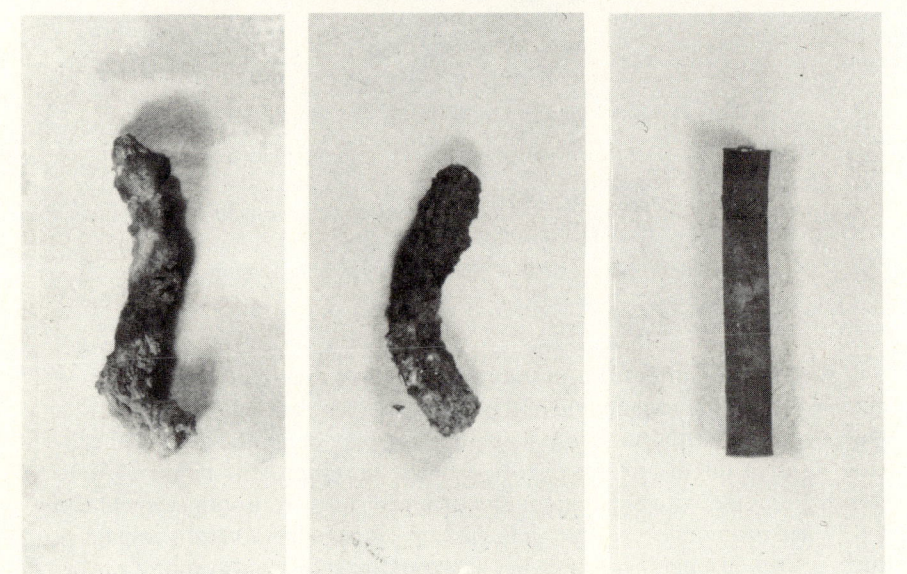

0 w/o Zr   0.52 w/o Zr   1.62 w/o Zr

Fig. 11.20 Effect of zirconium additions on the stability of cast uranium irradiated to 0.6 atom percent burnup. [Courtesy of Argonne National Laboratory.]

Figure 11.20 shows the marked improvement that can be brought about by small additions of zirconium.

The strength of uranium generally decreases with increasing temperature from an ultimate tensile strength of 524 MPa at 100°C to 35 MPa at the α–β transition temperature. The β-uranium is somewhat stronger than the α at the transition temperature and the γ phase shows very little strength. These changes are illustrated in Figure 11.21.

The ductility of the uranium rises from about 15 percent at room temperature to 30 percent at 100°C. Then it increases gradually to 35 percent at the

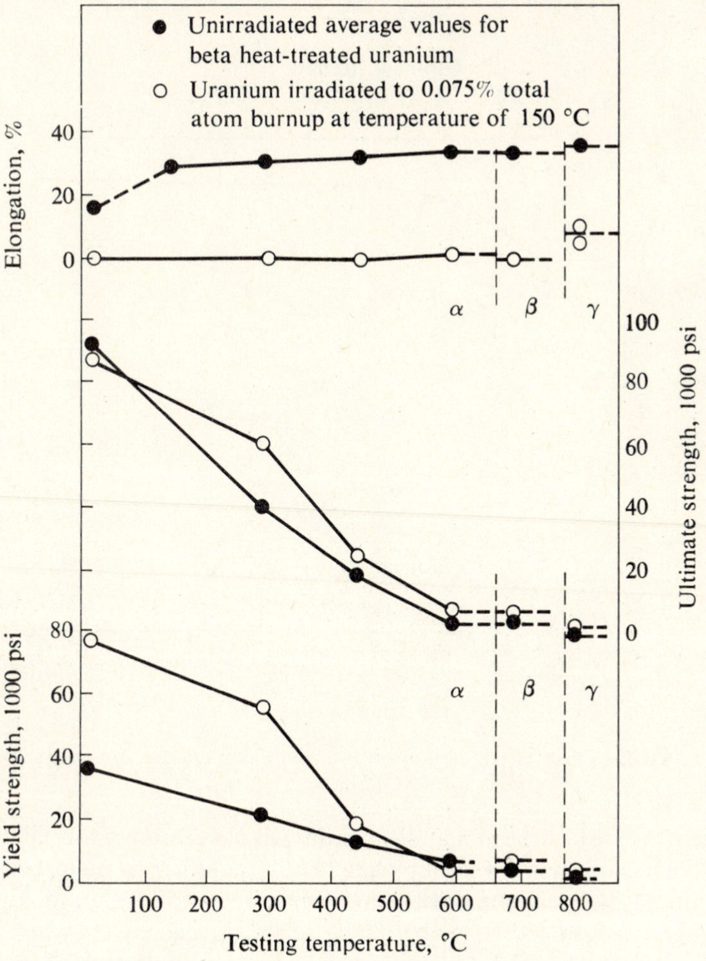

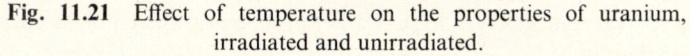

Fig. 11.21 Effect of temperature on the properties of uranium, irradiated and unirradiated.

$\alpha-\beta$ transition temperature and 37 percent at the upper extremity of the β region. It is somewhat surprising that the two lower-temperature phases show this much ductility, considering their complex lattice structures.

Figure 11.22 illustrates the effect of irradiation on the room temperature properties of uranium. Burnups greater than 0.02 percent result in loss of ductility. The ultimate and yield strengths approach each other at 0.1 percent burnup.

The typical as-cast structures of α-uranium are shown in Figures 11.23a and b. The photomicrographs are taken of the same area with polarized light but with the stage rotated 45°. Note that some areas have shifted from dark to light and that the subgraining in the central area is much more evident on the left. As

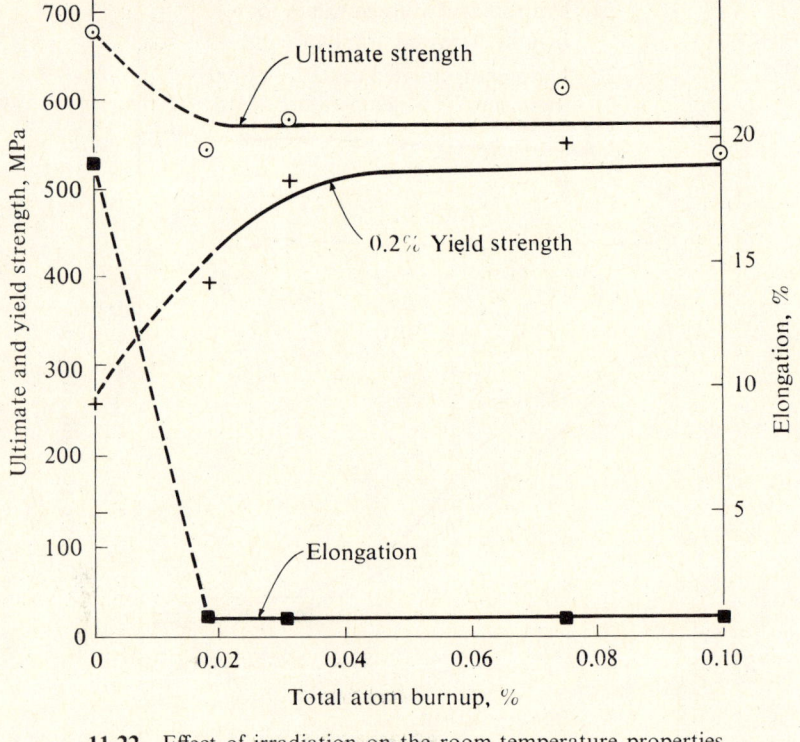

11.22   Effect of irradiation on the room-temperature properties
of uranium.

the stage is rotated each grain and subgrain will pass through a brilliant spectrum of colors, to which black-and-white photographs hardly do justice.

Figure 11.24 shows a microstructure where recrystallization took place at 625°C after rolling at 400°C.   Subgraining occurs during cooling where the anisotropic character of the shrinkage in neighboring crystals causes segments of the parent crystal to take on a slightly different orientation.   This is encouraged by thermal cycling.

As fuel temperatures have continually risen with the development of larger power reactors, metallic uranium has become less and less attractive as a reactor fuel.   Alloyed uranium, dispersion-type fuels, liquid metal fuels, and ceramic fuels are some of the approaches to providing fuel elements less subject to gross dimensional change.   These will be examined in some detail after a brief look at plutonium and thorium.

Plutonium

To all intents and purposes, plutonium is an artificial element produced by the transmutation of ^{238}U.   It does exist in uranium ore to the extent of 5 parts in a trillion, hardly enough to be commercially exciting.

(a)

(b)

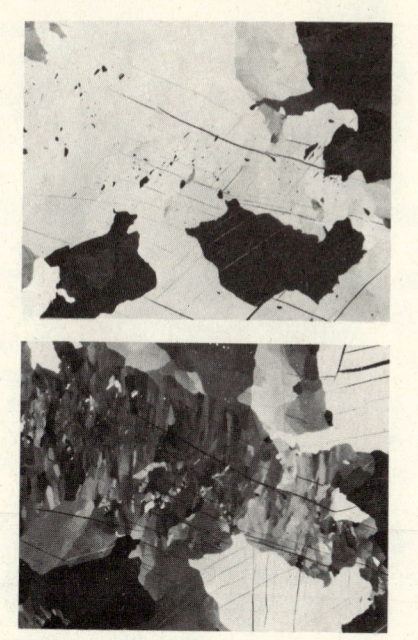

Fig. 11.23 High-purity cast uranium. (a) As electropolished; polarized light. (b) As electropolished; polarized light. Note subgraining in large grain in center of field. Stage rotated 45° from position in (a). Subgraining less evident in large central grain. Both 75X before 80 percent reduction. [From W. D. Wilkinson and W. F. Murphy, *Nuclear Reactor Metallurgy*. Princeton, N.J.: D. Van Nostrand Company, 1958.]

Plutonium is exceedingly toxic. The radioactivity concentration guides limit the maximum body concentration to 1500 Bq. *All* handling must be done in glove boxes. Figure 11.25 shows part of the ANL Fuel Fabrication Facility. Melting, casting, machining, rolling, and powder metallurgy operations are carried out entirely in glove boxes.

The use of metallic plutonium is complicated by its six allotropic forms. These are summarized in Table 11.6. The marked changes in density are accompanied by dimensional changes which are difficult to accommodate. The δ and δ' phases have negative coefficients of thermal expansion.

To be useful as a fuel, plutonium must be alloyed to provide a stable phase as a metal or a ceramic. It also may be used in a Pu-bearing liquid metal.

Thorium

Thorium is the most conventional of the three heavy metals being discussed. It solidifies at 1700°C with a body-centered cubic structure. At 1400°C it shifts to a face-centered cubic lattice which is stable to room temperature.

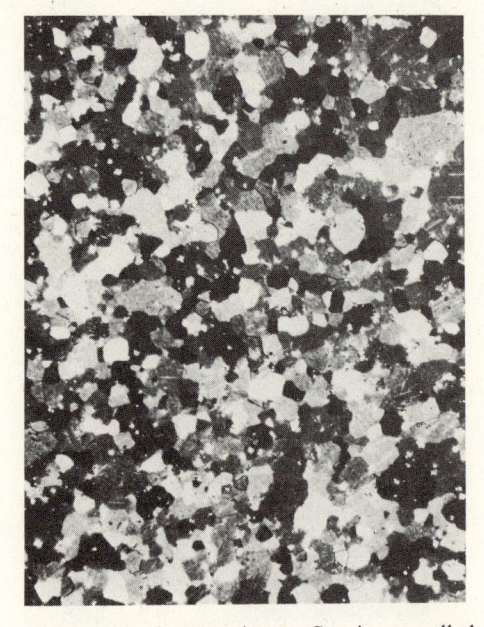

Fig. 11.24 Recrystallized α-uranium. Specimen rolled at 400°C, heated 1 h at 625°C, furnace cooled. Electropolished; polarized light. 100X before 70 percent reduction. [From W. D. Wilkinson and W. F. Murphy, *Nuclear Reactor Metallurgy*. Princeton, N.J.: D. Van Nostrand Company, 1958.]

Thorium is alpha active and many of its daughter products are gamma as well as alpha emitters. ^{220}Rn, which is sometimes called thoron, is a poisonous gas. It is undesirable to ingest thorium and one should guard against inhalation of ^{220}Rn.

The strength of thorium is quite dependent on the impurities and fabrication technique. Tensile strengths range from 138 to 276 MPa, with the yield strength being roughly 69 MPa less. At 700°C cold-rolled thorium will have a yield strength of about 55 MPa, while powder fabricated materials have a value of 35 MPa and cast thorium a value of only 14 MPa.

Recrystallization of severely cold-worked thorium will start at 500°C, but to be completed in 2 h it requires a temperature of 750°C. Cold reductions of 90 percent are possible, but thorium tends to seize in swaging or drawing dies, making jacketing in a soft metal like copper quite desirable.

Alloyed fuels

Although grain size and orientation in α-uranium can be partially controlled through heat treatment, its degradation in physical properties and tendency to swell with large amounts of irradiation tend to make it unattractive for high-

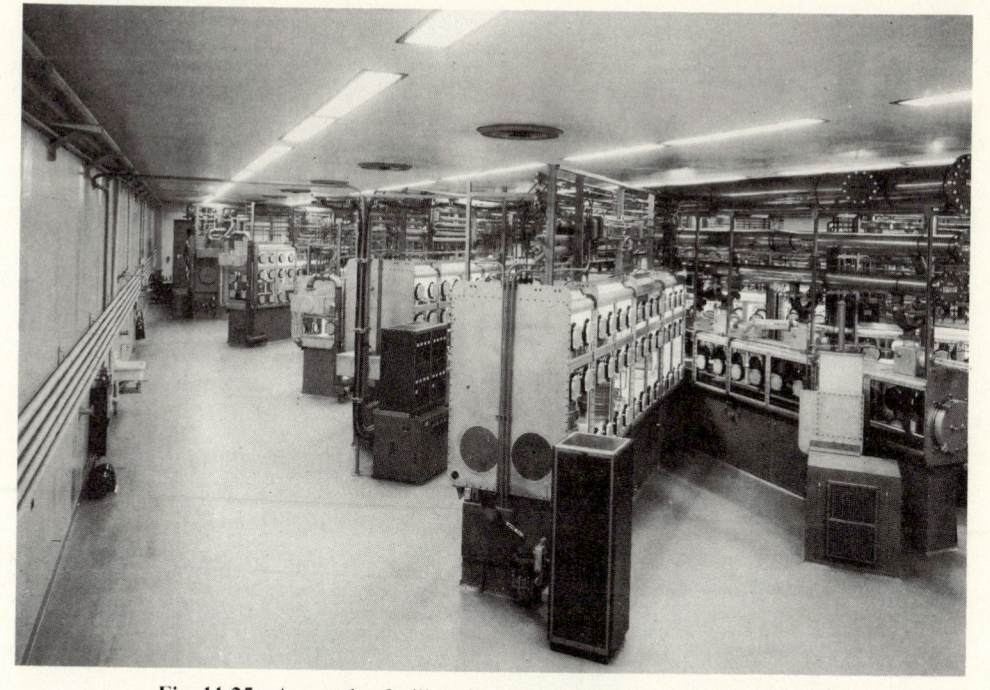

Fig. 11.25 An early facility for the fabrication of experimental plutonium fuels, at Argonne National Laboratory. Highly toxic plutonium fuels were fabricated within sealed glove boxes containing an inert atmosphere under negative pressure. By connecting these boxes with an enclosed conveyor system, material could be handled through all phases of fabrication while maintaining isolation from the outside environment. [Courtesy of Argonne National Laboratory.]

output reactors. The multiple phase changes in plutonium tend to eliminate its use in the pure metallic state. The development of suitable alloys systems for fuels may attempt to accomplish a number of the following objectives:

(1) Acquire a finer grain size.
(2) Raise melting point to improve high-temperature properties.
(3) Lower melting point for ease of casting.
(4) Improve corrosion resistance.
(5) Eliminate or reduce growth due to anisotropy.
(6) Reduce the tendency to swell.
(7) Improve working characteristics.
(8) Dilution of fissionable material.
(9) Produce a single phase that will be stable over the operating range of temperatures between startup and full power.
(10) Neutron economy. The alloying elements should have small absorption cross sections.

Table 11.6 Physical Properties of the Six Plutonium Allotropes

Phase	Crystal Lattice	Number of Atoms per Unit Cell	Transition Temperature to Next Higher Phase (°C)	Density (g/cm³)	Coefficient of Thermal Expansion (μ/m°C)	Volume Change on Transition (%)
Alpha	Monoclinic	16	112	19.8	46.4	$\alpha \to \beta$ 8.9
Beta	Body-centered Monoclinic	34	185	17.65	38.4	$\beta \to \gamma$ 2.4
Gamma	Face-centered orthorhombic	8	316	17.2	$(a = 34.7)$ $(b = 39.5)$ $(c = 83.4)$	$\gamma \to \delta$ 6.7
Delta	Face-centered cubic	4	451	15.9	— 8.8	$\delta \to \delta'$ −0.4
Delta Prime	Body-centered Tetragonal	2	480	16.0	−116 $(a = 305)$ $(c = -659)$	$\delta' \to \varepsilon$ −3.0
Epsilon	Body-centered cubic	2	640	16.51	+36.5	

Of the various fuel alloys, the following types have had considerable development with varying degrees of success:

(1) *Alpha-phase uranium alloys.* Small amounts of silicon, zirconium, or chromium all tend to produce a fine-grained α-uranium which may be fabricated in a manner similar to that for pure uranium in the α region. Figure 11.19 shows the effectiveness of zirconium in controlling gross dimensional stability.

An alloy with 5 weight percent zirconium and 1.5 weight percent niobium will have a somewhat distorted α structure after heat treating. It seems to be a supersaturated solid solution with improved corrosion resistance and radiation stability.

(2) *Gamma-phase uranium alloys.* Alloys containing 10 to 20 percent Mo or Nb will operate in the γ-phase region. They show better dimensional stability and improved resistance to aqueous corrosion. Figure 11.26 shows the phase diagram for the U–Nb system. Although an equilibrium condition at low temperatures would show an α plus γ mixture, all γ may be retained by a rapid quench. These alloys may be susceptible to phase change during cold work.

Alloys with 75 to 80 percent Nb and 15 to 25 percent U are all γ in the solid state. These alloys are refractory in nature and are difficult to fabricate.

(3) *Fissium alloys.* Fissium is the alloy content found in irradiated fuel which is reprocessed by melting. Many of the fission products with high cross sections oxidize and go into the slag or volatilize. Those left behind will actually improve the characteristics of uranium. Often the major alloy constituents are added to uranium in an unradioactive state to simulate true fissium. This makes the alloy easier to use for research purposes because of its nonradioactive nature.

In the EBR-2 reactor the fuel is U–5 percent fissium, whose composition is

U	95%	Rh	0.3%
Mo	2.5%	Ru	1.5%
Pd	0.5%	Zr	0.2%

Alloys of this type tend to have complex phase relationships but good radiation stability. They can be worked in the γ region, but the most probable methods of fabrication will be by injection casting in remote reprocessing facilities. Sixteen-inch pins have been satisfactorily cast by this technique (see Reference 3).

(4) *Ceramic fuels.* Ceramic fuels are nonmetallic oxides, carbides, nitrides, sulfides, and so on, of U, Pu, or Th. They tend to be hard and brittle at low temperatures but have very high melting points. To date the most widely used ceramic fuel is uranium dioxide. Mixed oxides are also in use. Thoria and UO_2 elements are used in the Indian Point Reactor blanket elements, and PuO_2–UO_2 combinations were used in the PRTR Reactor at Hanford.

Table 11.7 tabulates the densities and melting points for some of the uranium and plutonium ceramics.

The oxides, although less dense, show very good resistance to aqueous corrosion on exposure. This accounts for the wide use of UO_2 as a nuclear fuel in PWR and BWR power reactor cores. A cladding defect will not necessitate an unscheduled shutdown to remove the defective fuel assembly.

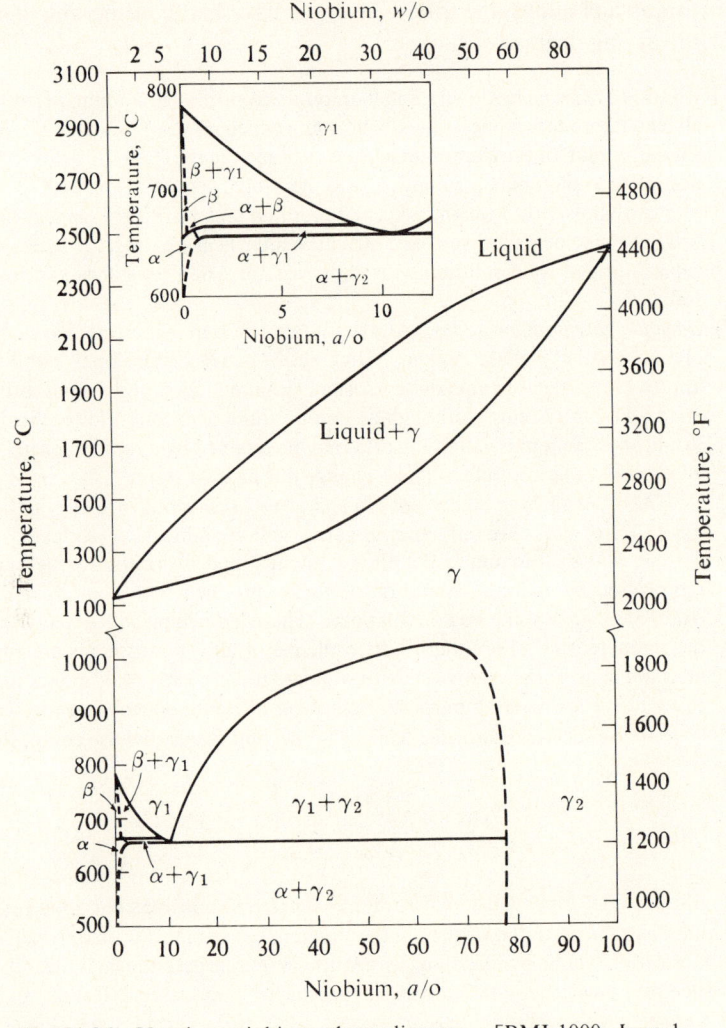

Fig. 11.26   Uranium–niobium phase diagram.   [BMI-1000, June 1, 1955.]

At the stoichiometric composition, uranium dioxide has the fluorite (CaF_2) crystal structure; here the uranium atoms occupy the positions of a face-centered cubic lattice with the oxygen atoms in the spaces in between. At elevated temperatures extra oxygen atoms may be accommodated interstitially. An excess of oxygen can result in observable precipitation of U_4O_9, which has a unit cell four times the size of UO_2. The bonding of UO_2 is often assumed to be predominantly ionic, but this has yet to be proved conclusively.

With its high melting point of 2800°C and its brittle nature below 1000°C, UO_2 is normally fabricated by powder metallurgical techniques. A large

Table 11.7 Uranium and Plutonium Ceramics

	Melting Point (°C)	Theoretical Density (g/cm³)	Typical Density (g/cm³)
UO_2	2750	10.97	10.5
PuO_2	2280	11.46	—
UC	2400	13.63	12.97
PuC	1654	13.6	—
UN	2630	14.32	13.52
PuN	>2500	14.2	—

surface area powder is compacted into pellets and then sintered to the desired densification (95 percent for a typical LWR fuel). Densities in the same range may be obtained by vibrational compaction; for example, loose powder is poured into the cladding tube and then vibrated at the tube's natural frequency to achieve the desired density.

The thermal conductivity of UO_2 is not especially high, and it is quite temperature dependent. Equation (11.12) expresses the thermal conductivity for UO_2 W/cm °C, where T is the temperature in degrees Celsius.

$$k = \frac{38.24}{402.2 + T} - 6.079 \times 10^{-13} (T + 273) \tag{11.12}$$

Figure 11.27 shows the thermal conductivity as a function of temperature. The value of $\int_0^T k \, dT$ is a measure of the ceramic's ability to transmit the heat that it generates internally. For GE 8 by 8 reload fuel, incipient melting occurs at a linear heat generation rate of 66.9 kW/m, which gives a value of the integral

$$\int_0^{T \text{ melt}} k \, dT = 93 \text{ W/cm}$$

The value of the integral decreases somewhat with burnup, as the melting temperature decreases with burnup at a rate of 32°C per 10^4 MWd burnup.

Experimental fuel assemblies have been run at linear heat generation rates that cause centerline melting in the fuel element. For the 8 by 8 assemblies mentioned previously, developmental assemblies have been run with centerline melting in the 72 to 89 kW/m LHGR range, and even as high as 190 kW/m. Figure 11.28 illustrates the structural changes in a rod that undergoes centerline melting. Voids tend to migrate up the thermal gradient to form a centerline void. The innermost fuel, which was molten, shows a nearly complete absence of voids. The adjacent material shows considerable porosity. Around this is a region with columnar grains. It has been postulated that one mechanism for

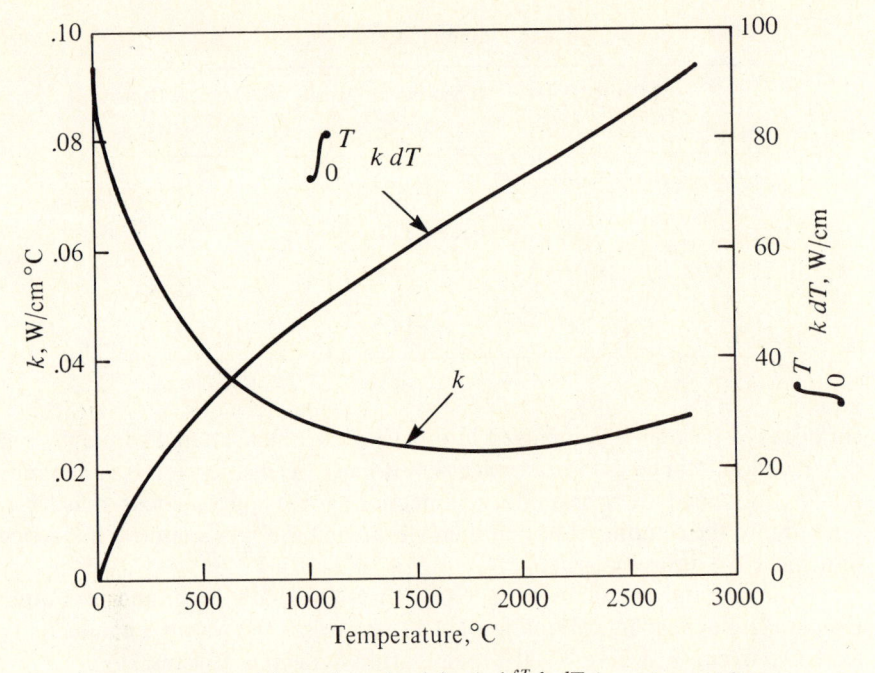

Fig. 11.27 UO_2 thermal conductivity and $\int_0^T k\,dT$ versus temperature.

this elongated grain structure is the transfer from the hotter side of a lenticular void to the cooler side by sublimation from the hot side and deposition on the cold side, leaving the columnar grain behind.   Less porous UO_2 tends merely to show large equiaxed grains in a similar region.   The varied grain structure in an irradiated fuel rod makes prediction of the property changes difficult.

UO_2, PuO_2, and ThO_2 are mutually soluble in all proportions, making the behavior and fabrication of mixed oxide fuels similar to that for a single oxide.   The ThO_2–PuO_2 combination is of interest because during processing the new fissionable [233]U that has been formed can be easily separated by chemical means from the thorium and plutonium.

Carbides of uranium are very reactive in an aqueous environment and thus are ruled out for pressurized or boiling water applications.   They are stable in sodium, and it is planned to use uranium carbide fuel for some sodium-cooled reactors.   The fuel behaves well under irradiation, being dimensionally stable, and it anneals out much damage under high operating temperatures.   Figure 11.29 shows the microstructure of a carbide ceramic fuel with 5.57 percent carbon.   Carbides which may be used in the dispersion-type elements are discussed in a subsequent section.

With the increased interest in the Th–[233]U fuel cycle there would be mixed oxide fuels, (Th, Pu) O_2 and (Th, [233]U) O_2, containing a preponderance of thorium among the heavy-metal atoms.   One might expect the performance of the thoria fuels to be superior to that for uranium dioxide because:

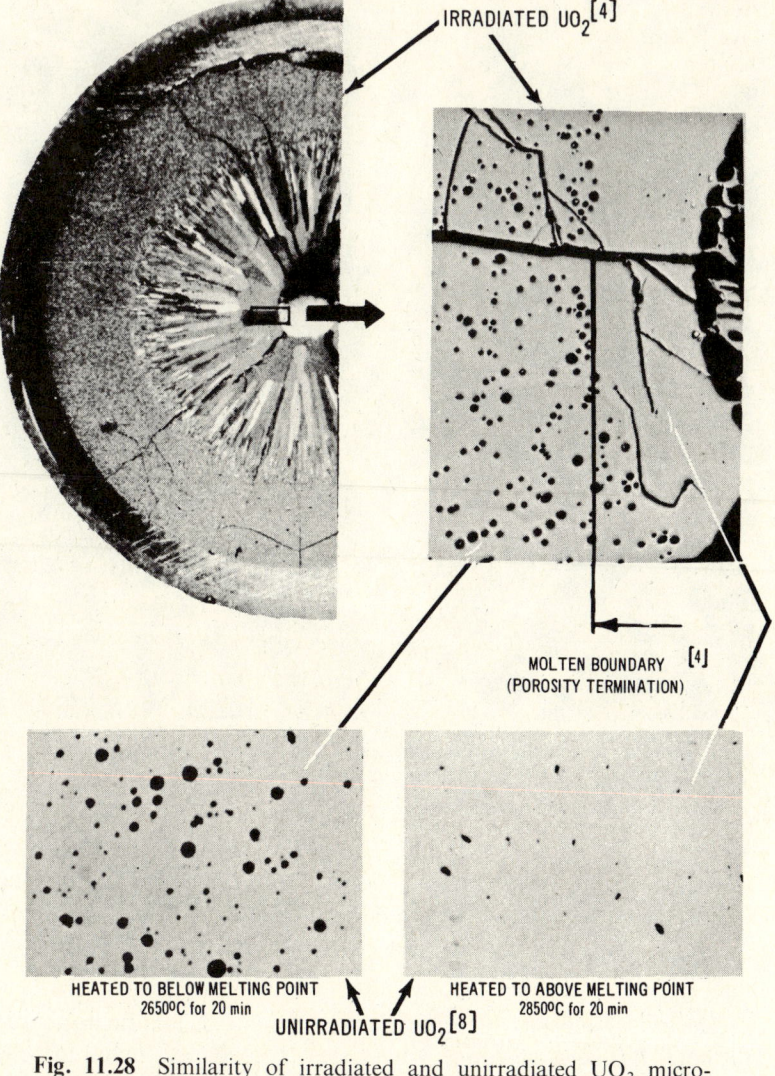

IRRADIATED UO$_2$[4]

MOLTEN BOUNDARY [4]
(POROSITY TERMINATION)

HEATED TO BELOW MELTING POINT
2650°C for 20 min

HEATED TO ABOVE MELTING POINT
2850°C for 20 min

UNIRRADIATED UO$_2$[8]

Fig. 11.28 Similarity of irradiated and unirradiated UO$_2$ micro-structure having been heated above and below the melting point. [From T. J. Pashos et al., "Irradiation Behavior of Ceramic Fuels," *Proceedings of the Third United Nations Conference on the Peaceful Uses of Atomic Energy*, Paper No. A/Conf. 28/P/240, September 1964.]

(1) For a given specific power level there should be lower fuel temperatures due to the superior thermal conductivity of ThO$_2$.

(2) Also due to the lower temperatures, there should be less fission gas release.

(3) The higher melting point should provide a greater safety margin against disruptive fuel failure under accident conditions.

(4) There is greater chemical stability, as ThO$_2$ is the only stable oxide of thorium.

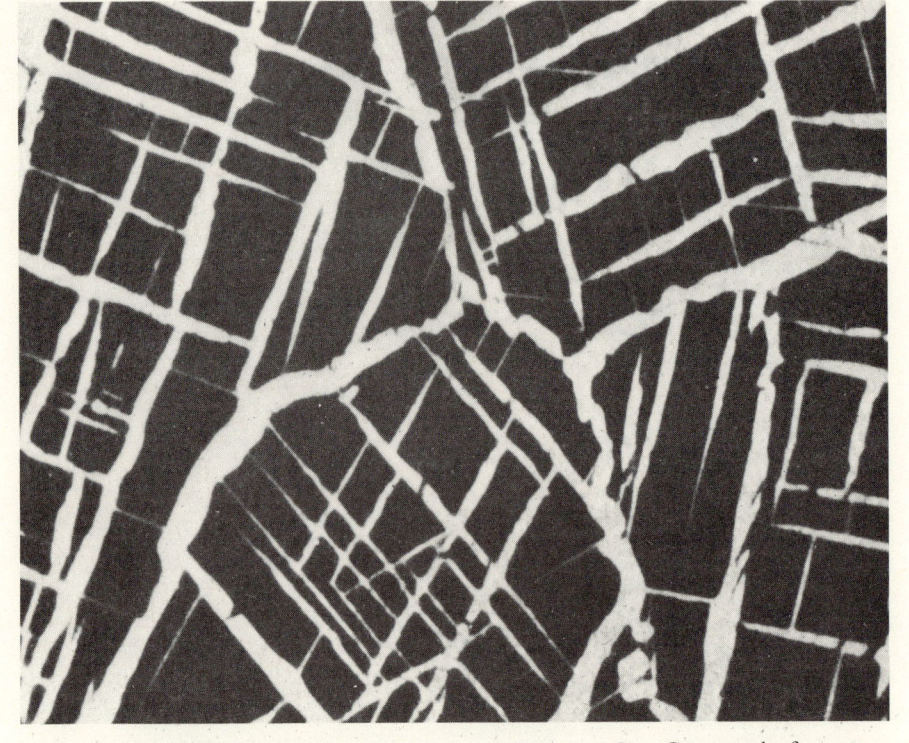

Fig. 11.29 Uranium–carbon alloy (5.75 percent C). Composed of 90 percent UC (dark) and 10 percent UC_2 (light Widmanstatten pattern). [From R. J. Gray et al., *Metal Progress* **74**, No. 1 (1958), p. 68.]

Fuel Densification

An increase in UO_2 density (densification) (Reference 31) in the hotter regions of LWR cores has been noted for some time. The changes noted in pellet dimensions have been small because the changes are localized in the central region of the pellet and are somewhat masked by other physical changes that occur at high temperature during the early part of the fuel cycle.

During a shutdown for refueling of the Ginna reactor during April 1972 and during a similar shutdown of the Point Beach reactor in September 1972, several rods were observed to have collapsed sections. These sections were at the top of the fuel rods and had lengths that varied from 1.25 to 10 cm. A little more than 3 percent of the rods showed collapsed sections, and none of the rods had more than a single collapsed section.

Fuel densification increases the percent of theoretical density of the UO_2 pellets from 90 to 95 percent up to 97 to 98 percent. In a 3.66-m column of pellets, if the density increased from 90 to 97 percent in an isotropic manner, there would be a decrease in column length of a bit more than 7.5 cm. Actually, the densification is anisotropic, the axial shrinkage being $1\frac{1}{2}$ times greater than that occurring in the radial direction. Thus, 7 to 10 cm gaps are quite possible. The densification is complete within 200 h of reactor operation. During subsequent operation cladding creep and oxide swelling may cause other problems.

As the column settles, mechanical interaction between the clad and the pellets may occur, preventing the settling of the pellet and those above it on the column below. Once the gap has been produced, outside water pressure can flatten the clad in the gap region, resulting in a flux spike. The thermal expansion of UO_2 is greater than that of Zircaloy, and the thermal response time for the fuel during a power change is shorter than that of the clad. Hence, the pellet temperature changes more quickly than the temperature of the cladding during a power change. If creep of the cladding has diminished the gap between itself and the fuel pellets, it is possible for the difference in thermal expansion to cause stresses exceeding the yield stress for the clad material. Since irradiation reduces ductility of the clad, the differential expansion may lead to clad failures.

The problems with cladding collapse that result from fuel densification and clad creep have occurred principally with unpressurized fuel rods in PWRs. The core pressure is roughly double that of a BWR, and the clad thicknesses tend to be less. Pressurizing the clad with a helium to pressures of 1.4 to 2.8 MPa and increasing fuel pellet density appear to have been successful in reducing the clad creep sufficiently to prevent the formation of fuel column gaps and subsequent tubing collapse.

There are three principal effects associated with fuel densification:

(1) An increase in the linear heat generation rate by an amount directly proportional to the decrease in pellet length.
(2) An increased local neutron flux and a local power spike in the axial gaps in the fuel column.
(3) A decrease in the clearance gap heat conductance between the pellets and the cladding. This decrease in heat transmission capability will increase the energy stored in the fuel pellet and will cause an increased fuel temperature.

All these effects must be evaluated for reactors in all modes of operation.

To minimize the effects of fuel densification on power plant operation, limits are established on the rate of change of power level. Also, limits are placed on permissible steam generator leakage to assure that potential leakage of radioactive material is minimized. It is required further that during a

loss of coolant accident the cladding temperature not be allowed to exceed 1200°C.

Other Major Causes of Fuel Defects

Internal hydriding of oxide fuels and cladding–pellet interactions have been major causes of fuel defects. The hydriding can be controlled by proper drying of fuel pellets or the inclusion in the fuel of a hydrogen getter.

Failures due to cladding–pellet interactions tend to occur under power ramping conditions. It is felt that fission product iodine and possibly cesium induce stress corrosion cracking. However, such failures have also been produced in low-burnup fuel with little I or Cs present. This may well be due to high local stresses which are induced over fuel cracks. Design features to counteract power ramp failures include:

(1) Increased clad thickness
(2) Increased clad–pellet gap, with pressurization to obviate cladding collapse
(3) Introduction of a layer of graphite or other lubricant between the fuel and the clad

The increased clad thickness, however, would have an adverse effect on neutron economy. Increasing the gap increases the resistance to heat flow, but the introduction of gas pressurization and/or graphite lubrication would tend to counteract that effect.

Dispersion-Type Alloys

In a dispersion-type fuel a fissile phase is embedded in a metallic or ceramic matrix. The fissile particles should be spaced far enough apart so that the areas damaged by fission fragments do not overlap and the matrix will retain its

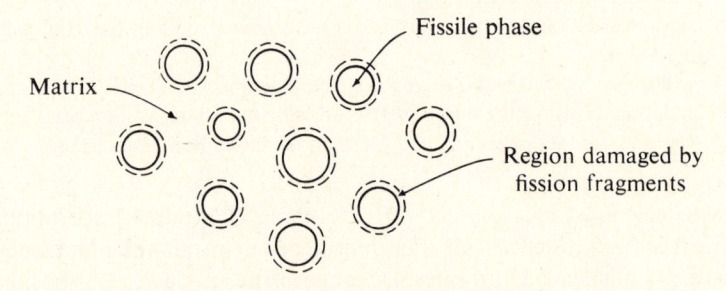

Fig. 11.30   Dispersion-type alloy showing the region in the matrix that is damaged by fission fragments.

Table 11.8 Examples of Dispersion-Type Fuel Systems

Dispersed Phase	Matrix Phase
UO_2	Stainless steel (austenitic and ferritic), Fe, Nichrome, SiC–Si, Nb, Al, Zr, Zircaloy, graphite, BeO, Al_2O_3, SiO_2
U	Mg, Th, Zr, Zircaloy, $ZrH_{1.65}$
UAl_4	Al
UZr_2	Zr, Zircaloy
UC	SS, Zr, Zircaloy
UN	SS, Zr
U_3Si	Zr
U_6Ni	Zr
U_2Ti	Zr
PuO_2	Mg, Al, Th, U, Zr, Fe, graphite

From E. A. Aitken, "Dispersion Fuel Elements," ASEE–AEC Summer Institute on Materials for Nuclear Reactors, Richland, Wash., August 1962.

strength effectively, as shown by Figure 11.30. The matrix will often have a better thermal conductivity than the fissile phase, tending to reduce the maximum temperatures in the fuel element. The retention of the mechanical properties of the matrix should assist in containing fission gases, allowing larger burnups without swelling. Table 11.8 shows some of the combinations that have been proposed for nuclear fuel elements.

Alloys of uranium with up to 45 weight percent Al can be produced by inexpensive casting techniques and are then amenable to plastic deformation. Figure 11.31 shows a phase diagram for this alloy system. There is a eutectic formed at 640°C between 13 weight percent U and 87 weight percent Al. Figure 11.32a shows an alloy with 10 weight percent U where there are primary η-phase grains (pure Al) in a matrix of UAl_4–η eutectic. Figure 11.32b shows a 15 weight percent U alloy where the primary crystals are now UAl_4. Alloys such as these may be hot-rolled with about a 3:1 reduction in thickness, cast clad in a 5 percent Si–95 weight percent Al alloy, followed by a final rolling to a fuel-plate thickness of less than 1.6 mm. Figure 11.33 shows the resultant structure. Note the uniform dispersion of the UAl_4 in the Al matrix in the final fuel plate. Elements of this type were developed at the Argonne National Laboratory's International Institute of Nuclear Science and Engineering as being suitable for production and use in research reactors in third-world nations. The simplicity of the production methods requires no heavy investment in special-purpose machinery.

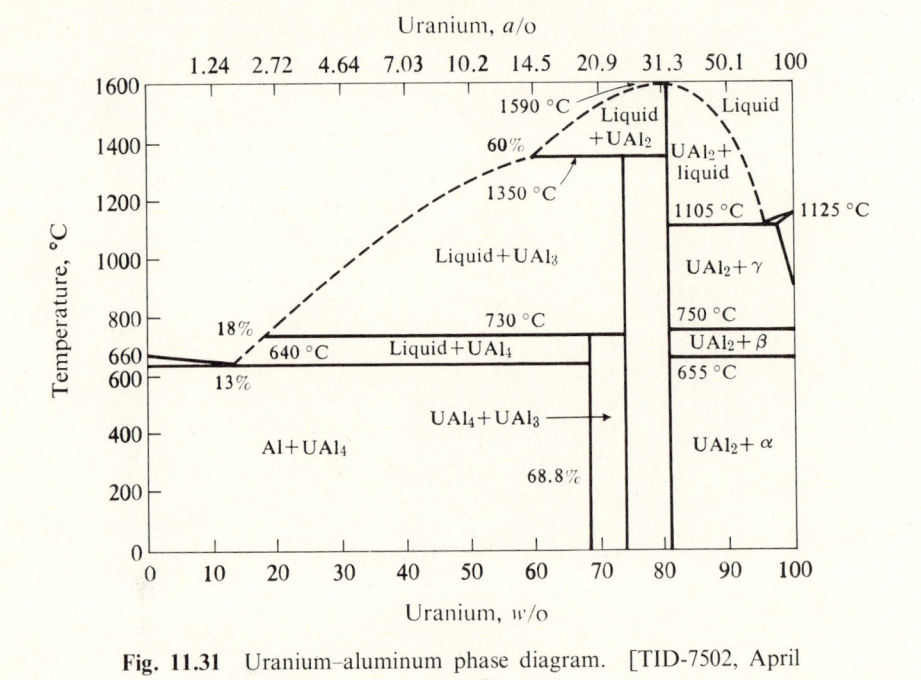

Uranium, a/o

Fig. 11.31   Uranium–aluminum phase diagram.   [TID-7502, April 1955.]

An opposite extreme is the Nichrome–UO_2 type of fuel element designed for the ill-fated Aircraft Nuclear Propulsion Project.   Here approximately 40 weight percent UO_2 is dispersed in a matrix of Nichrome (80 Ni–20 Cr).   Figure 11.34 shows a fuel stage of this material.   Ni, Cr, and UO_2 powders were blended in correct proportions, cold pressed, sintered, assembled with cover plates of Nb modified Nichrome, hot pressed, rolled, edge seal brazed, rolled to ribbon, spot

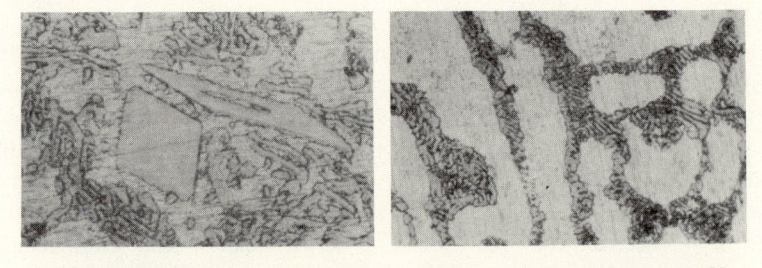

Fig. 11.32   Cast uranium–aluminum alloys.   (a) A 10 weight percent U hypoeutectic alloy where the primary crystals are nearly pure aluminum. (b) A 15 weight percent U hypereutectic alloy where the primary crystals are UAl_4.

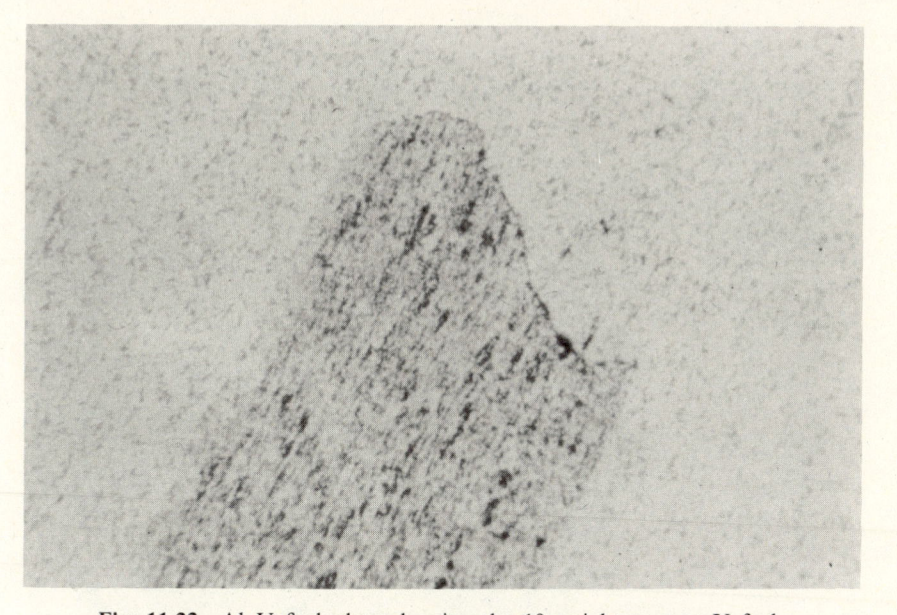

Fig. 11.33 Al–U fuel plate showing the 10 weight percent U fuel material surrounded by 5 weight percent Si–Al cladding. The fuel material is cast, hot-rolled, recast inside the cladding material, and rolled to a plate of approximately 1.6 mm thickness.

welded into an assembly, followed by a final brazing. The high temperature strength and oxidation resistance of these materials led to their use. The 100-h creep data shown in Figure 11.35 indicate the severe conditions under which these elements were intended to operate.

For even higher-temperature service the fissile phase can be embedded in a ceramic matrix. BeO, with up to 60 weight percent UO_2, is an example of such a material, which could be called a fueled moderator. Tubular elements of this type open up the possibility of all-ceramic reactors.

Fuel elements for the Peach Bottom High-Temperature Gas-Cooled Reactor (HTGR) were pyrolitic carbon-coated (Th, U) C_2 particles in a graphite matrix. Figure 11.36 shows the assembled elements. Figure 11.37 shows an individual fuel compact. In this reactor, purge gas was passed through the grooves on the outside of the annular elements removing a substantial fraction of the escaping fission products. The uranium and thorium in the fuel compacts were in the form of carbides. The composition of the compacts is shown in Table 11.9.

The carbide particles are 100 to 400 μm in diameter coated with pyrolytic carbon 50 to 60 μm thick. The PC prevents contact of the carbides with atmospheric moisture, which would lead to their rapid deterioration. It

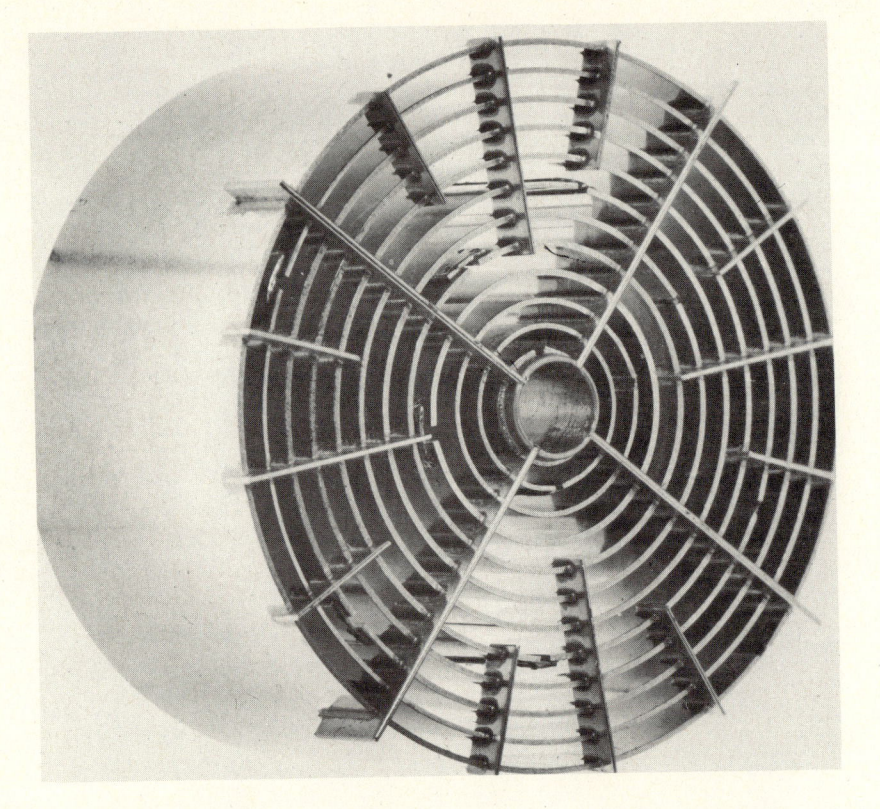

Fig. 11.34   Nichrome UO₂ fuel stage. [From E. A. Aitken, "Dispersion Fuel Elements," ASEE–AEC Summer Institute on Materials for Nuclear Reactors, Richland, Wash., August 1962.]

further helps contain fission fragments.   Figure 11.38 shows PC-coated (Th, U)C₂ particles.

Irradiation of these compacts at temperatures higher than anticipated in HTGR operation produces what is known as the *amoeba effect*.   When a temperature gradient exists across the carbide particle, the UC₂ tends to migrate toward the higher temperature.   It appears to "eat" the PC ahead of the particle and to precipitate carbon behind it, as shown by Figure 11.39.   The temperature gradient across this particle was estimated to be only 3°C.   Both UC₂ and (Th, U) C₂ particles show the same type of attack.

Multilayered fuel-particle coatings have subsequently been developed to improve metallic fission-product retention and to reduce fission-gas release. An inner porous layer acts as a buffer from recoiling fission fragments and provides voids for fission-gas retention.   An outer layer of pyrolytic carbon provides strength and retention of the noble fission gases.   It is, however, less successful

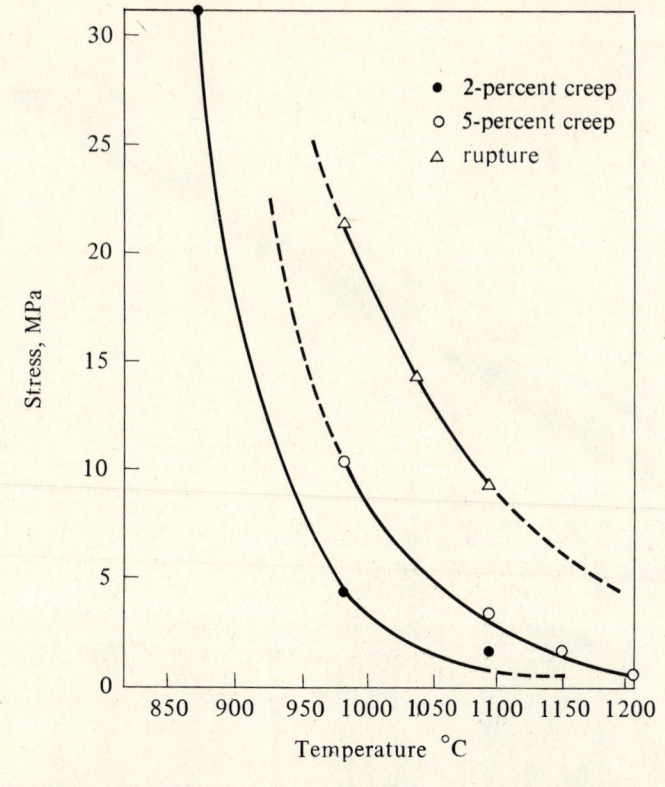

Fig. 11.35 Plot of 2 percent and 5 percent deformation and rupture for 80 Ni–20 Cr fuel ribbon in a 100-h period. [From E. A. Aitken, "Dispersion Fuel Elements," ASEE–AEC Summer Institute on Materials for Nuclear Reactors, Richland, Wash., August 1962.]

in stopping metallic fission products at high-temperature gas-cooled reactor temperatures. A layer of silicon carbide may be deposited on the porous carbon layer before the outer layer of pyrolytic carbon is laid down. The SiC layer provides for better retention of the metallic fission products. Figure 11.40 shows such a triple-coated UO_2 particle, both before and after irradiation to 4.6 percent burnup at 1250°C. The post-irradiation integrity of the SiC and pyrolytic carbon layers is evident. Improvement in the ability to retain fission gases has led to the abandonment of the purge-gas system used with the Peach Bottom HTGR.

The fuel being used in the 330-MW(e) Fort St. Vrain HTGR contains 100-μm UC_2 spheres and 400-μm spheres of ThC_2, each type of particle using a four-layer coating. An extra layer of pyrocarbon is deposited between the porous carbon layer and the silicon carbide layer previously described. The particles

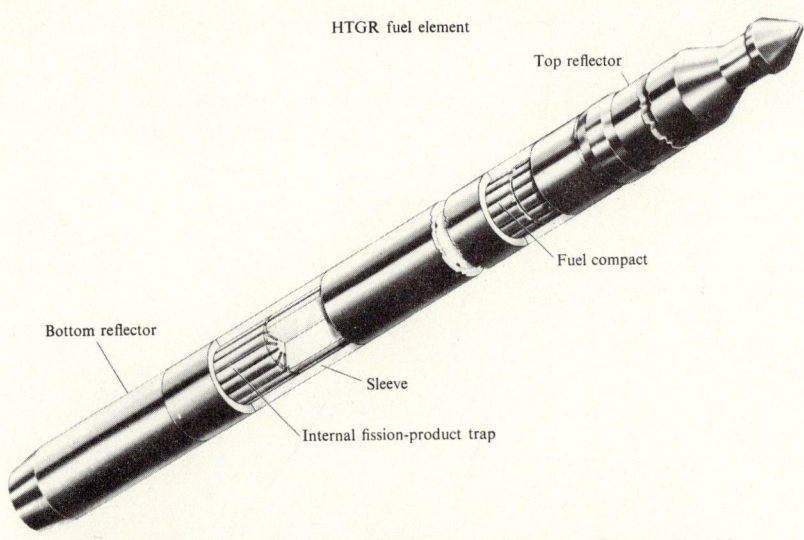

Fig. 11.36 Peach Bottom HTGR fuel element. [From W. V. Goeddel, *Nuclear Science and Engineering* **20** (1964), p. 201.]

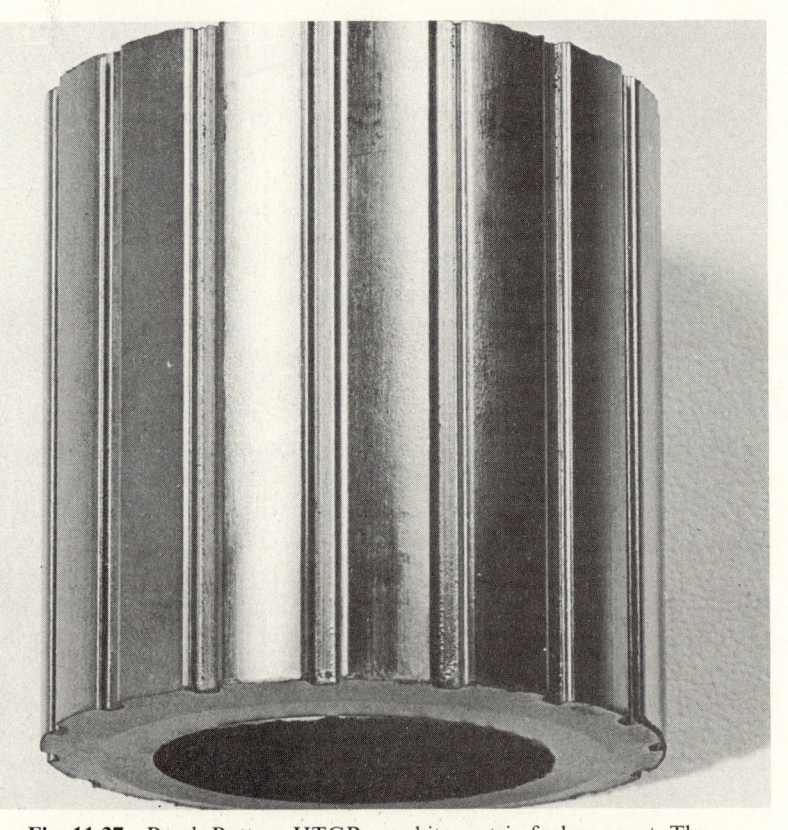

Fig. 11.37 Peach Bottom HTGR graphite matrix fuel compact. The compact is produced by hot pressing. The purge grooves are formed during the hot-pressing process; no machining is required. [From W. V. Goeddel, *Nuclear Science and Engineering* **20** (1964), p. 206.]

Table 11.9   HTGR Fuel Compact Data

Composition	
Uranium (93 % enriched)	2.8 wt %
Thorium	16.1 wt %
Total metal	18.9 wt %
Graphite	81.1 wt %
C/Th/U atom ratio	562:5.8:1
Temperature	
Maximum fuel compact temperature	1340°C
Average fuel compact temperature	980°C
Burnup (3-yr core life at 80 % load factor)	
Fissions/original fissile atom	0.6
MWd(th)/Mg ^{235}U + ^{232}Th	60 000
Fissions/cm^3	0.67×10^{20}

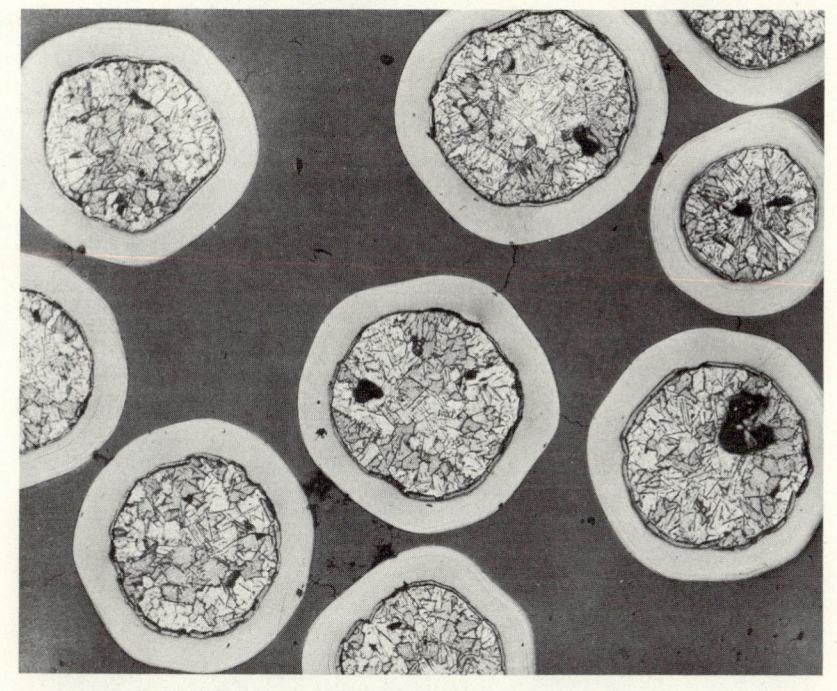

Fig. 11.38   Photomicrograph of pyrolytic carbon-coated (Th, U) C$_2$ particles.   75X before 50 percent reduction.   [From W. V. Goeddel, *Nuclear Science and Engineering* **20** (1964), p. 206.]

are embedded in a carbonaceous binder and formed into rods 1.27 cm in diameter × 5.08 cm long. These rods are inserted in holes drilled in the hexagonal graphite moderator blocks. The whole assembly, which is 0.356 m across flats and 0.787 m long, forms a massive fuel element. Figure 11.41 shows a fuel element and some of the quadruple-coated fuel particles. While the overall average burnup is expected to be 100 000 MWd/Mg, in the fuel particles with thier fully enriched uranium it will be 700 000 MWd/Mg and in the fertile ThC_2 particles it will be only 40 000 MWd/Mg. It is planned to replace one-sixth of the core at each annual refueling, so that a fuel element going through the full cycle will reside in the core for 6 yr, accounting for the large burnups.

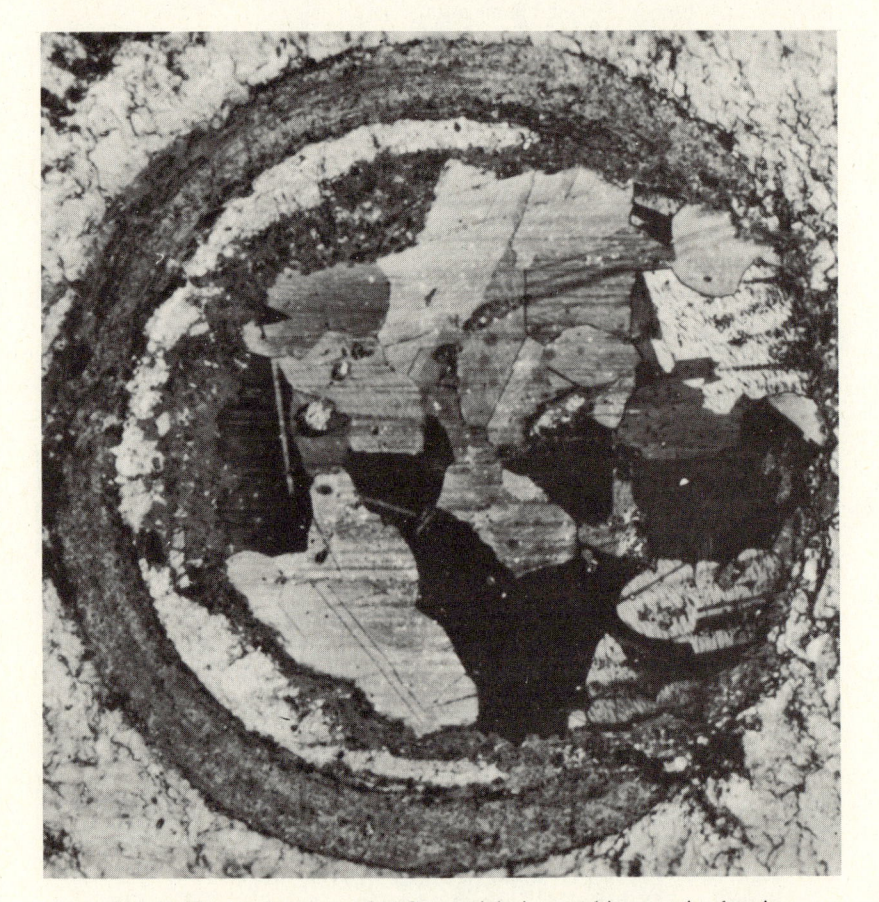

Fig. 11.39 Carbon-coated UC_2 particle in graphite matrix showing the "amoeba effect." Tested 150 h in temperature range from 1780 to 1920°C. The carbide particle has migrated in the direction of increasing temperature. 200X before 27 percent enlargement. [Courtesy of General Atomic Company.]

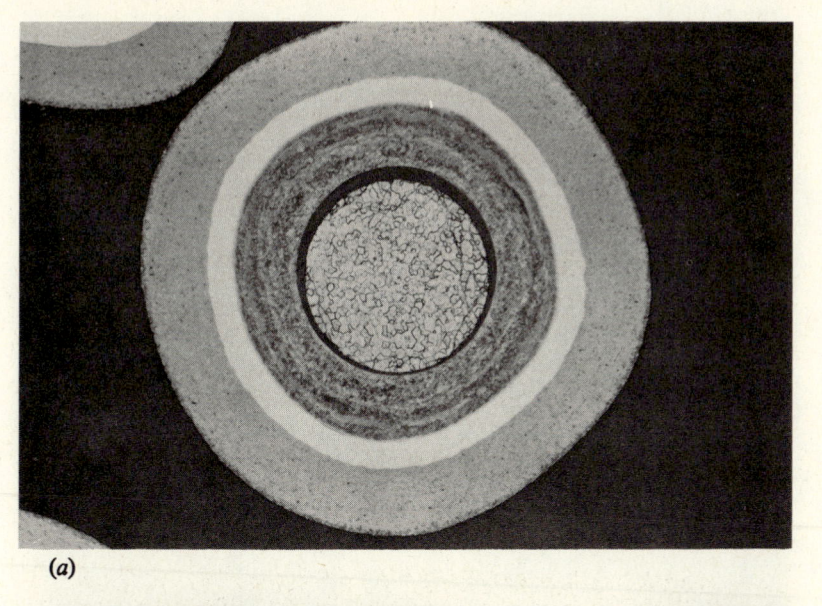

(a)

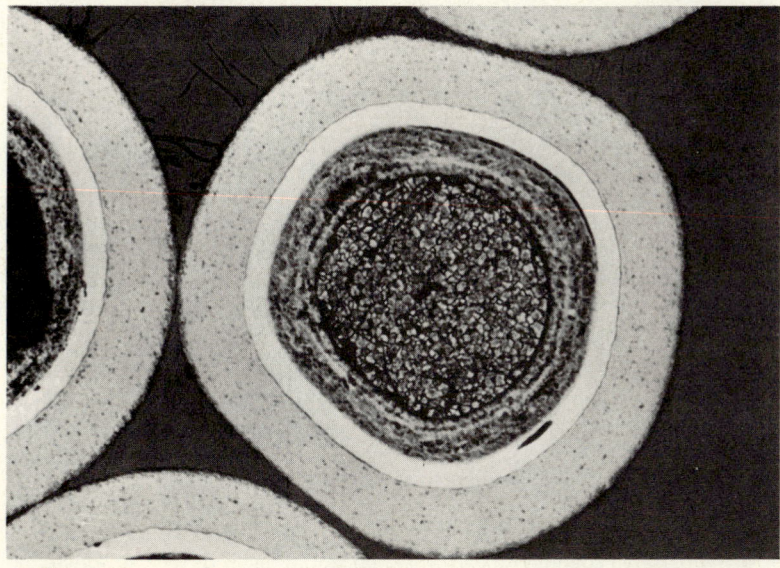

(b)

Fig. 11.40 Triple-coated UO_2 fuel particle. The outer layer is dense pyrolytic carbon over a layer of silicon carbide surrounding the inner layer of porous carbon. (a) Unirradiated and (b) irradiated to 4.6 atom percent burnup. Magnification 200X. [Courtesy of Oak Ridge National Laboratory, operated by Union Carbide Corp. for U.S. Atomic Energy Commission.]

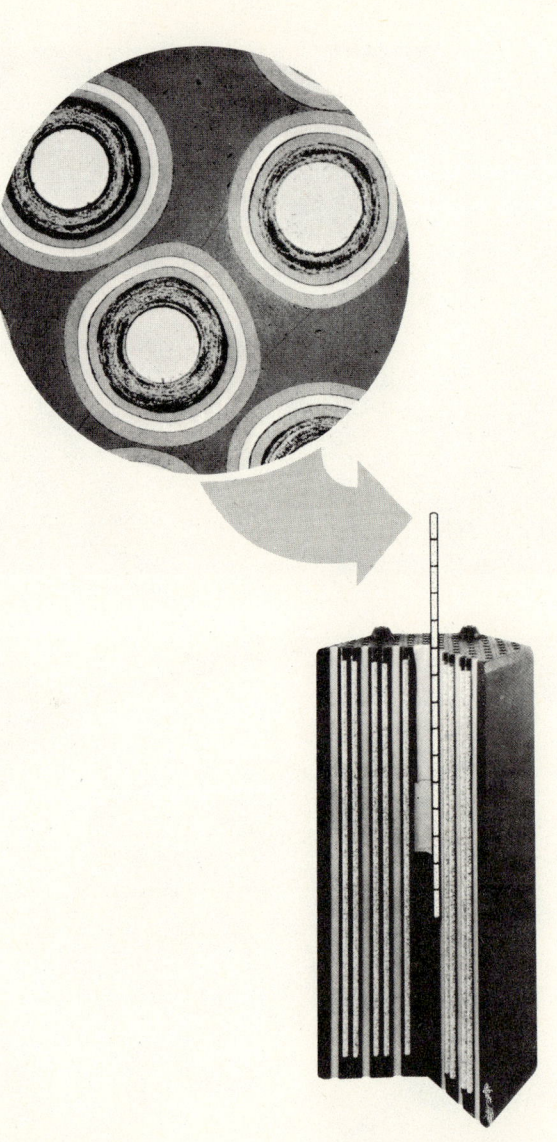

Fig. 11.41 Cross section of greatly magnified fuel particles with multilayered coatings (above) are bonded into fuel rods (below) which are inserted into a graphite block to form a single fuel element. [Courtesy of General Atomic Company.]

Metallic Fuels for Fast Breeders

Early fast reactors utilized metallic fuel (EBR-1, EBR-2, and the Fermi reactor), but as larger reactors were proposed, metallic fuels could not cope with the 10 atom percent burnups and 600 to 650°C sodium outlet temperatures. However, the 470°C outlet temperature for EBR-2 allowed continued use and development of the U–5 percent fissium driver fuel. The earlier fuels used a radial gap of only 0.152 mm and failed at about 3 atom percent burnup, when the fuel swelling caused contact between the fuel and clad. A redesigned fuel element with a 0.254-mm gap has been able to accommodate 10 atom percent burnup without significant probability of failure. A comparison of earlier Mark IA and the later Mark II fuel element designs is shown in Table 11.10

The in-core behavior of metal fuels is limited by two principal phenomena:

(1) Fuel swelling to cause cladding failure
(2) Fuel-clad metallurgical interaction

Metallic fuel generally uses liquid sodium as the bond in the fuel–clad gap to lower thermal resistance. Fuel swelling occurs due to fission bubble formation and growth. If the growth allows fuel–clad contact, the stresses will cause clad failure. When the volumetric swelling exceeds 25 percent, the voids begin to touch one another. By 30 percent they are completely interconnected and

Table 11.10 Design Features of the Mark-1A and Mark-II Driver-Fuel Elements

	Mark IA	Mark II
Enrichment (at% ^{235}U)	52.5	67.0
Fuel pin length (mm)	343	343
Fuel pin diameter (mm)	3.65	3.30
Fuel volume (m^3)	3.6×10^{-6}	2.9×10^{-6}
Fuel smeared density (%)	85	75
Fuel/clad radial gap (mm)	0.152	0.254
Cladding wall thickness (mm)	0.229	0.305
Cladding o.d. (mm)	4.42	4.42
Cladding material*	Type 304L (SA)	Type 316 (SA)
Element length (mm)	460	612
Plenum volume† (m^3)	0.67×10^{-6}	2.41×10^{-6}

* Solution-annealed Type 316 stainless steel is presently the reference cladding material. Some experimental Mark II elements were clad with solution-annealed Type 304L stainless steel.
† Preirradiated volume at 20°C.

From L. C. Walters and J. H. Kittel, *Nuclear Technology* **48**, No. 3 (May 1980).

most of the fission gases are released. Sufficient plenum volume must be provided to accommodate this released gas without creating undue stress in the cladding.

In the Mark IA elements only 17 percent swelling could take place prior to clad contact, and most of the fission gas was retained. In the Mark II pins the volumetric swelling before contact is 33 percent, which ensures interconnected porosity and gas release. At 10 atom percent burnup the gas pressure is 15 MPa. Clad dimples 13 mm above the top of the fuel preclude its lift-off into the gas plenum.

Fuel-clad metallurgical interactions are likely to occur during transient conditions of overpower or low Na flow, when the interface temperature may exceed the temperature at which a eutectic can be formed between fuel and clad. For 304L stainless steel and the U–5 percent Fs alloy, the eutectic temperature is 705°C. Many of the test elements have survived transients with temperatures in excess of 985°C. When failures did occur under transient conditions it was due to the eutectic penetration at temperatures well in excess of the eutectic temperature and it was felt that the clad–fuel metallurgical interactions were not the primary mode of breach for the Mark IA elements.

Looking toward the use of metal fuels in large fast power reactors, there is much interest in the U–Pu–Zr and the Th–U–Pu–Zr alloy systems. The U–15 weight percent–10 weight percent Zr alloy has a eutectic at 850°C. With an outlet temperature of 500°C, the fuel–clad interface temperature with uncertainty included is approximately 650°C. Thus, there would be little probability of exceeding the eutectic temperature under normal operating conditions. Even though a transient were to allow the temperature to exceed the eutectic temperature, the rate of eutectic penetration would be relatively slow. The addition of Zr not only raises the eutectic temperature, but also reduces the rate of growth of the interaction zone which develops because of interdiffusion between clad and fuel. Higher burnups appear achievable with this and other possible alloys, as the fission gas release at 30 volume percent swelling appears independent of the particular metal–fuel system.

Problems

1. Show that the maximum energy which can be transferred by a neutron of energy E to a knock-on atom of mass A is

$$T_m(E) = 4AE/(A + 1)^2$$

2. Calculate the number of atom displacements per scattering collision in iron for neutrons with an energy of 1 MeV. If one considers a fast mono-energetic flux of 10^{15} neutrons/cm^2 s for neutrons at this energy, calculate the number of displacements per iron atom per day.

3. In a fusion reactor the D-T reaction produces a 14-MeV neutron. Niobium is being considered as the material for the fusion chamber wall. Compare the number of atom displacements per neutron scattering collision for a 14-MeV neutron with the number produced by a 1-MeV neutron. Comment on the probable radiation damage to the fusion chamber wall.

4. In Reference 28, equations have been developed to fit experimental data on the percent swelling of solution-treated Type 316 stainless steel at fluences above 10^{22} neutrons/ cm^2 ($E > 0.1$ MeV).

$$\Delta V/V = (\phi t/10^{22})^{N(T)} F(T)$$

where

$$N(T) = \frac{2 + 3 \exp [0.05(T - 475)]}{1 + \exp [0.05(T - 475)]}$$

$$F(T) = \frac{\exp [0.09(T - 340)]}{1 + \exp [0.09(T - 340)]}$$

$$\times \left\{ \frac{0.022}{1 + \exp [0.05(T - 600)]} + \frac{0.06}{1 + \exp [0.06(T - 460)]} \right\}$$

where

$$\phi t = \text{fluence, neutrons/cm}^2 \ (E > 0.1 \text{ MeV})$$

$$\Delta V/V = \text{swelling, } \%$$

$$T = \text{temperature, } °C$$

Write a computer program to plot the percent swelling versus temperature between 300 and 600°C for fluences of 3×10^{22}, 6×10^{22}, and 1×10^{23} neutrons/cm². Compare the results to Fig. 11.6 for Type 304 stainless steel.

5. A sample of nickel is irradiated for 1 yr in a thermal flux of 2×10^{13} neutrons/cm² s at a temperature of 250°C.
   (a) What is the average energy transferred to the lattice by gamma emission upon capture of a thermal neutron?
   (b) What is the average number of Frenkel pairs produced per thermal neutron capture?
   (c) Calculate the concentration (atom fraction) of Frenkel pairs resulting from thermal neutron capture and recoil after gamma emission, assuming the temperature to be low enough that little recombination of vacancies and interstitials occurs.

References

1. Dienes, D. J., and D. H. Vineyard, *Radiation Effects in Solids*. New York: Interscience Publishers, 1957.
2. Holden, A. N., *Physical Metallurgy of Uranium*. Reading, Mass.: Addison-Wesley Publishing Company, 1958.

3. Wilkinson, W. D., *Uranium Metallurgy*, Vols. 1 and 2. New York: Interscience Publishers, 1962.

4. Wilkinson, W. D., and W. F. Murphy, *Nuclear Reactor Metallurgy*. Princeton, N.J.: D. Van Nostrand Company, 1958.

5. Strucken, E. F., "Anisotropic Growth in Metallic Uranium," ASEE–AEC Summer Institute on Materials for Nuclear Reactors, Richland, Wash., 1962.

6. Huntoon, R. T., "Properties of Uranium," ASEE–AEC Summer Institute on Materials for Nuclear Reactors, Richland, Wash., 1962.

7. Hudson, B., K. H. Westmacott, and J. J. Makin, *Dislocation Loops and Irradiation Growth in Alpha Uranium*, AERE R-3752, U.K. Atomic Energy Agency.

8. Kittel, J. H., "Uranium Alloys—Irradiation Behavior," ASEE–AEC Summer Institute on Materials for Nuclear Reactors, Richland, Wash., 1962.

9. Kaplan, G. E., "Metallurgy of Thorium," International Conference on the Peaceful Uses of Atomic Energy, Vol. 8, 1955.

10. Kemper, R. S., "Irradiation Effects—Metallic Uranium, Physical and Mechanical Properties," ASEE–AEC Summer Institute on Materials for Nuclear Reactors, Richland, Wash., 1962.

11. Aitken, E. A., "Dispersion Fuel Elements," ASEE–AEC Summer Institute on Materials for Nuclear Reactors, Richland, Wash., 1962.

12. Bates, J. L., "Thermal Conductivity of UO_2 Improves at High Temperatures," *Nucleonics* **19**, No. 6 (June 1961), pp. 83–87.

13. Walker, Jr., P. L., "Carbon—An Old But New Material," *American Scientist* **50**, No. 2 (June 1962), pp. 259–293.

14. Simnad, M. T., and L. P. Zumwalt, *Materials and Fuels for High Temperature Nuclear Energy Applications*. Cambridge, Mass.: The MIT Press, 1964.

15. Foster, A. R., "Radiation Damage to Hot Rolled Swedish Iron," ANL Special Nuclear Studies Institute 1960.

16. McDonnell, W. R., "Kinetics of Structural Changes During Beta Transformation of Uranium," *Nuclear Science and Engineering* **12**, No. 3 (March 1962), pp. 325–336.

17. Bolt, R. O., and J. G. Carroll, eds., *Radiation Effects on Organic Materials*. New York: Academic Press, Inc., 1963.

18. Blocher, J. M., et al., "Properties of Ceramic-Coated Nuclear-Fuel Particles," *Nuclear Science and Engineering* **20**, No. 2 (October 1964), pp. 153–170.

19. Goeddel, W. V., "Development and Utilization of Pyrolytic-Carbon-Coated Carbide Fuel for the High-Temperature Gas-Cooled Reactor," **20**, No. 2 (October 1964), pp. 201–218.

20. Gray, R. J., W. C. Thurber, and C. K. H. Du Bose, "Preparation of Arc-Melted Uranium Carbides," *Metal Progress* **74**, No. 1 (July 1958), pp. 65–70.

21. Fraser, A. S., I. R. Birss, and C. Cawthorne, "High Temperature Embrittlement of Stainless Steel Irradiated in Fast Fluxes," *Nature* **211**, No. 5046 (July 16, 1966), pp. 575–576.

22. Cawthorne, C., and E. J. Fulton, "Voids in Irradiated Stainless Steel," *Nature* **216**, No. 5115 (November 11, 1967), pp. 575–576.

23. Kaae, J. L., D. W. Stevens, and C. S. Luby, "Prediction of the Irradiation Performance of Coated Particle Fuels by Means of Stress-Analysis Models," *Nuclear Technology* **10**, No. 1 (January 1971), pp. 44–53.

24. Claudson, T. T., R. W. Baker, and R. L. Fish, "The Effects of Fast Flux Irradiation on the Mechanical Properties and Dimensional Stability of Stainless Steel," *Nuclear Application & Technology* 9, No. 1 (July 1970), pp. 10–23.

25. Reagan, P. E., E. L. Long, Jr., J. G. Morgan, and J. H. Coobs, "Irradiation Performance of Pyrolytic-Carbon- and Silicon-Carbide-Coated Fuel Particles," *Nuclear Application & Technology* 8, No. 5 (May 1970), pp. 417–431.

26. Bramer, S. E., H. Lurie, and T. H. Smith, "Re-entry Protection for Radioisotope Heat Sources," *Nuclear Application & Technology* 11, No. 2 (June 1971), pp. 232–245.

27. Hamilton, C. J., "Heavy Metal Buildup in the HTGR," *Transactions of the American Nuclear Society* 14, No. 1 (June 1971), p. 88.

28. Bates, J. F., and J. L. Straalsund, *A Compilation of Data and Representations of Irradiation-Induced Swelling of Solution-Treated Types of 304 and 316 Stainless Steel*, Hanford Engineering Development Laboratory Report No. HEDL-TME 71-139, UC-25, September 1971.

29. Ziebold, T. O., F. E. Foote, and K. F. Smith, "The Technology of Nuclear Reactor Safety," in *Materials and Metallurgy* (T. L. Thompson and J. G. Beckerley, eds.). Cambridge, Mass.: The MIT Press, 1973, Chap. 12.

30. Regulatory Staff, *Technical Report on Densification of Light Water Reactor Fuels*, U.S. Atomic Energy Commission, November 14, 1972.

31. Chubb, W., A. C. Hott, B. M. Argall, and G. R. Kilp, "The Influence of Fuel Microstructure on In-pile Densification," *Nuclear Technology* 26, No. 4 (August 1975), pp. 486–495.

32. Horak, J. A., and T. H. Blewitt, "Fast and Thermal Neutron Irradiation and Annealing of Cu, Ni, Fe, Ti, and Pd," *Nuclear Technology* 27, No. 3 (November 1975), pp. 416–438.

33. Boulton, J., "Materials and Design Interaction in Advanced Reactors," *Nuclear Tecynology* 40, No. 2 (September 1978), pp. 129–137.

34. Lawrence, L. A., D. C. Hata, and J. W. Weber, "The Effects of Stoichiometry on Cladding Attack in UO_2-PuO_2," *Nuclear Technology* 42, No. 2 (February 1979), pp. 195–206.

35. Walker, C. T., and S. Pickering, "Some Electron Microscope Analysis Results Concerning the Interaction Between Uranium-Plutonium Mixed Oxide," *Nuclear Technology* 42, No. 2 (February 1979), pp. 207–215.

36. Olander, D. R., *Fundamental Aspects of Nuclear Reactor Fuel Elements*, TID-26711-P1, Technical Information Center Office of Public Affairs, 1976.

37. Walters, L. C., and J. H. Kittel, "Development and Performance of Metal Fuel Elements for Fast Breeder Reactors," *Nuclear Technology* 48, No. 3 (May 1980), pp. 273–280.

Nuclear Heat Transfer

Heat transfer in a reactor system involves many quite conventional problems. However, the presence of fission and intense radioactivity due to both fission fragment decay and to neutron activation of clad, structure, moderator, coolant, and so on, causes intense heating. It is this heat that we wish to convert to a useful form of energy as effectively as possible. In Chapter 9 it was shown that the flux distribution was independent of power level. Consequently, the power extractable from a core is dependent on the magnitude of the heat transfer that can take place without damaging the reactor structure or the fuel elements.

Heat Transfer from Fuel Elements

Consider a fuel plate as an infinite slab with heat flow in a transverse direction through the element. This assumes that the problem is one-dimensional with temperature variation only in the transverse (x) direction. It is directly applicable for wide plates near the point of maximum temperature in the core where the axial temperature gradient is small.

The Fourier field equation, which describes the temperature distribution, can be written for a uniform value of k as

$$k\nabla^2 T + S = \rho c \frac{\partial T}{\partial t} \qquad (12.1)$$

where
$$k = \text{thermal conductivity, W/m °C}$$

$$T = \text{temperature, °C}$$

$$S = \text{source strength, W/m}^3$$

$$\rho = \text{density, kg/m}^3$$

$$t = \text{time, s}$$

$$c = \text{specific heat, kJ/kg °C}$$

The first term in Eq. (12.1) represents the heat conduction from a unit volume of material. The second term represents the source strength (in this case due to fission, but it can also represent electrical resistance heating or heating due to dissipation of alpha, beta, or gamma radiation).

In the steady-state condition, where $\partial T/\partial t = 0$, Eq. (12.1) reduces to

$$\nabla^2 T = -\frac{S}{k} \qquad (12.2)$$

If the temperature varies only in the x direction, Eq. (12.2) may be written

$$\frac{d^2 T}{dx^2} = -\frac{S}{k} \quad \text{(a constant per unit volume)} \qquad (12.3)$$

By integration of Eq. (12.3),

$$\frac{dT}{dx} = -\frac{Sx}{k} + C_1 \qquad (12.4)$$

A second integration gives

$$T = -\frac{Sx^2}{2k} + C_1 x + C_2 \qquad (12.5)$$

The boundary conditions will allow evaluation of the constants. If the surface temperatures on both sides of the fuel plate are equal, the temperature distribution is symmetrical (see Fig. 12.1). At the center of the plate $x = 0$, and $dT/dx = 0$. Therefore, $C_1 = 0$. Also, when $x = \pm w/2$, $T = T_s$, and the second constant of integration becomes

$$C_2 = T_s + \frac{Sw^2}{8k}$$

The temperature in the plate is then expressed as

$$T = -\frac{Sx^2}{2k} + T_s + \frac{Sw^2}{8k}$$

$$= T_s + \frac{S}{2k}\left(\frac{w^2}{4} - x^2\right) \qquad (12.6a)$$

The maximum temperature in the fuel is then

$$T_{max} = T_s + \frac{Sw^2}{8k} \qquad (12.6b)$$

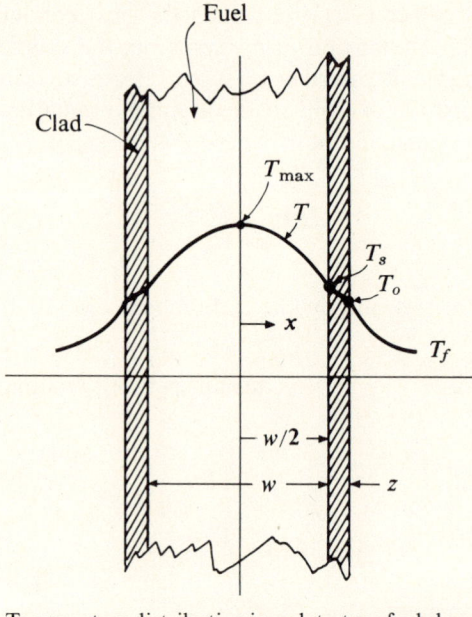

Fig. 12.1 Temperature distribution in a plate-type fuel element (treated as in infinite slab). Temperature drop through clad and convective film also shown.

The fission induced heating can be expressed as

$$S = 180 \times N \times \sigma_{\text{fis}} \times \bar{\phi} \times 1$$

$$\frac{\text{mev}}{\text{fission}} \quad \frac{\text{fis. nuclei}}{\text{cm}^3} \quad \frac{\text{cm}^2}{\text{fis. nuclei}} \quad \frac{\text{neutrons}}{\text{cm}^2 \text{ s}} \quad \frac{\text{fission}}{\text{neutron}}$$

$$\times\ 1.60210 \times 10^{-13} \times 10^6 \times 1$$

$$\frac{\text{J}}{\text{MeV}} \quad \frac{\text{cm}^3}{\text{m}^3} \quad \frac{\text{W s}}{\text{J}}$$

$$= 2.884 \times 10^{-5}\, N\sigma_{\text{fis}}\bar{\phi} \quad \text{W/m}^3 \tag{12.7}$$

The heat flux at the surface of the plate when $x = w/2$ is

$$\frac{q}{A} = -k\frac{dT}{dx} = \frac{Sw}{2} \tag{12.8}$$

The same heat flux passes through the cladding of thickness z.

$$\frac{Sw}{2} = \frac{k_{\text{cl}}(T_s - T_0)}{z} \tag{12.9}$$

The fuel's surface temperature can be expressed as a function of the outer cladding surface temperature.

$$T_s = T_0 + \frac{zSw}{2k_{cl}} \qquad (12.10)$$

Example 1   A fuel plate is fabricated from 0.3-cm-thick 1.5 percent enriched uranium. The cladding is 0.25-mm 304 stainless steel. The average thermal neutron flux is 2.5×10^{14} neutrons/cm^2 s. The surface temperature of the clad is to be 350°C.

   (a) What is the heat flux at the surface of the fuel?
   (b) What is the temperature at the fuel–clad interface?
   (c) What is the maximum fuel temperature?
   (d) Does any of the fuel transform to the β phase of uranium?

Assume an approximate average fuel temperature of 400°C for evaluation of the fission cross section.

$$\sigma_{fis} = 582 \frac{\sqrt{\pi}}{2} \sqrt{\frac{293}{673}} = 340 \text{ barns}$$

$$N_{235} = \frac{19 \times 6.023 \times 10^{23} \times 0.015}{238} = 7.21 \times 10^{20} \,^{235}\text{U nuclei/cm}^3$$

$$S = 2.884 \times 10^{-5} N_{235} \sigma_{fis} \phi$$
$$= 2.884 \times 10^{-5} \times 7.21 \times 10^{20} \times 340 \times 10^{-24} \times 2.5 \times 10^{14}$$
$$= 1.767 \times 10^9 \text{ W/m}^3$$

The surface heat flux is then

$$q/A = \frac{Sw}{2} = \frac{1.767 \times 10^9 \times 0.003}{2} = 2.6505 \times 10^6 \text{ W/m}^2$$

$$T_s = T_0 + \frac{zSw}{2k_{cl}} = 350 + \frac{0.00025 \times 2.6505 \times 10^6}{19}$$

$$= 350 + 34.9 = 384.9°\text{C}$$

$$T_{max} = T_s + \frac{Sw^2}{k}$$

$$= 384.9 + \frac{1.767 \times 10^9 \times 0.003}{34.8}$$

$$= 384.9 + 57.1 = 442°\text{C}$$

Since the α–β transition temperature is 663°C, phase change will cause no difficulty in this instance.

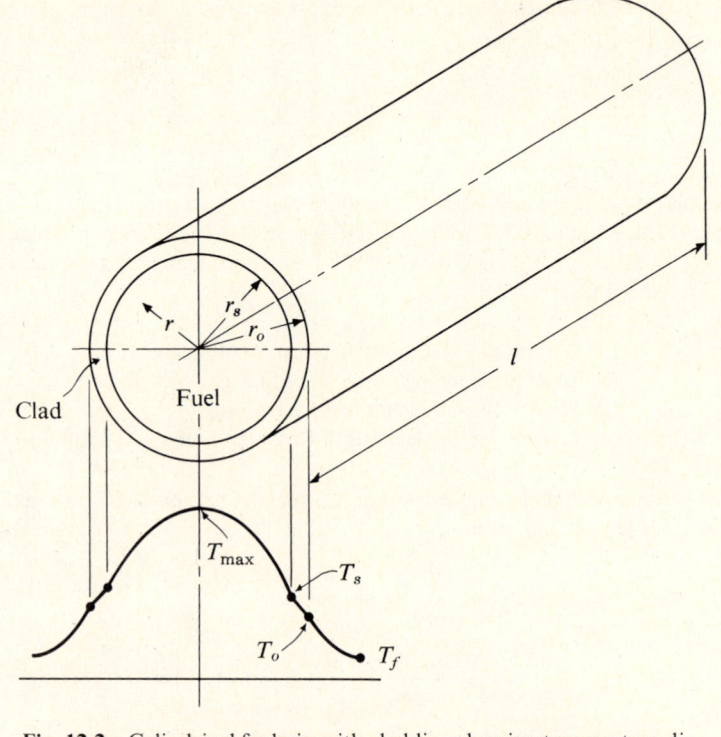

Fig. 12.2   Cylindrical fuel pin with cladding showing temperature distribution in pin, clad, and convective film.

An even more common type of fuel element is the pin or rod. If the axial temperature gradient is ignored, the problem is one-dimensional in the radial direction. Figure 12.2 shows a fuel pin with its cladding and temperature distribution. To solve this problem Eq. (12.1) must be converted to cylindrical coordinates. For steady state it reduces to

$$\frac{d^2 T}{dr^2} + \frac{(1/r)\, dT}{dr} = -\frac{S}{k}$$

(12.11a)

Rearranging, Eq. (12.11a) can be written as

$$r \frac{d^2 T}{dr^2} + \frac{dT}{dr} = -\frac{Sr}{k}$$

(12.11b)

Note that

$$\frac{d}{dr}\left(r \frac{dT}{dr}\right) = r \frac{d^2 T}{dr^2} + \frac{dT}{dr}$$

(12.12)

Substituting Eq. (12.12) into (12.11b) yields

$$\frac{d}{dr}\left(r\frac{dT}{dr}\right) = -\frac{Sr}{k}$$

(12.13)

Integrating, we obtain

$$r\frac{dT}{dr} = -\frac{Sr_s^2}{2k} + C_1$$

(12.14)

A second integration gives

$$T = -\frac{Sr^2}{4k} + C_1\ln r + C_2$$

(12.15)

For the cylinder of fuel the boundary conditions are:

(1) When $r = 0$, $dT/dr = 0$, because of symmetry.
(2) When $r = r_s$, $T = T_s$, at the fuel surface.

Therefore,

$$C_1 = 0$$

and

$$C_2 = T_s + \frac{Sr_s^2}{4k}$$

The temperature distribution in the element can then be expressed as

$$T = T_s + \frac{S}{4k}(r_s^2 - r^2)$$

(12.16)

The temperature gradient in a radial direction is

$$\frac{dT}{dr} = -\frac{rS}{2k}$$

(12.17)

The heat flux at the element surface is then

$$\frac{q}{A} = -k\left(\frac{dT}{dr}\right)_{r=r_s} = \frac{r_s S}{2}$$

(12.18)

Example 2   A 1.00-cm-diameter UO_2 fuel rod is clad with 0.025 cm of 304 stainless steel. The rod is to develop 16.4 kW/m at the point being considered in a BWR fuel channel. The boiling heat transfer coefficient is 7000 W/m^2 °C and the fluid saturation temperature is 280°C.

Compute the heat flux at the surface of the fuel and the temperature at the center of the fuel rod.

First, the source strength must be determined.

$$S = \frac{16.4 \text{ kW/m} \times 10^3 \text{ W/kW}}{(\pi/4)(0.01^2) \text{ m}^3/\text{m}} = 2.088 \times 10^8 \text{ W/m}^3$$

The heat flux at the fuel surface may then be evaluated.

$$q/A = \frac{r_s S}{2} = \frac{0.005 \times 2.088 \times 10^8}{2} = 5.22 \times 10^5 \text{ W/m}^2$$

Because they are in series, the heat flowing from the fuel surface, the heat flow through the clad, and the heat flow into the fluid are all equal. Let A_1 represent the area of the fuel surface and A_2 represent the outer-clad-surface area. To establish the temperature drop across the fluid film, the heat flow from the fuel is equated to the heat flow through the boiling film.

$$q/A \times A_1 = A_2 h_f (T_0 - T_f)$$

$$T_0 - T_f = \frac{(q/A)A_1}{h_f A_2} \tag{12.19}$$

$$T_0 = T_f + \frac{(q/A)A_1}{h_f A_2}$$

$$= 280 + \frac{5.22 \times 10^5 \times 1.000}{7000 \times 1.050}$$

$$= 280 + 71 = 351°C$$

For the clad, assuming that no heat source is present, integration of the heat flux equation

$$q = -kA(dT/dr)$$

gives

$$q = \frac{2\pi k_{cl}(T_s - T_0)}{\ln (r_0/r_s)}$$

The temperature distribution in the clad is given by

$$T = T_0 + \frac{S r_s^2}{2k_{cl}} \ln \frac{r_0}{r_s}$$

The temperature difference across the clad will be

$$T_s - T_0 = \frac{Sr_s^2}{2k_{cl}} \ln(r_0/r_s)$$

$$= \frac{2.088 \times 10^8 \times 0.005^2 \ln(1.050/1.000)}{2 \times 19.0} = 6.7°C$$

$$T_s = 351 + 6.7 = 357.7°C \qquad \qquad \textbf{(12.20)}$$

The centerline–surface temperature difference in the fuel is

$$T_{max} - T_s = \frac{Sr_s^2}{4k} \qquad \qquad \textbf{(12.21)}$$

The peak fuel temperature is then

$$T_{max} = T_s + \frac{Sr_s^2}{4k} = 357.7 + \frac{2.088 \times 10^8 \times 0.005^2}{4 \times 3.20}$$

$$= 357.7 + 407.8 = 765.5°C$$

The preceding example indicates the magnitude of the temperature that may exist at the center of a ceramic fuel element. Unfortunately, the thermal conductivity of UO_2 varies with both temperature and fuel burnup. Figure 12.3 indicates the changes in the effective thermal conductivity of uranium

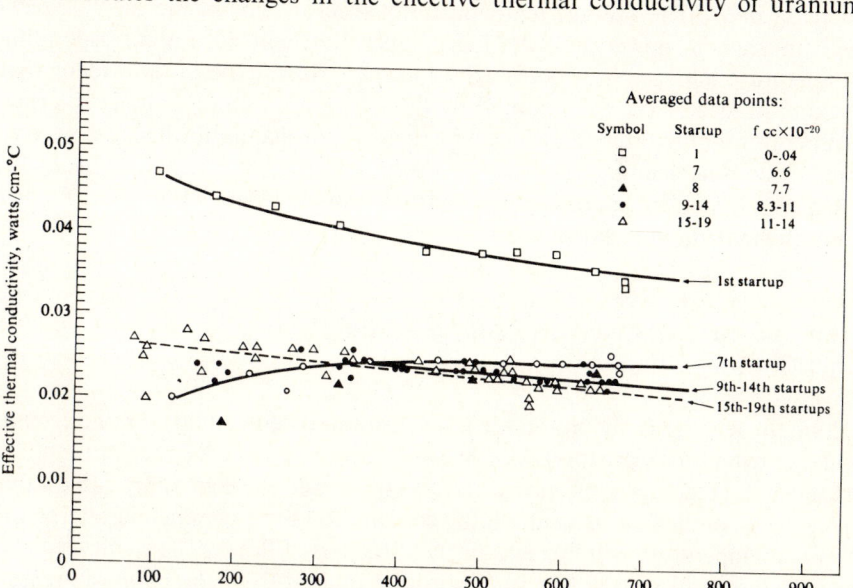

Fig. 12.3 Effective thermal conductivity of UO_2 showing the effect of various numbers of startups (see Reference 1).

dioxide after many startups. As the fuel suffers loss in conductivity, the temperature at the center of the elements may approach levels that can result in melting of some of the UO_2 at the centerline. The loss of conductivity occurs due to the microstructural changes and cracking during the heating and cooling cycles which the fuel must endure during its lifetime (see Fig. 11.28).

The example also does not account for heat transfer in the small space between fuel and cladding. This gap is normally filled with helium, and heat transfer is treated as conduction. Heat transfer across the gap will change somewhat with core life because as fission gasses build up, conductivity in the gap will decrease.

By adding Eqs. (12.19), (12.20), and (12.21) and setting $r = 0$, an expression for the temperature drop from the fuel centerline to the fluid can be obtained.

$$T_{max} - T_f = Sr_s^2 \left(\frac{1}{4k_f} + \frac{1}{2k_{cl}} \ln \frac{r_0}{r_s} + \frac{1}{2r_0 h_f} \right) \tag{12.22}$$

At any given point in core life, k and r will be fixed so that T_{max} becomes a function of h_f for a given S. In order to keep T_{max} from reaching too high a value, h_f would have to be increased a large amount since it affects only part of the equation. When reactor power (S) is increased, the temperature between fuel and coolant increases since it takes such a large change in h_f to offset an increase in S. The average temperature of reactor coolant increases linearly with increase in power, and the fuel temperature will have a corresponding temperature increase (not necessarily linear). Thus, it becomes evident that maximum reactor power is limited by the maximum allowable fuel centerline temperature. For pressurized water reactors the maximum linear power to limit centerline melting is about 65 kW/m. However, as was discussed in Chapter 10, accident conditions of departure from nucleate boiling may further limit the maximum linear power.

Temperature Distribution Along a Fuel Rod in a Cooling Channel

Along the length of a fuel rod there is a variation in neutron flux. For a rectangular or cylindrical core this variation can be approximated by a cosine function. In most real reactors a reflector will cause the flux to have an appreciable value at each end of the fuel element. If the coolant flows in a channel parallel to the rod, its temperature will increase in the direction of flow. The point of maximum temperature in the fuel will then be displaced from the point of greatest flux (and hence maximum energy release).

Figure 12.4 shows the neutron flux variation along the fuel rod. Note that the origin is at the midpoint of the rod, and that the ends of the rod are

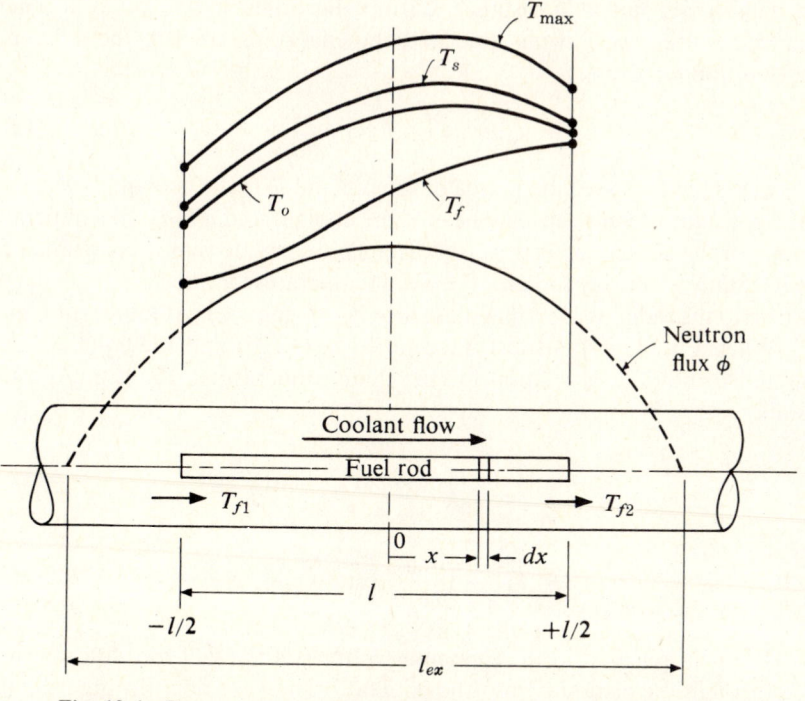

Fig. 12.4 Temperature and flux (neutron) variation along a fuel rod located in a fuel channel with flow parallel to the fuel rod. T_{max} is the temperature of the fuel at the centerline of the fuel rod, T_s is the fuel–clad interface temperature, T_0 is the temperature at the clad surface, and T_f is the coolant temperature.

at $\pm l/2$, where l is the rod length. l_{ex} is the extrapolated pin length. The flux, ϕ, may be expressed as a function of position and the maximum flux, ϕ_m.

$$\phi = \phi_m \cos \frac{\pi x}{l_{ex}} \tag{12.23}$$

The energy release rates are also position dependent, being proportional to the neutron flux.

$$S = S_0 \cos \frac{\pi x}{l_{ex}} \tag{12.24}$$

where S_0 is the energy release rate per unit volume at the point of maximum neutron flux.

Taking a point at a distance x from the origin and making an energy balance between the flowing fluid and the energy released in the differential length of rod dx gives

$$\dot{m}c_p\,dT_f = SA\,dx \tag{12.25}$$

where A is the cross-sectional area of the pin and $\dot{m}$ is the coolant flow rate. c_p is the constant pressure specific heat of the coolant and dT_f is the temperature change of the coolant as it travels a distance dx. Note that it is assumed that the heat flow is entirely radial. The axial temperature gradient is small in comparison to the radial value; thus, axial heat flow is ignored without serious error.

When Eq. (12.24) is substituted into Eq. (12.25) and the equation is integrated to point x, a relation for the fluid temperature, T_f, at any position results.

$$\dot{m}c_p \int_{T_{f1}}^{T_f} dT_f = S_0 A \int_{-1/2}^{x} \cos\frac{\pi x}{l_{ex}}\,dx \tag{12.26}$$

$$T_f = T_{f1} + \frac{AS_0 l_{ex}}{\dot{m}c_p \pi}\left(\sin\frac{\pi x}{l_{ex}} + \sin\frac{\pi l}{2l_{ex}}\right) \tag{12.27}$$

The heat released in the elemental volume of the fuel rod flows from the fuel, through the cladding, into the coolant. The heat flow through the fluid film between the coolant and clad is equated to the heat release to give the outer cladding temperature, T_0.

$$S_0 A \cos\frac{\pi x}{l_{ex}}\,dx = h_0(2\pi r_0)(T_0 - T_f)\,dx \tag{12.28}$$

Combining Eqs. (12.27) and (12.28) gives

$$T_0 = T_{f1} + \frac{S_0 A l_{ex}}{\dot{m}c_p \pi}\left(\sin\frac{\pi x}{l_{ex}} + \sin\frac{\pi l}{2l_{ex}}\right) + \frac{S_0 A \cos(\pi x/l_{ex})}{2\pi r_0 h_0} \tag{12.29}$$

The slope of the outer clad temperature in an axial direction is

$$\frac{dT_0}{dx} = \frac{S_0 A}{\dot{m}c_p}\cos\frac{\pi x}{l_{ex}} - \frac{S_0 A}{2r_0 h_0 l_{ex}}\sin\frac{\pi x}{l_{ex}} \tag{12.30}$$

Equating the slope to zero determines an expression for the position of the point of maximum surface temperature of the clad, $x_{0_{max}}$.

$$x_{0\,max} = \frac{l_{ex}}{\pi}\tan^{-1}\frac{2l_{ex}r_0 h_0}{\dot{m}c_p} \tag{12.31}$$

The temperature change across the cladding and the centerline temperature of the fuel can be determined by using the methods illustrated in Example 2 for the source strength at any given position. Problems of this type can be programmed conveniently for a digital computer (see Prob. 12.8).

Burnout in Water-Cooled Reactors

In both boiling water and pressurized water reactors the power-producing capability is limited by the reactor's ability to transfer fission energy from the fuel elements to the coolant for transport from the core. In both types of reactor there is a variation in heat flux in the axial and radial directions due to neutron flux variation. Flow rates along elements may be variable, due to dimensional differences in the flow passages. These may be due to manufacturing variations or to orifices placed in the flow channels. The orifices adjust flow in the channels so that there will be an equal enthalpy rise in each channel. In the BWR controlled phase change is intended, but in the PWR the water should remain liquid. In either case, film boiling that blankets the heat transfer surface with a layer of vapor must be prevented. Such a blanketing of the surface by vapor may result in "burnout." The very high temperatures produced in the cladding may produce rupture and release of fission products. At the point of maximum heating the heat flux must remain well below that required to cause burnout. The *factor of safety*, F_s, or *burnout ratio*, is the ratio of the burnout heat flux, $(q/A)_{BO}$ to the actual heat flux at a point in the flow channel.

$$F_s = \frac{(q/A)_{BO}}{q/A} \qquad (12.32)$$

Boiling

Pool boiling (natural-convection boiling)

Take a uniformly heated rod and immerse it in a pool of liquid at its saturation temperature and then increase the heat flux gradually. The flux (W/m^2) can be plotted, as shown in Fig. 12.5, versus the temperature differential between the rod surface temperature, T_s, and the liquid temperature, T_{sat}.

One can see that at very low heat fluxes from *a* to *b* there is only natural convection set up with evaporation at the liquid surface. Higher fluxes commence the formation of bubbles which leave the surface from favored nucleation sites and then collapse in the liquid. Somewhat higher fluxes from *b* to *c* result in the bubbles rising all the way to the surface. The agitation caused by the bubbles (they push the liquid back; it rushes in when they break away from the nucleation sites) produces high film coefficients. The bubble formation may become so intense, however, that the surface becomes blanketed with vapor

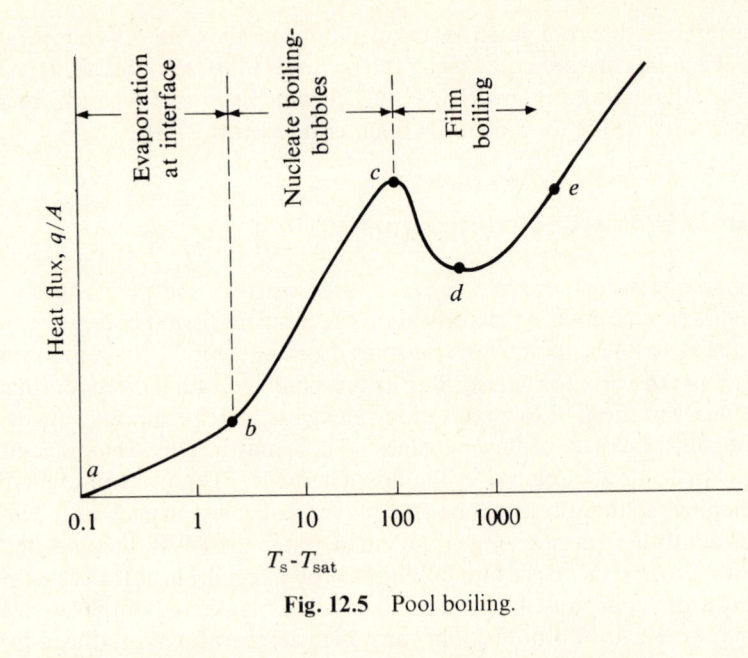

Fig. 12.5   Pool boiling.

and liquid cannot reach it.   Thus, the ability to remove heat is reduced.   This condition at c is known as the *burnout point* or the point of departure from nucleate boiling (*DNB*).   The flux must be reduced to continue a smooth increase in ΔT.   However, this is difficult with an electrically heated rod or a fission-heated fuel element in a reactor.   There will more likely be a large jump in ΔT from c to e, where stable film boiling will exist.   In the range from c to d boiling is unstable and the temperature will jump to a new larger value if the flux increases even slightly.   Beyond d the boiling is essentially all film boiling with radiation transporting the energy across the film.   A temperature at the point e may be well above the melting point of the cladding material on a fuel rod; hence, burnout ensues.

For pool boiling the burnout flux (W/m^2) can be evaluated from the Rohsenow and Griffith correlation (to be found in Reference 5).

$$\left(\frac{q}{A}\right)_{\text{BO}} = 451\, g^{1/4} \rho_v h_{fg} \left(\frac{\rho_l - \rho_v}{\rho_v}\right)^{0.6} \tag{12.33}$$

where g = acceleration of gravity in g's

ρ_v = vapor density, kg/m^3

ρ_l = liquid density, kg/m^3

h_{fg} = latent heat of vaporization, kJ/kg

For pool boiling the burnout flux is independent of the material used for the surface, as long as it is reasonably smooth. Knurled or threaded surfaces will have a somewhat higher burnout flux.

Boiling with forced convection

When boiling occurs with forced convection the mass velocity, the geometry of the system, and any subcooling of the entering water complicate the correlation of burnout data.

Figure 12.6 shows a schematic diagram of a burnout test where the tube is uniformly heated by electrical means. This simulates heating in the coolant passage of a reactor. If one picks a particular pressure, p, at the exit of the test channel, the only other variables remaining are the flow rate, the inlet temperature of the water, and the electrical input (which provides the heat flux in lieu of the fission heating in a reactor). For a given inlet temperature either the heat flux or the flow may be varied until burnout occurs. In general, with forced convection a higher heat flux is required to produce burnout than with pool boiling. This is because the forced convection sweeps away the bubbles as they are formed.

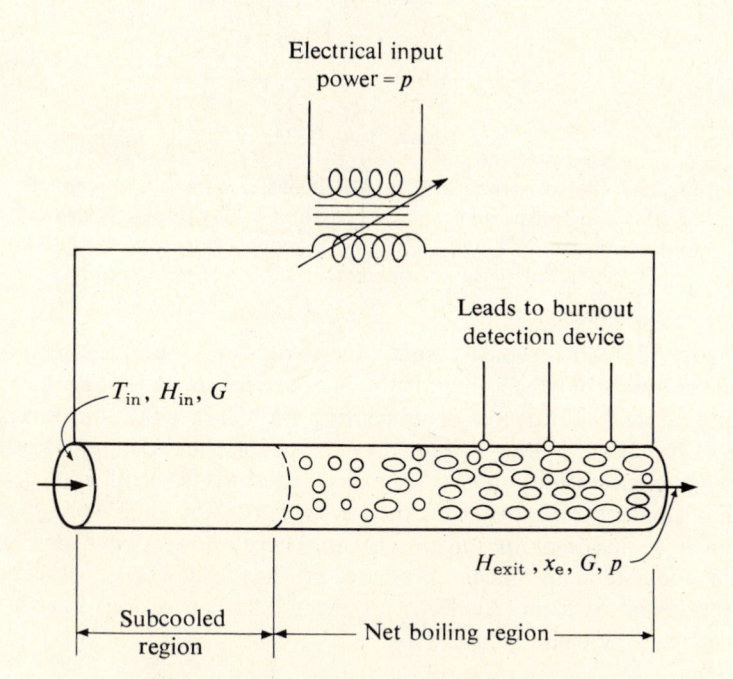

Fig. 12.6 Schematic diagram of burnout test. [From Reference 11.]

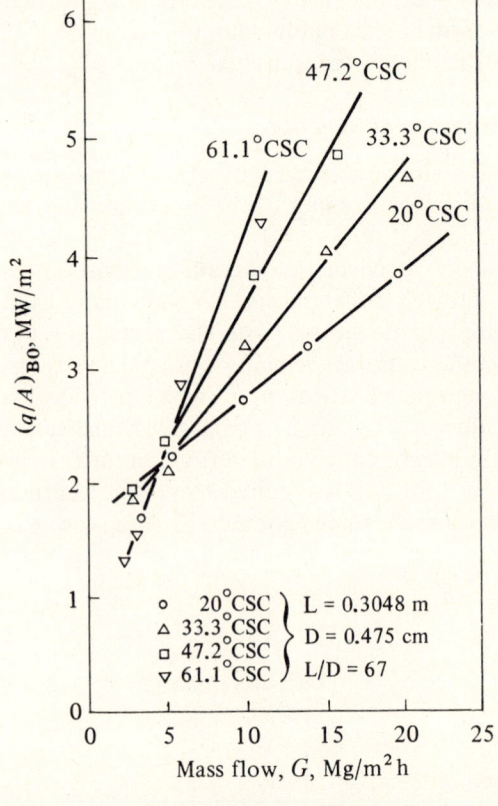

Fig. 12.7 Plot of burnout heat flux versus mass flow for various amounts of subcooling of the inlet water. Length of test section = 0.3048 m, tube diameter = 0.475 cm, $L/D = 67$, water pressure = 13.8 MPa. [Plotted in Reference 11 from data in Table 8 of Reference 14.]

Figure 12.7 shows typical results of tests run on a round test section using variable amounts of subcooling for the inlet water. Note that both increased flow and an increased degree of subcooling allow larger burnout fluxes. The utility of this set of curves is limited to the geometrical configuration tested. To correlate successfully data from many experiments of different geometry, Levedahl (Reference 12) proposed the steam energy flow (SEF) plot, where the burnout heat flux is plotted against steam energy flow (shown in Fig. 12.8).

To understand the meaning of the term steam energy flow, equate the energy transmitted from the electrically heated surface of Fig. 12.6 to the enthalpy gain of the passing fluid.

$$\frac{q}{A}\pi D_1 L = \frac{\pi D_2^2}{4} G(h_0 - h_{in}) \qquad (12.34)$$

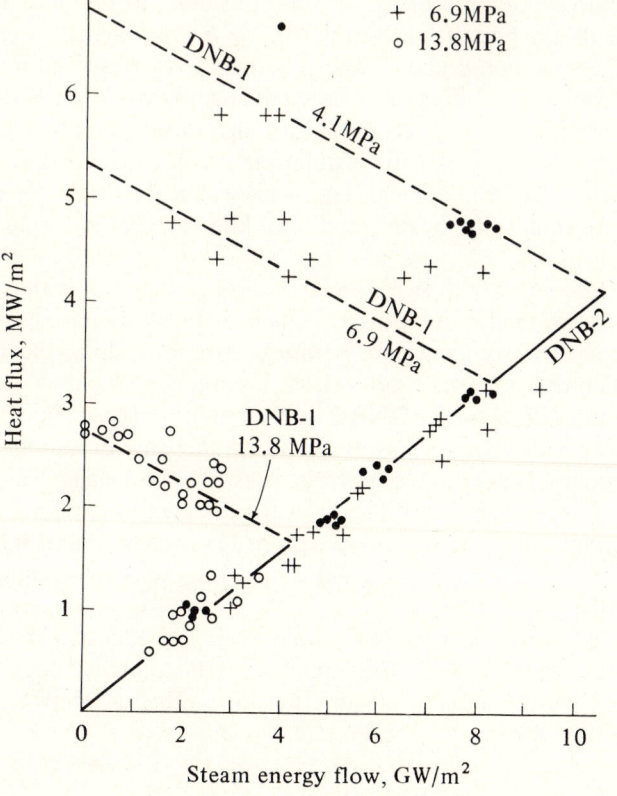

Fig. 12.8 SEF plot showing two burnout regimes for uniformly heated channels. [From Reference 12.]

where L = the section length from entrance to the point under consideration (often the exit), m

D_1 = reference diameter for heat flow, m

D_2 = equivalent diameter of flow passage, m; for a noncircular channel this is equal to four times the flow area divided by the heated perimeter

G = mass velocity, kg/m² h

h_0 = enthalpy at distance L from inlet, kJ/kg

h_{in} = inlet enthalpy, kJ/kg

Solving for the heat flux, the expression may be written

$$\frac{q}{A} = \frac{D_2^2}{4LD_1}[Gx_0 h_{fg} + G(h_f - h_{in})] \qquad \textbf{(12.35)}$$

where h_0 has been replaced by the enthalpy of the wet leaving steam ($h_0 = h_f + x_0 h_{fg}$) which has a quality of x_0. If one considers the enthalpy of saturated liquid as the datum point for SEF, the $G(h_{in} - h_f)$ represents the energy of the entering subcooled liquid and $(Gx_0 h_{fg})$ is the energy conveyed in the leaving steam above the datum level. For flow passing any point the SEF represents the departure of the enthalpy from the enthalpy of saturated liquid. For subcooled liquid its value is negative, while for the wet or superheated steam it will be positive. If the inlet condition of the water, the flow rate, and the heat flux along a uniformly heated rod are known, the exit quality can be computed.

Figure 12.8 shows that the burnout points align themselves and indicate that there may be two modes of burnout. Lines for these are marked DNB-1 and DNB-2 (departure from nucleate boiling). Burnout along the DNB-1 line represents failure due to the vapor release becoming so rapid as to blanket the surface. Points falling on the DNB-2 line represent burnout which occurs with accompanying hydrodynamic instability where there are pulsations in flow and pressure drop. DNB-2 data seem to be sensitive to orificing, surge volumes, and length of lines. It is very difficult for the designer to apply DNB-2 data as useful design information. Wiley (Reference 11) suggests that it signifies to the experimenter that he or she is generating ambiguous information which is difficult for the designer to utilize.

It should be noted that the SEF, as plotted in Fig. 12.8, uses the SEF of saturated liquid as the zero point. Thus, any part of the total SEF due to subcooling $[G(h_f - h_{in})]$ should be laid out as a negative value, as shown in Fig. 12.9.

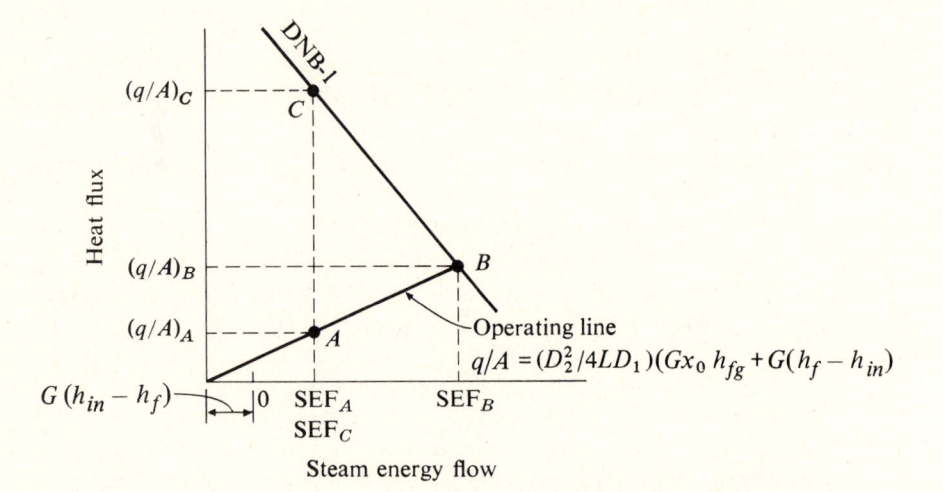

Fig. 12.9   Steam energy flow plot showing an operating line and a
DNB-1 line.

Use of DNB-1 Data in Design

In design it is important that the heat flux be kept safely below the burnout value indicated along the DNB-1 line. Figure 12.9 shows an operating line and a DNB-1 line on an SEF plot. The operating line represents the locus of points for a given mass velocity, G, as the heat flux is raised to burnout. There is some uncertainty as to whether the burnout ratio should be taken as the ratio of $(q/A)_B/(q/A)_A$ or $(q/A)_C/(q/A)_A$.

The heat flux burnout ratio is defined as

$$F_{S_1} = \frac{(q/A)_C}{(q/A)_A} \tag{12.36}$$

This has significance only if the DNB-1 line is determined solely by local conditions. The $(q/A)_C$ condition could not be reached for the flow rate used for the operating line AB.

A ratio that gives a better indication of the safety factor attainable with a given flow rate is called the *power burnout ratio*. It is defined as

$$F_{S_2} = \frac{(q/A)_B}{(q/A)_A} \tag{12.37}$$

Figure 12.10 shows the axial heat flux variation along a reactor fuel rod for two power levels, P and P'. The power level P represents the power such that the SEF at the exit represents the design point A of the previous Fig. 12.9. The power P' is a level such that burnout would occur at the exit, B, with the same flow rate as at A. At any other point A' the operating line will show burnout at B'. Depending on the shape of the power distribution curve and the slope of the DNB-1 curve, it is possible for the power burnout ratio to reach unity initially at a point other than the channel exit. For a core with control rods inserted from the top the peak flux may be distorted with the peak heat flux closer to the inlet because of the rod insertion. At the end of the core life, when the rods are fully removed, the peak may move toward the exit because of the lesser poison buildup in the upper end of the core. In this region the flux was previously depressed. Figure 12.11 shows a case where the burnout will occur at B'' for a power level P''.

Burnout fluxes can be evaluated from any one of several correlations. The Westinghouse equations (Reference 14) for 13.8 MPa water are useful for the design of pressurized water cores. The particular design equation (12.38) shown here incorporates a factor of 65 percent so that the burnout fluxes predicted

will be well below the actual values of burnout flux (MW/m^2). The design equation for round tubes is

$$(q/A)_{\text{BO}} = 0.574 \left(\frac{h_{\text{BO}}}{10^3}\right)^{-2.5} \left(1 + \frac{G}{10^7}\right) e^{-0.0012L/D} \qquad \textbf{(12.38)}$$

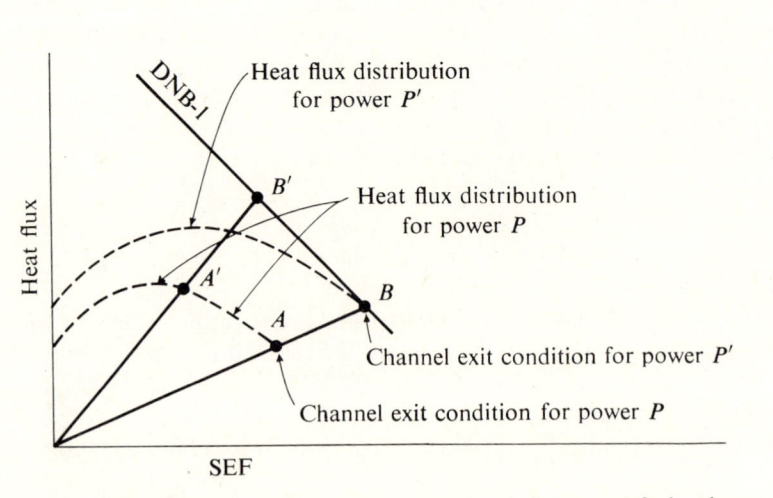

Fig. 12.10   Nonuniform axial power distribution for reactor fuel rod operating at power levels P and P'. The power level P is the design power level and P' is a level such that burnout will occur at the channel exit.

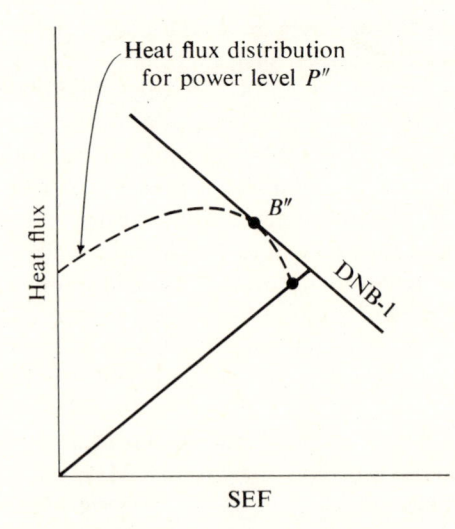

Fig. 12.11   SEF plot showing outlet flux peak causing burnout to occur to exit.

The equation is valid over the following range of variables:

Pressure	$12.8 \leq P \leq 14.8$ MPa
Enthalpy	$116 \leq h_{BO} \leq 2326$ kJ/kg
Mass velocity	$0.94 \times 10^6 \leq G \leq 37.5 \times 10^6$ kg/m^2 h
Length/diameter	Data based on values $21 \leq L/D \leq 365$
	Extrapolation is possible.

The design burnout flux may be further reduced below the predicted value by the uncertainty factor, F_{S_1}. Reference 14 recommends a value of 1.1 to allow for lack of transient burnout and pressure drop data, plus another factor of 1.1 to allow for meager data on nonuniformly heated channels. These combine to give a total uncertainty of approximately $F_{S_1} = 1.2$.

Example 3 In a vertical, 4-m-long, uniformly heated tube with 2.5 cm inside diameter, subcooled water enters at 250°C and leaves at a pressure of 4.1 MPa with a quality of 20 percent. The mass flow rate is 4167 kg/s m^2. Determine:

(a) The steam energy flow
(b) The heat flux
(c) The power burnout ratio, F_{s2}

$$SEF = G x_0 h_{fg} + G(h_f - h_{in})$$

$$= 4167 \frac{kg}{s\,m^2} \times 10^3 \frac{W\,s}{kJ} \times 0.2 \times 1706.5 \frac{kJ}{kg}$$

$$+ 4167 \times 10^3 (1094.4 - 898.5)$$

$$= 1.422 \times 10^9 + \underline{0.816 \times 10^9} = 2.238 \times 10^9 \text{ W/m}^2$$

$$\underset{\substack{\text{offset for} \\ \text{subcooling}}}{}$$

$$(q/A)_A = \frac{D_2^2}{4 L D_1} [G x_0 h_{fg} + G(h_f - h_{in})]$$

In this instance, equivalent diameter of the flow passage, D_2, is equal to the reference diameter for heat flow, D_1, so we may write (see Fig. 12.12)

$$(q/A)_A = \frac{D}{4L} \times SEF = \frac{0.025}{4 \times 4} (2.238 \times 10^9)$$

$$= 3.50 \times 10^6 \text{ W/m}^2$$

$$F_{s_2} = (q/A)_B / (q/A)_A = \frac{5.2 \times 10^6}{3.5 \times 10^6}$$

$$= 1.49$$

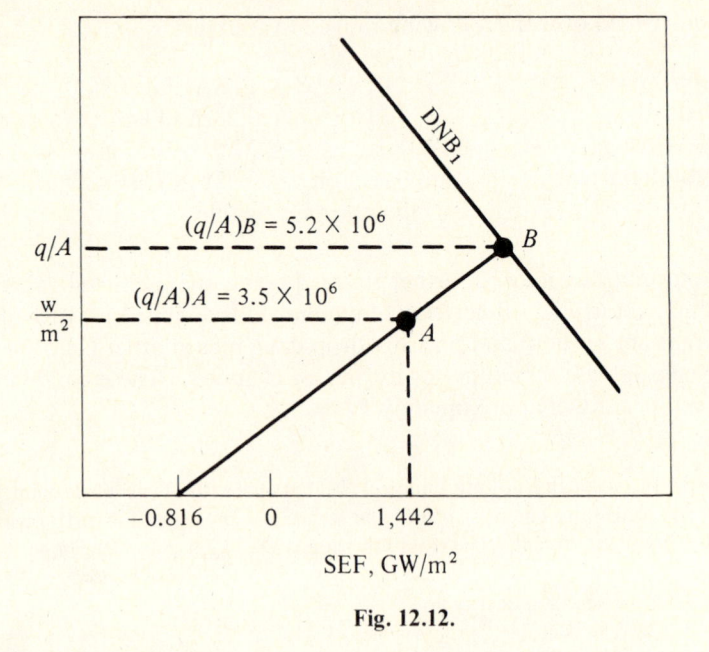

$(q/A)B = 5.2 \times 10^6$

q/A

$\dfrac{\mathrm{w}}{\mathrm{m}^2}$

$(q/A)A = 3.5 \times 10^6$

DNB₁

B

A

−0.816   0   1,442

SEF, GW/m²

Fig. 12.12.

Nuclear Superheat

The generation of saturated steam by both **PWR** and **BWR** reactors limits the attainable thermal efficiency. It also makes the large amounts of moisture formed during the expansion of the steam an aggravating problem. The design of reactors to accomplish boiling and superheating in the same core would permit attainment of better thermal efficiencies along with reduced moisture toward the end of the expansion, but it does present several other problems:

(1) The higher fuel and cladding temperatures associated with heat transfer to the superheated steam. This also may induce deposition of solids on the heat transfer surfaces in the superheat section.

(2) Maintaining the proper balance between power production in the boiling and the superheating sections of the core.

(3) Preventing excessive heat transfer from the superheated steam passages back to the cooler boiling water. This temperature difference will also cause some problems with differential expansion.

(4) The high concentration of radiolytically decomposed oxygen (30 ppm) precludes usage of most of the low-neutron-cross-section materials for cladding and tubing in contact with the superheated steam. Cladding must operate at temperatures as high as 675°C to produce 538°C steam.

Figure 12.13 shows the type of element that was used in the Puerto Rican Boiling Nuclear Superheat Reactor (BONUS). Here the stainless steel pressure tube

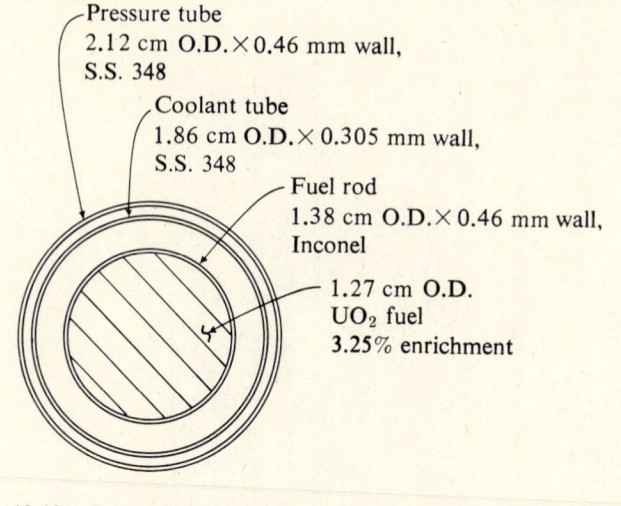

Pressure tube
2.12 cm O.D. × 0.46 mm wall,
S.S. 348

Coolant tube
1.86 cm O.D. × 0.305 mm wall,
S.S. 348

Fuel rod
1.38 cm O.D. × 0.46 mm wall,
Inconel

1.27 cm O.D.
UO_2 fuel
3.25% enrichment

Fig. 12.13 Cross section of BONUS superheater element. [From *Nuclear Reactor Technology* **6**, No. 2 (1963), p. 76.]

and the stainless steel coolant tube are separated by an insulating space. Steam flows between the coolant tube and the Inconel clad UO_2 fuel. The details of the core arrangement are shown in Fig. 12.14. The central region of the core provides the majority of the steam, while 6 percent of the steam is

Table 12.1 Performance Data for the BONUS Reactor

	Boiler Region	Superheater Region
Volume of active region (liters)	1126	1118
Net heat generation in fuel [MW(th)]	37.0	13.0
Heat loss to moderator due to r and n heating [MW(th)]	1.1	0.4
Thermal insulation heat loss to coolant [MW(th)]	—	1.2
Net heat transfer to coolant [MW(th)]	32.9	11.4
Uranium content of region (tons)	2.81	1.79
Fuel enrichment (%)	1.85 and 0.71	3.5
Average specific power [MW(th)/ton U]	13.2	7.25
Average heat flux (W/m²)	321 700	205 000
Maximum heat flux (W/m²)	1 018 700	747 500
Steam flow (kg/h)		
Core region	450 300	—
Moderator	—	29 100
Superheater passages	—	470 400

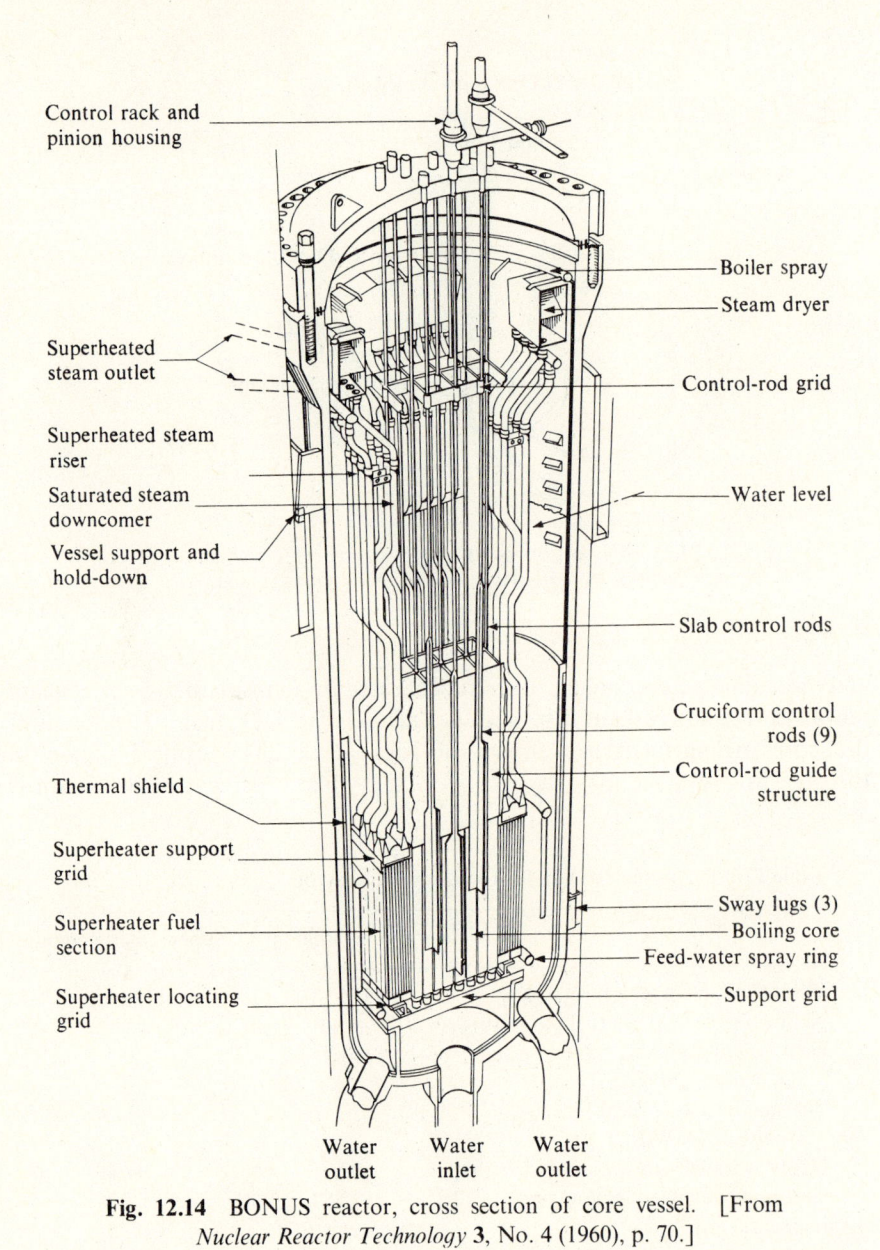

Control rack and
pinion housing

Boiler spray

Steam dryer

Superheated
steam outlet

Control-rod grid

Superheated steam
riser

Saturated steam
downcomer

Water level

Vessel support and
hold-down

Slab control rods

Cruciform control
rods (9)

Control-rod guide
structure

Thermal shield

Superheater support
grid

Sway lugs (3)

Boiling core

Superheater fuel
section

Feed-water spray ring

Superheater locating
grid

Support grid

Water
outlet

Water
inlet

Water
outlet

Fig. 12.14   BONUS reactor, cross section of core vessel. [From
Nuclear Reactor Technology **3**, No. 4 (1960), p. 70.]

generated in the moderator surrounding the pressure tubes in the superheat
region.

It is interesting to note the comparison of the heat transfer and power
density characteristics of the reactor core boiler region and the superheat
region, as shown shown by Table 12.1   Because of the lower film coefficients in the
superheat region, the heat flux and the power density are significantly less.

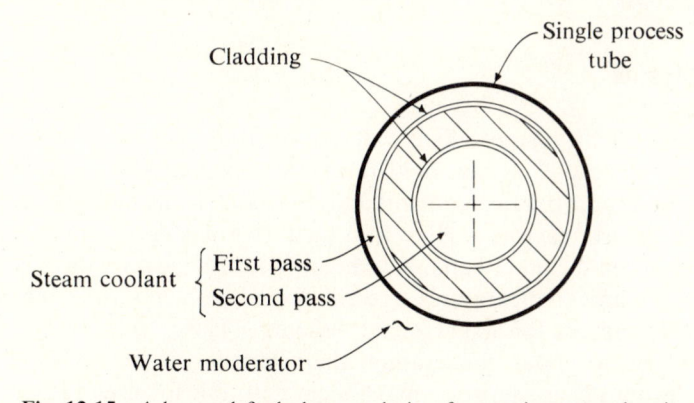

Fig. 12.15 Advanced fuel element design for use in a superheating reactor. [From *Nuclear Reactor Technology* **6**, No. 2 (1963), p. 77.]

Annular fuel elements are used in some of the designs intended for nuclear superheat. Figure 12.15 shows such an element. The steam in the first pass loses some heat to the water moderator and its final temperature is produced in the second pass.

Figure 12.16 shows a ring of superheating tubes embedded in a large annular fuel element. There is water on both sides of the annulus. Such complex fuel elements can be fabricated effectively by vibrational compaction of UO_2 powder.

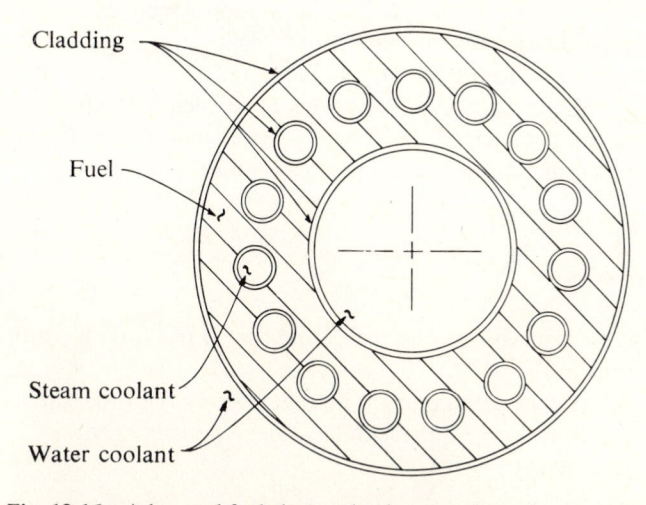

Fig. 12.16 Advanced fuel element having superheated steam passages embedded in the annular fuel element. [From *Nuclear Reactor Technology* **6**, No. 2 (1963), p. 77.]

Liquid Metals

In the cooling of fast reactors, light water, heavy water, and organic liquids are unsatisfactory because of their moderating characteristics. Gases with their low atom density are possible coolants and studies are being made of helium and steam-cooled fast reactors. However, their thermal conductivity is poor. Liquid metals, on the other hand, have high thermal conductivities. Although their high melting temperatures present some problems, they enable the use of low pressures for liquid metal systems. Their high atomic weights give liquid metals poor moderating characteristics which, when coupled with the previous characteristics, make them desirable for use in fast reactors. Appendix G indicates some of the properties of liquid sodium, NaK (a eutectic mixture of sodium and potassium), lithium, and water. The liquid metals all have much better conductivities than water because of their free electrons, which transport the majority of their thermal energy in conduction. Sodium has by far the best conductivity and is the principal coolant for fast reactors.

This is not to say that sodium does not have some problems. It becomes radioactive due to neutron capture with ^{24}Na which is activated, being a β, γ emitter with a 15-h half-life. Also sodium reacts with water violently. To prevent radioactive sodium contacting water in the event of a rupture in a steam generator, a secondary sodium loop is used, where the radioactive Na transfers heat to the secondary sodium, which in turn transports the energy to the steam generator. In addition, sodium reacts readily with oxygen. Oxygen contents below 30 ppm should be maintained and free surfaces should be covered with an inert gas such as helium or argon. Deposits of NaO_2 on heat transfer surfaces add a significant resistance to heat transfer and in extreme cases they may plug narrow heat transfer passages.

A number of heat transfer correlations have been made for liquid metals. Most involve the product of the Reynolds (Re) and Prandtl (Pr) numbers, which is in turn dimensionless and known as the *Peclet number*.

$$\text{Pe} = \text{Re} \cdot \text{Pr} = \frac{DV\rho}{\mu} \frac{C_p \mu}{k} = \frac{DV\rho C_p}{k} \qquad (12.39)$$

Note that the viscosity drops out, indicating that it has little bearing on liquid metal heat transfer.

For circular tubes with constant heat flux along the tube, the Lyon–Martinelli correlation states the dimensionless *Nusselt number* (Nu) is

$$\text{Nu} = \frac{hD}{k} = 7 + 0.025\text{Pe}^{0.8} \qquad (12.40)$$

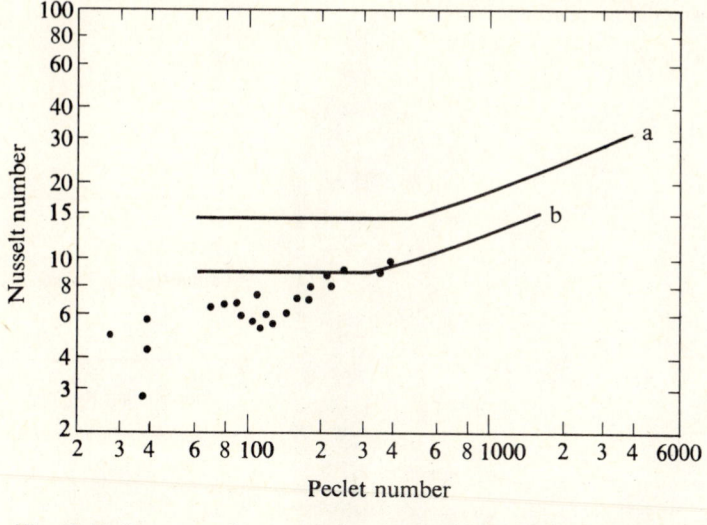

Fig. 12.17 Heat transfer correlation for liquid metal flow parallel to rod bundles. For curve a, $S/D = 1.5$ and for curve b, $S/D = 1.2$. [From M. M. El-Wakil, *Nuclear Heat Transport*. Scranton, Pa.: *International Textbook Co.*, 1971, Fig. 10.4, p. 269.]

If there is a uniform wall temperature, Seban and Shimazaki recommend

$$Nu = 5.0 + 0.025Pe^{0.8} \qquad (12.41)$$

For flow parallel to rod bundles a graphical correlation is given in Fig. 12.17 for fully developed turbulent flow and uniform heat flux. Curve a is for $S/D = 1.5$ and curve b is for $S/D = 1.2$. S is the distance between tubes and D is the tube O.D. The poor correlation of the experimental points at low Peclet numbers may be due to oxide contamination.

PWR Superheat

The development of a once-through steam generator (OTSG) for use with PWRs permits approximately 15.6°C superheat in the steam supplied to the high-pressure turbine (Reference 21). This means less moisture must be separated after the expansion through the high-pressure turbine with an accompanying improvement in cycle thermal efficiency and heat rate.

Figure 12.18 shows the arrangement of the heat transfer surfaces in the Babcock and Wilcox OTSG-type steam generator (similar to those installed at TMI). The primary reactor coolant enters at the top at approximately 15.2 MPa and 317°C, and passes downward through 16 000 steam generator tubes (1.588 cm O.D. × 0.86 mm wall of alloy 600 (Ni–Cr–Fe) with an exposed

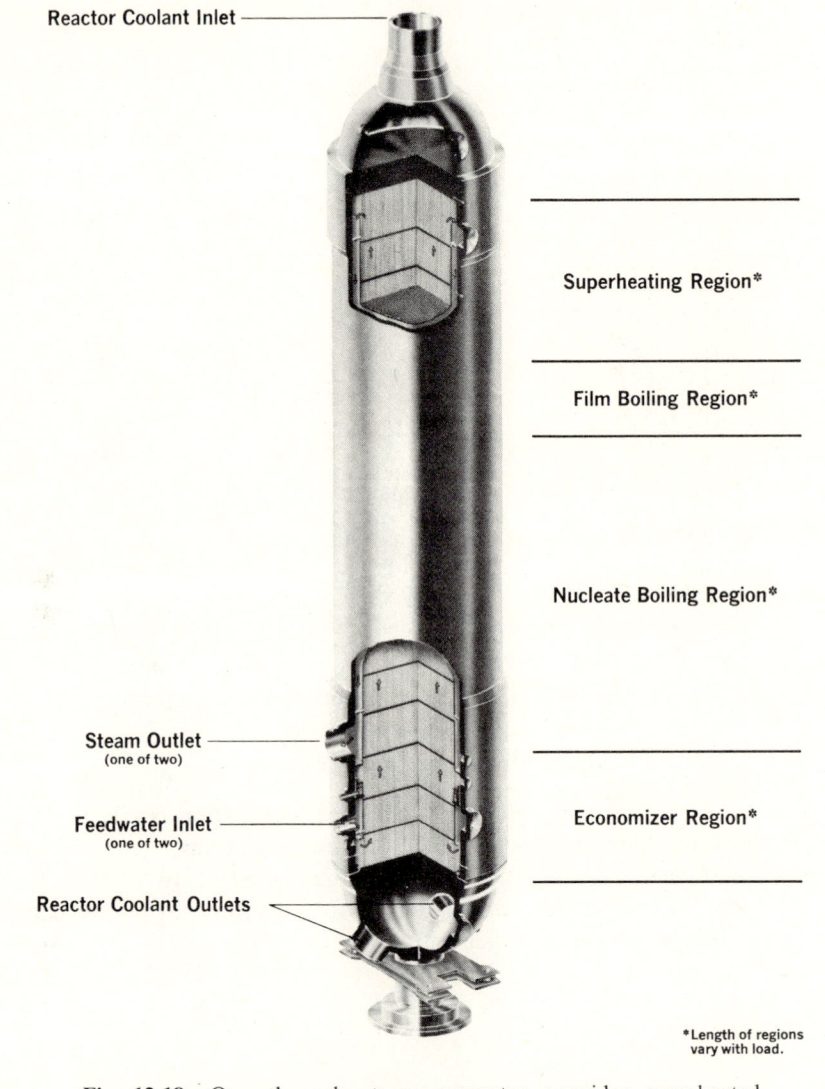

Reactor Coolant Inlet

Superheating Region*

Film Boiling Region*

Nucleate Boiling Region*

Steam Outlet
(one of two)

Feedwater Inlet
(one of two)

Economizer Region*

Reactor Coolant Outlets

*Length of regions
vary with load.

Fig. 12.18 Once-through steam generator provides superheated steam to the high-pressure turbine in a PWR system. [Courtesy of Babcock and Wilcox Co.]

length of about 15.8 m). The feed-water enters the shell side at the bottom and passes upward through the unique broached-plate tube supports of the type shown in Fig. 12.19. These guide the subcooled flow entering at about 238°C upward through an economizer region, where the water rises to its boiling temperature (280°C), a nucleate boiling region where the steam acquires a 100 percent quality; and finally, the flow reaches a superheat region from which

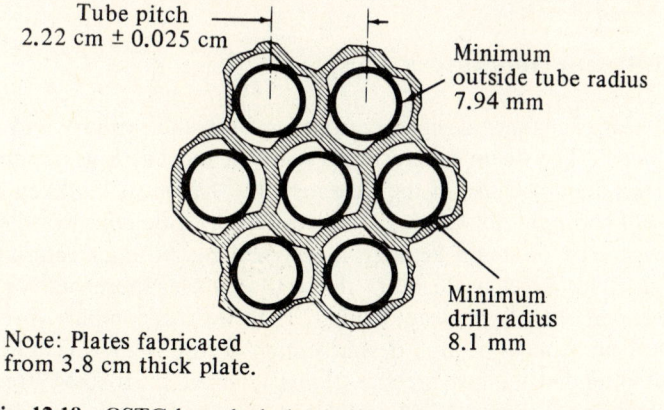

Tube pitch
2.22 cm ± 0.025 cm

Minimum
outside tube radius
7.94 mm

Minimum
drill radius
8.1 mm

Note: Plates fabricated
from 3.8 cm thick plate.

Fig. 12.19 OSTG broached-plate tube supports. [Courtesy of Babcock and Wilcox Co.]

the steam leaves at about 313°C and 6.4 MPa. Under conditions of changing load, the OTSG changes the proportion of tube length devoted to boiling and superheating as the amount of feedwater flow is controlled to maintain a fixed turbine throttle pressure. For a plant with a designed superheat of 28°C at 100 percent load, the superheat increases to 33°C when load is reduced to 60 percent of the rated value.

The first plant utilizing this type of OTSG to go on line was Duke Power Company's Oconee Station, which went into operation in 1974. Table 12.2 shows how the design and measured performances compared for Oconee Unit 1.

Table 12.2 Oconee Unit 1 Performance—Design Versus Measured Performance at 100 Percent Power Level

	Design	Measured
Primary side		
Inlet temperature (°C)	318.8	316.8
Outlet temperature (°C)	290	308.7
Secondary side		
Steam flow (10^6 kg/h)	2.58	2.42
Feedwater temperature (°C)	242	237.8
Steam pressure (MPa)	6.38	6.39
Steam temperature (°C)	299*	313.3
Steam superheat (°C)	19.4*	33.3

* Design minimum.
From W. O. Parker, Jr., et al., "Post Operational Test and Examination—Oconee Unit 1," American Power Conference. April 21-23, 1975.

LMFBR Steam Generators

LMFBR heat exchanger design is critical in both the primary and secondary sodium loops. The steam generator presents the most serious design challenge for heat exchange because of the extreme care that must be taken due to the reactivity of sodium and water. The first design to be considered will be the *hockey-stick* type of steam generator designed by Atomics International for the Clinch River Breeder Reactor. Here the twin evaporators are separated from the superheater by a steam drum. Then we will compare three different units based on a once-through design similar to the B&W OTSG for PWRs. These are intended for fast breeder plants to produce 1000 MW(e) or more. The B&W design for an LMFBR and the one for the French Super Phénix both use helically coiled tubes to reduce the number of tube-sheet weld connections. The Westinghouse design uses a large number of straight tubes and incorporates third-fluid leak detection capability.

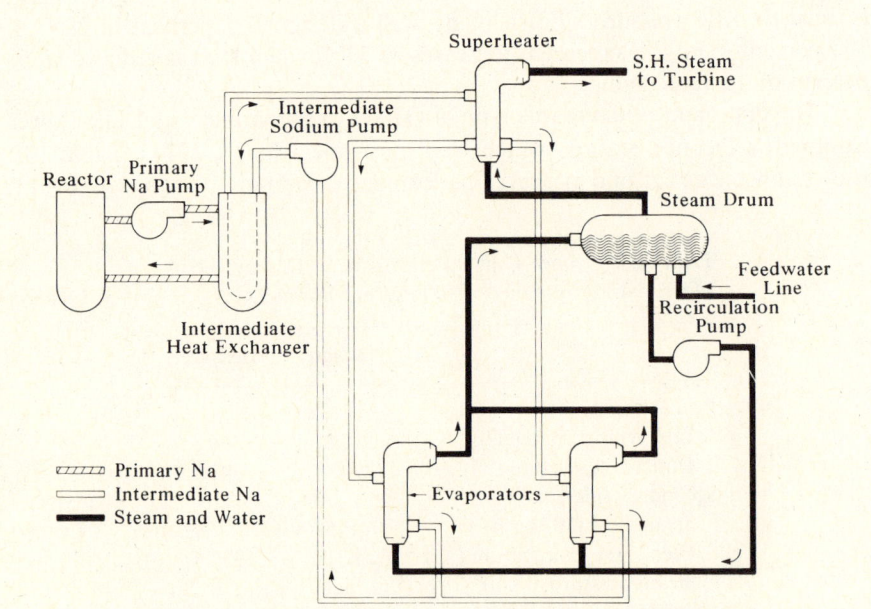

Fig. 12.20 Schematic diagram for the heat transport system on the Clinch River Breeder Reactor. Notice that this is but one of three identical loops operating from the single core. [Courtesy of Project Management Corporation and Westinghouse Advanced Reactors Division for the Clinch River Breeder Reactor Project.]

Clinch River Breeder Reactor steam generator

The reactor heat transport system for the Clinch River Breeder Reactor will be described with particular emphasis on the steam generator. The reactor core generates 975 MW(th), and each of three identical transport circuits handles 325 MW(th). In each circuit the radioactive sodium in the primary loop is coupled to the steam generator indirectly by an intermediate sodium loop, which is nonradioactive. Table 12.3 shows some of the design values of the heat transport system for the CRBR.

In each loop dry steam is generated in a pair of evaporators operated in parallel. Since the recirculation rate is 2:1, a steam drum must separate the water and steam before the steam passes through a single superheater, as shown in Fig. 12.20. The design chosen for the evaporators and superheater is called the *hockey-stick* design. It is essentially a simple counterflow arrangement with the sodium flowing downward on the shell side of the tubes and the water and/or steam flowing upward on the inside of the tubes. It is interesting to note that this design was developed allowing for departure from nucleate boiling (DNB)

Table 12.3 Clinch River Breeder Reactor Heat Transport System —Design Values

Primary loop	
Thermal power [MW(th)]	975
Reactor outlet temperature (°C)	535
Reactor inlet temperature (°C)	388
Primary sodium flow rate (total kg/h)	18.80×10^6
Primary pump flow rate at 535°C (m³/s)	2.13
Primary pump head at rated flow (m Na)	123
Intermediate loop	
Hot-leg temperature (°C)	502
Superheater–evaporator crossover temp. (°C)	458
Cold-leg temperature (°C)	344
System flow rate (total kg/h)	17.39×10^6
Pump flow rate at 344°C (m³/s)	1.86
Intermediate pump head at rated flow (m Na)	100.6
Steam/water loop	
Superheater outlet temperature (°C)	485
Superheater outlet pressure (MPa)	10.5
Feedwater temperature (°C)	232
Recirculation ratio	2:1
Steam flow rate (total kg/h)	1.515×10^6

Courtesy of Breeder Reactor Corp.

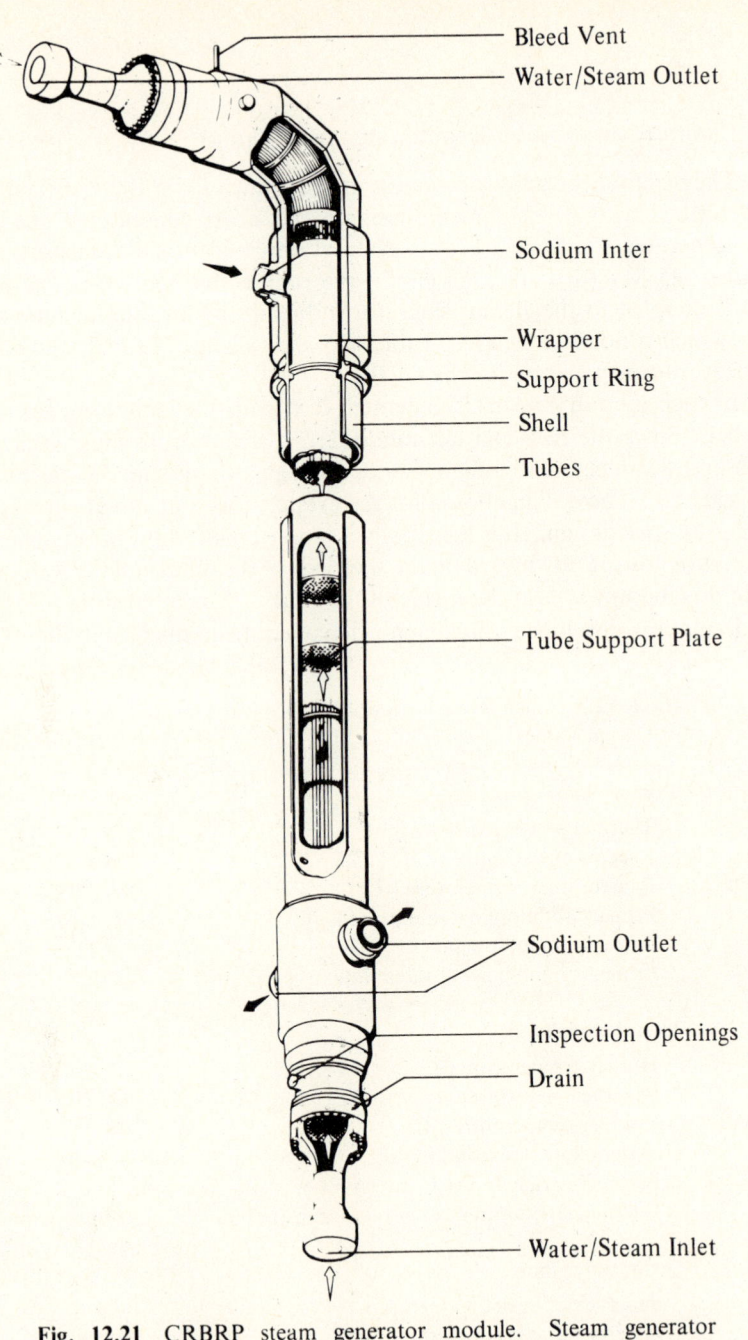

Bleed Vent

Water/Steam Outlet

Sodium Inter

Wrapper

Support Ring

Shell

Tubes

Tube Support Plate

Sodium Outlet

Inspection Openings

Drain

Water/Steam Inlet

Fig. 12.21 CRBRP steam generator module. Steam generator modules are identical for evaporators and superheaters, except that orifice inserts are added to the evaporators for hydraulic stability. [Courtesy of Project Management Corporation and Westinghouse Advanced Reactors Division for the Clinch River Breeder Reactor Project.]

Table 12.4 CRBR Steam Generator Design Data

	Evaporator	Superheater
Overall height (m)	19.8	19.8
Shell O.D. (m)	1.32	1.32
Number of heat transfer tubes	757	757
Heat transfer tubes		
O.D. (cm)	1.588	1.588
Thickness (cm)	0.277	0.277
Triangular pitch (cm)	3.10	3.10
Shell side sodium velocity (m/s)	1.83	3.66
Water/steam side exit velocity (m/s)	16.8	66.4
Steam quality at evaporator outlet (%)	50	—

Courtesy of Clinch River Breeder Reactor Project.

to occur in the evaporator modules. It is designed so that the heat transfer tube wall temperature variations and tube scale exfoliation or accelerated tube corrosion at DNB would not be excessive. The only difference in the superheaters and the evaporators is that inlet water orifice inserts are used in the evaporators to assure hydraulic stability. Figure 12.21 shows a cutaway for a steam generator module. The design characteristics for the hockey-stick evaporator/superheater units are given in Table 12.4.

B&W once-through, helical coil LMFBR steam generator

Babcock & Wilcox has designed a once-through, counterflow, helical coil steam generator to be part of a proposed 1000-MW(e) LMFBR plant, as shown in Fig. 12.22. In this steam generator the sodium enters through a single nozzle in the upper head and flows downward through a cylindrical flow distributor to be dispersed evenly before continuing down through the nested tube coils. The water enters through four nozzles in the lower head and flows upward within the coiled tubes. The lowest coils function as an economizer, heating the water to its saturation point. The midsection functions as an evaporator wherein the wet steam has its quality go from 0 to 100 percent. The upper coils add superheat to the steam, which leaves through four nozzles in the upper head.

Table 12.5 shows a comparison of this B&W unit with the French Super Phénix and a Westinghouse design, both of which will be described shortly. By use of two generators in each of the three loops it produces 437.5 MW(th) per generator, compared to 203 from its U.S. competitor. This results in a shell I.D. of 3.61 m, which is the largest shell permitted by rail clearances, thus minimizing the number of units required.

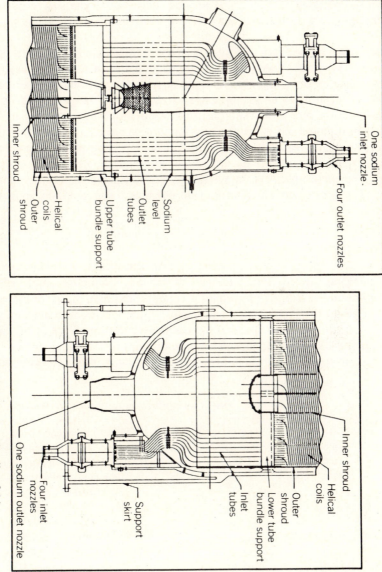

Fig. 12.22 B&W 1000-MW(e) LMFBR steam generator design showing details of (a) the upper head arrangement and (b) the lower head arrangement. [From G. Grant, ASME Paper No. 80-C2/NE-29, August 1980.]

Table 12.5 Comparison of Once-Through LMFBR Steam Generators

	Super Phénix*	Westinghouse†	B&W‡
Plant output [MW(e)]	1200	1000	1000
Number of loops and units/loop	4 and 1	3 and 3	3 and 2
Thermal output/generator	750	293	437.5
Na inlet temperature (°C)	525	482	482
Na outlet temperature (°C)	345	318	321
Na flow/generator (kg/s)	3273	1399	2125
Steam outlet temperature (°C)	490	457	457
Steam outlet pressure (MPa)	18.4	15.69	15.67
Tube outside diam. (cm)	2.5	2.1	3.19
Tube inside diam. (cm)	1.98	1.1	2.32
Number of tubes	—	3366	240
Tube length (m)	—	23.4 (total) 20.6 (active)	110.3 (av.)
Dry mass (Mg)	180	—	347

* From P. Baqué et al., ASME Paper No. 80-C2/NE-28, August 1980.
† From C. R. Adkins and D. J. Bongaards, ASME Paper No. 80-C2/NE-27, August 1980.
‡ From G. Grant, ASME Paper No. 80-C2/NE-29, August 1980.

The construction material, except for bolts and feedwater orifices is $2\frac{1}{4}$ Cr–1 Mo steel, which has good properties up to 590°C, good fabricability, and good corrosion resistance in a Na–H_2O environment. The heavy wall thickness provides an extra factor of safety for leak prevention. If a leak should occur, the material is resistant to impingement damage and resists the attack of the resultant caustic solution, which may induce stress corrosion cracking. Electroslag remelt is used to produce the tubes and tube sheets, enhancing the integrity of the Na–H_2O boundary. The high tube side mass velocity assures a high heat transfer coefficient, which will not change markedly during the transition from nucleate to film boiling. (Note the small drop in the heat flux at the DNB point for the Super Phénix, as shown in Fig. 12.24.) At full load the steam reaches 100 percent quality at approximately two-thirds of the bundle height and exits superheated. As the point moves up and down with varying loads, the temperature fluctuations and resulting thermal stresses are small.

The long average tube length and large tube diameter permit a relatively small number of tubes. These can be welded to four upper and four lower tube sheets, which have a diameter of 0.572 m and a thickness of 14 cm. These are located outside the shell between the shell and the nozzles.

The Na has a velocity of 0.85 m/s through the coiled tubes. This velocity is chosen to minimize flow-induced vibrations in the coiled tubes. Shrouds at

the inner and outer diameters of the coil bundle prevent the sodium from by-passing the heat transfer surfaces.

Creys–Malville Super Phénix LMFBR steam generator

The Super Phénix generators (four per reactor) are also of the helical coil type, rather similar to the B&W design. Figure 12.23 shows the overall configuration of the French unit.

In Table 12.5 it may be noted that this steam generator uses an Na inlet temperature 43°C higher and steam temperature and pressure that are 33°C and 2.7 MPa higher, respectively. The French unit also uses more smaller

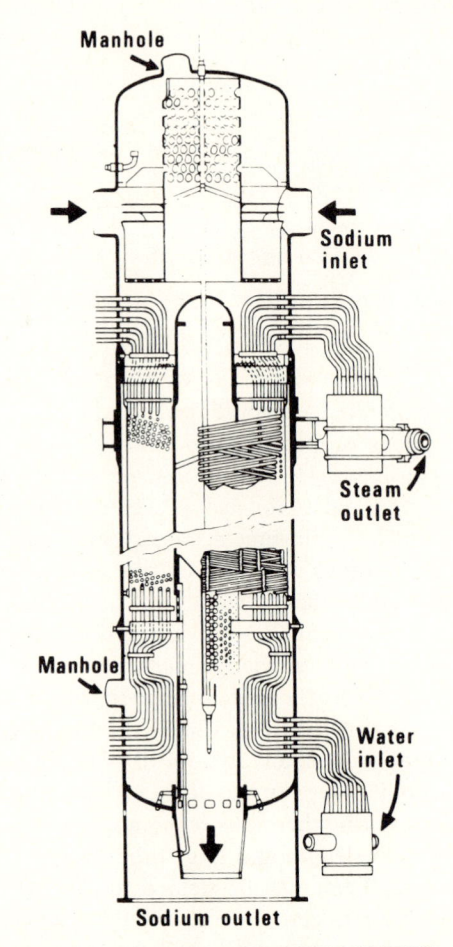

Fig. 12.23   French Super Phénix 1200-MW(e) LMFBR steam generator. [From M. Banal et al., *International Nuclear Engineering* **23**, No. 232 (June 1978), p. 48.]

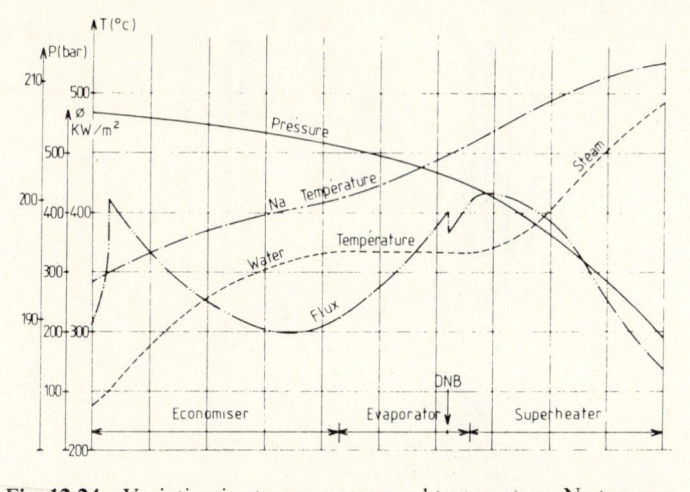

Fig. 12.24 Variation in steam pressure and temperature, Na temperature, and heat flux passing upward through the economizer, evaporator, and superheater sections of the Super Phénix steam generator. [From P. Baqué et al., ASME Paper No. 80-C2/NE-28, August 1980.]

tubes of Incolloy 800. At the design temperature this alloy has an ultimate tensile strength of 100 MPa, compared to about 50 MPa for $2\frac{1}{4}$ Cr–1 Mo steel.

Figure 12.24 is of particular interest, as it shows the variation in H_2O temperature and pressure, Na temperature, and heat flux, as one passes up through the economizer, evaporator, and superheater sections of the steam generator. Note the small, sharp drop in the heat flux at the DNB point, where there is no longer a liquid film on the wall. The heat flux picks up again, as the last few percent of moisture is dried out, due to the increasing temperature difference. However, once the steam starts to superheat, its temperature rises faster than that of the Na, the ΔT decreases, and the heat flux falls.

Westinghouse double-walled, straight-tube LMFBR steam generator

The Westinghouse LMFBR steam generator design is of the once-through straight-tube type, with built-in third-fluid leak detection capability. This design should make Na–H_2O interactions virtually nonexistent. Figure 12.25 shows the prestressed, double-walled tubes, which have four longitudinal grooves running the full tube length and open to the space between the upper steam and sodium tube sheets. The interface grooves are 0.5 mm deep by 1.3 mm wide and are cut on the inner surface of the outer tube. The third fluid is helium. The H_2O pressure is greater than the helium pressure, which in turn is higher than the Na pressure. Thus, steam leaks into the He may be detected by hygrometers or by an increase in pressure. If there is a leak on the sodium

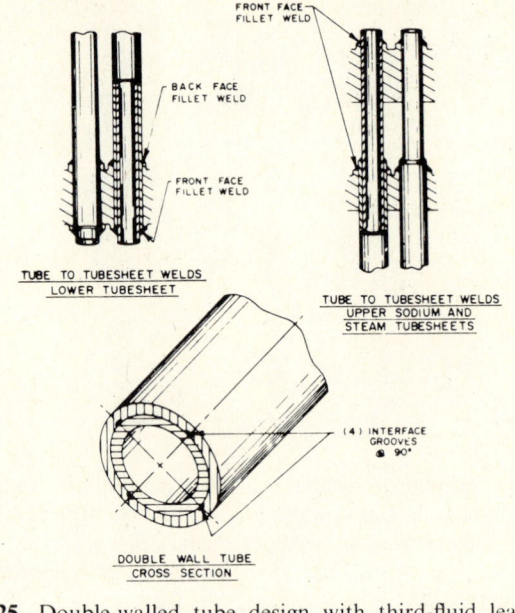

Fig. 12.25 Double-walled tube design with third-fluid lead detection capability for the proposed 1000-MW(e) Westinghouse LMFBR steam generators. [From C. R. Adkins and D. Bongaards, ASME Paper No. 80-C2/NE-27, August 1980.]

side, He will show up in the sodium leaving the core. The gas is detected by sparging and a gas chromatography. In sparging, an inert gas, say argon, is bubbled up through the sodium and will pick up any leaking He.

With this design the reliability of the plant should be excellent. By precluding $Na–H_2O$ reactions decay heat can still be removed through the steam generator. Further, it removes the probability of pressure loadings on the intermediate heat exchanger (IHX) as a potential failure mode. As steam and sodium leaks would be separately detected, there should be less downtime, with no necessity for the cleanup of reaction products, thus contributing to high plant availability. High availability is important in reducing fuel cycle costs and for the production of bred fuel.

To reduce problems with differential expansion of the straight tubes in the event of tube blockage, a large number of closely spaced tube-support plates (39) are used. These not only preclude tube buckling, but also have alternating flow-hole patterns to ensure cross-flow and mixing. To provide for bundle-to-shell differential expansion, a convoluted shell expansion joint is provided in the heat exchanger shell.

Problems

1. If the fuel element used in Example 2 operates over a period of 18 reactor startups, estimate the maximum fuel temperature in the element. Use the effective UO_2 thermal conductivity in Appendix G for 650°C. Will the UO_2 at the center be molten?

2. A 3 percent enriched metalic uranium fuel rod of 1.9 cm diameter is clad with 0.635 mm of Zircaloy 2. The outer clad temperature is not to exceed 315°C. What is the highest thermal flux that will not allow any transition to β-phase uranium at the centerline of the fuel rod?

3. A fuel plate is fabricated with 1.5 mm of 10 weight percent U–90 weight percent Al alloy as the fuel. It is clad with 0.3 mm of aluminum on each side. Assume that the thermal conductivity of the fuel material is 175 W/m °C and that its density is 3.0×10^{-3} kg/m^3. The thermal conductivity of the clad is 225 W/m °C. The film coefficient on the outside of the plate is 4250 W/m^2 °C and the coolant temperature is 175°C. The fuel is 5 percent enriched and the peak thermal flux is 4×10^{14} neutrons/cm^2 s. Determine:
   (a) The peak heat flux at the plate surface
   (b) The temperatures at the point of peak flux on the outer clad surface and at the centerline of the fuel.
   (c) The kW produced per plate. The plates are 7.5 cm wide and 1.2 m long with a truncated-cosine variation in neutron flux along the rod. The peak neutron flux is four times the value at the end of the plate.

4. In the pebble-bed reactor concept, spheres of uranium carbide (UC) are coated with pyrolytic graphite. For spherical pellets, develop the expression shown below for the temperature in the pellet, T, in terms of the temperature at the fuel–coating interface, T_s, the conductivity of the uranium carbide, k_f, and the rate of internal heat generation, S. The radius at any point is r and the outer radius is r_s.

$$T = T_s + \frac{S}{6k_f}(r_s^2 - r^2)$$

5. A fuel pellet of the type described in Prob. 12.4 has a diameter of 2.25 cm for the UC and a pyrolytic graphite coating 3 mm thick. The material properties are as follows:

Uranium carbide: density = 13.0×10^{-3} kg/m^3
   avg. thermal conductivity = 21.0 W/m °C
   uranium enrichment = 5 percent

Pyrolytic graphite: density = 1.75×10^{-3} kg/m^3
   avg. thermal conductivity = 86.5 W/m °C

The temperature of the outer surface of the graphite is 535°C and the thermal flux is 5×10^{13} neutrons/cm^2 s. Assume an average UC temperature of 850°C. Determine:
   (a) The rate of internal heat generation, $S(W/m^3)$
   (b) The temperature drop across the graphite
   (c) The maximum temperature at the center of the sphere

6. Fuel elements for the German Pebble-Bed Reactor (AVR) are fueled graphite injection-molded inside a 6-cm-diameter sphere with a 1-cm wall thickness. If the fueled portion has an internal heat generation rate of 93.2 W/cm^3 and the surface temperature of the outer shell is 950°C, determine the peak temperature at the center of the fuel and the heat flux at the outer surface of the sphere.

   Assume a mean thermal conductivity for the fueled graphite of 0.145 W/cm °C and a value for the graphite shell of 2.0 W/cm °C.

7. The fuel pins for EBR-2 are 0.366 cm in diameter by 36.2 cm long. The fuel is ^{235}U–^{238}U–fissium alloy. The clad is a 304 stainless steel tube with 0.0442 cm O.D. and 0.23 mm wall thickness. The intervening annulus of 0.15 mm is sodium filled to give a good clad–core bond for heat transfer.

   The fuel is discharged after 2 percent burnup. After 15 days the β-γ activity is releasing 3.8×10^6 W/m^3. Reprocessing by melt refining reduces the activity to one-third of the value after the cooling period. Assume that 75 percent of the energy is dissipated in the fuel and the balance escapes to the surroundings.

   If after refabrication of newly clad reprocessed elements they lie horizontally in still air at 27°C, compute the maximum temperature in the center of the pin.

   The pins are subsequently placed in a subassembly containing 91 elements in a hexagonal configuration that is about 4.7 cm across flats. Would you suggest force cooling? Explain.

8. In a fuel assembly for a proposed 1150-MW(e) PWR, water enters a fuel channel at 285°C. The UO$_2$ fuel pellets have a diameter of 0.90 cm and are clad with 304 stainless steel, which has a thickness of 0.25 mm. The fuel rod length is 4 m. The effective flow diameter for each rod is 1.43 cm. The average water velocity is 3.5 m/s. Assume a truncated-cosine-shaped energy release rate. The peak value is 10^8 W/m^3, which is 2.5 times the value at each end of the rod. Take the outside film coefficient as 27 500 W/m^2 °C and the mean specific heat of the pressurized water as 4800 J/kg °C.

   It is desired to determine the temperature variation along the rod at the following points:

   (a) In the fluid
   (b) At the outer clad surface
   (c) At the clad–fuel interface
   (d) At the centerline of the fuel

   Write a computer program to give the desired temperatures at positions 0.25 m apart along the length of the rod. Plot the results.

9. Water flows at 5×10^6 kg/h m^2 through an annular passageway whose inner surface is uniformly heated. For the annulus the I.D. is 2 cm and the O.D. is 3 cm with a length of 4 m. Subcooled water at 250°C enters the passageway and leaves at 13.8 MPa with a quality of 10 percent. Determine:

   (a) The steam energy flow
   (b) The heat flux
   (c) The heat flux burnout ratio

10. A BWR pressure tube is to have a mass flow rate of 10^7 kg/m^2 h. Subcooled water at 200°C enters the tube and the pressure of steam leaving the channel is 6.9 MPa. The channel contains a 19-tube cluster of 1.588-cm-O.D. fuel rods within its 10-cm I.D.

The length of the passageway is 3.66 m. Design is based on a heat flux burnout ratio (F_{s_1}) of 1.30. Using Fig. 12.8, estimate the SEF and the quality of the channel exit. Also determine the power burnout ratio (F_{s_2}) at that point.

11. Determine the outlet steam and sodium temperatures for a superheater using the hockey-stick design. The design data are given below.

Inlet steam (flows upward through tubes)	
Flow rate (kg/h)	5.05×10^5
Quality (%)	100
Pressure (MPa)	11.0
Exit steam	
Pressure (MPa)	10.5
Temperature (°C)	To be determined
Sodium (flows downward on tube side)	
Flow rate (kg/h)	5.8×10^6
Inlet temperature (°C)	502
Exit temperature (°C)	To be determined
Velocity (m/s)	3.66
Superheater design parameters	
Shell I.D. (m)	1.32
Number of tubes	757
Tube material	304 SS
Tube O.D. (cm)	1.588
Tube wall thickness (cm)	0.277

References

1. Daniel, R. C., and I. Cohen, *In-pile Effective Thermal Conductivity of Oxide Fuel Elements to High Fuel Depletions*, WAPD-246, U.S. Atomic Energy Commission, April 1964.

2. Glasstone, S., and A. Sesonske, *Nuclear Reactor Engineering*. New York: D. Van Nostrand Company, 1963.

3. Sutherland, W. A., *Heat Transfer to Superheated Steam*, GEAP-4528, U.S. Atomic Energy Commission, May 1963.

4. El-Wakil, M. M., *Nuclear Power Engineering*. New York: McGraw-Hill Book Company, 1962.

5. Rohsenow, W. M., and H. Y. Choi, *Heat, Mass, and Momentum Transfer*. Englewood Cliffs, N.J.: Prentice-Hall, Inc., 1961.

6. Hausner, H. H., and J. F. Schumar, *Nuclear Fuel Elements*. New York: Reinhold Publishing Corporation, 1959.

7. Kreith, F., *Principles of Heat Transfer*. Scranton, Pa.: International Textbook Co., 1965.

8. Lustman, B., and F. Kerze, *Metallurgy of Zirconium*. New York: McGraw-Hill Book Company, 1955.

9. Holden, A. N., *Physical Metallurgy of Uranium*. Reading, Mass.: Addison-Wesley Publishing Co., Inc., 1958.

10. McAdams, W. H., *Heat Transmission*. New York: McGraw-Hill Book Company, 1954.
11. Wiley, John S., "Burnout Limits for Boiling Water Reactors," *Power Reactor Technology* 7, No. 1 (Winter 1963–1964), pp. 14–26.
12. Levedahl, W. J., "Application of Steam Energy Flow to Reactor Design," *Transactions of the American Nuclear Society* 5, No. 1 (June 1962), pp. 149–150.
13. Janssen, E., and J. A. Keriven, *Burnout Conditions for Nonuniformly Heated Rod in Annular Geometry, Water at 1000 psia*, GEAP-3755, U.S. Atomic Energy Commission, June 1963.
14. DeBertoli, R. A., et al., *Forced Convection Heat Transfer Burnout Studies for Water in Rectangular Channels and Round Tubes at Pressures 500 psia*, WAPD-188, U.S. Atomic Energy Commission, October 1958.
15. Walker, P. L., Jr., "Carbon—An Old But New Material," *American Scientist* 50, No. 2 June 1962), pp. 259–293.
16. *Nuclear Reactor Technology*, 3, No. 4 (1960); pp. 68–74; 4, No. 3 (1961), pp. 71–85; 6, No. 2 (1963), pp. 75–80; 1, No. 3 (1964), pp. 276–312.
17. Burkett, M. N., and W. P. Eatherly, "Fueled-Graphite Elements for the German Pebble-Bed Reactor (AVR)," in *High Temperature Nuclear Fuels* (A. N. Holden, ed.). New York: Gordon and Breach, Science Publishers, Inc., 1966.
18. El-Wakil, M. M., *Nuclear Heat Transport*. Scranton, Pa.: International Textbook Co., 1971.
19. Muenchow, H. O., and R. C. Armstrong III, *A 1000-MWe LMFBR Steam Generator*, ASME Paper No. 71-NE-13, 1971.
20. *Technical Progress Report 1974*, Clinch River Reactor Project, Knoxville, Tenn., 1974.
21. Parker, W. O., Jr., J. C. Deddens, D. W. Berger, and K. L. Hladek, "Postoperational Test and Examination—Oconee Unit 1," American Power Conference, Chicago, April 21–23, 1975.
22. Marinelli, V., "Critical Heat Flux: A Review of Recent Publications," *Nuclear Technology* 34, No. 2 (July 1977), pp. 135–171.
23. Adkins, C. R., and D. J. Bongaards, *Straight Double Wall Tube Steam Generator with Third Fluid Leak Detection Capability of LMFBR Application*, ASME Paper No. 80-C2/NE-27, August 1980.
24. Baqué, P., *The Creys Malville FBR Superphenix Steam Generators*. ASME Paper No. 80-C2/NE-28, August 1980.
25. Grant, G., *Design and Development of a Once-Through Helical Coil Steam Generator for Large Steam Plants*, ASME Paper No. 80-C2/NE-29, August 1980.

13

Nuclear Reactors

The previous chapters have dealt with various phases of reactor design such as determination of critical size, reactor kinetics, heat transfer, and materials problems. This chapter surveys some of the current and projected types of reactors. Although the list is far from exhaustive, it should give the reader a reasonable view of the spectrum of reactor types, particularly those useful for power generation in the next two decades.

By 1966 installed domestic nuclear electric generating capacity amounted to 1 percent of the U.S. total and that year for the first time commitments to new nuclear generating capacity exceeded commitments for new fossil-fired capacity. By 1970 the installed capacity amounted to 7000 MW(e) (2 percent of the total). By the end of the 1970s installed nuclear capacity had risen to in excess of 50 000 MW(e) (about 10 percent of the total), but predictions for future growth had been severely curtailed. Figure 1.2 shows that a 1974 prediction of somewhat less than 500 000 MW(e) had been cut back to less than 200 000 MW(e) by 1979. Several factors contributed to this reduction: escalating costs due to increased safety and environmental concerns resulting from Three Mile Island, increased emphasis on conservation and load management, plus rampant inflation. Still, during the 1980s 45 U.S. plants are scheduled for completion. They have a capacity in excess of 48 000 MW(e) and at an estimated average cost of \$1000/kW(e), this will amount to an investment of \$48 $\times$ 10^9.

Fuel Burnup

Before looking at nuclear economics and power costs, fuel burnup will be considered. Fuel burnup may be expressed as:

(1) Megawatt days per metric ton of heavy metal, MWd/Mg
(2) Percent atom burnup, or
(3) Fissions per cm^3

Fuel is considered as atoms of the fertile or fissile isotopes of thorium, uranium, or plutonium. Fuel material includes fuel plus any alloying elements. In (Th, UO_2) the thorium and uranium atoms are fuel, but the oxygen in this mixed oxide may be counted only as fuel material.

If all the atoms in a gram of ^{235}U could be fissioned, there would be an energy release of 0.949 MW-days (MWd) of energy. This energy release is computed as

$$\frac{6.023 \times 10^{23} \ (\text{atoms } ^{235}U/\text{g atom})}{235 (\text{g } ^{235}U/\text{g atom})} \times 200 (\text{MeV}/^{235}U \text{ atom})$$

$$\times \ 4.44 \times 10^{-23} \ (\text{MW h/MeV}) \times (\tfrac{1}{24})(\text{days/h})$$

$$= 0.949 (\text{MWd/g } ^{235}U)$$

It should be remembered here that every atom of ^{235}U that absorbs a neutron does not produce fission and, hence, in considering the actual consumption of the fissile atoms, some are destroyed without fission due to (η, γ) capture. Although 0.949 MWd is produced per gram of ^{235}U fissioned, only 0.839 g is fissioned per gram of ^{235}U destroyed. Thus, the energy released per gram of uranium destroyed is 0.796 MWd/g.

If all of the atoms in a ton of fuel were fissioned, nearly 10^6 MWd of energy would be produced. However, because of the buildup of fission products that reduce reactivity, the release of fission gases that contribute pressure on the cladding, a tendency of the fuel to swell, and radiation damage to the cladding, the burnup is limited to a fraction of the total number of heavy atoms. The higher the percent burnup is, the higher fuel enrichment that must be used. For the Canadian heavy water-moderated and cooled reactors, which use natural uranium, a burnup of only 7500 MWd/ton is typical; this points up the importance of on-line refueling for this reactor concept. The ceramic oxide fuels used with light-water reactors may have enrichments of 2 to 4 percent with average burnup as high as 30 000 MWd/ton. The fueled graphite for the HTGR is designed for burnups of 70 000 to 100 000 MWd/ton. The Clinch River Breeder Reactor Project, which is the 350-MW(e) demonstration LMFBR, is designed for an enrichment of 33 percent resulting in a peak fuel burnup of 110 000 MWd/ton.

Example 1 Fuel near the center of the CANDU–BLW reactor undergoes a burnup of 8500 MWd/ton. The thermal flux is 3×10^{14} neutrons/cm^2 s and the mean oxide temperature is 1000°C.

(a) Determine the atom percent burnup.

$$\frac{8500 \dfrac{MWd}{ton} \times \dfrac{6.023 \times 10^{23}}{235} \dfrac{atoms\ ^{235}U}{g\ ^{235}U}}{0.949 \dfrac{MWd}{g\ ^{235}U\ burned} \times 10^6 \dfrac{g\ fuel}{ton} \times \dfrac{6.023 \times 10^{23}}{238} \dfrac{atoms\ fuel}{g\ fuel}}$$

$$= 0.00907 \frac{atoms\ ^{235}U}{atom\ fuel}$$

atom percent burnup = 0.907

(b) How long will it take to achieve this burnup?

$$N_{235} = 10.5 \frac{g\ UO_2}{cm^3} \times \frac{6.023 \times 10^{23} \dfrac{molecules\ UO_2}{g\ mole\ UO_2}}{270 \dfrac{g\ UO_2}{g\ mole\ UO_2}}$$

$$\times\ 0.00714 \frac{atoms\ ^{235}U}{molecule\ UO_2}$$

$$= 1.67 \times 10^{20} \frac{atoms\ ^{235}U}{cm^3}$$

$$\sigma_{fis} = 582 \frac{\sqrt{\pi}}{2} \sqrt{\frac{293}{1273}} = 250\ barns$$

$$\frac{1}{10.5} \frac{cm^3}{g\ UO_2} \times 10^6 \frac{g\ UO_2}{ton\ UO_2} \times 1.67 \times 10^{20} \frac{atoms\ ^{235}U}{cm^3} \times 250$$

$$\times\ 10^{-24} \frac{cm^2}{atom\ ^{235}U} \times (\phi t) \frac{neutrons}{cm^2} \times 200 \frac{MeV}{neutron} \times 4.44$$

$$\times\ 10^{-23} \frac{MWh}{MeV} \times \frac{1\ day}{24\ h} = 8500 \frac{MWd}{ton}$$

$$\phi t = 5.95 \times 10^{21} \frac{neutrons}{cm^2}\ (thermal\ fluence)$$

$$t = \frac{5.95 \times 10^{21} \dfrac{neutrons}{cm^2}}{3 \times 10^{14} \dfrac{neutrons}{cm^2\ s} \times 3600 \times 24 \dfrac{s}{day}} = 229.5\ days$$

A look at the answer for part (a) of Example 1 indicates that too simple an analysis has been used and that the burnup exceeds the ^{235}U content for natural uranium. However, the conversion of ^{238}U to ^{239}Pu and its subsequent fission account for a significant fraction of the 8500 MWd/ton.

Nuclear Power Costs

The capital cost of a nuclear power plant includes both direct and indirect costs. The annual cost is usually taken as a fixed percentage of the total capital investment. The direct component of the capital investment consists of the nuclear steam supply, turbines, pumps, electrical equipment, structures, and so on. The indirect costs include land, administration, architect–engineer fees, contingencies, and interest during construction.

Capital cost

Table 13.1 shows a breakdown of capital costs for a hypothetical 1200-MW(e) plant whose expected date of operation could be 1984. Note that, although the ranges of annual escalation rates (inflation) used have proved to be low, the enormous magnitude of the investment and its division between the various categories is shown effectively. The indirect costs account for approximately one-third of the total. Also, since the time span for such a project is about 10 years, interest during construction for the "expected" case amounts to 165.4×10^6. This is 20.8 percent of the total cost.

The annual fixed charge rate is taken as a percentage of the total capital investment. For private investment this has risen to a range of 16 to 22 percent, while for government ownership a range of 8 to 11 percent might be more appropriate. For a light-water reactor (LWR) plant, assuming an 11.47 percent cost of capital and a return on equity of 6.72 percent, a 16.5 percent annual fixed-charge rate (Reference 78) could be divided as follows:

Cost of capital	7.7%
Depreciation	4.3%
Federal income tax	3.4%
Insurance	0.4%
Revenue tax	0.7%
	16.5%

For public ownership there is a lower return on capital and no taxes, thus the lower fixed-charge rate.

Operating and maintenance cost

The operating and maintenance cost tends to be the smallest of the three categories. It is composed of costs of operation, maintenance, supervisory personnel, liability insurance, and the materials required for maintenance and repair. Current O & M cost runs on the order of $7 per installed kW per year. No great difference is expected between the various reactor types.

Table 13.1 Capital Cost Evaluation for a Hypothetical 1200-MW(e) PWR to Begin Commercial Operation in January 1984

System and Project Parameters

Construction schedule	Yr/Month	Elapsed Months	
Baseline date for evaluation	74–07	0	
Begin plant contruction	77–07	36	
Order turbine generator	76–05	22	
Ship turbine generator	81–10	87	
End plant construction	83–07	108	
Begin commercial operation	84–01	114	
Annual escalation charges (%)	Low	Expected	High
Turbine generator	6.00	7.00	8.00
Balance of plant			
Equipment	4.50	5.00	5.50
Labor	9.00	10.00	11.00
Material	4.50	5.00	5.50
Interest during construction (%)	7.50	8.50	10.00
Total plant cost (millions of dollars)			
Nuclear steam supply	60.00	60.00	60.00
Turbine generator	36.00	36.00	36.00
Balance of plant			
Equipment	1.70	2.10	2.50
Material	29.20	36.50	43.80
Labor	40.80	51.00	61.20
Electrical facilities			
Equipment	16.60	20.70	24.80
Material	24.60	30.80	37.00
Labor	46.90	58.60	70.30
Other facilities			
Equipment	7.80	9.80	11.80
Material	4.40	5.50	6.60
Labor	11.80	14.80	17.80
Environmental costs	30.00	35.00	40.00
Architect–engineer	55.00	60.00	65.00
Indirect costs	25.00	30.00	35.00
Total escalation	169.85	178.19	186.52
Total interest during construction	152.24	165.40	178.56
Total plant capital			
Cost at commercial operation	770.11	794.64	819.18

Adapted from J. B. Hoge, *Power Engineering* **78**, No. 12 (December 1974), pp. 36–37.

Fuel cycle cost

The fuel cycle cost includes mining and milling of the ore, conversion of the uranium to UF_6, reconversion and fabrication of fuel elements, shipping spent fuel, reprocessing, waste management, plutonium credit, and a fuel inventory carrying charge.

From 1977 through 1980 by presidential order there was no fuel reprocessing in the United States, while the various fuel cycle options were studied extensively. Thus, Table 13.2 shows the range of costs for both once-through operation and thermal fuel recycling. The NUREG/CR-1041 study was based on 1979 dollars and considered fuel costs between 1980 and 2020. Figure 13.1 shows the expected variation in the levelized fuel costs over the four decades. The ranges of various costs shown in Table 13.2 indicate the expected changes in the levelized fuel costs during the period, exclusive of the effects of inflation. If inflation were 6 percent per year, by 2020 the fuel cycle cost with once-through operation would be 88 mills/kWh and it would be 65 mills/kWh for thermal recycling. In terms of 1979 dollars, these costs are 9.7 mills/kWh and 7.6 mills/kWh.

Real fuel costs will increase during this period due to higher uranium costs resulting from reduced ore quality. In 1950 the average ore mined and milled in the United States contained 0.32 percent U_3O_8. By 1980 this had dropped

Table 13.2 LWR Fuel Cycle Costs (mills/kWh) for Median Growth Cases Based on 1979 Dollars

	Once-Through			Thermal Recycle		
	1980–84	*2000–04*	*2020*	*1980–84*	*2000–04*	*2020*
Mining and milling	2.982	3.356	6.211	2.982	2.281	2.603
Enrichment	1.842	2.203	2.256	1.948	1.890	1.623
UO_2 fuel fabrication	0.657	0.580	0.595	0.703	0.506	0.428
Reprocessing	—	—	—	—	1.003	0.803
Conversion, transportation, and other costs	0.212	0.047	0.192	0.161	0.182	0.141
Spent fuel and waste disposal		0.451	0.437	0.025	0.536	0.539
Mixed-oxide fuel fabrication	—	—	—	—	0.144	0.236
Fissile exchange*	—	—	—	−0.052	−0.515	1.374
Total	5.693	6.638	9.691	5.766	5.957	7.569

* Fissile exchange refers to the credits and costs associated with using mixed-oxide fuel instead of uranium fuel in LWRs. The cost of prior mixed-oxide fuel fabrication and plutonium credit from subsequent fuel reprocessing are netted in order to more closely relate the actual cost incurred to the electricity generated.
From L. L. Clark and A. D. Chockie. *Fuel Cycle Cost Projections*, Battelle Pacific Northwest Laboratory. NUREG/CR-1041. December 1980).

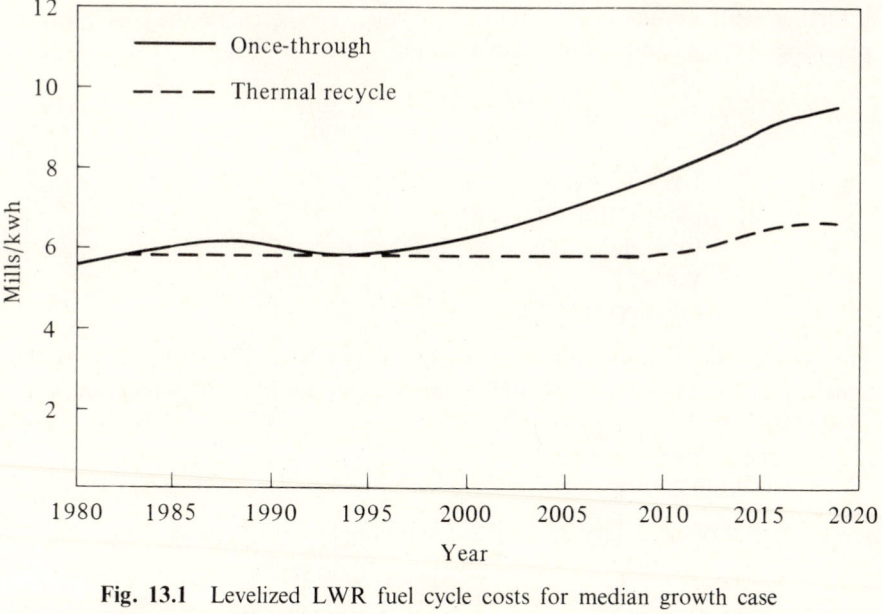

Fig. 13.1 Levelized LWR fuel cycle costs for median growth case using a once-through fuel cycle and thermal recycle (1979 dollars). [From L. L. Clark and A. D. Chockie, *Fuel Cycle Cost Projections*, Battelle Pacific Northwest Laboratory, NUREG/CR-1041, December 1980].

to 0.16 percent from underground mines and 0.12 percent from open-pit mines. This decreasing trend is expected to continue. The lower cost with thermal recycling results from credit from the recovered plutonium and uranium. There is then a reduction in the fresh ore requirements which will slow the decrease in ore quality.

For an 1100-MW(e) reactor, operating with a 70 percent capacity factor over a 30-yr lifetime with an initial fuel cycle cost of 6 mills/kWh and an 8 percent annual escalation in fuel cost, the accumulated total fuel cost will be 3.48×10^9. This is the order of 3.5 times the initial cost of the plant and serves to point out the extreme importance of effective fuel cycle management.

Plant size and inflation

One of the most significant factors in the cost of nuclear power is plant size. Some costs increase more or less linearly with size, while others are less sensitive to the rating of the plant. Overall, there tends to be a decrease in unit cost for higher ratings. In estimating unit costs, it is also important to estimate the effects of inflation, so the estimated date of operation also must be known. The empirical equation given below starts with a 1980 estimated capital cost per installed kilowatt of capacity of $620/kW for a 1000-MW(e) LWR plant.

It further assumes that an annual inflation rate, r, will escalate the cost between planning and the competion of construction.

$$C = \frac{9.67 \times 10^4}{R^{0.8}} \left(1 + \frac{r}{100}\right)^t \tag{13.1}$$

where C = LWR unit capital cost, \$/kW(e)
   R = plant rating, MW(e)
   t = time elapsed between 1980 and expected date of operation, years
   r = inflation rate, %

Figure 13.2 shows how unit cost decreases with size. It also indicates the drastic increase in cost if the annual inflation rate should be 10 percent between 1980 and 1990.

Cost of electricity

The cost of electricity, e, in mills/kWh may be expressed as

$$e = 1000 \frac{\phi I + O + F}{E} \tag{13.2}$$

where ϕ = annual fixed charge rate, yr^{-1}
   I = capital investment for plant, \$
   O = operating expense, \$/yr
   F = annual fuel cost, \$/yr
   E = electrical energy produced, kWh(e)/yr

The net annual energy generation may be expressed as

$$E = 8760 \times \qquad L \qquad \times \qquad P_r$$

$$\frac{h}{yr} \qquad \frac{kWh(e)\ actual}{kWh(e)\ rated} \qquad kW(e)\ rated$$

$$= 24 \times \frac{1}{1000} \times \frac{1000}{1} \times \eta \times B \times U$$

$$\frac{h}{day} \quad \frac{ton}{kg\ U} \quad \frac{kW(th)}{MW(th)} \quad \frac{kW(e)}{kW(t)} \quad \frac{MWd(th)}{ton} \quad \frac{kg\ U}{yr}$$

$$E = 8760\ LP_r = 24\eta BU \tag{13.3}$$

where L = capacity or load factor, kWh(e) actual/kWh(e) rated

   P_r = rated capacity of plant, kW(e)

   η = plant thermal efficiency, kW(e)/kW(th)

   B = fuel burnup at discharge, MWd(th)/ton

   U = nuclear fuel consumption, kg U fed to reactor/yr

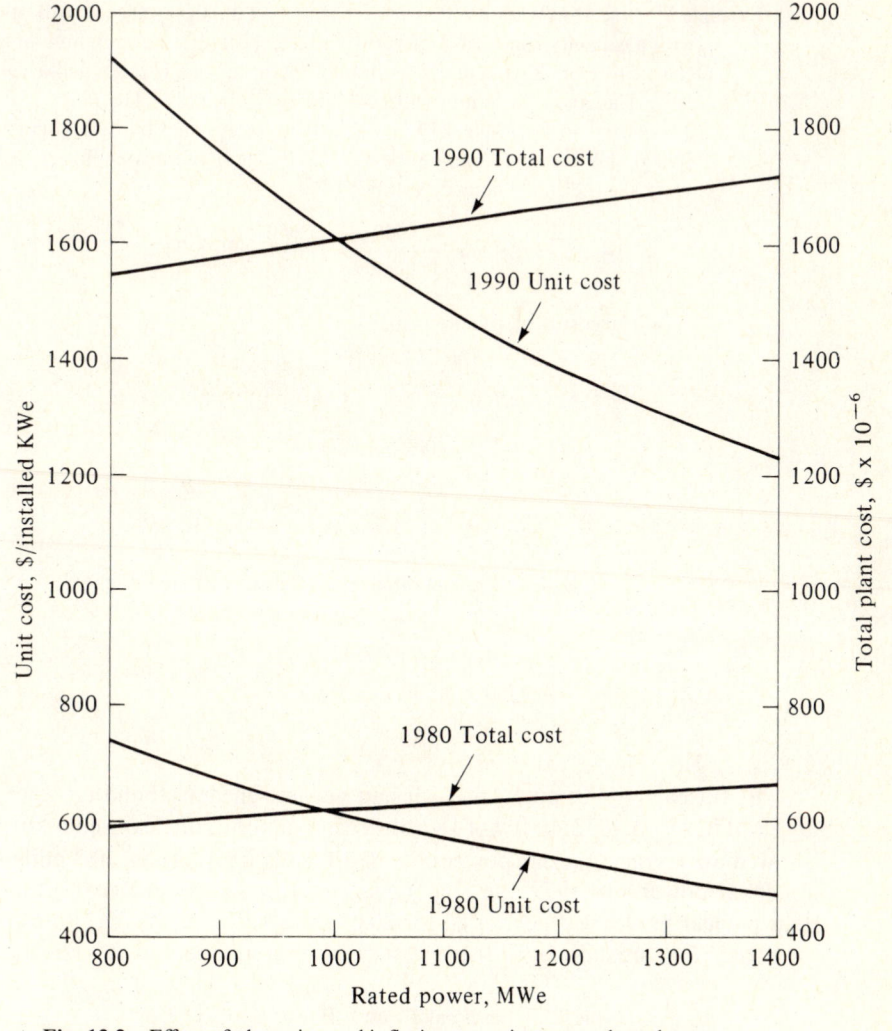

Fig. 13.2   Effect of plant size and inflation on unit costs and total costs. An inflation rate of 10 percent was used for this plot.

The annual fuel costs, F, is

$$F = C_f U \qquad (13.4)$$

where C_f is the total fuel cycle cost, \$/kg U fed to reactor.

Now Eq. (13.2) may be rewritten:

$$e = \frac{1000}{8760L}\left(\phi\frac{I}{P_r} + \frac{O}{P_r}\right) + \frac{1000}{24}\left(\frac{C_f}{\eta B}\right) \qquad (13.5)$$

Example 2   A 1200-MW(e) LWR is expected to cost $900 000 000. It will operate with a capacity factor of 70 percent and is expected to have an annual fixed-charge rate of 18 percent. The annual operating cost is estimated at $9.5 × 10^6. The cost of uranium will be $1350 per kilogram. The plant heat rate is expected to be 10 900 kJ/kWh and the average fuel burnup ought to be 30 000 MWd/ton. Calculate the cost of electricity in mills/kWh.

The thermal efficiency will be

$$\eta = \frac{3600}{\text{heat rate}} = \frac{3600}{10\,900} = 0.3303$$

The cost of electricity is then

$$e = \frac{1000}{8760 \times 0.70} \left(\frac{0.18 \times 900 \times 10^6}{1200 \times 10^3} + \frac{9.5 \times 10^6}{1200 \times 10^3} \right)$$

$$+ \frac{1000 \times 1350}{24 \times 0.3303 \times 30\,000}$$

$$= 22.01 + 1.29 + 5.68 = 28.98 \text{ mills/kWh}$$

For a government-operated plant the fixed charge might be 9 percent. The cost of electricity in that case would then be

$$e = \frac{1000}{8760 \times 0.70} \left(\frac{0.09 \times 900 \times 10^6}{1200 \times 10^3} \right) + 1.29 + 5.68$$

$$= 17.98 \text{ mills/kWh}$$

By the early 1970s nuclear power had become cheaper than that generated by fossil fuels. A 1972 survey of 18 utilities having both nuclear and fossil units showed an average nuclear power cost of 8.1 mills/kWh versus 10.3 mills/kWh for fossil power. In 1979, the Department of Energy (Reference 77) reported that nuclear power was generated for 20.67 mills/kWh versus 22.35 mills/kWh by coal. The breakdown of these costs is shown in Table 13.3. The average

Table 13.3 Baseload Electric Power Costs During 1979

	Power Cost (mills/kWh)	
	Coal	Nuclear
Fuel cost (actual)	12.48	3.73
Operation, maintenance and supplies (actual)	2.26	4.11
Capital cost (estimated)	7.61	12.83
Total	22.35	20.67

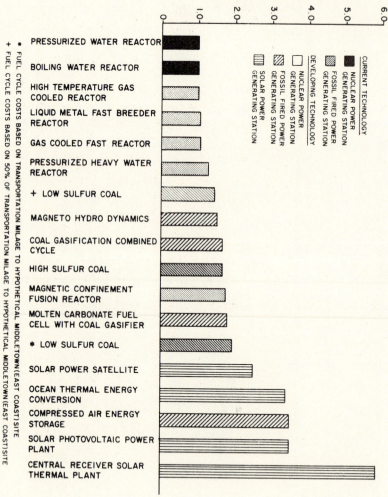

Fig. 13.3 Economic comparison of current and developing technologies for power generation alternatives having a 2001 startup and normalized to 1200 MW(e). The study was based on constant 1978 dollars. [From R. E. Allen, "Economics of the Nuclear Option: Can It Compete with Alternatives?" *Proceedings of the ANS Topical Meeting: A Technical Assessment of Nuclear Power and Its Alternatives*, Los Angeles, February 27–29, 1980.]

Table 13.4 Economic Comparison of Nuclear Power Generation Alternatives Having 2001 Startup Based on Constant 1978 Dollars

	Current Technologies		Developing Technologies				
	Pressurized Water Reactor	Boiling Water Reactor	High-Temperature Gas-Cooled Reactor	Pressurized Heavy-Water Reactor	Gas-Cooled Fast Reactor	Liquid-Metal Fast Breeder Reactor	Magnetic Confinement Fusion Reactor
Plant size [MW(e)]	1139	1190	1330	1162	917	1390	660
Assumed capacity factor (%)	70	70	70	70	70	70	65
Capital cost (mills/kWh)	10.0	9.9	9.6	11.2	12.8	12.4	27.8
Fuel cycle cost (mills/kWh)	7.8*	7.8*	6.4*	8.2†	4.1‡	3.6‡	4.9
Operating and maintenance cost (mills/kWh)	1.9	1.9	1.6	2.4	2.7	2.1	5.3
Total cost (mills/kWh)	19.7	19.6	17.6	21.8	19.6	18.1	38.0
$/kW (capital cost)	581	571	559	646	744	721	1612

* Throwaway cycle.
† Natural uranium.
‡ Plutonium has zero value.
From R. E. Allen, "Economics of the Nuclear Option: Can It Compete with Alternatives?" *Proceedings of the ANS Topical Meeting: A Technical Assessment of Nuclear Power and Its Alternatives*, Los Angeles, February 27–29, 1980.

capacity factor in 1979 dropped from 65 percent to 59.8 percent, largely because of shutdowns to accomplish post–TMI checking and modifications.

Figure 13.3 shows an economic comparison of current and developing technologies for electric generating systems for a 2001 startup date in terms of 1978 dollars. On this economic basis the nuclear alternatives would be cheaper than coal, MHD, ocean thermal conversion, and the various solar conversion schemes. Table 13.4 shows a breakdown of the power costs (1978 dollars) for the various nuclear alternatives with this projected 2001 startup. It is interesting to note that the high-temperature gas-cooled reactor has not only the lowest power cost, but also the least capital cost.

Pressurized Water Reactors

In a pressurized water reactor (PWR) the fission energy released in the core increases the enthalpy of high-pressure water (~ 14 MPa). This hot water generates steam at a lower pressure in the secondary loop of the system (see Fig. 13.4). The steam then produces power in a conventional turbine. Now it is condensed, and the feedwater is pumped back into the steam generator. Steam pressures in the secondary loop have increased from 4.1 MPa to nearly 7.0 MPa as reactor plants have progressed from Shippingport, which went critical in 1957 with a capacity of 60 MW(e), to Connecticut Yankee, which went critical in 1967 and which operates currently at a power level of 575 MW(e). TVA's Bellefonte 1 and 2 PWRs will each have a rating of 1213 MW(e) and are expected online during 1983 and 1984.

The primary pressure is maintained by electrical heaters in the pressurizer. Water in this tank is maintained at the saturation temperature corresponding to the desired pressure level.

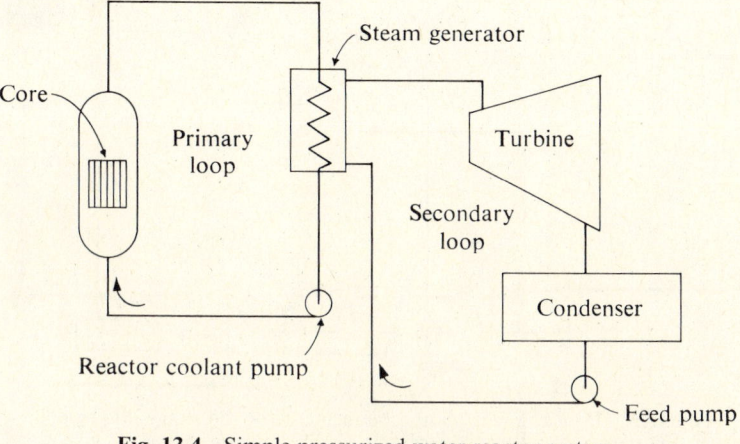

Fig. 13.4 Simple pressurized water reactor system.

The fact that no superheat is available means that considerable moisture will form during expansion through the turbines. Moisture separators between the high- and low-pressure turbine casings help to eliminate some of the problems due to excess moisture in the final stages of the expansion. This difficulty with moisture may be further alleviated by using live steam for reheating after the moisture separation. A system using reheating and several stages of regenerative feedwater heating is shown in Fig. 13.5. Notice that all of the primary working fluid stays inside the reactor containment.

The Connecticut Yankee Plant has a power cycle similar to that of Fig. 13.5. Water at 14.8 MPa is heated from 285°C to 306.4°C in the core. The primary coolant flow rate is 46×10^6 kg/h. When operating at 590 MW(e) a total of 3.45×10^6 kg/h is pumped to the four steam generators. At this power level steam is supplied to the throttle at 4.4 MPa and one-fourth percent moisture. There is approximately a 0.35 MPa pressure drop between the steam generator and the throttle. From each steam generator 0.61 m lines lead to a 0.91 m manifold and thence to the two turbine stop-trip valves via 0.76 m diameter lines. The turbine is a tandem compound unit with the double-flow high-pressure casing which has one impulse and seven reaction stages on each end. Two double-flow low-pressure casings have 10 reaction stages per end. The blade length for the final row of blades is 1.12 m.

The reactor core contains 75 300 kg of uranium dioxide pellets. They are 1.52 cm long and 0.974 cm in diameter. They are contained in stainless steel

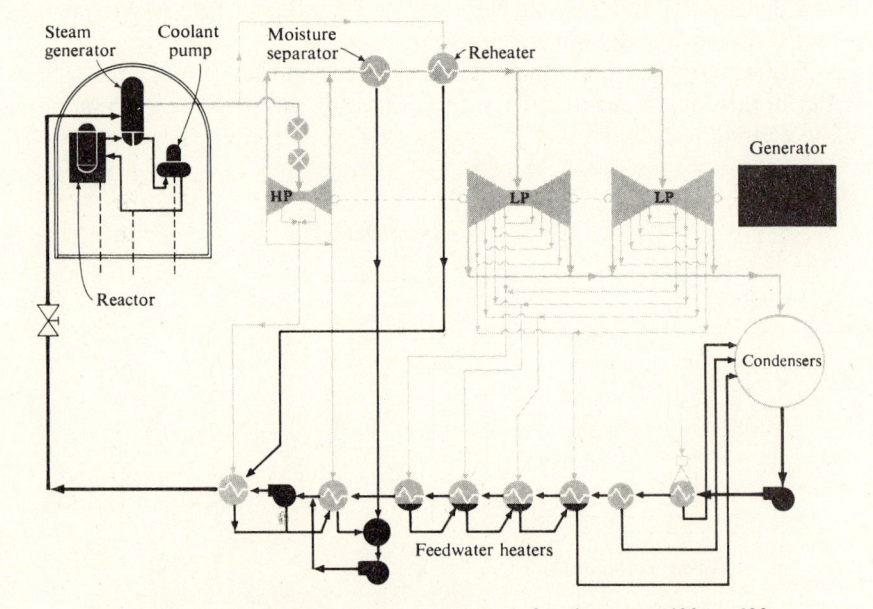

Fig. 13.5 Typical PWR plant arrangement for the range 400 to 600 MW(e). [Courtesy of Westinghouse Hanford Company.]

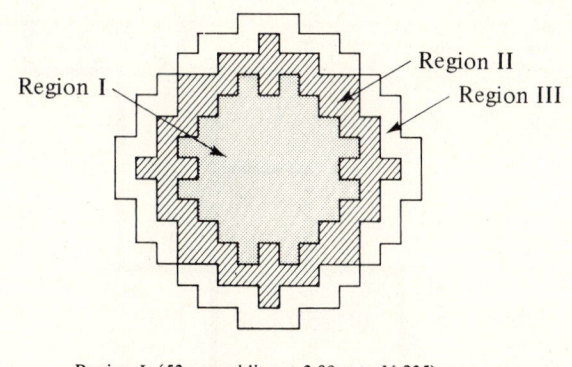

Region I   Region II
                      Region III

Region I (53 assemblies at 3.00 w/o U-235)

Region II (52 assemblies at 3.24 w/o U-235)

Region III (52 assemblies at 3.67 w/o U-235)

Fig. 13.6   Region boundaries in the first cycle. [From Connecticut Yankee Atomic Power Company, *Facility Description and Safety Analysis*, Vol. I, Report No. NYO-3250-5.]

tubes with 1.07 cm O.D. and 0.042 cm wall thickness. The active fuel length in the tubes is 3.09 m; the rods are arranged in a 15 by 15 array to form an assembly. There are 157 of these assemblies in the core. There are only 204 fuel rods per assembly, with 20 of the remaining positions being available for control rods; one position is available as a central instrumentation sheath. The control rods are 85 percent Ag, 15 percent In, 5 percent Cd, clad with stainless steel. They have an O.D. of 1.12 cm.

The first core loading had three regions of varying enrichment, as shown by Fig. 13.6. The outermost 52 assemblies had a 3.67 percent enrichment, the intermediate region contained 52 assemblies with 3.24 percent ^{235}U, and the central region had a 3 percent enrichment. This variable enrichment tends to provide a flatter flux variation across the core. As refueling occurs the outer elements will move inward and new fuel will be added to Region III. The Region I fuel will be removed and stored as spent fuel, awaiting possible reprocessing at a future date. The division of power between the various fuel assemblies in the three regions is shown in Fig. 13.7. These figures are calculated for 15 900 MWd/ton burnup with the control rods withdrawn from the core. Under these conditions 36.1 percent of the power is developed in the central region, 36.4 percent in the intermediate region, and 27.4 percent in the outer region, indicating the effectiveness of the variable enrichment in equalizing the power generation across the core.

Reactivity is controlled by both control rods and by the injection of boric acid into the core (chemical shim). The use of boron as a uniformly distributed poison promotes a more uniform flux distribution by requiring less control rod insertion. The cold core with no power and no Xe or Sm buildup has an effective

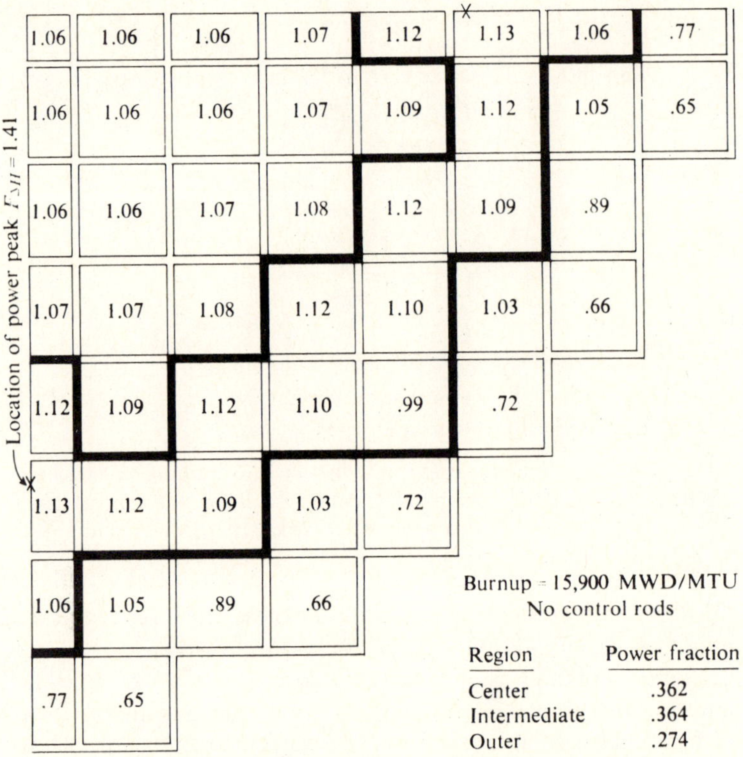

1.06	1.06	1.06	1.07	1.12	1.13	1.06	.77
1.06	1.06	1.06	1.07	1.09	1.12	1.05	.65
1.06	1.06	1.07	1.08	1.12	1.09	.89	
1.07	1.07	1.08	1.12	1.10	1.03	.66	
1.12	1.09	1.12	1.10	.99	.72		
1.13	1.12	1.09	1.03	.72			
1.06	1.05	.89	.66				
.77	.65						

Location of power peak $r_{3/I} = 1.41$

Burnup = 15,900 MWD/MTU
No control rods

Region	Power fraction
Center	.362
Intermediate	.364
Outer	.274

Fig. 13.7 Assembly to average power distribution. [From Connecticut Yankee Atomic Power Company, *Facility Description and Safety Analysis*, Vol. 1, Report No. NYO-3250-5.]

multiplication factor of 1.28. When the core is hot at full power with equilibrium Xe and Sm buildup the effective multiplication factor drops to 1.163. Initially, a boron concentration of 2470 ppm will give a shutdown margin of 9 percent with the control rods fully inserted. At full power with equilibrium xenon and samarium the boron concentration will be reduced to 1800 ppm. As the fuel burnup proceeds, the boron concentration will be gradually reduced to zero. The control rod worth is designed to be 0.0642 at the beginning of core life: at the end of core life this will have increased to 0.0732.

Improvements in core design have permitted marked increases in the water temperature leaving the reactor core. One factor in this improvement is the reduction of the enthalpy rise hot channel factor. This is the ratio of the enthalpy rise of the water in the hottest channel to the enthalpy rise in the average channel. In 1960 this factor was 3.6. It decreased to a value of 1.8 at the end of 1966. Further reduction to 1.6 is anticipated. This permits higher water temperatures in the core with a corresponding improvement in the thermal

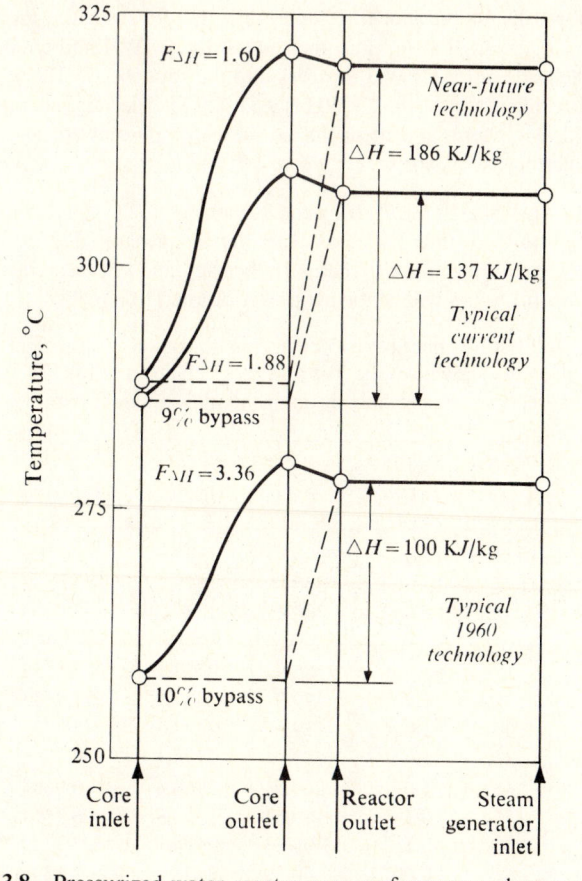

Fig. 13.8 Pressurized water reactor core performance advancements since construction of the 1960 plant. [Courtesy of Westinghouse Electric Corporation.]

efficiency of the steam cycle and power costs. Figure 13.8 shows the water temperature variation, as it passes through the core, for 1960 technology, the technology of the mid-1960s, and for "near future" technology.

Example 3 In a pressurized water reactor (PWR) 75×10^6 kg/h of water at 150 bars enter the core at 300°C and are heated to 330°C. In the steam generators heat is transferred from the pressurized water to water in the secondary loop, generating saturated steam at 70 bars. The steam expands to 10 bars through the high-pressure turbine. Moisture is separated and live steam reheat increases the steam temperature to 260°C before entering the low-pressure turbine, where it expands to 1 bar, at which point a fraction is bled to a closed feedwater heater (CFWH), and the remaining steam continues its expansion to the condenser pressure of 0.075 bar. Condensate from the reheater is

trapped into the moisture separator. Liquid from the moisture separator passes through a trap into the shell of the CFWH and condensate from the shell side of the CFWH throttles into the condenser. The terminal temperature difference for the CFWH is 9.63°C. The electric generator is 95 percent efficient, each segment of the turbine is 85 percent efficient, and the pump efficiency is 75 percent.

(a) Sketch the cycle on the *T–S* plane.
(b) How many kg/h of steam must be generated?
(c) What are the thermal efficiency and the heat rate for the plant?
(d) What will be the net power output [MW(e)]?

First it is necessary to determine the state-point properties at the different points shown on the *T–s* and block diagrams (Fig. 13.9). Consider first the isentropic expansion through the high-pressure turbine.

Point	P (MPa)	T (°C)	h (kJ/kg)	s (kJ/kg K)	x
a	15.0	300	1337.3	3.2260	SC*
b	15.0	330	1518.1	3.5332	SC
c	15.0	300	1337.3	3.2260	SC
1	7.0	285.9	2772.1	5.8133	1.0000
2i	1.0	179.9	2427.9	5.8133	0.8262
2	1.0	179.9	2479.5	5.9275	0.8518
3	1.0	179.9	2778.1	6.5865	1.0000
4	1.0	260.0	2964.6	6.9664	SH†
5i	0.1	99.63	2528.9	6.9664	0.9351
5	0.1	99.63	2594.3	7.1394	0.9641
6i	0.0075	40.29	2226.2	7.1394	0.8551
6	0.0075	40.29	2281.4	7.3155	0.8781
7	0.0075	40.29	168.79	0.5764	0
8	7.0	40.3+	178.2	—	CL‡
9	7.0	90	382.3	1.1861	CL
10	1.0	179.9	762.81	2.1387	0
11	0.1	99.63	417.46	1.3026	0
12	7.0	285.9	1267.0	3.1211	0

* SC = subcooled.
† SH = superheated.
‡ CL = compressed liquid.

$$s_{2i} = s_1$$

$$5.8133 = 2.1387 + x_{2i}(4.4478)$$

$$x_{2i} = 0.8262$$

The ideal enthalpy at the end of the initial expansion is

$$h_{2i} = 762.81 + 0.8262(2015.3) = 2427.9 \text{ kJ/kg}$$

T
15.0 MPa
b
a, c
12
7.0 MPa
1
4
10
12′
1.0 MPa
3
2i
2
11
10′
0.1 MPa
9
5
5i
8
8i
11′
0.0075 MPa
7
6i
6
b
1
1 kg
m
(1 − m)
H.P.
turbine
L.P.
turbine
Generator
c
a
4
2
(1 − n)
5
6
(1 − n − p)
3
12
T
(n + p)
p
12′
Condenser
n
10
T
10′
7
9
8
11
T
11′

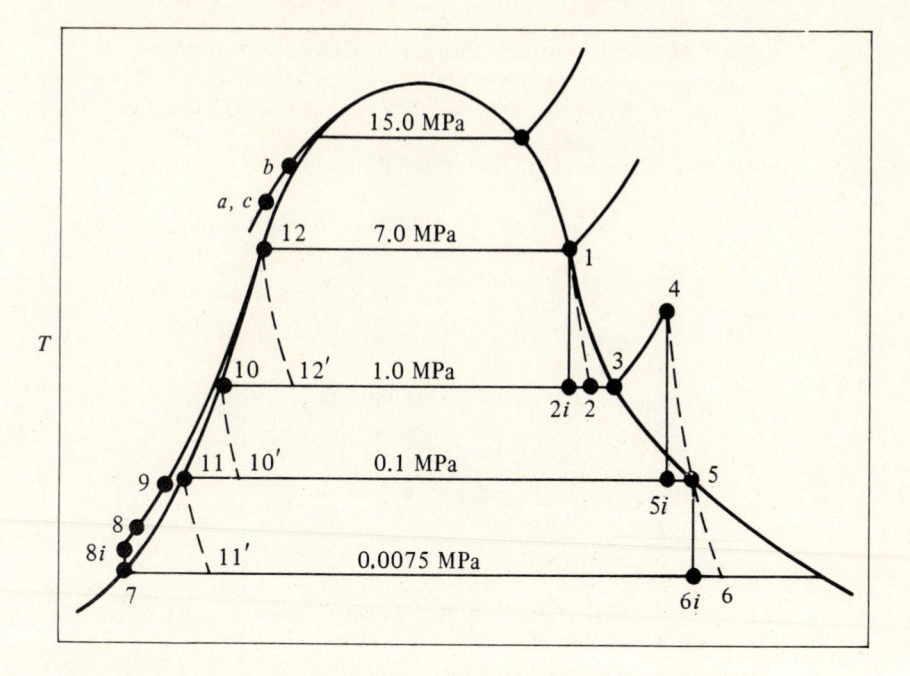

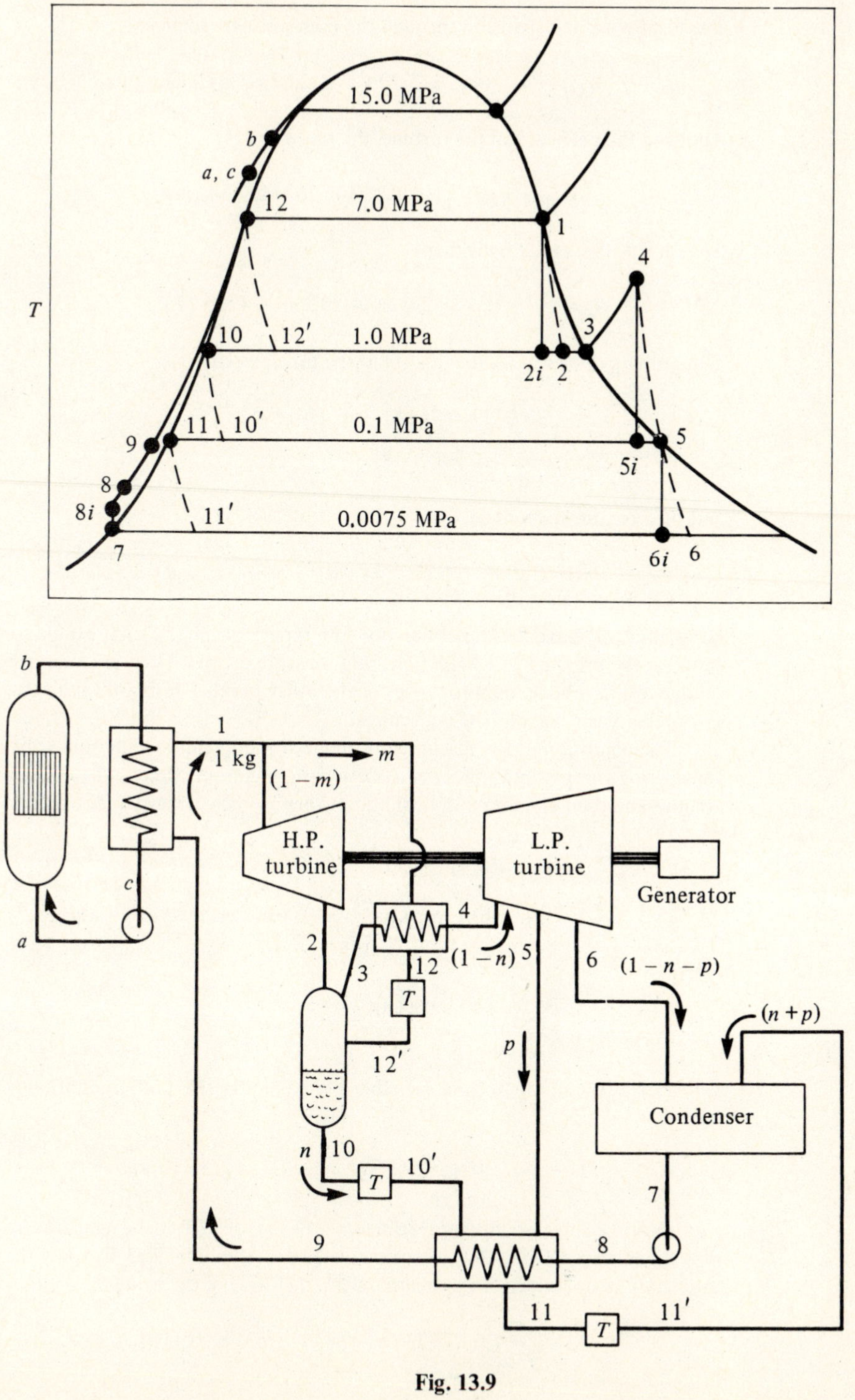

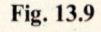

Fig. 13.9

The ideal work of expansion through the high-pressure turbine is

$$W_{12i} = h_1 - h_{2i} = 2772.1 - 2427.9 = 344.2 \text{ kJ/kg}$$

Applying the efficiency of the turbine, the real work is

$$W_{12} = \eta_T(W_{12i}) = 0.85(344.2) = 292.6 \text{ kJ/kg}$$

The turbine exit enthalpy is then

$$h_2 = h_1 - W_{12} = 2272.1 - 292.6 = 2479.5 \text{ kJ/kg}$$

This enthalpy determines the quality at the turbine exit

$$2479 = 762.81 + x_2(2015.3)$$

$$x_2 = 0.8518$$

and the exit entropy will be

$$s_2 = 2.1387 + 0.8518(4.4478) = 5.9275 \text{ kJ/kg}$$

Dry steam will leave the top of the moisture separator and be superheated to 260°C at the pressure of 1 MPa (ignoring pressure drop). Then in a manner similar to that above, the properties of the other points passing through the low-pressure turbine may be determined.

The condition at the discharge of the feed pump can be determined by considering the pumping process to be ideally incompressible and constant volume, and then dividing this ideal pump work by the efficiency of the pump.

$$W_{78} = - v_7(p_8 - p_7)/\eta_p$$

$$= -\frac{1.0079 \times 10^{-3} \text{ m}^3/\text{kg} (7.0 - 0.0075) \text{ MPa} \times 10^3}{0.75} \frac{\text{kN}}{\text{m}^2 \text{ MPa}}$$

$$\times 1 \frac{\text{kJ}}{\text{kN m}} = -9.4 \text{ kJ/kg}$$

Considering the pump to be irreversible and adiabatic, the discharge enthalpy will be

$$h_8 = h_7 - W_{78} = 168.79 - (-9.4) = 178.2 \text{ kJ/kg}$$

With a terminal temperature difference (TTD) of 9.63°C, the temperature of the feedwater leaving the closed feedwater heater will be 9.63°C below the saturation temperature of the steam on the shell side of the heater.

$$T_9 = T_{11} - \text{TTD} = 99.63 - 9.63 = 90°C$$

The value of h_9 may then be determined either from a compressed liquid table or approximated by adding the ideal incompressible adiabatic work necessary to raise saturated liquid at 90°C to the 7.0-MPa pressure. In this case the table was used.

In passing through the traps the saturated liquid is throttled to a lower-pressure region and a fraction of vapor flashes, while the enthalpy is unchanged ($h_{12} = h_{12}'$, $h_{11} = h_{11}'$, and $h_{10} = h_{10}'$).

Next it will be necessary to determine the fractions of each kilogram of secondary steam generated:

(1) Bled for reheating, m
(2) Separated as moisture to be throttled to the shell of the CFWH, n
(3) Bled for feedwater heating, p

Energy balances around the moisture separator and the reheater both contain m and n, thus requiring the solution of the two equations simultaneously. The balance around the moisture separator gives

$$(1 - m)h_2 + mh_{12} = (1 - n)h_3 + nh_{10}$$

$$(1 - m)2479.5 + m(1267.0) = (1 - n)2778.1 + n(762.81)$$

$$m = 1.6621n - 0.2463$$

The balance for the reheater is

$$(1 - n)(h_4 - h_3) = m(h_1 - h_{12})$$

$$m = \frac{(1 - n)(h_4 - h_3)}{h_1 - h_{12}} = \frac{(1 - n)(2964.6 - 2778.1)}{2772.1 - 1267.0}$$

$$= (1 - n)(0.1239)$$

Equating the two expressions for m, we get

$$(1 - n)(0.1239) = 1.6621n - 0.02463$$

$$n = 02073$$

$$m = (1 - 0.2073)(0.1239) = 0.09822$$

A balance around the CFWH will yield the value of p.

$$n(h_{10} - h_{11}) + p(h_5 - h_{11}) = h_9 - h_8$$

$$p = \frac{h_9 - h_8 - n(h_{10} - h_{11})}{h_5 - h_{11}}$$

$$= \frac{382.3 - 178.2 - 0.2073(762.8 - 417.46)}{2594.3 - 417.46}$$

$$= 0.609$$

The total heat addition for the cycle is

$$Q_s = \dot{m}_{pri}(h_b - h_a) = \frac{75 \times 10^6 (1518.1 - 1337.3)}{3600 \times 1000}$$

$$= 3767 \text{ MW(th)}$$

The secondary steam flow rate is

$$\dot{m}_{sec} = \frac{Q_s}{h_1 - h_9} = \frac{3767 \times 3.6 \times 10^6}{2772.1 - 382.3} = 5.674 \times 10^6 \text{ kg/h}$$

The cycle heat rejection is

$$Q_r = \dot{m}_{sec}[h_7 - (1 - n - p)(h_6) - (n + p)(h_{11})]$$

$$= \frac{5.674 \times 10^6 (168.79 - 0.7318 \times 2281.4 - 0.2682 \times 417.46)}{3600 \times 1000}$$

$$= -2542 \text{ MW(th)}$$

The net electrical power for the cycle is then

$$W_{net} = \eta_{gen}(Q_s + Q_r) = 0.95(3767 - 2542) = 1163.8 \text{ MW(e)}$$

The thermal efficiency for the plant is then

$$\eta_{th} = W_{net}/Q_s = 1163.8/2542 = 0.3090$$

The heat rate would be

$$\text{heat rate} = 3600/\eta_{th} = 3600/0.3090 = 11\,650 \text{ kJ/kWh}$$

The preceding example indicates the general type of analysis used to calculate the thermal performance of light-water reactor systems (LWRs). However, an actual plant would have a thermal efficiency of 31 to 34 percent, largely because it would use several additional stages of feedwater heating (see Fig. 13.5).

PWR superheat

One of the most interesting innovations for PWRs is the once-through steam generator (OTSG) developed by Babcock and Wilcox (see Chapter 12 for a description of the OTSG). The addition of 28 to 33°C superheat to steam supplied by the steam generators reduces the moisture in the later stages of the high-pressure turbine. This helps alleviate blade erosion and, if properly exploited in the cycle, will improve the heat rate and the thermal efficiency.

Figure 13.10 shows a sketch of the Oconee-type Babcock and Wilcox nuclear steam supply system. Two OTSGs, each having two circulating pumps,

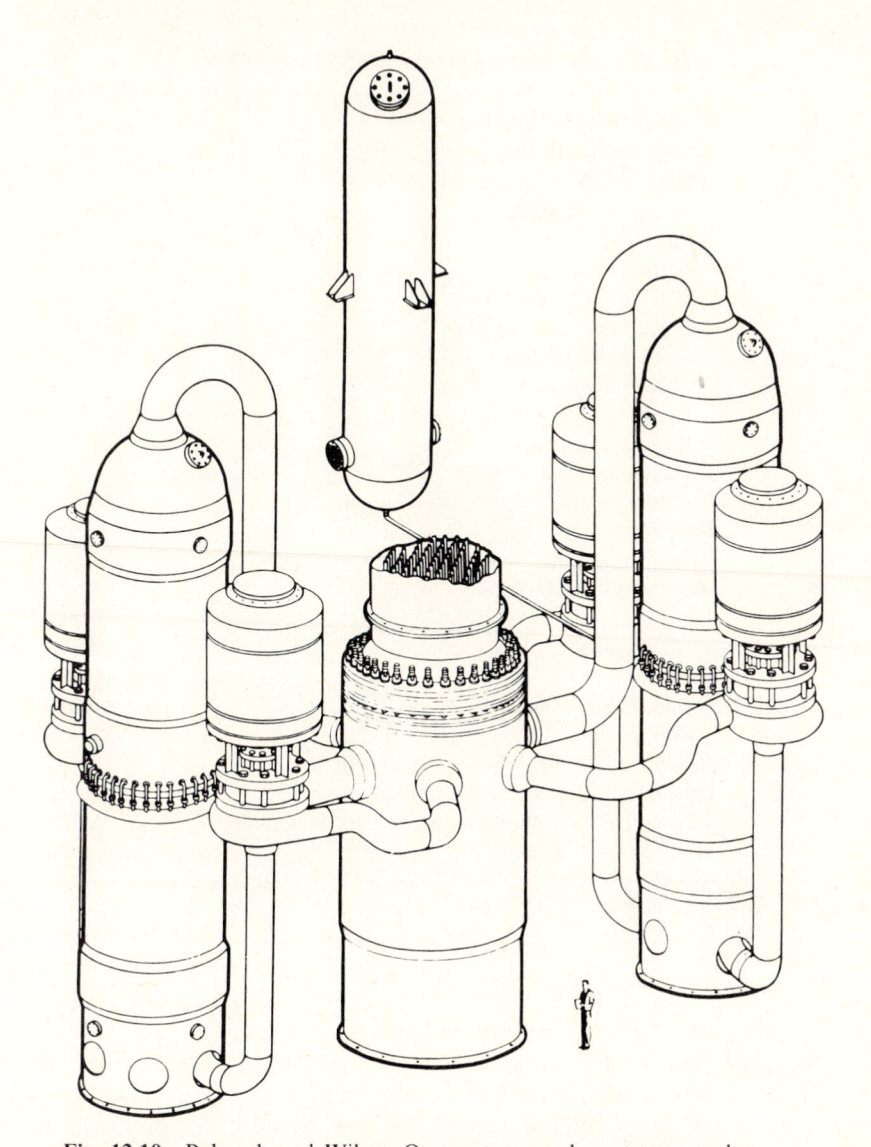

Fig. 13.10 Babcock and Wilcox Oconee-type nuclear steam supply system. The reactor is shown at the center; and it supplies energy for the once-through steam generators, each of which has a pair of circulating pumps. The pressurizer is shown at the top center. [From W. O. Parker, Jr., et al., "Postoperational Test and Examination— Oconee Unit 1," American Power Conference, Chicago, April 21–23, 1975.]

Table 13.5 3600-MW(th) Standard NSS Design Data

Number of fuel assemblies	205
Core power [MW(th)]	3600
NSS power [MW(th)]	3618
Linear power (kW/m)	17.8
Steam pressure (MPa)	7.31
Superheat (°C)	27.8
Net plant output [MW(e)]*	1244
Net plant heat rate (kJ/kWh)*	10 458
Active fuel length (m)	3.63
Number of control rod drives	
U core	72
Pu core	77
maximum	89
Reactor coolant system flow (10^6 kg/h)	68.3
Reactor vessel inlet temperature (°C)	300
Reactor vessel outlet temperature (°C)	331.5
Reactor vessel I.D. (m)	4.62
Reactor coolant hot leg I.D. (m)	0.965
Reactor coolant cold leg I.D. (m)	0.711
Pump suction I.D. (m)	0.813
Number of steam generator tubes	16 000
Steam generator height (m)	23.0
Reactor vessel–steam generator spacing (m)	10.4
Core flooding tanks, number/volume (unit/m³)	51.0
Low-pressure injection pumps, number/flow (unit/min)	2/18 900
High-pressure injection pumps, number/flow (unit/min)	3/2650

* Estimated at 51 mm Hg backpressure, seven FW heaters.
Courtesy Babcock & Wilcox.

produce steam for units as great as the 3600-MW(th) units for TVA's Bellefonte 1 and 2. Table 13.5 gives the principal design and performance parameters for a Babcock and Wilcox unit of this type, as an example of the large PWR units currently available.

Boiling Water Reactors

In the boiling water reactor, steam is generated directly within the core. This steam is the working fluid for the power cycle, as shown in Figure 13.11, which uses a single cycle. In this cycle all steam generation takes place within the core. All BWRs built in the United States since Dresden 1 have been of the single-

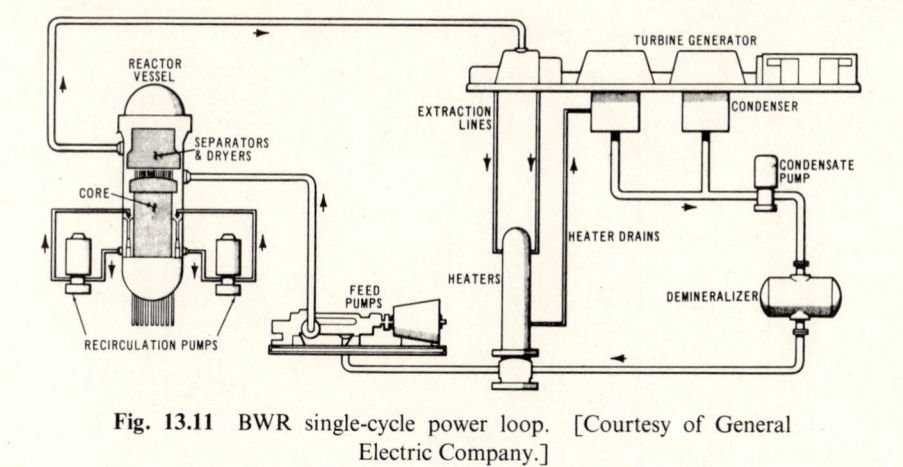

Fig. 13.11 BWR single-cycle power loop. [Courtesy of General Electric Company.]

cycle design. The dual-cycle generates secondary steam at a lower pressure outside the core using heat taken from high-pressure water returning from the steam drum to the core. This dual cycle will be discussed further after a more thorough consideration of the single cycle. These plants have grown from Oyster Creek 1's 640 MW(e), which has operated since 1969, to Dresden 2's 800 MW(e) (1970 operation), Brown's Ferry 1's 1065 MW(e) (1974 operation) to Hartsville A1's 1233 MW(e) (1986 expected operation).

The core contains assemblies of Zircaloy-2 clad fuel rods in an 8 by 8 array. These rods have an active fuel length of approximately 3.66 m and contain pellets of UO_2. Figure 13.12 shows such an assembly. Each assembly is contained in a can-like fuel channel with the rods being supported between upper and lower tie plates. Eight of the rods are threaded into the bottom support plate and are bolted to the upper support plate, while end plugs on the remaining rods fit into holes in the tie plates. Assemblies are inserted into or removed from the core by the upper tie plate handle.

Four assemblies are arranged in a module surrounding a cruciform control blade, as shown in Fig. 13.13. These blades enter the core through the bottom of the core vessel, and each blade contains 84 0.476-cm-diameter stainless steel tubes filled with boron carbide powder as the principal neutron absorbing material. The positioning of the control rods in the lower part of the core helps to effect an axial flux flattening by counterbalancing the effect of the increasing number of steam voids near the top of the core. Certain of the fuel rods may contain gadolinium (Gd_2O_3) as a burnable poison. Once the high-absorption gadolinium isotopes have disappeared, these rods function as ordinary fuel rods. The modules are arranged roughly in a right cylinder with flow entering at the bottom of each assembly through a nose piece and then passing upward between the fuel rods where the steam is generated. An orifice at the base of each assembly controls the flow of water to that assembly.

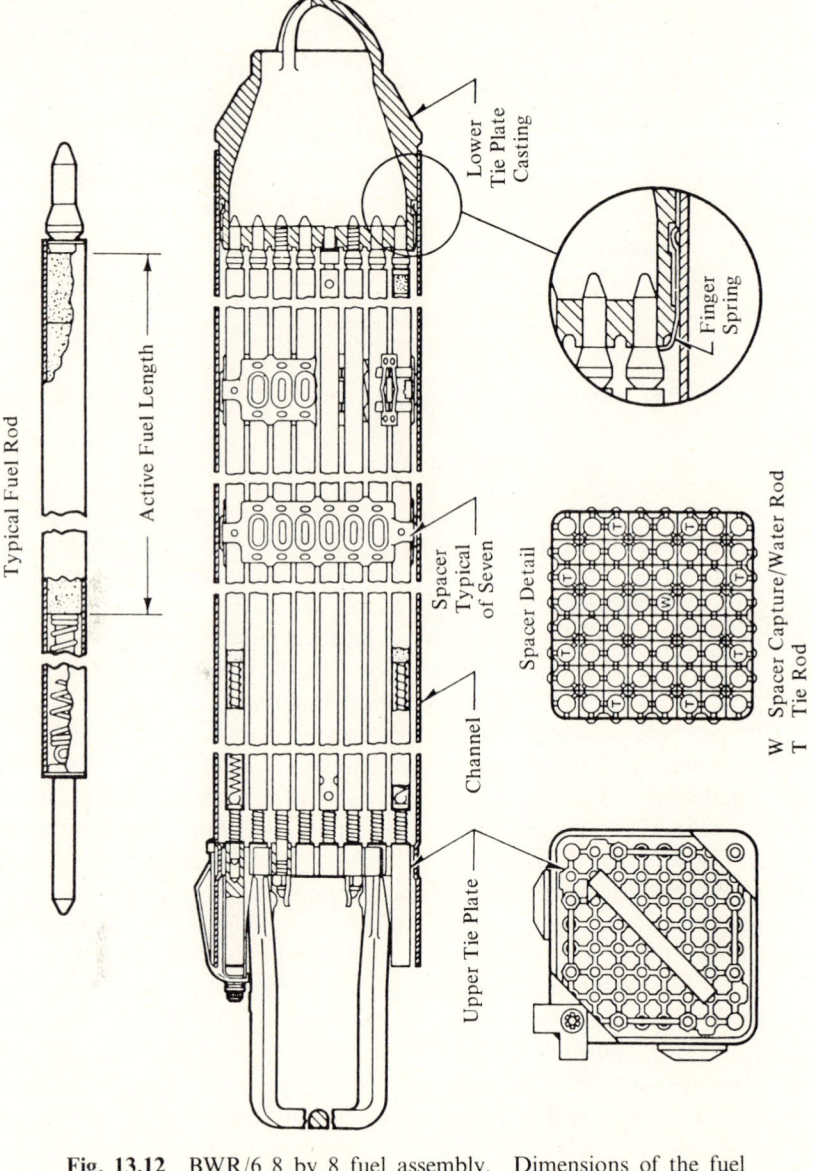

Fig. 13.12 BWR/6 8 by 8 fuel assembly. Dimensions of the fuel channel are 14 cm × 14 cm × 4.24 m long. [Courtesy of General Electric Company.]

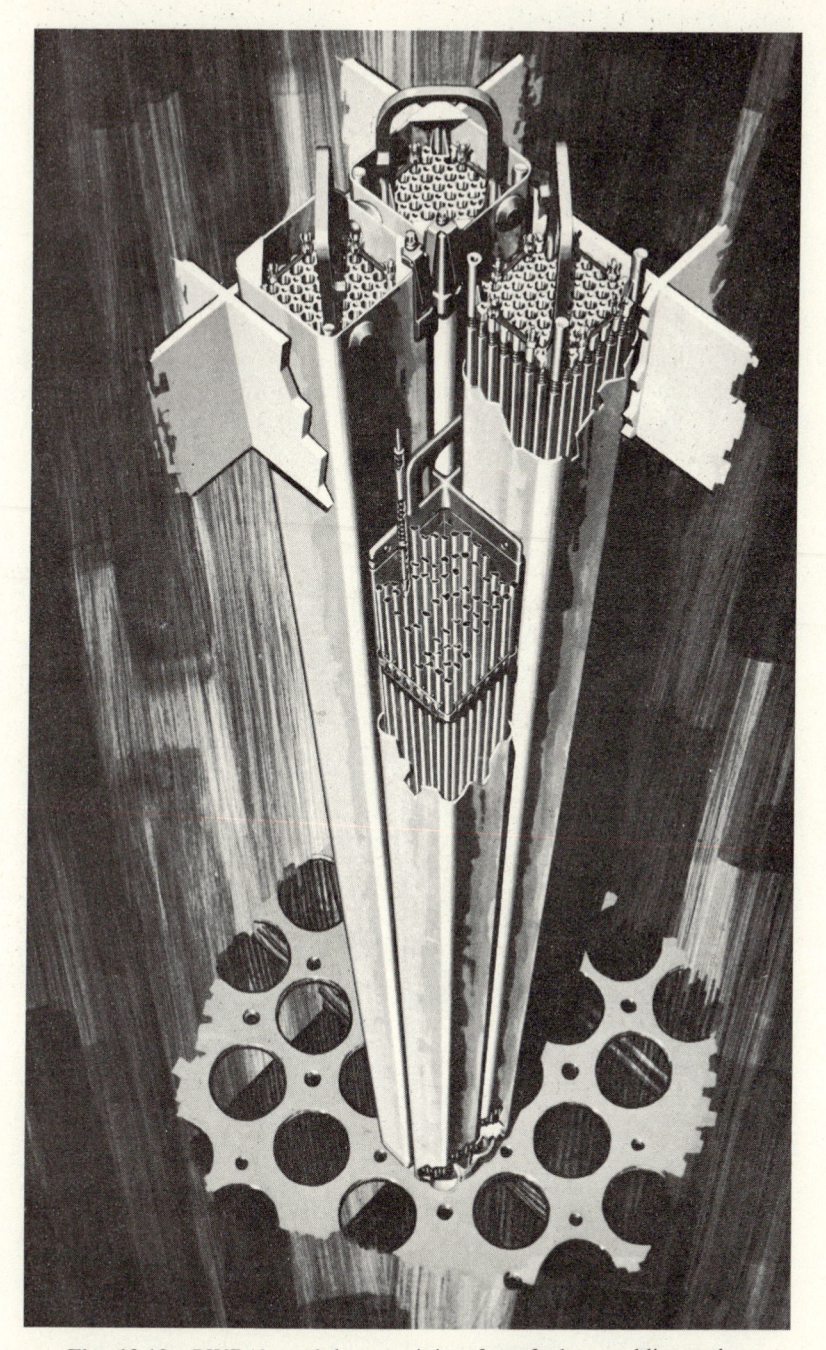

Fig. 13.13 BWR/6 module containing four fuel assemblies and a cruciform control rod. [Courtesy of General Electric Company.]

As a control blade is withdrawn, water in the vacated space provides more effective neutron moderation, causing a local flux peaking.   Figure 13.14 indicates that several (four or five) different fuel enrichments may be used to offset the flux peaking adjacent to the water gap for the control rod.   The average fuel enrichment for initial cores will be from 1.6 to 2.2 percent, depending on the cycle requirements.   Reload fuel will have an enrichment of 2.4 to 2.8 percent.

The BWR fuel is in the form of UO_2 pellets stacked in a 3.66-m column contained by free-standing Zircaloy-2 cladding.   In the 8 by 8 fuel assemblies the Zircaloy tube is 4.06 m long, has an O.D. of 1.25 cm, and has a wall thickness of 0.86 mm.   The active pellet height is 3.75 m, on top of which there is a 0.3-m fission gas plenum.   There is a 0.23-mm gap between the tube I.D. and the fuel

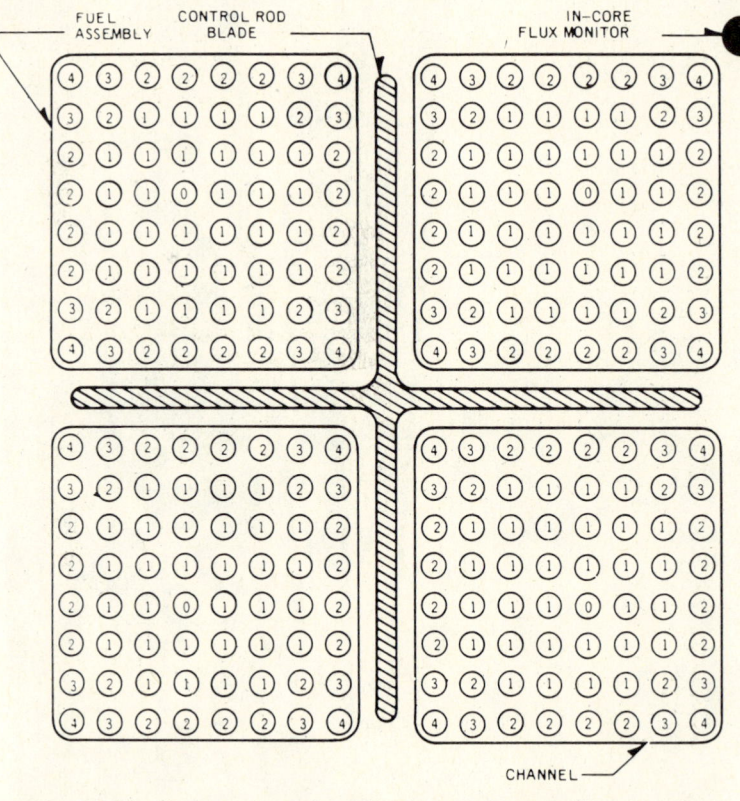

Fig. 13.14   Cross section of a BWR/6 module containing four fuel assemblies and a cruciform control rod.   Note that 1, 2, 3, and 4 indicate the location of four different fuel rod enrichments in each fuel assembly.   0 represents a water rod.   Several of the interior high-enrichment rods in each assembly contain Gd_2O_3 as a burnable poison.

[Courtesy of General Electric Company.]

pellets. A plenum spring maintains a downward pressure on the fuel pellets to keep them in place during pre-irradiation handling and then during operation to be sure that pellet-to-pellet contact is maintained, if there should be any fuel shrinkage. The plenum volume is backfilled with helium and provides a reservoir to accommodate gaseous fission products released during operation.

Fuel rod failure occurs when cladding perforation or rupture occurs, allowing the release of fission gases. During reactor transients, fuel damage may be due to local overheating because of insufficient cooling or excessive strain from relative pellet–cladding expansion. Fuel damage due to local heating is expected to occur as the film boiling regime develops. The onset of film boiling occurs at a linear heat generation rate of 52.5 kW/m but actual damage is not expected to occur until well into the film boiling region. Thus, the design value of 44 kW/m for the maximum linear heat generation rate provides a safety margin greater than the ratio of 16 to 13.4 would indicate. It might be noted here that a minimum critical heat flux ratio greater than 1.90 is used in the design.

A value of 1 percent plastic strain is a fairly conservative value below which fuel damage is not expected to occur. Actually, available data indicates that the threshhold for damage to Zircaloy cladding is in excess of the 1 percent limit. The linear heat generation rate to produce a plastic strain of this amount is approximately 82 kW/m. In the 7 by 7 BWR/5 fuel, it has been determined that localized strains due to ridging at pellet interfaces and pellet cracking can result in a small, but significant number of clad failures during normal operation. This can be related to the random occurrence of local cladding strains that are greater than average, combined with the variable ductibility of irradiated cladding. For the 8 by 8 BWR/6 fuel, the following changes have been instituted to reduce the localized strain:

(1) A reduction in the pellet L/D ratio from 2:1 to 1:1, which reduces ridging.
(2) The fuel pellets are chamfered, again reducing ridging.
(3) The maximum linear heat generation is reduced from 61 kW/m to 44 kW/m, reducing both thermal distortion and ridging.
(4) The heat treatment procedures for the clad have been modified to reduce the clad variability.

At any one time less than 1 percent of the fuel rods in the core should experience the peak linear heat generation rate (LHGR). The long-term LHGR for elements seeing this peak generation rate is of the order of 26 to 33 kW/m. The local design peak exposure for these pellets is 45 000 MWd/ton with a 30 000-MWd/ton maximum assembly exposure. The core residence time for a fuel assembly is 4 yr when partial refueling occurs annually, and approximately $4\frac{1}{2}$ yr when an 18-month refueling cycle is used.

Figure 13.15 shows the core internals in a schematic fashion, and in particular, shows the jet pumps that provide recirculation of water through the core. As water passes up through the core, steam is generated and its buoyancy causes it to rise. The vapor percentage (quality) at the end of the fuel channel is of the

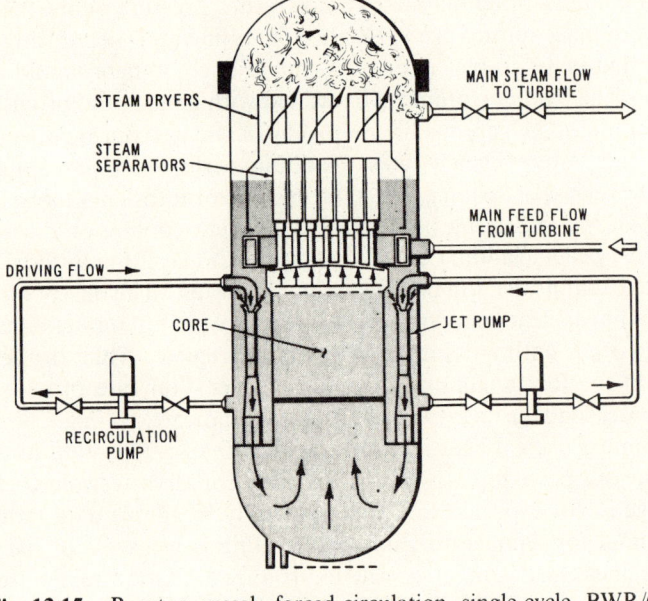

STEAM DRYERS

STEAM
SEPARATORS

DRIVING FLOW →

CORE

RECIRCULATION
PUMP

MAIN STEAM FLOW
TO TURBINE

MAIN FEED FLOW
FROM TURBINE

JET PUMP

Fig. 13.15   Reactor vessel, forced-circulation single-cycle BWR/6.
Company.]

order of 9 percent. Steam separators and driers in the top of the core vessel
separate the steam and water. Dry steam passes to the turbine and water is
recirculated. About one-third of the total flow is taken by the recirculation
pumps and discharged to the throat of the diffusers, thus inducing more flow
through the diffusers to augment the tendency of natural circulation to produce
flow. Figure 13.16 offers a more detailed view of the reactor internals. Between
75 percent and 100 percent power, the output level may be maintained by variation of the recirculation flow rate while the control rods remain stationary.
Up to 30 percent power may be generated using only natural circulation.

Control of single-cycle BWRs may be accomplished by either of the modes
shown in Fig. 13.17. In Fig. 13.17a, the turbine is slaved to the reactor. Pressure is held at a fixed level by an initial pressure regulator. The turbine governor
then adjusts the position of the control rods or the recirculation rate to change
the amount of steam being generated. As indicated previously, a 25 percent
change in power level may be accomplished by modulating the recirculation
flow through the core while the control rods remain stationary. To raise the
power level, the flow rate through the core is increased, reducing the void fraction and increasing the reactivity. This, in turn, raises the thermal output.
The increased power output will stabilize when the temporary excess of reactivity is balanced by the higher void fraction. To reduce the power level,
the recirculation flow rate is reduced.

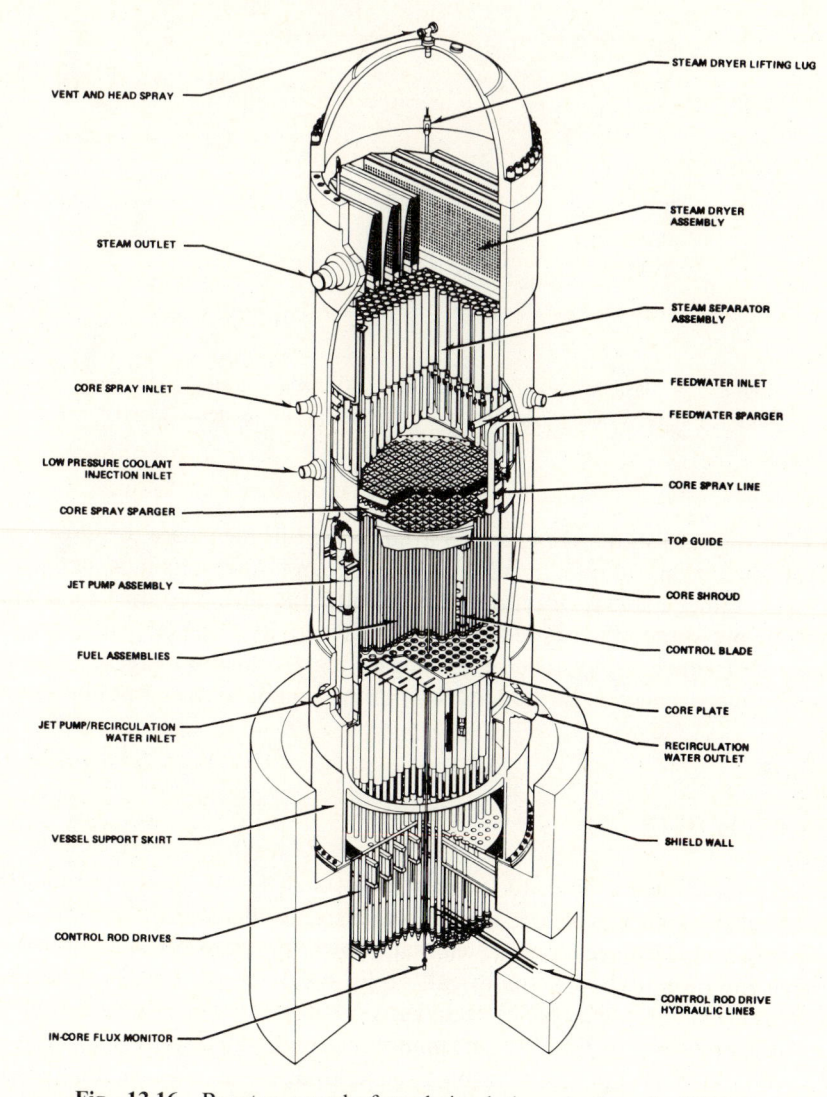

VENT AND HEAD SPRAY

STEAM OUTLET

CORE SPRAY INLET

LOW PRESSURE COOLANT
INJECTION INLET

CORE SPRAY SPARGER

JET PUMP ASSEMBLY

FUEL ASSEMBLIES

JET PUMP/RECIRCULATION
WATER INLET

VESSEL SUPPORT SKIRT

CONTROL ROD DRIVES

IN-CORE FLUX MONITOR

STEAM DRYER LIFTING LUG

STEAM DRYER
ASSEMBLY

STEAM SEPARATOR
ASSEMBLY

FEEDWATER INLET

FEEDWATER SPARGER

CORE SPRAY LINE

TOP GUIDE

CORE SHROUD

CONTROL BLADE

CORE PLATE

RECIRCULATION
WATER OUTLET

SHIELD WALL

CONTROL ROD DRIVE
HYDRAULIC LINES

Fig. 13.16   Reactor vessel, forced-circulation single-cycle BWR/6.
[Courtesy of General Electric Company.]

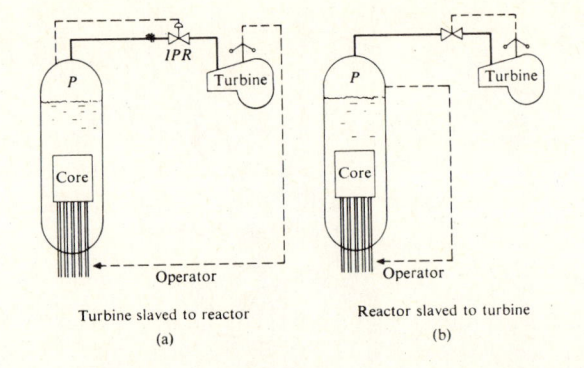

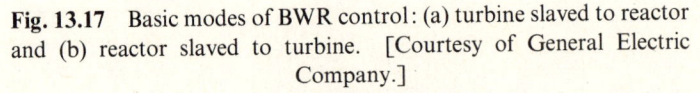

Fig. 13.17   Basic modes of BWR control: (a) turbine slaved to reactor and (b) reactor slaved to turbine. [Courtesy of General Electric Company.]

Figure 13.17b shows the reactor slaved to the turbine where the governor controls the turbine admission valves. A change in flow results in a change in pressure. Either manual or automatic repositioning of the control rods will return the pressure to the desired level. Since reactivity is sensitive to changes in pressure (reduced void fraction accompanying a pressure increase will enhance the effectiveness of the moderator), this method of control is not popular.

Dual-Cycle BWR

In the dual-cycle BWR (Fig. 13.18) only about half the steam is generated in the core at the higher pressure. On its way back to the core the recirculated water is subcooled by generating steam at a lower pressure. The 200-MW(e) Dresden 1 Station of the Commonwealth Edison Company uses this cycle and generates primary steam at 6.9 MPa and secondary steam at 3.6 MPa for maximum flow (this rises to 6.55 MPa at minimum flow). High-pressure steam is admitted to the first stage of the turbine and secondary steam is admitted at the ninth stage. Some of the operating conditions are shown for the cycle in Table 13.6.

A very interesting characteristic of this system is its ability to respond to load changes without requiring any motion of the control rods. This is accomplished by allowing the turbine governor to regulate the admission of secondary steam to the turbine. The increased flow of secondary steam creates more subcooling in the recirculation line. The cooler water entering the core reduces the void fraction somewhat, increasing the effectiveness of the water as a moderator. This, in turn, raises the reactivity of the core, causing it to be slightly supercritical. The output rises until the void fraction increases sufficiently to

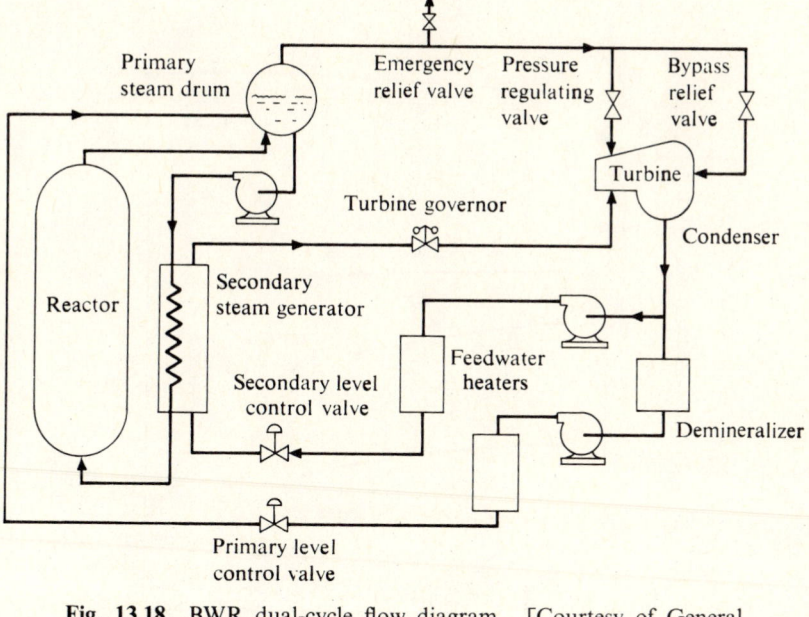

Fig. 13.18 BWR dual-cycle flow diagram. [Courtesy of General Electric Company.]

Table 13.6 Dresden 1 Rated Operating Conditions

Power level	630 MW(th)
	180 MW(e)
Recirculation flow*	1.13×10^6 kg/h
Primary steam flow	0.635×10^6 kg/h
Secondary steam flow	0.544×10^6 kg/h
Inlet core subcooling	123 kJ/kg
Average core exit quality	5%
Peak heat flux (at overpower)†	1.1×10^6 W/m²
Minimum burnout ratio (at overpower)	2.1
Maximum void fraction at hot	
channel exit (at overpower)	65%
Thermal efficiency	
Test value at rated power	29.75%
Test value at 60% power	29.45%

* During system tests recirculation was measured at 12.3×10^6 kg/h.
† Core loading of 452 fuel assemblies having a peaking factor of 3.64.

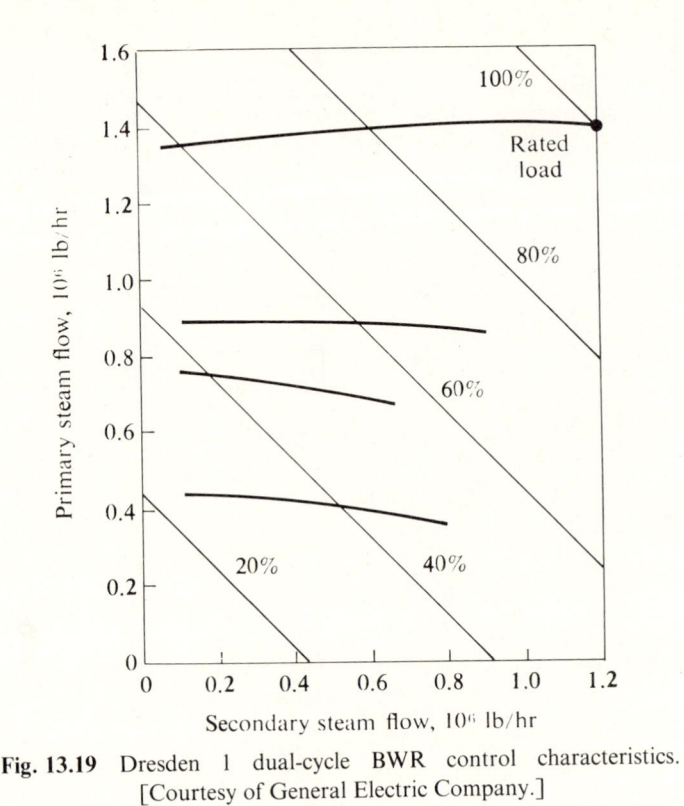

Fig. 13.19 Dresden 1 dual-cycle BWR control characteristics.
[Courtesy of General Electric Company.]

reduce the effective multiplication factor to unity. The increased output tends
to raise the primary steam pressure and open the primary admission valves
wider. Reduced load demands reverse this sequence. Figure 13.19 shows the
relation between primary and secondary steam flows for the Dresden 1 station
for various fixed control rod settings. Load changes of 40 percent may be
accommodated with a particular control rod setting.

Comparison of Dual- and Single-Cycle BWRs

The dual cycle has the advantage of inherent load-following capability by
regulation of secondary steam flow to the turbine. This has been matched by
the single-cycle units with the advent of flow control applied to the core recirculation rate. The dual cycle has the ability to produce more power from a given
size of core vessel because of the generation of the secondary steam outside the
core proper. This could be a distinct advantage for siting conditions where
transportation of a large core vessel is a serious problem.

The simplicity of the single cycle allows a lower capital investment for
piping, heat exchangers, pumps, and so on. It also has a somewhat better

Table 13.7 Design Progress—BWR Technology

Characteristics	Dresden 1[a] (1960)*	Dresden 2[a] (1970)*	Brown's Ferry 1[b] (1974)*	Hartsville A1[b] (1986)*
Power rating [MW(e) (net)]	200	714	1080	1233
Type reactor	Dual cycle	Single cycle	Single cycle	Single cycle
Systems and hardware				
Containment	Dry sphere	Pressure suppression	Pressure suppression	Pressure suppression
Secondary steam generators	4	0	0	0
Recirculation loops	4	2 (jet pumps)	2 (jet jumps)	2 (jet pumps)
Number of fuel bundles	488	724	764	748
Control rods	80	177†	185	177
Rods/MW(e)	0.4	0.25	0.171	0.147
Steam separation	External steam drum	Internal	Internal	Internal
Power density				
Core average (th) [kW(th)/liter]	28	37†	51	54
Refueling time (days)	33	15–20	—	4–7
Fuel exposure (MWd/ton)	8000	15 000–20 000‡	19 500–25 000	27 800
Plant heat rate (Btu/kWh)	11 816	11 300	10 950	10 657

* Year of operation (or expected operation).
† Based on initial rating of 714 MW(e) (net).
‡ Equilibrium core is 20 000 MWd/ton.
[a] From V. A. Elliott, "Boiling Water Reactor," *Mechanical Engineering* **89**, No. 1 (January 1967), pp. 19–26.
[b] Courtesy of Tennessee Valley Authority.

thermal efficiency because of the direct generation of all steam at the maximum cycle temperature. The loss in availability as heat is transferred from the re-circulated water to the secondary steam costs the cycle about 1 percent of its thermal efficiency at rated load. A study of comparable cycles working between 6.9 MPa and 3.8 cm Hg gave an efficiency of 30.9 percent for the single-cycle plant, as compared to 29.7 percent for the dual-cycle plant.

Table 13.7 shows an interesting comparison of the Dresden 1 and 2, Brown's Ferry 1, and the Hartsville A1 plants, which indicates the progress in BWR design. Table 13.8 shows an economic comparison between the BWR plant that is planned for Hartsville and the low- and medium-sulfur coal-fired alternatives. The advantage is clearly in favor of the nuclear alternative.

Table 13.8 Hartsville Project—Economic Comparison of Alternative Baseload Alternatives

		Coal-Fired	
	Nuclear	Low Sulfur	Medium Sulfur*
Net power heat rate (kJ/kWh)	10 654	9573	9609
Fuel cost (¢/10^6 kJ)	33	152	104
Investment ($/kW) (net)†	519	488	500
Annual use (h)	7000	7000	7000
Generating cost (mills/kWh)			
Investment	6.4	6.0	6.1
Fuel	3.5	14.5	10.1
Operation and maintenance	1.2	1.5	4.4
Total	11.1	22.0	20.6

* Includes estimates of SO_2 removal equipment capital and operating costs.
† All capital-cost estimates include cooling towers.
Courtesy of Tennessee Valley Authority.

Gas-Cooled Reactors

Gas cooling for a reactor core will allow substantially higher temperatures for the working fluid than is possible when water is the cooling medium. Carbon dioxide was used in the European Magnox Reactors, but as cycle temperatures have risen helium has come into favor. It is inert with respect to materials of construction and ^{4}He has no neutron absorption (there is 0.000 13 percent ^{3}He which undergoes an (n, p) reaction to form tritium). The majority of gas-cooled plants built thus far generate steam at conditions equivalent to modern fossil practice, but the German KSH plant, described later in this chapter, uses the helium directly in a gas turbine cycle.

The High-Temperature Gas-Cooled Reactor in the United States started with the 40-MW(e) prototype at Peach Bottom, Pennsylvania, which went critical in 1966. This has been followed by a 330-MW(e) plant completed in 1976. It is this 330-MW(e) Fort St. Vrain plant which will be described in some detail.

The plant is graphite moderated, helium cooled, and fueled with coated microspheres of thorium and fully enriched uranium embedded in a carbon-aceous binder. The hexagonal fuel element which was described in Chapter 11 and shown in Fig. 11.41 contains 210 fuel cavities and 100 coolant passages. These fueled graphite blocks are stacked in columns which are arranged seven columns to each of 37 refueling regions. Figure 13.20 shows the stacking of the fuel elements in a segment of the core. There are 2 B_4C control rods for each of the refueling regions and the He flow is orificed to each region to attain a

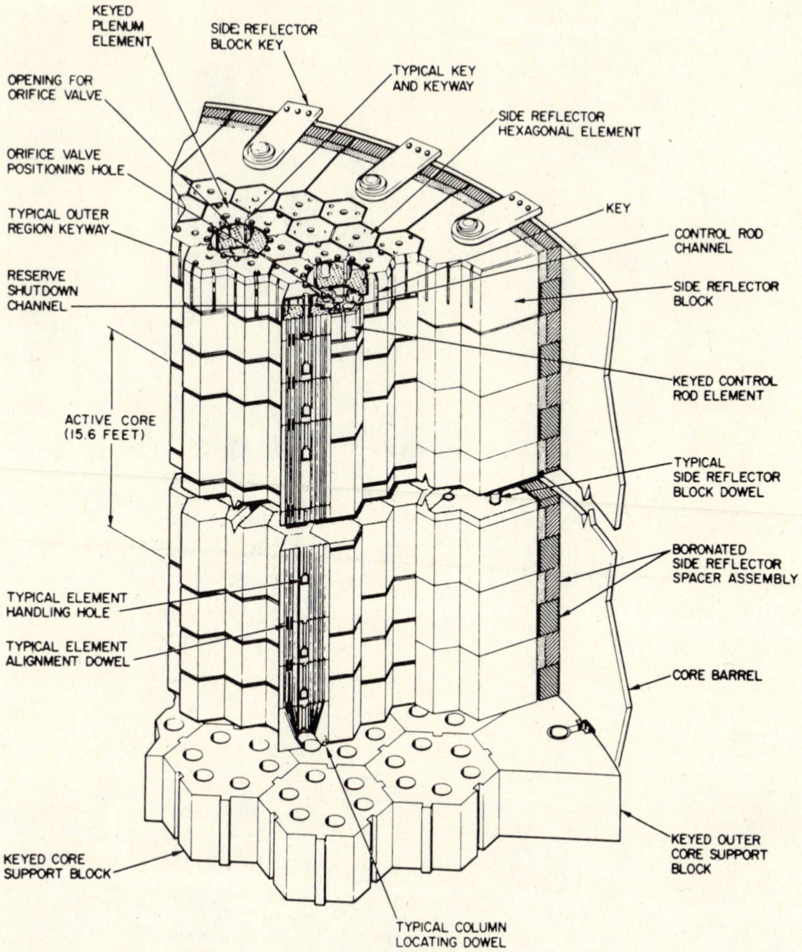

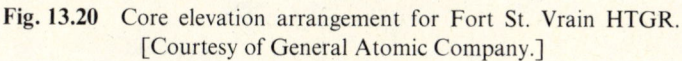

KEYED
PLENUM
ELEMENT

SIDE REFLECTOR
BLOCK KEY

TYPICAL KEY
AND KEYWAY

OPENING FOR
ORIFICE VALVE

SIDE REFLECTOR
HEXAGONAL ELEMENT

ORIFICE VALVE
POSITIONING HOLE

KEY

TYPICAL OUTER
REGION KEYWAY

CONTROL ROD
CHANNEL

SIDE REFLECTOR
BLOCK

RESERVE
SHUTDOWN
CHANNEL

KEYED CONTROL
ROD ELEMENT

ACTIVE CORE
(15.6 FEET)

TYPICAL
SIDE REFLECTOR
BLOCK DOWEL

BORONATED
SIDE REFLECTOR
SPACER ASSEMBLY

TYPICAL ELEMENT
HANDLING HOLE

TYPICAL ELEMENT
ALIGNMENT DOWEL

CORE BARREL

KEYED CORE
SUPPORT BLOCK

KEYED OUTER
CORE SUPPORT
BLOCK

TYPICAL COLUMN
LOCATING DOWEL

Fig. 13.20   Core elevation arrangement for Fort St. Vrain HTGR.
[Courtesy of General Atomic Company.]

uniform discharge temperature for the hot helium. In the helium circuit the
core heats helium from 404 to 777°C.

Figure 13.21 shows a schematic diagram for the plant power cycle. Hot
helium passes from the core down to two steam generators each containing
six modules. The modules are once-through steam generators producing superheated steam at 16.5 MPa and 538°C. The internal piping for one of these
modules is shown in Fig. 13.22. After expansion through the high-pressure
turbine steam returns to the reheat section of the steam generator modules via
single-stage turbines driving the four helium circulators for the core. After
reheat to 538°C steam at 4.14 MPa passes to an intermediate turbine and then

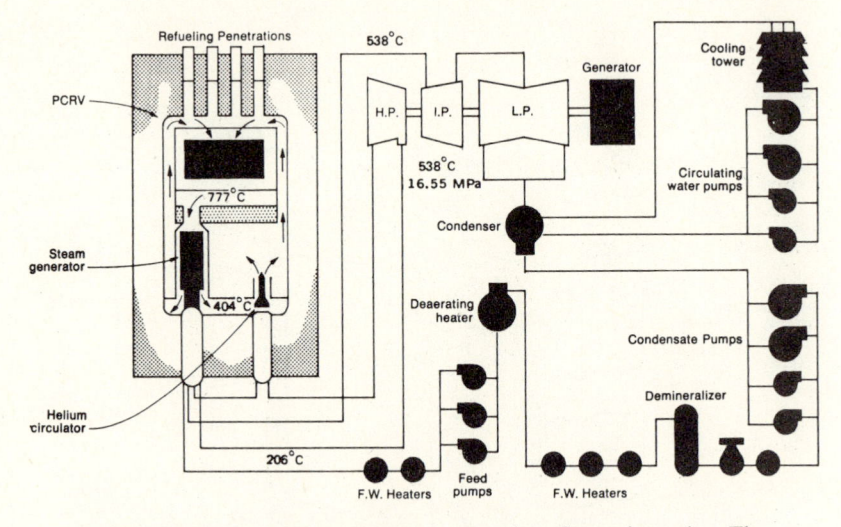

Fig. 13.21 Fort St. Vrain HTGR power plant flow schematic. The temperatures and pressures will differ somewhat in the larger HTGRs. [Courtesy of General Atomic Company.]

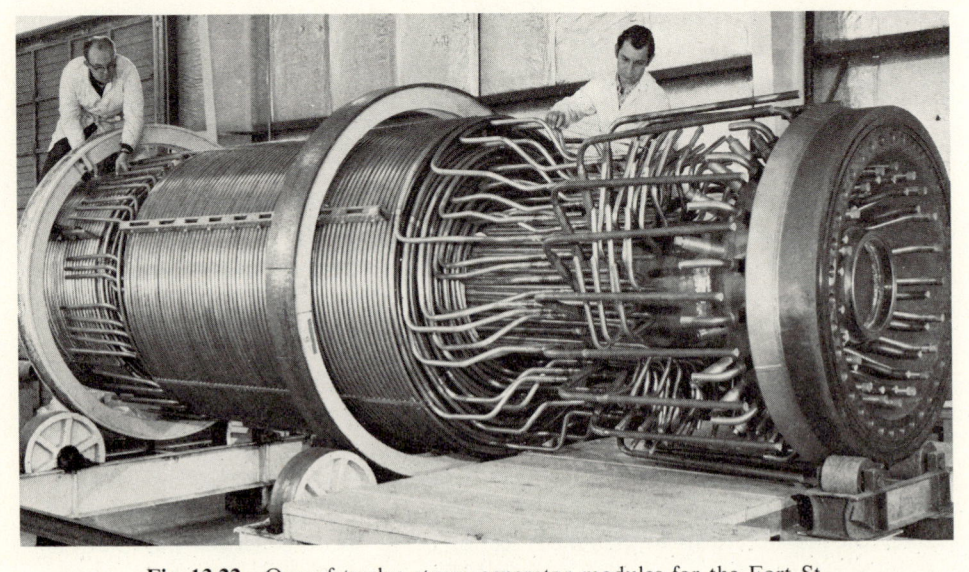

Fig. 13.22 One of twelve steam generator modules for the Fort St. Vrain HTGR. [Courtesy of General Atomic Company.]

through a double-flow low-pressure turbine before being condensed. The condensate passes through three low-pressure feedwater heaters, a deaerating heater, and two high-pressure heaters before returning to the economizer section of the steam generator modules at 206°C.

The entire primary circuit consisting of core, steam generator modules, and helium circulators is contained in a Prestressed Concrete Reactor Vessel (PCRV) which is more or less a hexagonal prism, 18.6 m across flats and 32.3 m high. The internal cavity has an internal diameter of 22.9 m and an internal height of 9.45 m. The upper and lower heads have a nominal thickness of 4.6 m. The vessel provides not only containment of the 4.83-MPa helium, but also radiological shielding. Steel tendons pass through tubes in the concrete and when properly tensioned they will place the concrete under enough compression to more than offset the tensile stress due to the internal helium pressure. The arrangement of the reactor within the PCRV is shown in Fig. 13.23. Penetrations in the top head provide for refueling and control rod drives.

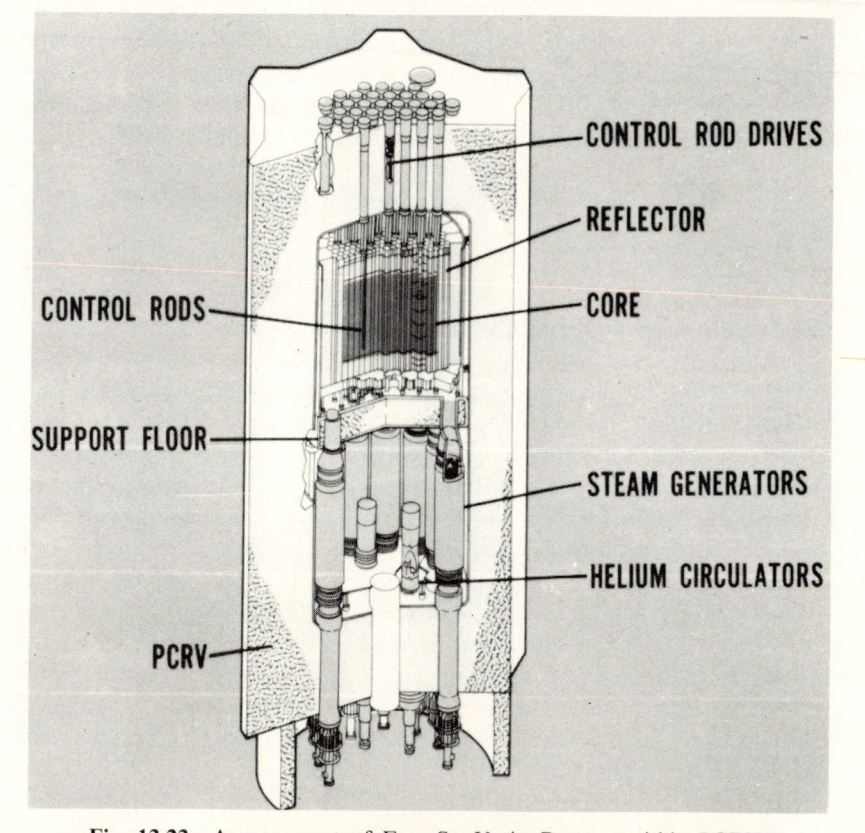

Fig. 13.23 Arrangement of Fort St. Vrain Reactor within PCRV. [Courtesy of General Atomic Company.]

Control over the plant output is by a reactor-follow-turbine system. The governor on the turbine controls steam flow. It is then desired to maintain the steam pressure and temperature at the throttle and the reheat steam temperature. The steam pressure at the throttle is maintained by variation of feedwater flow, the throttle steam temperature is controlled by variation of helium flow, and the reheat steam temperature is controlled by variation of control rod position in the core.

The ^{235}U–^{232}Th–^{233}U fuel cycle being employed is an interesting one. As burnup continues an increasing fraction of the fission occurs in the ^{233}U which has been bred from the fertile thorium. Since the value of η for ^{233}U is higher than that for ^{235}U, this helps the conversion ratio, which is expected to be 0.62 for the equilibrium core. In the all-ceramic core there is no structural metal with its parasitic capture of neutrons and the coolant has virtually no absorption. If sufficient ^{233}U becomes available for the fissile UC_2 microspheres, a higher conversion ratio should be possible.

Plans were fairly well developed for two 770-MW(e) HTGRs for the Delmarva Power and Light Company's Summit Station and also for two 1140-MW(e) units for Philadelphia Electric's Fulton 1 and 2, when the contracts were canceled late in 1975—apparently due to escalating costs. General Atomic announced that it was accepting no further orders for steam-cycle HTGRs. For the time being, it is unclear whether there will be any further units of this type built in the United States or whether development will proceed to direct-cycle HTGRs and VHTRs (very high temperature reactors) for process heating.

Direct-Cycle Gas Turbine Cycles

The next step beyond the steam-generating HTGR will be a high-temperature gas-cooled reactor using a direct helium gas turbine cycle. Figure 13.24 shows a simple regenerative gas turbine cycle coupled to a dry cooling tower. The high temperature of the helium entering the precooler (283°C) makes the use of the dry cooling tower quite feasible. An intermediate water loop transports the rejected energy between the precooler and the tower, and in the case illustrated, air leaves the tower at 116°C.

Most of the components for a direct-cycle HTGR may be contained in vaults within the PCRV, as shown in Fig. 13.25. Although efficiencies could be improved by multistage compression with intercooling, many initial studies have been based on the simple regenerative cycle. An exception is the 25-MW(e) KSH plant in Germany, which incorporates three stages of compression with intercooling (see Fig. 13.26).

Loads between 50 and 100 percent may be varied by controlling the density of the helium within the system. Pressure ratios and temperatures around the circuit are affected little by the addition or subtraction of helium. At loads less

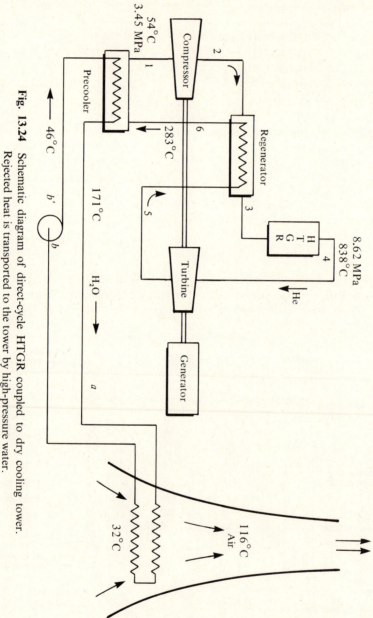

Fig. 13.24 Schematic diagram of direct-cycle HTGR coupled to dry cooling tower. Rejected heat is transported to the tower by high-pressure water.

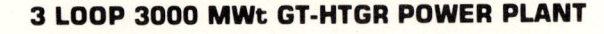

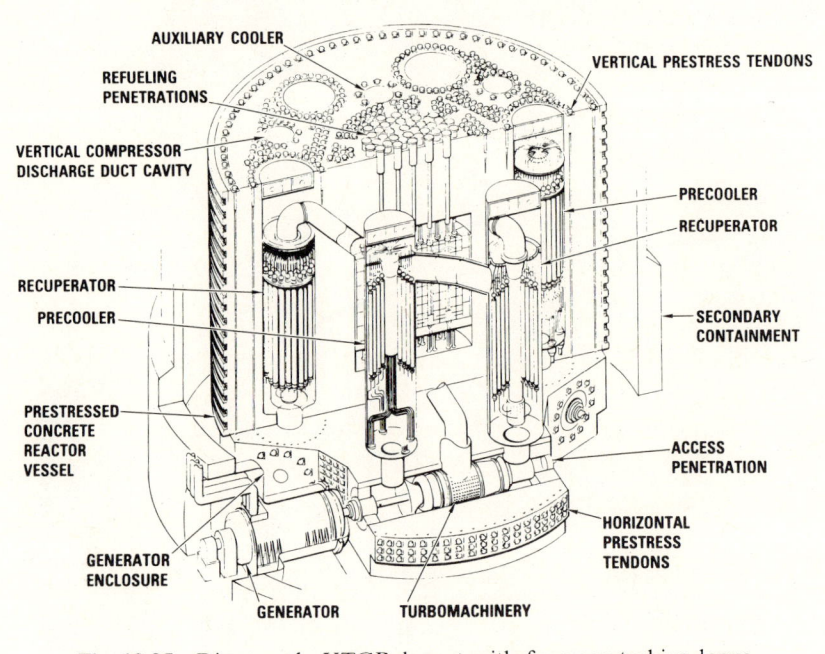

Fig. 13.25 Direct-cycle HTGR layout with four gas turbine loops. Turbines, compressors, precooler, and recuperator are all contained within vaults in the PCRV. [Courtesy of General Atomic Company.]

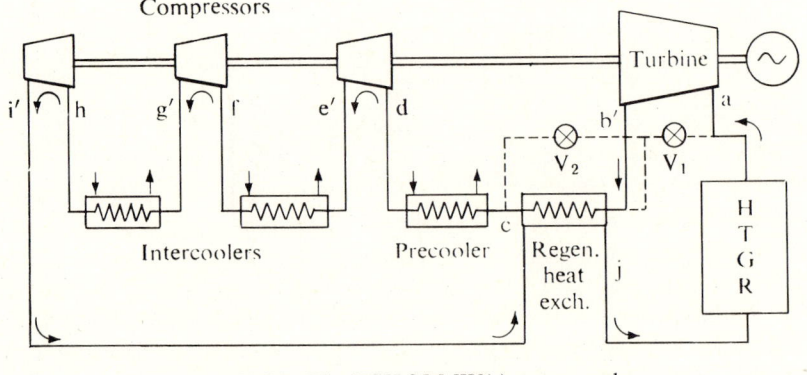

Fig. 13.26 The KSH 25-MW(e) power cycle.

than 50 percent some of the hot gas can be bypassed around the turbine through the valve (V1) shown in Fig. 13.26. As this gas is much hotter than the turbine exhaust, it also is necessary to bypass some gas through valve V2 around the regenerator. On the KSH plant, the reactor power required for 50 percent electrical output is 51 percent of the full reactor power when the reduction is by inventory depletion, but it is 92 percent when the reduction is accomplished by utilizing bypass control.

Serious consideration is being given to gas cooling for fast breeder reactors. The direct cycle offers much in the way of capital cost reduction as the double energy interchange of the sodium-cooled fast reactor systems is eliminated. A proposed helium-cooled fast breeder is discussed later in this chapter along with other breeders.

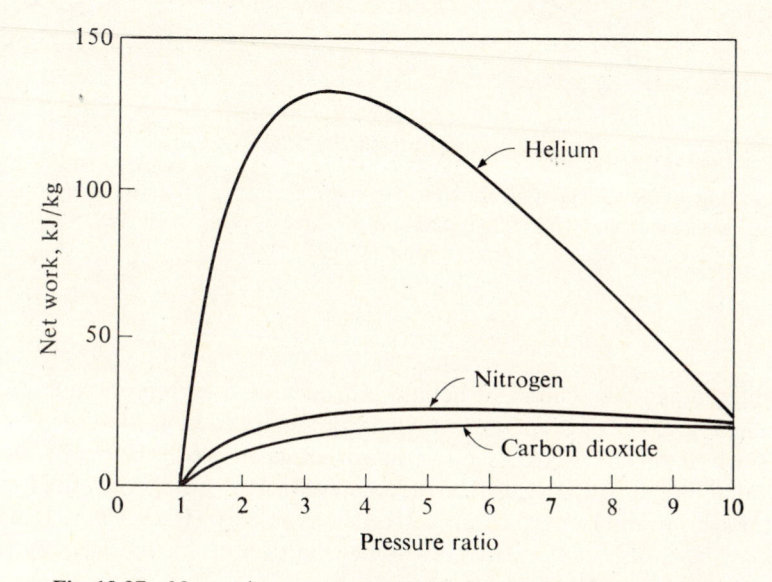

Fig. 13.27 Net work versus pressure ratio for gas-cooled reactor direct cycle. Compressor efficiency, 80 percent; turbine efficiency, 85 percent.

The choice of working fluid for a gas-cooled reactor must consider the questions of neutron activation, required pressure levels, cost, and the properties of the gas as they affect the cycle performance. The early Magnox reactors in England and France, as well as their successors, the advanced gas-cooled reactors, use carbon dioxide as the coolant. The high-temperature gas-cooled reactors are using helium. Figures 13.27 and 13.28 show how work per kilogram and thermal efficiencies vary with the ratio of high-side to low-side pressure in gas turbine cycles either with or without a heat exchanger for three different

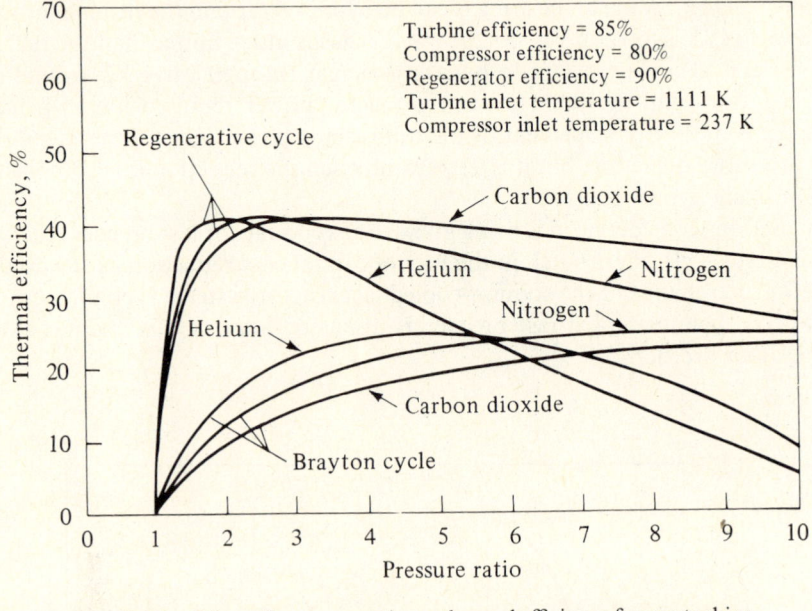

Turbine efficiency = 85%
Compressor efficiency = 80%
Regenerator efficiency = 90%
Turbine inlet temperature = 1111 K
Compressor inlet temperature = 237 K

Fig. 13.28 Effect of pressure ratio on thermal efficiency for gas turbine cycles with and without a regenerator for three working fluids (He, N_2, and CO_2).

working fluids. The choice of helium, nitrogen, and carbon dioxide gives a considerable range of properties which are summarized in Table 13.9.

Carbon dioxide was chosen for the earlier gas-cooled reactors because of its availability at low cost, its inertness in contact with materials used in the reactor system, and lack of serious activation problems. However, as temperatures increase, dissociation becomes somewhat of a problem. Helium, being

Table 13.9 Properties of Working Fluids Used in Gas-Cooled Reactor Comparisons

Gas	Molar Mass	k	(kJ/kg K)
He	4.003	1.667	5.191
N_2	28.016	1.4	1.038
CO_2	44.01	1.285	0.852

inert and monatomic and having no neutron absorption cross section, is attractive at high temperatures. It should be noted in Fig. 13.27 that the maximum work per kilogram for helium occurs at a much lower pressure ratio than CO_2. The nitrogen was chosen as an inert diatomic gas with properties intermediate between the other two.

For an ideal Brayton cycle, the thermal efficiency will continuously increase with pressure ratio. However, when the turbine and compressor efficiencies are considered, and fixed values of turbine and compressor inlet temperatures are used, the efficiency will eventually peak. In Fig. 13.28 this is evident only for the helium, but will eventually occur for the N_2 and CO_2 as well. If a regenerator is incorporated into the ideal cycle with given values of turbine and compressor inlet temperatures, the highest efficiency occurs at a pressure ratio of 1. The incorporation of efficiencies for the turbine, compressor, and regenerator makes the efficiency drop to zero at a pressure ratio of unity, but the efficiency peaks at pressure ratios much lower and at values much higher than for the Brayton cycle, with comparable efficiencies for the turbine and compressor. For helium the regenerative cycle efficiency peaks at a pressure ratio of 2.0 and the work per kilogram peaks at a pressure ratio of 3.25. In the KSH plant, although the cycle is somewhat more complex, the ratio of turbine inlet pressure to exhaust pressure is 2.54, which indicates a design compromise between maximum work and efficiency. Note that the Brayton cycle peaked at an efficiency just under 25 percent at a pressure ratio of 4.75 for the cycle conditions studied compared with 41 percent with the regenerator. The advantage of the regenerator is evident. To attain high thermal efficiencies a large regenerator is required. Some recent studies indicate that if the desired thermal efficiency is reduced somewhat to 30 or 35 percent rather than striving for the 40 to 50 percent which is attainable, the savings in regenerator investment may produce minimum power cost.

Example 4 A 1200-MW(e) direct-cycle HTGR (see Fig. 13.24) admits helium to the gas turbine(s) at 85 bars and 1127°C. At the inlet to the compressor the pressure is 34 bars and the temperature of the helium is 50°C. The compressor and turbine efficiencies may be taken as 85 percent and the regenerator effectiveness as 80 percent. There is assumed to be a 2 percent pressure loss as the helium passes through each heat exchanger. Compute the thermal efficiency for the cycle and the required helium flow rate (see Fig. 13.29).

The various pressures are

$$p_3 = 0.98 \times 85 = 83.3 \text{ bars}$$

$$p_4 = 0.98 \times 83.3 = 81.63 \text{ bars}$$

$$p_6 = 34/0.98 = 34.69 \text{ bars}$$

$$p_5 = 34.69/0.98 = 35.40 \text{ bars}$$

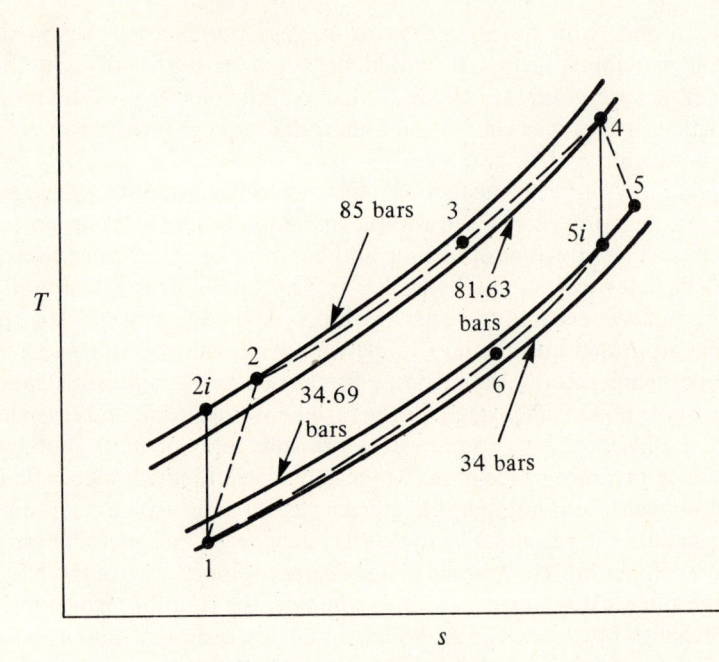

Fig. 13.29

The ideal temperatures at the end of isentropic compression and expansion are

$$T_{2i} = T_1(p_2/p_1)^{(k-1)/k}$$
$$= (50 + 273)(85/34)^{0.667/1.667} = 466.0 \text{ K}$$

$$T_{5i} = T_4(p_5/p_4)^{(k-1)/k}$$
$$= (1127 + 273)(35.40/81.63)^{0.667/1.667} = 1002.2 \text{ K}$$

Applying the turbine and compressor efficiencies allows the determination of the actual compressor and turbine discharge temperatures.

$$W_{12} = W_{12i}/\eta_c = \frac{c_p(T_1 - T_{2i})}{\eta_c} = c_p(T_1 - T_2)$$

$$T_2 = T_1 + (T_{2i} - T_i)/\eta_c = 323 + 0.85(1400 - 1002.2)$$
$$= 491.2 \text{ K}$$

$$W_{45} = \eta_e W_{45i} = c_p(T_4 - T_{5i})\eta_e = c_p(T_4 - T_5)$$

$$T_5 = T_4 - (T_4 - T_{5i})\eta_e = 1400 - (1400 - 1002.2) \times 0.85$$
$$= 1061.9 \text{ K}$$

The regenerator effectiveness establishes the actual temperature rise of the high-pressure gas as a fraction of the maximum possible temperature rise

$(T_5 - T_2)$. The low-pressure turbine exhaust gas has an equal temperature drop.

$$\eta_{reg} = \frac{\Delta h_{actual}}{\Delta h_{max}} = \frac{c_p(T_3 - T_2)}{c_p(T_5 - T_2)} = \frac{c_p(T_5 - T_6)}{c_p(T_5 - T_2)}$$

$$T_3 = T_2 + \eta_{reg}(T_5 - T_2)$$
$$= 491.2 + 0.80(1061.9 - 491.2) = 947.8 \text{ K}$$

$$T_6 = T_5 - \eta_{reg}(T_5 - T_2)$$
$$= 1061.9 - 0.80(1061.9 - 491.2) = 605.34 \text{ K}$$

Because of the regenerative heating, the helium needs only to be heated from T_3 to T_4 as it passes through the HTGR core.

$$Q_a = c_p(T_4 - T_3) = 5.2(1400 - 947.8) = 2351.4 \text{ kJ/kg}$$

The sum of the turbine and compressor works gives the net work for the cycle.

$$W_{net} = c_p(T_4 - T_5) + c_p(T_1 - T_2)$$
$$= 5.2(1400 - 1061.9) + 5.2(323 - 491.2)$$
$$= 1758.1 - 874.6 = 883.5 \text{ kJ/kg}$$

This results in a thermal efficiency of

$$\eta_{th} = W_{net}/Q_a = 883.5/2351.4 = 0.3757$$

The helium flow rate required may be found by dividing the power (kJ/h) by the net work per kg.

$$\dot{m}_{He} = \frac{P \times 3.6 \times 10^6}{W_{net}} = \frac{1200 \times 3.6 \times 10^6}{883.5} = 4.89 \times 10^6 \text{ kg/h}$$

Pressurized Heavy-Water Reactors

The Canadian Deuterium Uranium (CANDU) reactor uses heavy water (D_2O) as the moderator, natural uranium oxide as the fuel, and pressurized heavy water as the coolant. The cool moderator is unpressurized and contained in a cylindrical calandria vessel, which is traversed by concentric calandria and pressure tubes. The pressure tubes contain bundles of fuel rods that are cooled by the pressurized heavy water. Figure 13.30 shows a simplified CANDU pressurized heavy-water flow diagram. Heat from the coolant generates light-water steam in boilers, and this steam drives turbines in a cycle similar to those used with PWRs. Refueling takes place while the reactor is on-line, thus eliminating annual shutdowns of several weeks for this purpose, as required by most other reactor types.

The steam generating heavy-water reactor (SGHWR) generates steam from light water pumped into vertical pressure tubes where steam of about 16

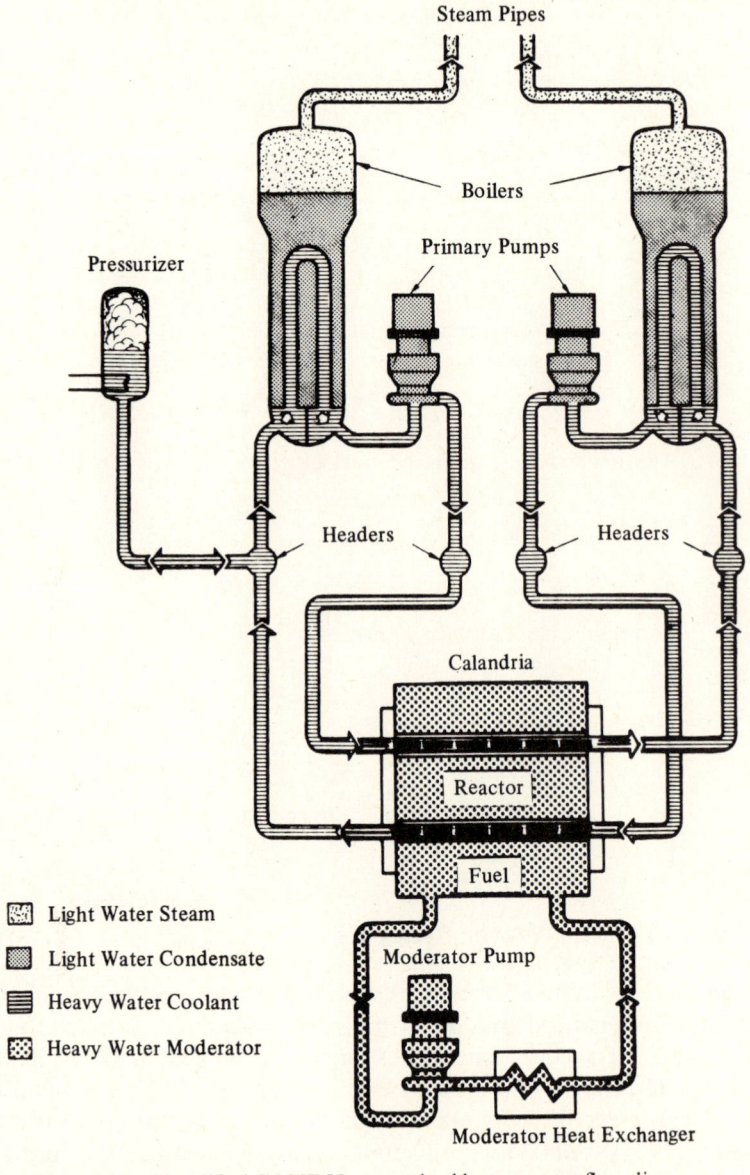

Fig. 13.30 Simplified CANDU pressurized heavy-water flow diagram.
[Courtesy of Atomic Energy of Canada Limited.]

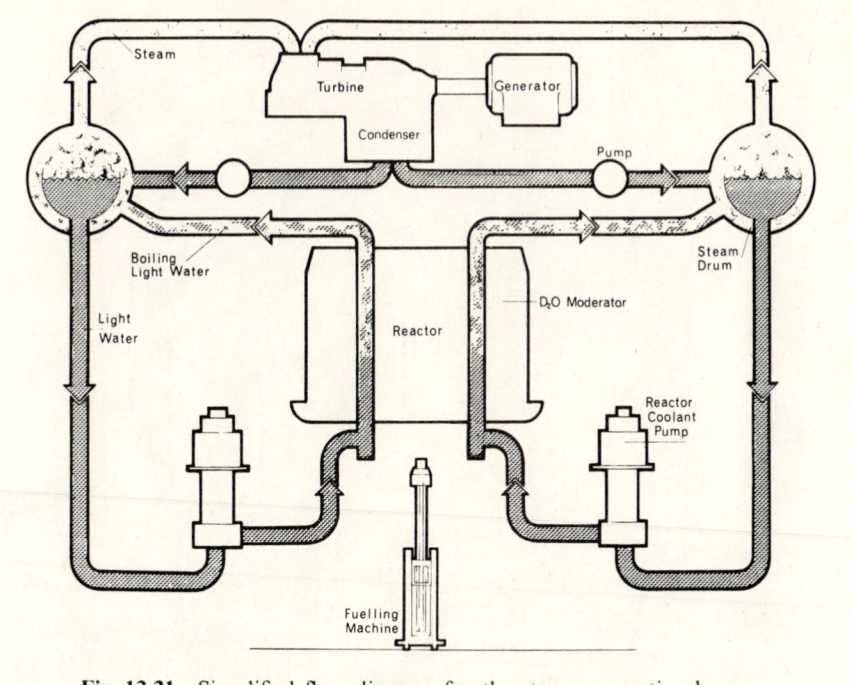

Fig. 13.31 Simplified flow diagram for the steam generating heavy-water reactor (SGHWR) of the type used for the Gentilly 1 nuclear power station. [Courtesy of Atomic Energy of Canada Limited.]

percent quality leaves the tube and then has the vapor separated from the liquid in a steam drum. Figure 13.31 shows a simplified flow diagram for the 250-MW(e) Gentilly 1 station. Steam enters the tubes at 6.34 MPa, leaves the tubes at 5.55 MPa and arrives at the turbine at 5.17 MPa and 266°C. Between the high- and low-pressure turbine, there are moisture separators and live steam reheaters similar to those previously described for the PWR. The cycle uses seven stages of feedwater heating and attains a thermal efficiency of 34.45 percent. This reactor has served as a prototype for a significant number of SGHWRs being built in Great Britain.

The reactor described here in more detail is Gentilly 2, which is the first of a standardized type 600-MW(e) plant developed for use both in Canada and abroad. The Rio Tercero plant located in Argentina's Cordoba Province is also of this type.

Figure 13.32 shows the general layout of the plant's nuclear steam supply, and Fig. 13.33 shows the reactor in more detail with its myriad feeder tubes connecting the boilers to the reactor. The calandria vessel is traversed by 380 fuel channel assemblies. Each fuel channel assembly consists of a zirconium–2.5 percent niobium alloy pressure tube 10.34 cm (4.07 in.) I.D. × 6.30 m (20 ft 8 in.)

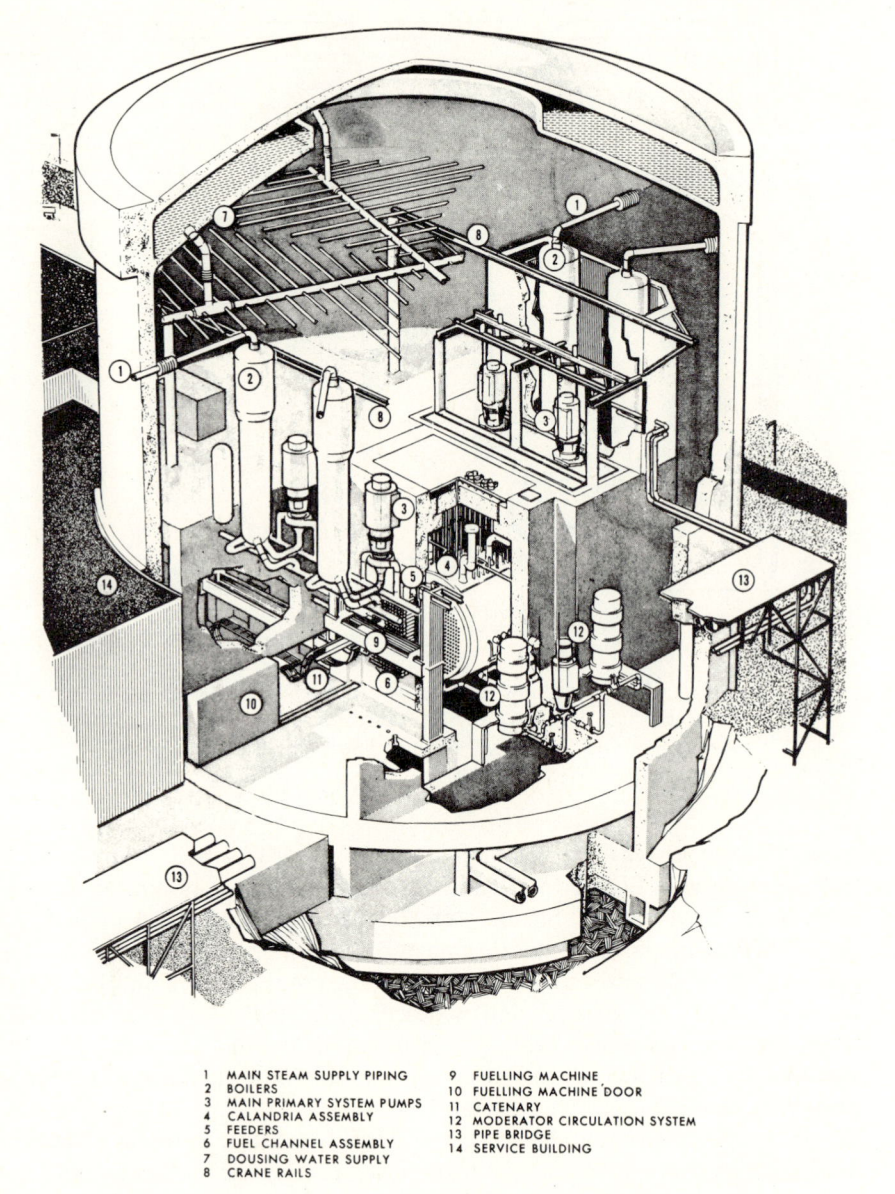

1	MAIN STEAM SUPPLY PIPING	9	FUELLING MACHINE
2	BOILERS	10	FUELLING MACHINE DOOR
3	MAIN PRIMARY SYSTEM PUMPS	11	CATENARY
4	CALANDRIA ASSEMBLY	12	MODERATOR CIRCULATION SYSTEM
5	FEEDERS	13	PIPE BRIDGE
6	FUEL CHANNEL ASSEMBLY	14	SERVICE BUILDING
7	DOUSING WATER SUPPLY		
8	CRANE RAILS		

Fig. 13.32 General layout of the nuclear steam supply system for a CANDU pressurized heavy-water reactor. [Courtesy of Atomic Energy of Canada Limited.]

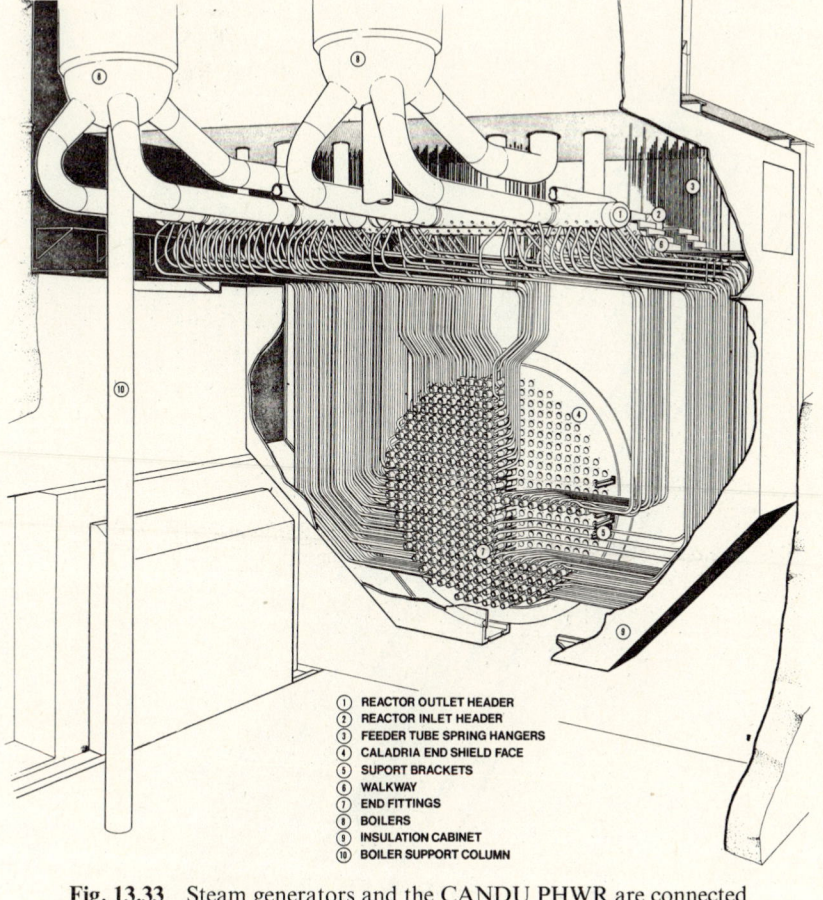

1. REACTOR OUTLET HEADER
2. REACTOR INLET HEADER
3. FEEDER TUBE SPRING HANGERS
4. CALADRIA END SHIELD FACE
5. SUPORT BRACKETS
6. WALKWAY
7. END FITTINGS
8. BOILERS
9. INSULATION CABINET
10. BOILER SUPPORT COLUMN

Fig. 13.33 Steam generators and the CANDU PHWR are connected by myriad feeder tubes. [Courtesy of Atomic Energy of Canada Limited.]

containing 12 fuel bundles, each with 37 tubes and weighing approximately 23.7 kg, of which 18.5 kg is uranium in the form of UO_2 with a density of 10.6 g/cm^3. The UO_2 pellets have a length of 1.64 cm and a diameter 1.215 cm, and are clad with Zircaloy-4 having a nominal O.D. of 1.308 cm and a wall thickness of 0.38 mm.   Each bundle can produce a nominal maximum power of 830 kW. A 37-tube fuel bundle is shown in Fig. 13.34.   Each pressure tube is contained within a calandria tube that separates it from the cool moderator.   The D_2O moderator enters the calandria at 43.3°C (110°F) and leaves at 71°C (160°F). A gap of 8.6 mm separates the calandria tube from the pressure tube to provide an effective barrier to heat flow.   2143 MW(th) are generated by the core at rated conditions.   Of the 2024.4 MW(th) generated in the fuel channels, only

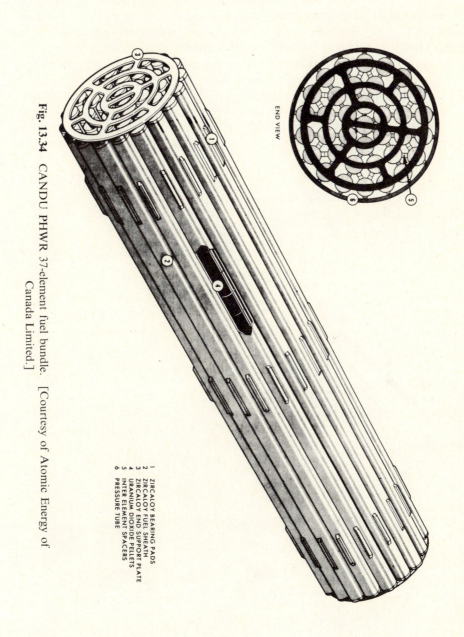

END VIEW

Fig. 13.34 CANDU PHWR 37-element fuel bundle. [Courtesy of Atomic Energy of Canada Limited.]

1 ZIRCALOY BEARING PADS
2 ZIRCALOY FUEL SHEATH
3 ZIRCALOY END SUPPORT PLATE
4 URANIUM DIOXIDE PELLETS
5 INTER ELEMENT SPACERS
6 PRESSURE TUBE

3 MW(th) are lost to the moderator. The moderator picks up 118.5 MW(th) from all forms of energy deposition, and the shield system picks up 5.9 MW(th).

Two heat transfer loops are used with water flow in opposite directions in adjacent tubes, as indicated in Fig. 13.30. Each loop has its own pump and U-tube steam generator. A total of 18.1×10^6 kg/h of dry saturated steam at 6.36 MPa is supplied to the 1800-rpm tandem compound steam turbine, which employs a moisture separator and reheat between the double-flow high and two double-flow low-pressure turbine casings. The twin condensers maintain a pressure of 3.8 cm Hg absolute. There are three low-pressure feedwater heaters, a deaerator, and a final high-pressure stage of feedwater heating. The temperature of the returning feedwater is 187°C. A steam bypass will dump steam directly to the condenser in the event of a loss of electrical load without shutting down the reactor.

Control of the CANDU-type reactor strives to maintain the operating power level and the power distribution such that the failure rate for the fuel will be minimized.

In-core flux detectors are used by the regulating system to limit the operating power level to 115 percent of normal in any bundle where the acceptable critical power ratio is 130 percent. Spatial and bulk regulation are both possible using *liquid control rods*. These are vertical tubes containing variable amounts of light water. Acting in unison, bulk regulation is obtained; and, when operated differentially, spatial control is possible. Six vertical tubes are located along two planes in the core. The two on the axial centerline have three chambers, and the four outer assemblies each have two chambers, giving a total of 14 such adjusting chambers.

There are also 21 absorber rods, or *adjusters*, individually driven and normally fully inserted for flux flattening. They may be withdrawn in groups to override xenon buildup following a power reduction or a reactor scram, and will provide about a half hour of poison override time. These rods are stainless steel tubes of varying thickness to provide optimal neutron absorption.

Four mechanical control absorbers provide a supplement to the liquid control rods. Neutron absorption is provided by a cadmium–stainless steel sandwich in the form of a 11.4 cm-diameter tube, which is pulled out by a cable-and-winch arrangement and is held out by an electromagnetic clutch. When deenergized, the clutch releases a cable and gravity, assisted by a spring, inserts the rod in approximately 2 to 3 s. The rods may be stopped at an intermediate position to supplement the zone control absorbers.

At the start of core life, gadolinium or boron solutions may be added as a shim to compensate for the excess reactivity in new fuel or the lack of fission poisons after a lengthy shutdown. These are in the form of deuteroboric acid or gadolinium nitrate.

Fast shutdown may be accomplished by the injecton of a high-pressure gadolinium nitrate solution and/or the insertion of 28 mechanical absorber rods containing cadmium. Either can provide full shutdown capability.

Figure 13.35 shows how the in-core reactivity control mechanisms and flux detectors come down vertically among the calandria tubes. The core not only has 28 vertical flux detector tubes, but it also has three horizontal detectors.

The reactor and all of the primary loop are contained within a prestressed concrete envelope. In the event of a loss of coolant accident, there is a pressure suppression water dousing system and an emergency core cooling system supplying ordinary water cooling to the reactor fuel.

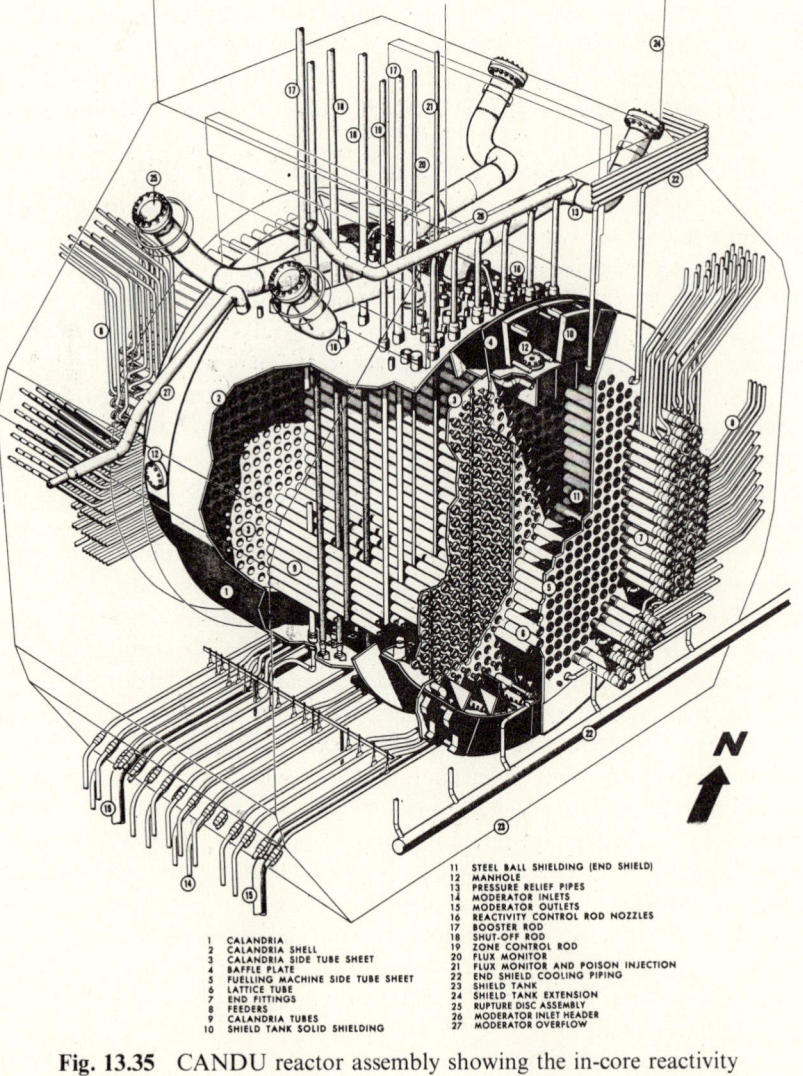

1	CALANDRIA	11	STEEL BALL SHIELDING (END SHIELD)
2	CALANDRIA SHELL	12	MANHOLE
3	CALANDRIA SIDE TUBE SHEET	13	PRESSURE RELIEF PIPES
4	BAFFLE PLATE	14	MODERATOR INLETS
5	FUELLING MACHINE SIDE TUBE SHEET	15	MODERATOR OUTLETS
6	LATTICE TUBE	16	REACTIVITY CONTROL ROD NOZZLES
7	END FITTINGS	17	BOOSTER ROD
8	FEEDERS	18	SHUT-OFF ROD
9	CALANDRIA TUBES	19	ZONE CONTROL ROD
10	SHIELD TANK SOLID SHIELDING	20	FLUX MONITOR
		21	FLUX MONITOR AND POISON INJECTION
		22	END SHIELD COOLING PIPING
		23	SHIELD TANK
		24	SHIELD TANK EXTENSION
		25	RUPTURE DISC ASSEMBLY
		26	MODERATOR INLET HEADER
		27	MODERATOR OVERFLOW

Fig. 13.35 CANDU reactor assembly showing the in-core reactivity control mechanisms and flux detectors coming down vertically among the calandria tubes. [Courtesy of Atomic Energy of Canada Limited.]

This reactor system develops low cost power due to:

(1) The neutron economy because of the heavy-water moderator and the use of Zircaloy-4 as the major structural metal in the core
(2) The use of natural uranium as a low-cost "throwaway" fuel
(3) The replacement of the heavy-walled core vessel of the usual light-water reactor by the lighter calandria
(5) Greater availability due to on-line refueling

MSBR: A Thermal Breeder Reactor

The molten salt breeder reactor (MSBR) concept depends on molten fluoride salts containing ^{7}LiF, BeF_2, ThF_4, and $^{233}UF_4$ being pumped through a graphite-moderated core where heat released by fission will raise the salt temperature to 700°C ($\sim$1300°F). The heated salt then passes through an intermediate heat exchanger, where its enthalpy is reduced by transfer of heat to a secondary salt, sodium fluoroborate, which in turn may transfer the energy to steam which can be generated at modern fossil conditions (say 24 MPa and 528°C).

Initially, a two-salt system was favored with a fuel salt containing only fissile UF_4 and no fertile ThF_4 (see Table 13.10). The blanket salt would

Table 13.10 Typical MSBR Salt Compositions and Properties

	Two-Fluid Core		Single-Fluid Core	Secondary Salt
	Fuel Salt	Blanket Salt		
Composition (mole %)				
^{7}LiF	68.75	71.0	71.6	—
BeF_2	31.0	0	16.0	—
UF_4 (fissile)	0.25	0	0.4	—
ThF_4	0	29.0	12.0	—
$NaBF_4$	—	—	—	92.0
NaF_2	—	—	—	8.0
Liquidus temp (°C)	449	566	499	385
Density (kg/m³)			3284 (at 704°C)	1874 (at 531°C)
Molar mass	46.3	121.3	64	104
Viscosity (kg/m s) × 10³			6.78 (at 704°C)	1.03 (at 482°C)
Specific heat (W/kg °C)			1.34	1.42
Thermal conditions (W/m °C)			1.30	0.467

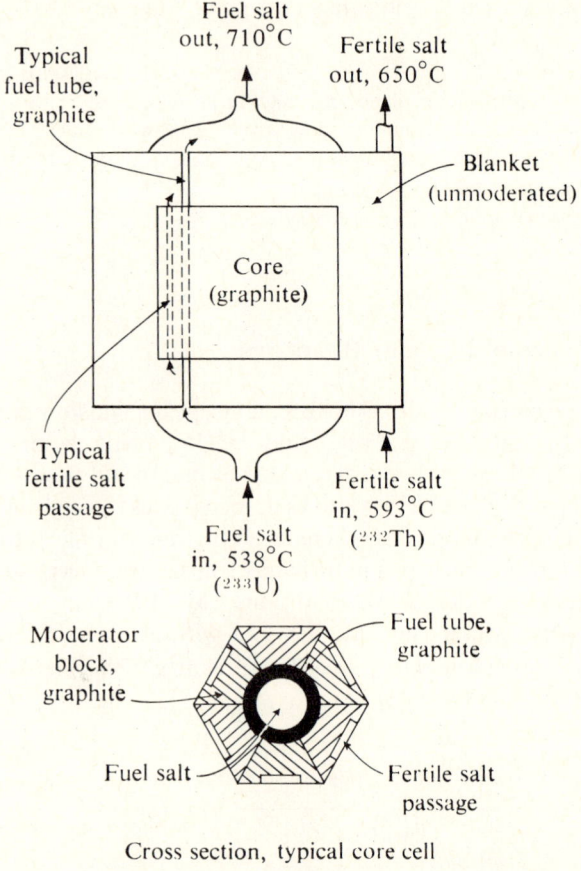

Fuel salt out, 710°C

Fertile salt out, 650°C

Typical fuel tube, graphite

Blanket (unmoderated)

Core (graphite)

Typical fertile salt passage

Fuel salt in, 538°C (^{233}U)

Fertile salt in, 593°C (^{232}Th)

Moderator block, graphite

Fuel tube, graphite

Fuel salt

Fertile salt passage

Cross section, typical core cell

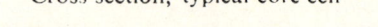

Fig. 13.36 MSBR-I conceptual design for a thermal, two-stream reactor core. [From H. F. Bauman and P. R. Kasten, *Nuclear Applications* **2**, No. 4 (1966).]

contain only ThF_4 and no UF_4. Some of the blanket salt would also circulate through the core in separate passageways cut in the moderator blocks. The fuel salt would pass up through graphite fuel tubes, as shown in Fig. 13.36. The fertile salt in the central core region would enhance the conversion of thorium to ^{233}U.

The two-fluid concept appeared to offer a simple fuel reprocessing scheme as shown in Fig. 13.37. The fuel salt would be fluorinated to convert the UF_4 to UF_6 (a gas), allowing the separation of the uranium. The remaining salt could then have the carrier fluorides separated from the fission products by vacuum distillation. The blanket salt would only be fluorinated to remove the bred uranium, since there would be few fission products if the uranium content were kept low. As indicated previously, the graphite fuel tubes would keep

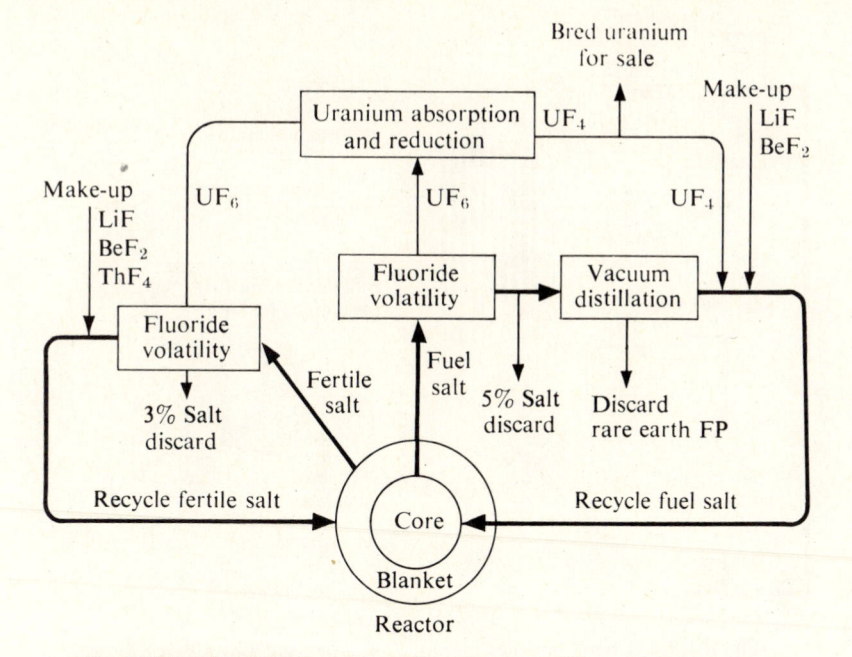

Fig. 13.37 MSBR fuel processing flow diagram. [From H. F. Bauman and P. R. Kasten, *Nuclear Applications* **2**, No. 4 (1966).]

the two-salt streams from mixing. A breeding ratio of 1.07 to 1.08 was predicted for such a system with low fuel costs and small fuel inventories.

As data were accumulated on the dimensional instability of graphite under long-term irradiation, it raised the question about the ability of the graphite piping to stand up under large fluences. This concern led to the consideration of a single-salt, two-region core with an easily replaceable graphite assembly in the core. The blanket region in the single-salt core is obtained by increasing the salt fraction from 13 percent to 35 to 40 percent. This will make k_∞ for the blanket region less than 1.0. Figure 9.15 shows how, as the salt content is increased, the decrease in moderation reduces the resonance escape probability. The thermal utilization factor improves as the salt fraction becomes larger and there is less parasitic neutron capture by the graphite. As the other two terms in the four-factor equation do not change, k_∞ will peak at approximately 4 percent salt content and drop below unity at just over 18 percent salt. In a single-salt core the reflector region having a 37 percent salt content will have a subcritical k_∞ of 0.392 and the core region with a 13 percent salt fraction will have $k_\infty = 1.034$. Figure 13.38 shows a schematic diagram of a single-fluid two-region MSBR.

To effect breeding ^{233}Pa with its 27.4-day half-life must be held up outside the core so that it can decay to ^{233}U. With ^{233}Pa, $\sigma_a = 43$ barns neutron

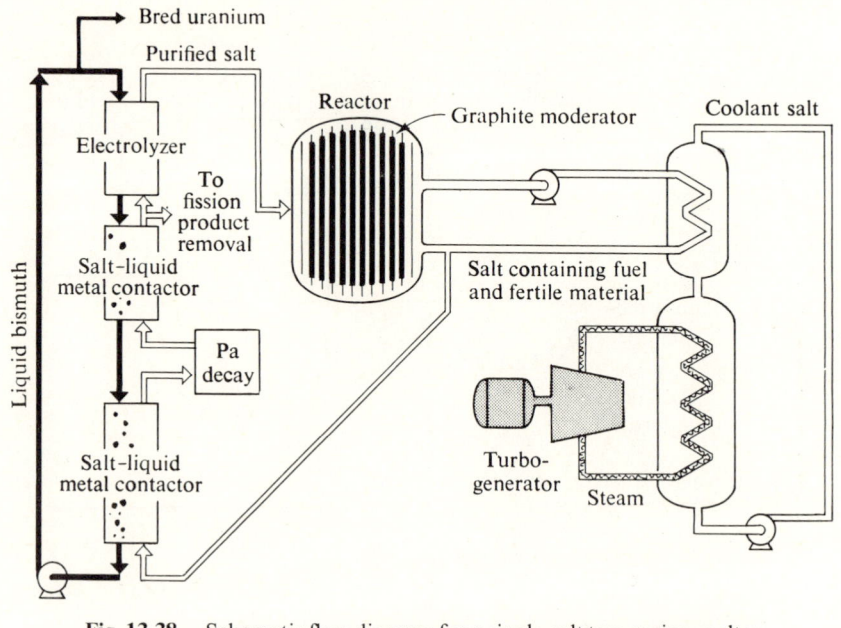

Fig. 13.38   Schematic flow diagram for a single-salt two-region molten salt breeder reactor. [From M. W. Rosenthal, P. R. Kasten, and R. B. Briggs, *Nuclear Applications & Technology* (now *Nuclear Technology*) **8**, No. 2 (February 1970), p. 110.]

capture in the core would produce ^{234}Pa, which decays to ^{234}U, which is not thermally fissionable. It also reduces the thermal utilization factor and in turn k_∞ and the excess reactivity of the core. A liquid-to-liquid extraction process shows promise for removal of Pa and U from the molten salts. The Pa is then trapped in salt in a decay tank and the U may be transferred back to the carrier salt by electrolysis for return to the reactor.

Also essential to breeding will be the reduction of ^{135}Xe by a factor of 10. This gaseous fission poison has an absorption cross section of 2.72×10^6 barns. Much of the reduction is accomplished by sparging with helium (bubbling helium through the molten salt). Also, there must be a marked reduction in the porosity of the graphite used in the core. This will prevent a significant quantity of ^{135}Xe from diffusing into the pores. It will be necessary to reduce the porosity below 10^{-8} cm^2/s. Impregnating the graphite surfaces with pyrolytic carbon shows promise, but more development and irradiation testing are required to produce the required characteristics in reactor-sized graphite blocks.

Capital costs for the MSBR are expected to be comparable to those of LWRs. There is an increased cost for remote maintenance due to the circulation of fission products through the primary loop, the fuel reprocessing facilities, and the off-gas system. Offsetting this, the cost of a turbogenerator unit is

Table 13.11 Fuel Cycle Cost Breakdown for a Single-Fluid Molten Salt Breeder Reactor*

	Mills/kWh
Fissile inventory	0.26
Thorium inventory	0.01
Carrier salt inventory	0.04
Thorium and carrier salt makeup	0.05
Processing plant fixed charges and operating cost	0.30
Credit for sale of bred material	−0.09
Graphite replacement cost (4-yr interval)	0.10
Net fuel cycle cost	0.67

* At 10% per year inventory charge on material, 13.7% per year fixed-charge rate on processing plant, $13/g ^{233}U, $11.2/g ^{235}U, $12/kg ThO$_2$, $120/kg ^{7}Li, $26/kg carrier salt (including ^{7}Li).

From M. W. Rosenthal, P. R. Kasten, and R. B. Briggs, *Nuclear Applications & Technology* (now *Nuclear Technology*) **8**, No. 2 (February 1970), p. 111.

reduced by $17 × 10^6 for a 1000-MW(e) plant due to the high thermal efficiency achieved in the steam cycle.

It is the fuel cycle cost that is expected to provide the major cost reduction for this system. Table 13.11 shows a cost breakdown for a single-fluid MSBR. Note that the cost of graphite is a significant item. Using the Th–^{233}U fuel cycle, only thorium makeup is required and there is a credit for the bred uranium, which is available for sale.

On balance the inherent safety characteristics of this system appear to be an asset. On the negative side the accumulation of fission products in the primary system, the reprocessing plant, the off-gas system, and fuel storage tanks dictate provisions for containment and removal of decay heat under all circumstances. Positively, however, the system has a number of virtues:

(1) The operating conditions of the salt at low pressures and at temperatures more than 538°C below the boiling temperature.
(2) The ability to drain the salt to tanks with redundant cooling systems.
(3) Continuous fission product removal reduces the need for excess reactivity.
(4) The molten salt has an inherent negative temperature coefficient of reactivity associated with heating.

Table 13.12 indicates some of the principal characteristics expected of a one-fluid two-region MSBR. The high outlet temperature of salt leaving the core (704°C) will permit the generation of supercritical steam to produce a

Table 13.12 Characteristics of One-Fluid, Two-Region Molten Salt Breeder Reactors

Fuel—fertile salt (mole %)	72 ^{7}LiF, 16 BeF$_2$, 12 ThF$_4$, 0.3 UF$_4$
	Melting point, 499°C
Moderator	Graphite (bare)
Salt volume fractions (%)	Core, 13; blanket, 40
Core temperatures (°C)	Inlet, 566; outlet, 704
Reactor power [MW(e)]	1000–2000
Steam system	17.2 MPa, 538°C, 44% net cycle efficiency
Breeding ratio	1.05–1.07
Specific fissile fuel inventory* [kg/MW(e)]	1.0–1.5
Doubling time (compound interest)* (yr)	15–25

* The lower values are associated with the higher reactor powers.
From M. W. Rosenthal, P. R. Kasten, and R. B. Briggs, *Nuclear Applications & Technology* (now *Nuclear Technology*) **8**, No. 2 (February 1970), p. 111.

cycle thermal efficiency of 44 percent. A breeding ratio of 1.05 to 1.07 is predicted along with a doubling time of 15 to 25 years. Design studies look forward to the eventual construction of such a reactor.

Fast Reactors

Fast reactors use no moderator to promote the slowing down of fission neutrons. There is, however, some degradation of neutron energies, with the result that the median energy for the neutron spectrum might lie between 0.1 and 0.5 MeV. This energy loss occurs mainly due to inelastic scattering collisions with atoms of the coolant, structural materials, and the fuel material. The coolant for fast reactors is usually sodium, which has excellent heat transfer characteristics. The fuel is enriched uranium or plutonium surrounded by a blanket region containing thorium, natural uranium, or depleted uranium.

The neutron spectrum for a particular reactor is affected by the core size and the combination of materials used. Oxide or carbide fuels will cause a shift of the spectrum to lower energies because of the moderating effect of the oxygen or carbon that is present. Figure 13.39 shows the spectra calculated for the EBR-I, EBR-II, and PBR (proposed, but never built) reactors. EBR-I was a 1-MW, 6-liter, fast core reactor which used highly enriched uranium, NaK coolant, and stainless steel cladding. EBR-II has a 50-liter core fueled by 50 percent enriched uranium, sodium as the coolant, and stainless steel for cladding. The Power Breeder Reactor (PBR) would have had an 800-liter core

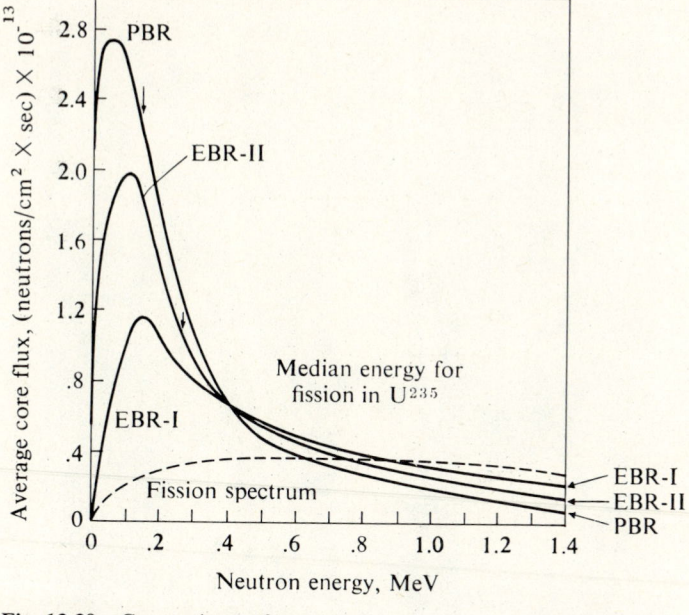

Fig. 13.39 Comparison of neutron spectra for various sizes of fast reactors. [From D. Okrent, R. Avery, and H. H. Hummel, "A Survey of the Theoretical and Experimental Aspects of Fast Reactor Physics," *Proceedings of the Second United Nations International Conference on the Peaceful Uses of Atomic Energy*, Vol. V, 1955, p. 347.]

and would have been fueled by 15 to 20 percent enriched uranium or a $Pu–U^{238}$ fuel. Its output was designed to be greater than 600 MW(th).

Clinch River Breeder Reactor

The Clinch River Breeder Reactor (CRBR), which is shown in Fig. 13.40, has been designed as a 975-MW(th) and 375-MW(e) (gross) demonstration plant. The reactor site is 10 mi south of the Oak Ridge National Laboratory and 30 mi west of Knoxville, Tennessee. It is intended as a prototype for full-scale, commercial, liquid-metal, fast breeder reactors (LMFBRs) having an output of 1000 MW(e) or greater.

When the contract was signed on July 25, 1973, the cost estimate was 600×10^6 with the utility industry pledging 250×10^6 toward the project. Inflation, stretch-out of the project to a 1988 startup, and more stringent safety and environmental requirements have all contributed to a 1980 cost estimate of 2.88×10^9. It should be noted that this includes research and development, construction funds, and five years of operation. As the costs rose so dramatically, the government was forced to shoulder the increase. The Department of

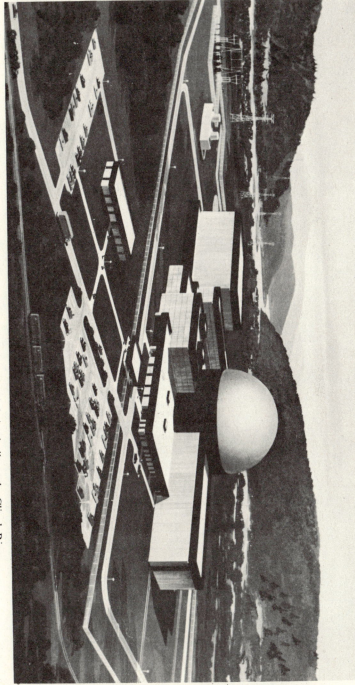

Fig. 13.40 The Clinch River Breeder Reactor Plant that is being built on the Clinch River near the town of Oak Ridge, Tennessee. [Courtesy of Project Management Corporation and Westinghouse Advanced Reactors Division for the Clinch River Breeder Reactor Project.]

Energy became responsible for direction of the project with support from the utility industry through its Project Management Corporation.

Presidential concern over proliferation of nuclear weapons and the desireability of fuel reprocessing has led to a reevaluation of fuel cycles and their diversion resistance. President Carter attempted to stop the CRBR Project, at least pending the outcome of these studies. However, congressional continuing appropriations have allowed the design work to proceed toward completion. In addition, fabrication of long lead-time components is underway, with some already finished and awaiting construction.

This plant is designed as a loop system where all of the primary system is not immersed in a pool of molten sodium, as it is in EBR II and the French Phénix reactors. The entire primary loop is contained, however, in shielded inerted vaults below the elevation of the operating floor level within the reactor containment building, as shown in Fig. 13.41. This diagram indicates the relationship of the fuel handling, fuel storage, and radwaste facilities located in the reactor service building. It also shows the reactor containment building, which houses the reactor core, the primary sodium pump, and the intermediate heat exchanger. The steam generator building houses the intermediate sodium pumps (one per cell and three cells), the evaporators (two per cell and three cells), steam drums (one per cell and three cells), and the superheaters (one per cell and three cells). The turbine building houses the turbogenerator and the associated condensers. In choosing the loop system, it was felt that there was no clear cost or safety advantage of one system over the other, but the mechanical and thermal decoupling of the primary units would allow for easier maintenance and permit changes in the system to be accomplished with less difficulty. Plant maintainability is enhanced by adequate sizing of the containment with particular attention to the location of equipment to allow removal and replacement and hands-on accessibility with shielding and atmospheric separation of redundant systems. Designs provide separate cleaning and maintenance facilities, casks for the removal of radioactive equipment, redundancy of components, the ability of in-place tube plugging for the intermediate heat exchangers and steam generators, removal of pump internals, and removal of spent fuel from the containment to allow complete accessibility during normal operation.

Plant reliability and availability are enhanced by such things as design simplicity, redundancy, and operational flexibility. Examples of the simplification are the elimination of main loop isolation valves, elimination of the fast rod insertion system, and the use of identical components where possible (e.g., steam generators and sodium pumps).

The heat transport system, as shown in Fig. 13.42, supplies 1.5×10^6 kg/h of superheated steam at 485°C and 10.7 MPa to a turbine having one high-pressure, double-flow cylinder and three low-pressure, double-flow cylinders. There are three low-pressure closed feedwater heaters in the feedwater heating chain ahead of the deaerating open feedwater heater, followed by three additional

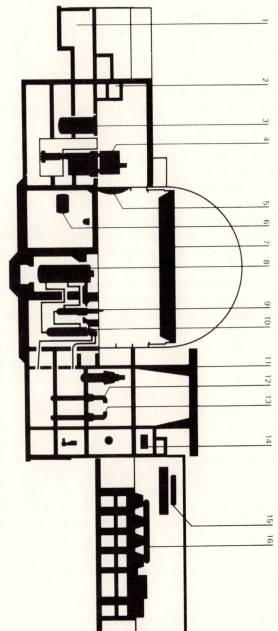

Fig. 13.41 Plant layout and containment for CRBRP. [Courtesy of Clinch River Breeder Reactor Project.]

1 Radwaste Area
2 Refueling Control Room
3 Ex-Vessel Storage Tank
4 Ex-Vessel Transfer Machine
5 Refueling Hatch
6 Overflow Tank

7 Polar Crane
8 Reactor
9 Primary Pump
10 Intermediate Heat Exchanger
11 Gantry Crane
12 Evaporator (2 per cell)

13 Superheater (1 per cell)
14 Auxiliary Heat Removal Equipment
15 Deaerator
16 Turbine Generator

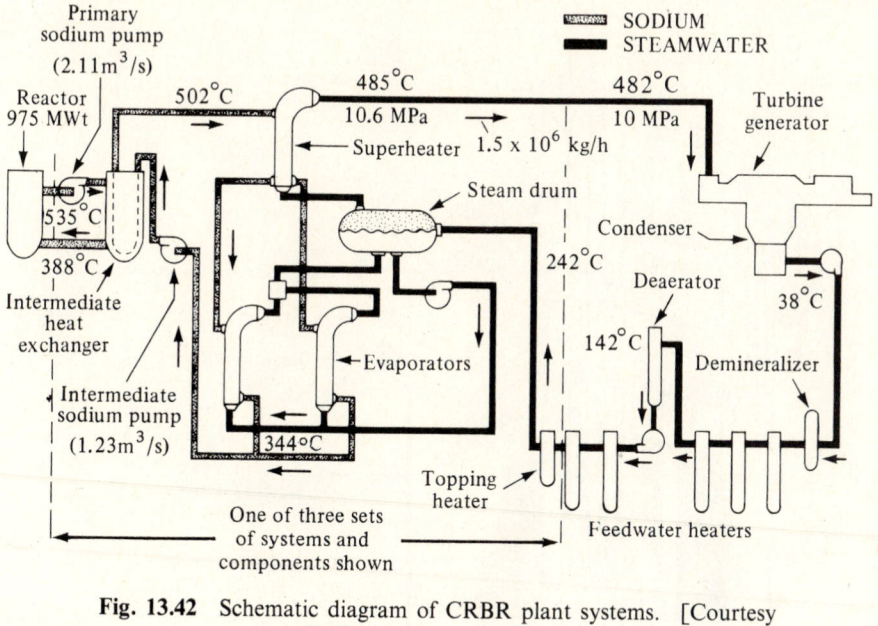

Fig. 13.42 Schematic diagram of CRBR plant systems. [Courtesy of Breeder Reactor Corp.]

closed feedwater heaters. The feedwater returns to the steam generators at 233°C. Notice that the temperature of the superheated steam is ~55°C less than for most modern fossil plants. The lowered steam temperatures will lessen problems with creep and should enhance reliability. The expected gross heat rate is 9368 kJ/kWh, which corresponds to a thermal efficiency of 38.4 percent.

Figure 13.43 shows the reactor assembly. The cylindrical core contains fuel assemblies each having 271 wire-wrapped rods per assembly. These rods have an outside diameter of 0.584 cm and the assembly has an overall length of 4.27 m, as shown in Fig. 13.44. The core region is 0.914 m high and contains enriched oxide fuel; a 0.356 m axial blanket region is located both above and below the core proper. A plenum distance of 1.22 m is provided in each fuel rod above the upper blanket region to accommodate the fission gases that will be released. Current heterogeneous core design calls for a fuel burnup of 110 000 MWd/Mg, but ultimately it is hoped to reach 150 000 MWd/Mg. The high burnup accounts for the plenum length being *four times* that previously described for the BWR. The mixed oxide (UO_2 and PuO_2) fuel is fabricated into pellets of 85 percent theoretical density. This low density helps to provide for improved fission gas retention within the pellets.

The original homogeneous core design is shown in Fig. 13.45a. The central core region contained 108 inner core elements with an initial cycle

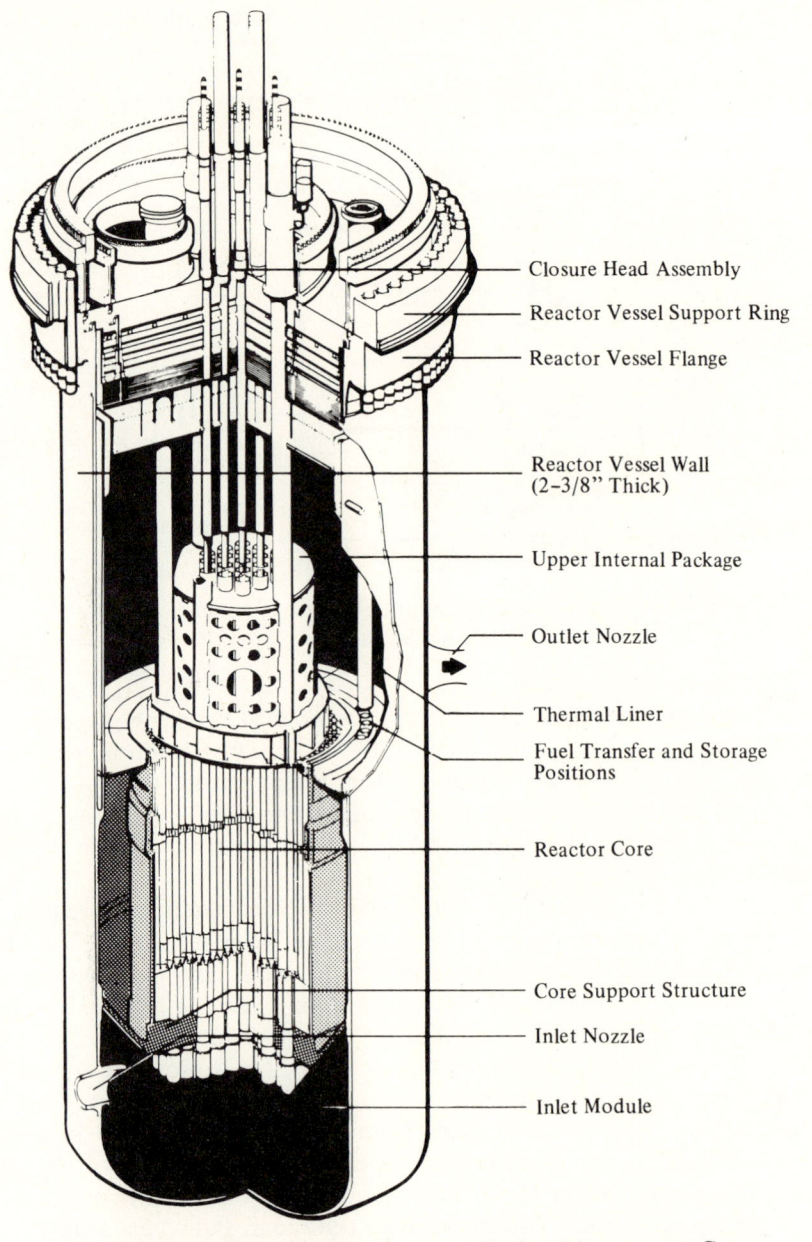

Closure Head Assembly

Reactor Vessel Support Ring

Reactor Vessel Flange

Reactor Vessel Wall
(2–3/8" Thick)

Upper Internal Package

Outlet Nozzle

Thermal Liner

Fuel Transfer and Storage
Positions

Reactor Core

Core Support Structure

Inlet Nozzle

Inlet Module

Fig. 13.43 CRBRP reactor. [Courtesy Project Management Corporation and Westinghouse Advanced Reactors Division for the Clinch River Breeder Reactor Project.]

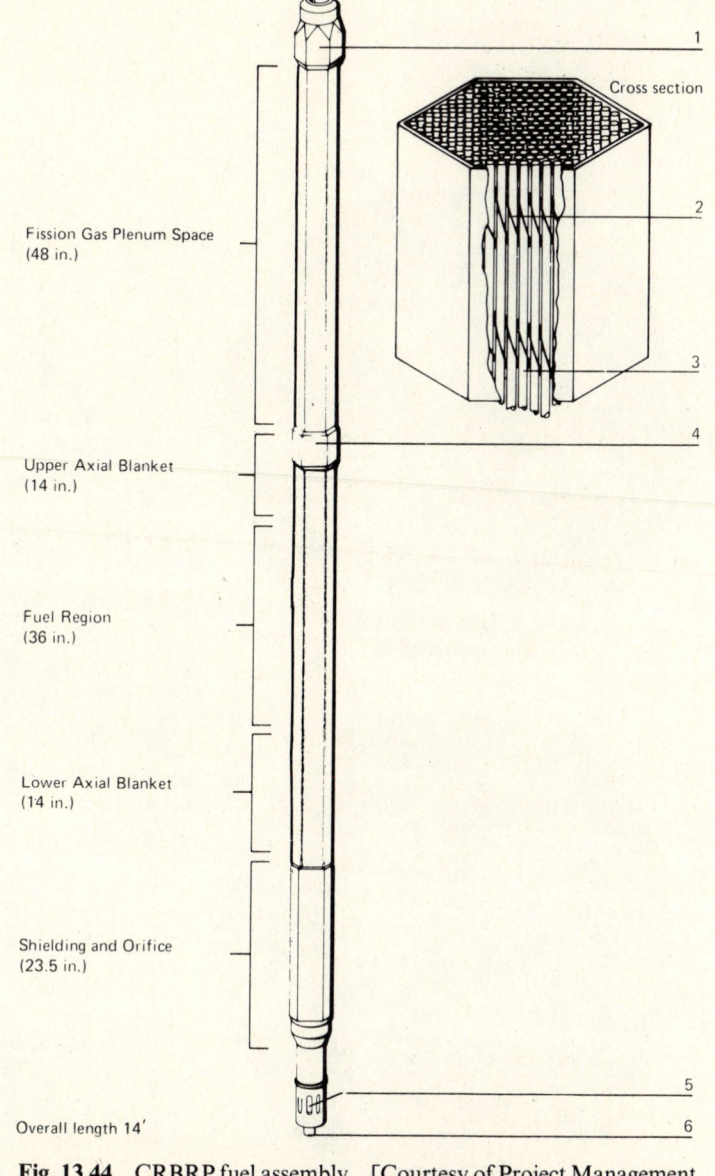

Cross section

Fission Gas Plenum Space
(48 in.)

Upper Axial Blanket
(14 in.)

Fuel Region
(36 in.)

Lower Axial Blanket
(14 in.)

Shielding and Orifice
(23.5 in.)

Overall length 14′

1

2

3

4

5

6

Fig. 13.44   CRBRP fuel assembly.   [Courtesy of Project Management
Corporation and Westinghouse Advanced Reactors Division for the
Clinch River Breeder Reactor Project.]

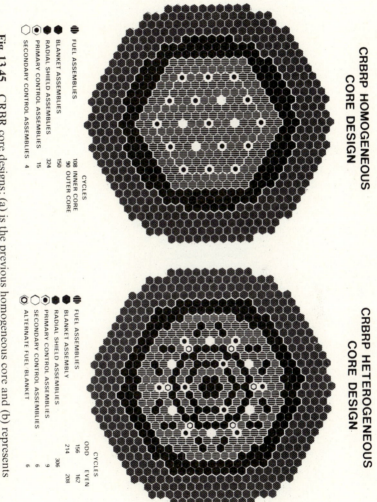

CRBRP HOMOGENEOUS
CORE DESIGN

		CYCLES
⬤	FUEL ASSEMBLIES	108 INNER CORE
		90 OUTER CORE
⬢	BLANKET ASSEMBLIES	150
⬤	RADIAL SHIELD ASSEMBLIES	324
◉	PRIMARY CONTROL ASSEMBLIES	15
⬡	SECONDARY CONTROL ASSEMBLIES	4

CRBRP HETEROGENEOUS
CORE DESIGN

		CYCLES	
		ODD	EVEN
⬤	FUEL ASSEMBLIES	156	162
⬢	BLANKET ASSEMBLY	214	208
⬤	RADIAL SHIELD ASSEMBLIES	306	
◉	PRIMARY CONTROL ASSEMBLIES	9	
⬡	SECONDARY CONTROL ASSEMBLIES	6	
⬡	ALTERNATE FUEL BLANKET	6	

Fig. 13.45 CRBR core designs: (a) is the previous homogeneous core and (b) represents the current improved heterogeneous core. [Courtesy of Project Management Corporation and Westinghouse Advanced Reactors Division for the Clinch River Breeder Reactor Project.]

enrichment of 18.7 percent surrounded by an outer zone of 90 assemblies with 27.1 percent enrichment. This core would contain 15 primary control assemblies and four secondary control assemblies, either group of which can completely shut down the core if its rod of the greatest worth is hung up in a completely withdrawn position. The control material was to be boron carbide. Surrounding the core were 150 radial blanket assemblies containing depleted uranium in oxide form. About 80 percent of the flow is through the fuel and 12 percent through the blanket assemblies. Surrounding the blanket assemblies were 324 radial shield assemblies, requiring approximately 3 percent of the sodium flow for adequate cooling.

It became apparent that a core design change would be necessary when it was discovered that the homogeneous core could meet all the design requirements except the 1.2 goal for breeding gain. Rather than increase fuel pin diameter, it was decided to shift to the heterogeneous core configuration shown in Fig. 13.45b. Annular rings of blanket assemblies are moved into the core, being interspersed among the fuel assemblies. This change produces a breeding gain of 1.24 in the equilibrium cycle. In the initial one-year cycle there are 156 fuel assemblies, 214 blanket assemblies, 306 radial shield assemblies, nine fully enriched (90 percent ^{10}B) primary control assemblies, and six fully enriched secondary control assemblies. During the even cycle in the second year six of the inner blanket assemblies are replaced by fresh fuel assemblies. At the end of the second year all fuel assemblies, the inner blanket assemblies, and some radial blanket assemblies are discharged from the reactor.

The new design not only improved the breeding ratio, it reduced fuel costs by reducing the original 198 fuel assemblies to 156 or 162, depending on whether one considers the first or second year of the complete cycle. This is expected to result in a savings of $\$8 \times 10^6$. The ratio of peak to average flux is also reduced. Note in Table 13.13, which is based on a two-year fuel lifetime, that there is a 10 percent increase in fuel rod power, only a 6 percent increase in burnup, and a 20 percent reduction in fuel elements. There is only a single fuel enrichment of ~ 33 percent used in this core configuration.

The CRBR Preliminary Safety Analysis Report (PSAR) indicates a lower calculated fuel sodium void reactivity worth (see Table 13.13) and improved incoherency in voiding, both of which contribute to improved safety margins for core-disruptive accidents. Transient behavior proved to be both sluggish and energetically benign. Although there is a reduction in the Doppler constant, the studies showed no nonallowable or unstable conditions with the reduced Doppler feedback.

This heterogeneous core design has been found very flexible. It can accommodate the ^{233}U–thorium cycle for proliferation resistance using either oxide or carbide fuels. Breeding ratios are reduced to ~ 1.10 for oxide fuels and blankets, but could rise to 1.18 by a shift to carbide fuel and blanket material. For Pu–U fuel its diversion resistance can be improved during shipping and handling by fission product spiking (^{95}Zr, ^{95}Nb, ^{103}Ru, ^{106}Ru, etc.).

Table 13.13 Comparison of CRBR Performance Using Homogeneous and Heterogeneous Cores

	Homogeneous	Heterogeneous
Breeding ratio with low 240Pu fuel		
Initial cycle	1.15	1.29
Equilibrium cycle	1.08	1.24
Peak fast flux (neutrons/cm^2 s)	4.2×10^{15}	3.4×10^{15}
Peak fuel burnup (MWd/Mg)	104 000	110 000
Peak liner kW/m in fuel	4.69	5.21
Maximum positive sodium void reactivity (fuel assemblies, \$)	4.00	2.31
Doppler constant ($-T \, dk/dT \times 10^4$)		
Fuel	55.9	25.8
Inner blankets	—	—
Radial blankets	7.0	11.8
Axial blankets	4.4	2.6

Courtesy of Breeder Reactor Corp.

Table 13.14 shows a comparison of the CRBR to the larger French Super Phénix, which is discussed in the next section.

Super Phénix Breeder Reactor

The Super Phénix (Creys–Malville) 1200-MW(e) fast breeder reactor is being built near Lyon, France. The project is an international undertaking in which the French hold a 51 percent interest, with the balance held by West Germany, Italy, Belgium, and the Netherlands. Commercial operation is planned for the end of 1983. This plant is the direct descendent of the highly successful 250-MW(e) French Phénix Reactor, which has been in operation since July 1974. As Super Phénix represents the world's most ambitious fast breeder project to date, it will be described in some detail.

A schematic diagram for the plant is shown in Fig. 13.46. The primary sodium is pumped through the core, leaving with a temperature of 545°C. It then transfers heat to a secondary sodium loop, which in turn generates steam at 487°C and 17.7 MPa. The secondary loop prevents the radioactive primary sodium from coming in contact with water in the steam generators. Note in the diagram that the entire core, the intermediate heat exchangers, and the primary circulating pumps are all totally immersed in molten sodium. This protects the core from large fluctuations in temperature due to low power demand

Table 13.14 Comparison of the Clinch River Breeder Reactor and the French Super Phénix

	CRBR*	Super Phénix†
Overall plant		
Thermal power [MW(th)]	975	3000
Net electrical power [MW(e)]	350	1200
Net overall plant eff. (%)	35.9	40
Plant capacity factor	0.75	0.75
No. primary loops	3	4
Containment		
Diameter (m)	56.7	64
Height (m)	78	84
Reactor		
Fuel material	PuO_2–UO_2	PuO_2–UO_2
Cladding material	316 SS	316 SS
Fuel rod diameter (cm)	0.58	0.85
No. fuel rods per assembly	217	271
No. core assemblies	156/162	364
No. blanket assemblies	214/208	233
Fuel assembly dimensions		
Total length (m)	4.27	5.4
Pin length (m)	2.85	2.7
Fuel region length (m)	0.91	1.0
Upper and lower axial blankets (m)	0.36	0.3
Fis. gas plenum		
Upper (m)	1.22	0.16
Lower (m)	—	0.85
Breeding ratio	1.29/1.24	1.24
Doubling time (yr)	23	
Peak liner power rating (W/cm)	5.21	4.8 initial core
		4.5 equil. core
Peak fuel burnup (MWd/t)	110 000	70 000
Primary heat transport system		
Reactor outlet temp. (°C)	535	545
Reactor inlet temp. (°C)	388	395
Pump flow rate (kg/s)	1820	4 × 4100
Pump head at design flow (m Na)	137	63
Intermediate heat transport system		
Hot-leg temp. (°C)	502	525
Cold-leg temp. (°C)	344	345
Pump flow rate (kg/s)	1613	3300
Pump head at design flow (m Na)	125	28
Steam generator-turbine system		
Superheater outlet temp. (°C)	485	487
Superheater outlet press. (MPa)	10.5	17.7
Steam flow rate (kg/s)	927	4 × 340

* Courtesy of Breeder Reactor Corp.
† From M. Banal, Creys-Malville special issue of *Nuclear Engineering International* **23**, No. 272 (June 1978).

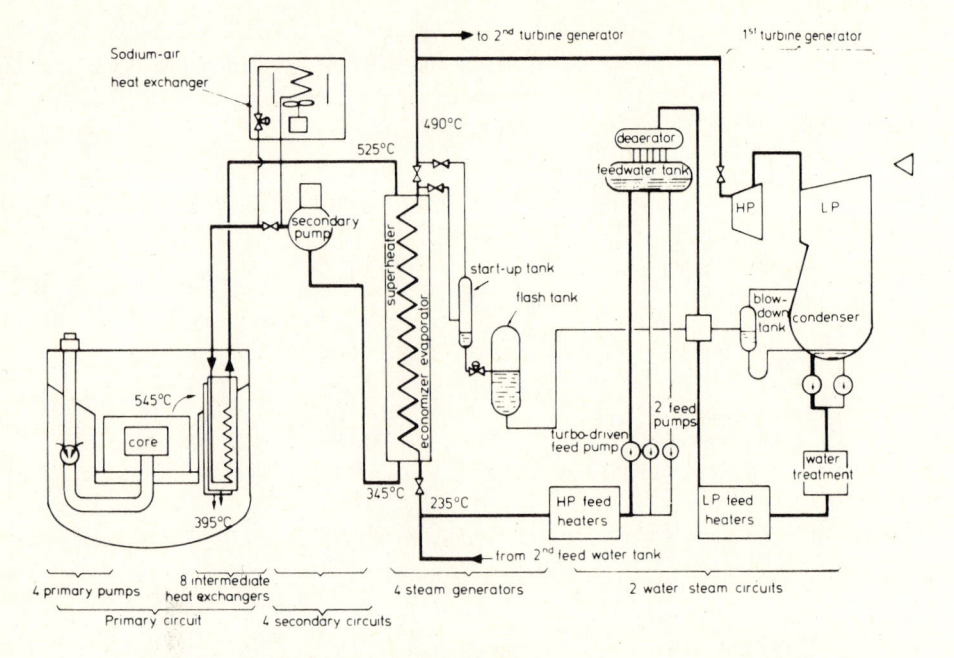

Fig. 13.46   Schematic diagram of the primary, secondary, and water–steam loops for the Super Phénix fast breeder reactor. [From M. Banal et al., Creys–Malville special issue of *Nuclear Engineering International*, **23**, No. 232 (June 1978), p. 47.]

or changed conditions in the secondary loop. It also ensures against loss of sodium, except for the possibility of a rupture of the 21-m-diameter core vessel, which is in turn contained in a 22.4-m-diameter safety vessel.

The arrangement of the primary circuit components within the main vessel is shown in Fig. 13.47. Four primary pumps take 395°C sodium and force it through the fuel and blanket assemblies, where it is heated to 545°C and then forced through four intermediate heat exchangers. A cross section of the core is shown in Fig.13.48. The fuel, fertile blanket elements, steel reflector assemblies, and radial shielding assemblies are arranged in four concentric zones, as follows:

(1) 193 internal core assemblies, having a 15 percent Pu enrichment
(2) 171 outer core assemblies, having an enrichment of 18 percent (the higher enrichment of this outer region tends to flatten the radial flux distribution)
(3) 197 steel reflector assemblies, which reduce neutron leakage and contribute to shielding
(4) 1076 steel radial shielding assemblies, which reduce the radiation dose to other primary loop components

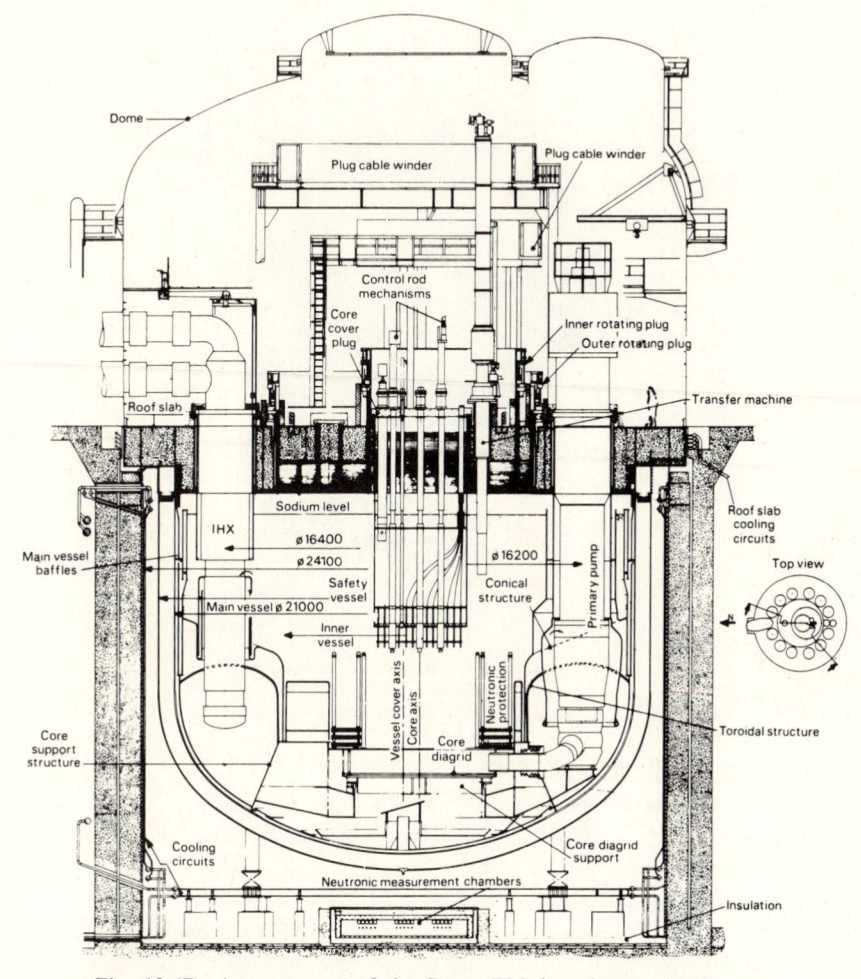

Fig. 13.47 Arrangement of the Super Phénix primary circuit components within the main vessel, which in turn is protected by the safety vessel and the dome, and all of which is within the reactor containment building. [From M. Banal et al., Creys–Malville special issue of *Nuclear Engineering International* **23**, No. 272 (June 1978), p. 52.]

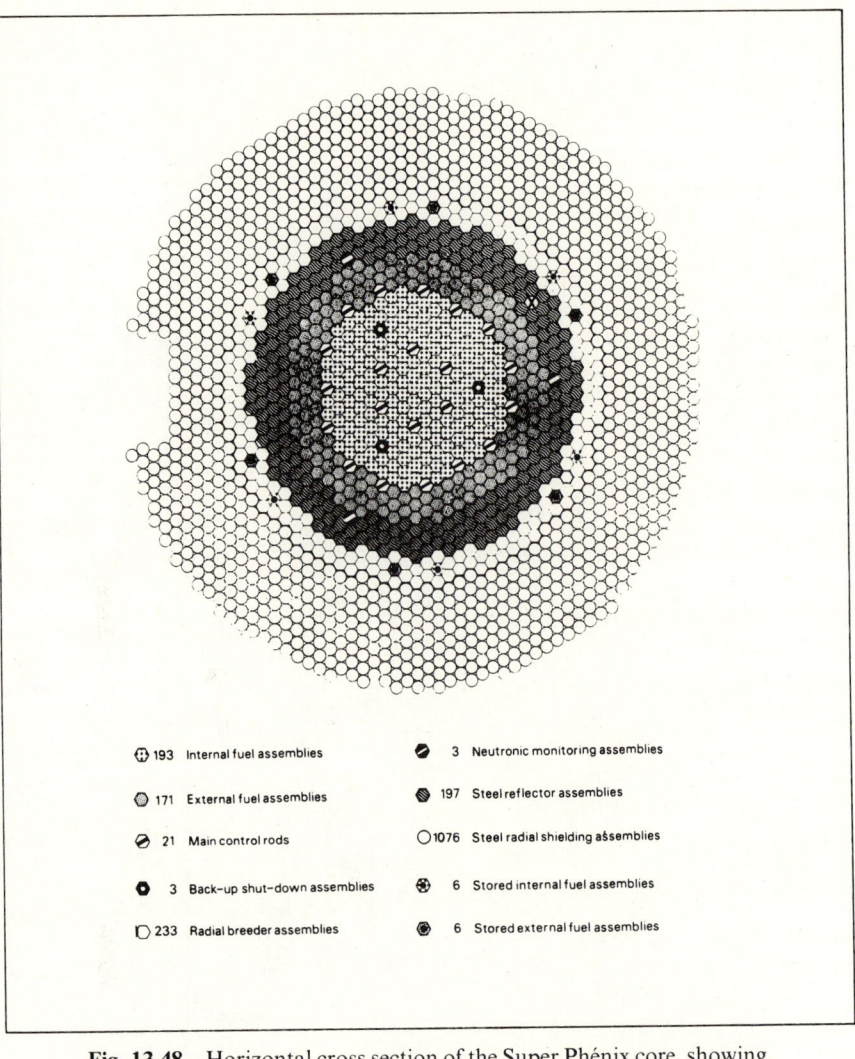

⊕ 193	Internal fuel assemblies	◑ 3	Neutronic monitoring assemblies
◎ 171	External fuel assemblies	◉ 197	Steel reflector assemblies
⊘ 21	Main control rods	○ 1076	Steel radial shielding assemblies
⬢ 3	Back-up shut-down assemblies	⊛ 6	Stored internal fuel assemblies
◗ 233	Radial breeder assemblies	◉ 6	Stored external fuel assemblies

Fig. 13.48 Horizontal cross section of the Super Phénix core, showing the arrangement of the fuel, radial blanket elements, control assemblies, the steel reflector assemblies, and the radial shielding assemblies. [From M. Banal et al., Creys–Malville special issue of *Nuclear Engineering International* **23**, No. 272 (June 1978), p. 54.]

Reactivity is controlled by 21 multipurpose main control rods. These each contain a bundle of 31 pins of B_4C (enriched to 90 percent ^{10}B), performing the following functions:

(1) Scram and rapid shutdown controlled by the reactor protective system
(2) Reactivity shim to offset power and temperature reactivity effects, as well as burn-up reactivity depletion
(3) Power control

There are also three backup, shutdown assemblies, which have only a safety function. They are held out of the core in the ready position by an electromagnetic coupling. A scram or rapid shutdown will cut the current to the magnets and the rods will fall into the core.

A fuel pin and the arrangement of 271 of these pins in an assembly inside a wrapper tube is shown in Fig. 13.49. The pin length is 2.7 m. The pin assembly in the wrapper tube, head structure, and bottom fitting have an overall length of 5.4 m. The 1-m-long central section of each pin contains the hollow PuO_2–UO_2 mixed-oxide fuel pellets. On top and below the fuel is 0.3 m of axial blanket pellets of depleted UO_2. Above and below the axial blanket sections are gas plenums to collect the fission gases. The location of the principal plenum at the cooler lower end of the pins reduces the gas pressure and the volume required. A spring at the top of the pin keeps the entire column in place.

The fuel pins have a bottom steel plug which fits into a grid plate. They are spaced from one another by a 1.2-mm-diameter wire wound around the outside of the cladding, being welded to the upper and lower plugs. This spacing system extends the entire pin length, discourages pin vibration, encourages turbulence in the sodium to promote mixing and heat transport, and allows some swelling in the bundle without causing excessive stress levels or reductions of cooling and flow.

The fuel, radial blanket, control, and shutdown assemblies have the same external hexagonal shape (17.3 cm across flats) and identical top and bottom fittings. They are fed at a pressure of 5 bars by the primary sodium pumps through six radial inlets near the assembly base. To increase the average ΔT across the core, the sodium flow is adjusted by diaphragms placed inside the base fitting. There are six flow zones for the fuel assemblies, three for the radial blanket assemblies, and two for the control rods.

The reflector and radial shielding assemblies have the same annular cross section with an O.D. of 17.0 cm and an I.D. of 10.0 cm. Since their power level is low, a natural circulation flow is sufficient.

At a 75 percent load factor the reference fuel cycle length is 13 months. At each refueling one-half of the fuel assemblies, control assemblies, and backup shutdown assemblies are reloaded. For the first refueling a special procedure is being used where the unloaded half of the fuel assemblies are saved to be reloaded at the start of the third cycle. This will bring the first core to the full burnup level. For the radial blanket assemblies the first, second, and third rows

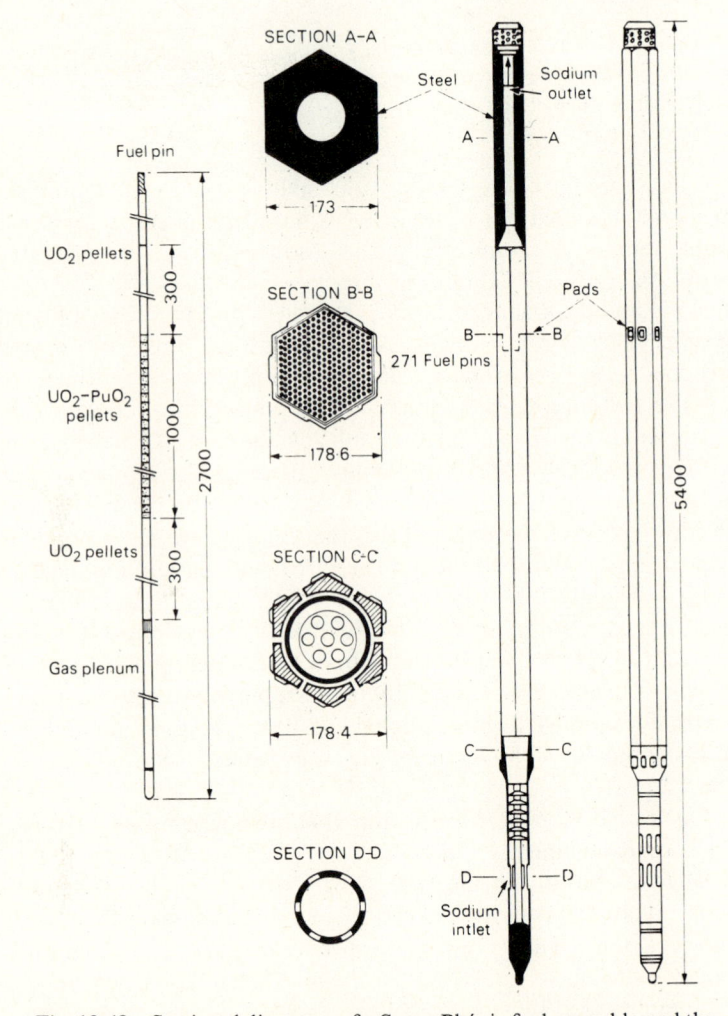

Fig. 13.49 Sectional diagrams of a Super Phénix fuel assembly and the arrangement of the pellets within one of its fuel pins. [From M. Banal et al., Creys–Malville special issue of *Nuclear Engineering International* **23**, No. 272 (June 1978), p. 55.]

will be removed after the third, fourth, and fifth cycles. The core should attain a breeding ratio of 1.24.

The main safety features of Super Phénix are essentially the same as for its already operating predecessor, Phénix, with the exception of more stringent requirements for resistance to earthquakes and missiles. The essential features are as follows:

(1) The pool-type primary circuit is protected by surrounding the main vessel with a slightly larger diameter safety vessel. In the event of a rupture of the main vessel, the sodium depth in the safety vessel is still deep enough to allow natural convection cooling by the intermediate heat exchangers (IHXs).

(2) The great 3200-Mg sodium mass with its high inherent thermal inertia allows design of the components to function close to optimum conditions, the temperature transients being low.

(3) Although a loss of flow and failure to scram type of accident has a low probability of occurrence, the primary containment is designed to withstand such an event and to provide adequate public protection.

To minimize the fuel inventory the linear heat generation rate should be as high as possible without exceeding the melting point of the PuO_2–UO_2 in the interior of the fuel pellets. To maintain an adequate margin of safety, the maximum linear heat generation rate will be 480 W/cm for the initial core and about 450 W/cm for the equilibrium core.

To permit reasonable size of heat transfer surface and high-temperature steam generation, the maximum rated cladding temperature has been set at 620°C. The hot-spot temperature should be less than 700°C. The resulting overall net thermal efficiency of the plant is 40 percent, which is close to the efficiencies of large fossil plants.

Good fuel cycle economy is closely linked to a high burnup. The contributions to fuel cycle cost of fabrication, transportation, and reprocessing are all inversely proportional to the burnup at discharge. For Super Phénix the guaranteed peak burnup value is 70 000 MWd/ton. Some fuel assemblies in the smaller Phénix have exceeded 75 000 MWd/ton.

Gas-cooled fast breeder reactor

A 350-MW(e) demonstration gas-cooled fast breeder design has been proposed and a flow diagram is shown in Figure 13.50. It is essentially a high-temperature gas-cooled reactor (HTGR) without fueled graphite, using assemblies of mixed-oxide fuel rods, much like those for LMFBRs. The hot helium generates steam in three main loops, each with its own independent steam generator, an electric motor driven, single-stage, radial flow, He circulator, and a gravity closing isolation valve. Each main loop has an auxiliary cooling loop with an electrically driven circulator, heat exchanger, and gravity opening isolation valve.

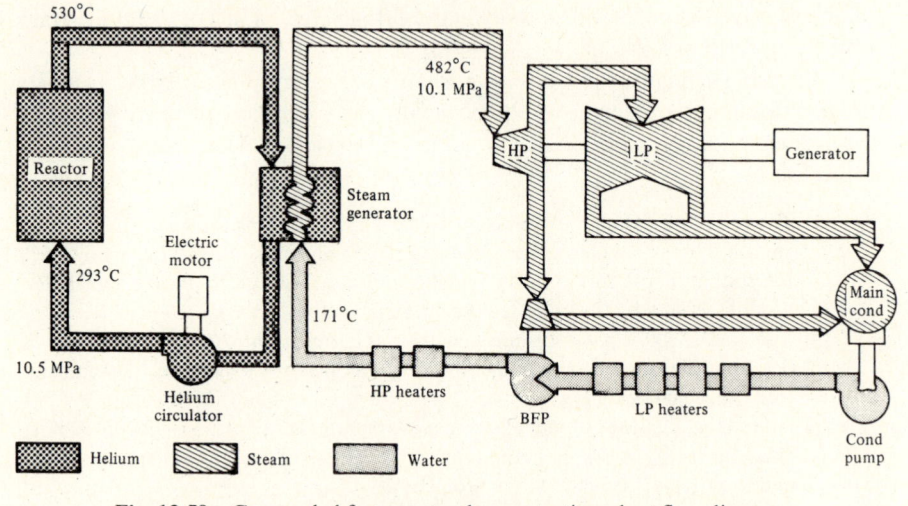

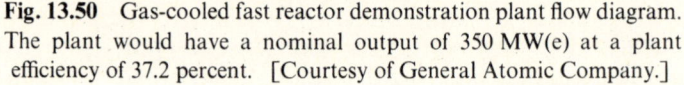

Fig. 13.50   Gas-cooled fast reactor demonstration plant flow diagram. The plant would have a nominal output of 350 MW(e) at a plant efficiency of 37.2 percent.   [Courtesy of General Atomic Company.]

The prestressed concrete reactor vessel (PCRV), which encloses the entire system, is reinforced with steel rods and is prestressed by circumferential wrapping and longitudinal tendons, much like what is shown in Fig. 13.25.   The central reactor cavity contains fuel assemblies (150), blanket assemblies (162), control assemblies (15), shutdown assemblies (4), and radial shield assemblies (138).   Cavities in the surrounding wall contain three steam generators, helium circulators, and the auxiliary cooling systems.   The auxiliary cooling loops provide for backup for the main cooling loops in case of emergencies and for shutdown cooling.   It should be noted that flow in the design shown has been changed to upward flow to facilitate emergency cooling by natural circulation. The overall height of the PCRV is 24.4 m and the outside diameter is 25.6 m.

Helium at 293°C and 10.5 MPa enters the core and flows upward, leaving at 524°C.   Steam is generated at 10 MPa and 482°C.   Each helium circulator requires 10.7 MW(e).

The fuel assemblies are 2.87 m in length and consist of 265 fuel rods, each 8 mm in diameter and located on an 11.5-mm triangular pitch.   A hexagonal flow duct channels the helium flow upward through the rod bundle.   The exit nozzle is used for fuel handling and contains a fixed area replaceable orifice. The fueled region in the rods is a 1.2-m long stack of mixed-oxide (U, Pu) O_2 pellets with a Pu enrichment ~ 20 percent.   Upper and lower axial blankets (0.6 m each) are located above and below the fueled region and consist of depleted UO_2 pellets.   The blanket design is similar, except for a smaller number

of larger rods. These could be of either depleted UO_2 or ThO_2. A breeding ratio of 1.4 is expected.

The reactor would be refueled through the bottom of the vessel, replacing one-third of the fuel annually. At the time of refueling, the orifices in the outlet of the fuel and blanket assemblies would be changed. This will compensate for changes in power distribution when new fuel is placed adjacent to spent fuel.

The reactor control elements are fuel elements with the 37 center rods replaced by a B_4C control rod. There are 15 control rods attached to top-mounted control rod drives; electromagnets permit gravity trip. There is also an independent set of four electrically driven shutdown rods providing a redundant means of scramming the reactor.

Testing and development work have been done on GCFR vented fuel rods and some reactor physics tests were done in addition to the proposed design. But since there has been a cutoff in U.S. government funding, it would appear that any further development work in the United States must await an increased demand for nuclear power. Proponents believe the GCFR should have a low fuel-cycle cost and that it could provide a very significant source of energy.

Light-water breeder reactor

The light-water breeder reactor, sometimes known as the seed-blanket aqueous reactor, is a thermal breeder. It makes use of much of the well-established PWR technology, thus avoiding some of the technical and licensing problems of less developed concepts.

It uses a highly enriched fuel in a seed region surrounded by a fertile blanket. Because of the higher thermal value of η for ^{233}U, the Th$-^{233}U$ fuel cycle is chosen. To improve the neutron economy, the seed region is moved in and out of the blanket region to vary the neutron leakage, as shown by Fig. 13.51, rather than by using control rods.

The flux is tailored in the two regions so that each can perform its own function best. In the seed region the neutron spectrum is made as thermal as possible to maximize the value of η and encourage a large fissile–atom burnup. In the ThO_2 blanket region there is a minimum amount of water, to make the blanket flux as hard as possible. This encourages resonance captures by fertile nuclei and hence breeding. It also encourages fast fission of fertile nuclei and (n, 2n) reactions.

The first demonstration of this concept has been carried out in the light-water breeder reactor (LWBR) project at the Shippingport nuclear plant. It was converted from a small PWR core to an LWBR core, which achieved full power with this demonstration core in December 1977. The net plant output is 50 MW(e). Figure 13.52 shows a cross section of the core. Note the modular nature of the core, which will allow it to be easily adapted to larger sizes simply

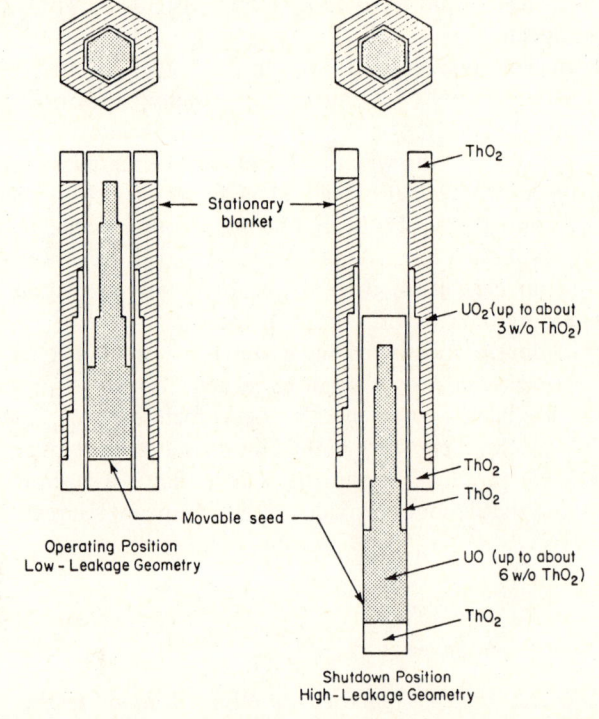

Fig. 13.51 LWBR seed-blanket module, showing the variable geo-
metry control concept. [From *Status and Prospects of Thermal
Breeders and Their Effect on Fuel Utilization*, Technical Reports Series
No. 195, International Atomic Energy Agency, Vienna, 1979.]

by increasing the number of modules. In fact, this seed-blanket concept has
been visualized as a replacement core for existing LWRs. Some of these might
be prebreeders used to produce the necessary ^{233}U required for new LWBRs.
Table 13.15 gives some of the design parameters and core loading information
for a conceptual 1000-MW(e) light-water breeder reactor.

The advantages for this type of reactor are:

(1) Control may be provided by motion of the seed region.
(2) A significant fraction of the energy is generated in the blanket region.
(3) There is a strongly negative temperature coefficient of reactivity, which is typical
of small highly enriched cores.
(4) Refueling can be accomplished by insertion of fresh seeds, leaving the blanket
until it is depleted to its metallurgical limit.
(5) High power density in the highly enriched seed regions because of their low-
resonance capture.

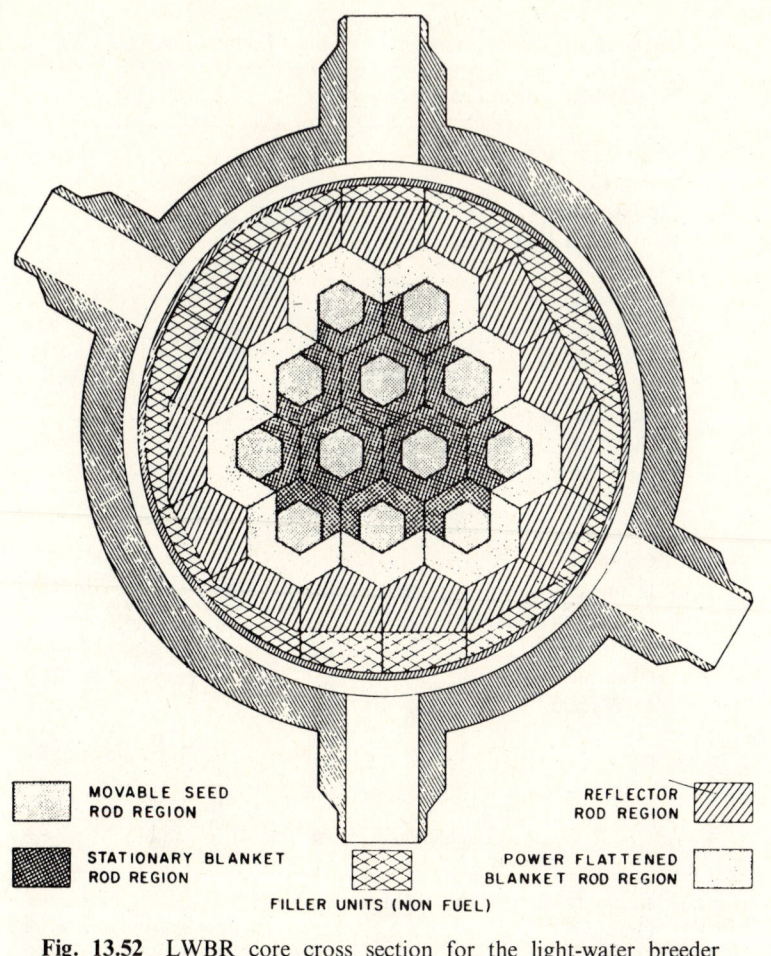

| | MOVABLE SEED ROD REGION | | REFLECTOR ROD REGION |
| | STATIONARY BLANKET ROD REGION | | POWER FLATTENED BLANKET ROD REGION |

FILLER UNITS (NON FUEL)

Fig. 13.52 LWBR core cross section for the light-water breeder reactor at Shippingport, Pennsylvania. [From *Status and Prospects of Thermal Breeders and Their Effect on Fuel Utilization*, Technical Reports Series No. 195, International Atomic Energy Agency, Vienna, 1979.]

Table 13.15 Core parameters for a Base-Technology 1000-MW(e) LWBR

Typical Mechanical Features

Module configuration	Hexagonal
Number of modules	74
Module pitch	0.438 m
Full core enclosing diameter	4.778 m
Fuel height	3.20 m
Axial blanket height	0.23 m
Seed rods	
Composition	$^{233}UO_2$–ThO_2
Diameter	7.77 mm
Number per module	619
Cladding	Zircaloy
Clad thickness	0.559 mm
M/W ratio	1.72
Blanket rods	
Composition	$^{233}UO_2$–ThO_2
Diameter	14.50 mm
Number per module	444
Cladding	Zircaloy
Clad thickness	0.705 mm
M/W ratio	2.98

	Movable Seed Fuel	
Control Mechanism	Initial 3-yr Core	Near-equilibrium 2-yr Core
Material inventory—initial charge		
Uranium, fissile	4 000 kg	4 300 kg
Uranium, other	400 kg	1 000 kg
Thorium-232	188 000 kg	186 000 kg
Burnup data		
Refueling interval	$\frac{1}{2}$ or 1 yr	$\frac{1}{2}$ or 1 yr
Refueling fraction	$\frac{1}{6}, \frac{1}{5}, \frac{1}{4}$, or $\frac{1}{3}$	$\frac{1}{6}, \frac{1}{5}, \frac{1}{4}$, or $\frac{1}{3}$
Fuel exposure, MWd/Mg heavy metal		
Seed	24 700	16 300
Blanket	12 600	8 470
Specific power (MW/Mg fis. inventory)	956	812

From *Status and Prospects of Thermal Breeders and Their Effect on Fuel Utilization*, Technical Reports Series No. 195, International Atomic Energy Agency, Vienna, 1979.

Pumped Storage for Peaking Power

Pumped storage plants are being constructed in order to have large nuclear power stations baseloaded continually and at the same time to give a utility system a readily available large block of peaking power. During off-peak hours a pump-turbine operates as a motor-driven pump. Water is discharged to an upper reservoir and stored until peaking power is required. During that part of the day when peak load occurs on the system, flow is reversed and water flows back to the lower reservoir. The pump-turbine now functions as a turbine driving an electric generator to provide the peaking power.

The Northfield Mountain Pumped Storage Project on the Connecticut River at Northfield, Massachusetts, is an excellent example of such a plant. Here four Francis-type pump-turbines deliver water against a head, depending on the difference in reservoir levels, at 223 to 251 m. When generating the net operating heads are 219.5 to 251 m. Each machine has a nominal capacity of 250 MW, giving the plant a total output of 1000 MW.

The massive size of the project is indicated by the dimensions of the power-house cavern which contains the machinery. It is 100 m long by 21.4 m wide by 36.6 m high. It is reached through an access tunnel which is 7.9 m by 7.9 m and 747 m long. The pressure shaft carrying the water from the upper reservoir down to the turbine is 9.45 m in diameter and 259 m long, while the tailrace tunnel out to the river is a 10-m-diameter horseshoe 1585 m long.

At this site and in the adjacent state forest, in excess of $4 000 000 have been spent by Northeast Utilities and the state and federal governments in a joint development of new recreational facilities. Included are boat launching areas, facilities for fishing, swimming, and camping, as well as trails for hiking, snow-mobiling, and horseback riding. With a growing population, increased recreational opportunities are equally as important as power if we are to achieve an increased quality of life.

Fusion Reactors

With experimental work on the containment of plasmas proceeding in a hopeful manner much attention is being given to other aspects of the development of fusion power such as engineering, environmental problems, and economics. This is a measure of the newfound confidence in the ultimate solution of the containment problem and urgency for the development of a power source not so dependent on limited resources, such as oil, coal, thorium, or uranium.

Engineering

The probable fuel choice for the first reactors will be deuterium and tritium, which when joined by fusion produce 17.6 MeV shared as 14.1 MeV by a neutron

and 3.5 MeV by an alpha particle. It will be burned in a closed torroidal system with the neutrons escaping from the plasma through a vacuum wall into a coolant surrounded by more coolant and moderator, neutron shielding, and finally the coils of a superconducting magnet. Figure 13.53 shows a conceptual drawing of such a Molten Lithium Fusion Breeder Reactor (MLFBR) less containment domes and hot cells for handling. The inset shows the blanket cross section.

The most severe radiation damage problem occurs in the vacuum chamber wall, where a flux the order of 3.7×10^{15} neutrons/cm^2 s with an energy of 14 MeV passes through the wall. Here the problem occurs at the boundary of the plasma, where in a fission reactor both fuel and structure occupy the volume where neutrons are produced. In the fusion reactor the designer has control over the surface area-to-volume ratio.

Figure 13.54 shows the size of a 5000-MW(th) torroidal reactor. The superconducting magnets are rated at 100 kG at the coils, would cost approximately 200×10^6, and store 10^5 MJ of energy which require 6 h to charge or discharge in a noncatastrophic manner due to their inductance. The superconducting Nb$_3$Sn–NbTi coils must be maintained superconducting at 4K by boiling He.

The deuterium required for fuel is available from seawater and is almost inexhaustible, but the tritium must be bred from ^{6}Li(n, α)T and ^{7}Li(n, αn')T reactions, where the former reaction has a $1/v$ cross section below 0.3 MeV and the latter has a threshold of 2.8 MeV. The ^{6}Li reaction produces about two-thirds of the tritium, with the balance coming from ^{7}Li. Steiner has shown that with the ORNL blanket design (see Fig. 13.55) that a breeding ratio of 1.33 is possible using Li as the coolant and Nb as the structural material. Breeding would be required during the initial phases of an expanding fusion economy, but once equilibrium is established, the breeding ratio may be reduced.

The heat developed in the coolant is transported from the core to a lithium–potassium heat exchanger where the potassium boils and acts as the topping fluid in a binary steam cycle having possible thermal efficiencies as great as 58 percent. The tritium will diffuse into the potassium cycle, where it must be trapped out and kept from either leaking or getting into the steam cycle. One challenge is to keep the radioactive tritium from leakage less than 0.0001 percent per day.

Environmental concerns

There are three principal environmental concerns with fusion reactors:

(1) High neutron flux and activation of the structure
(2) Creation, separation, and confinement of tritium
(3) Waste disposal

Activity due to the flux of approximately 4×10^{15} neutrons/cm^2 s for 14-MeV fusion neutrons will amount to a decay power of ~ 0.2 percent [10 MW(t)]

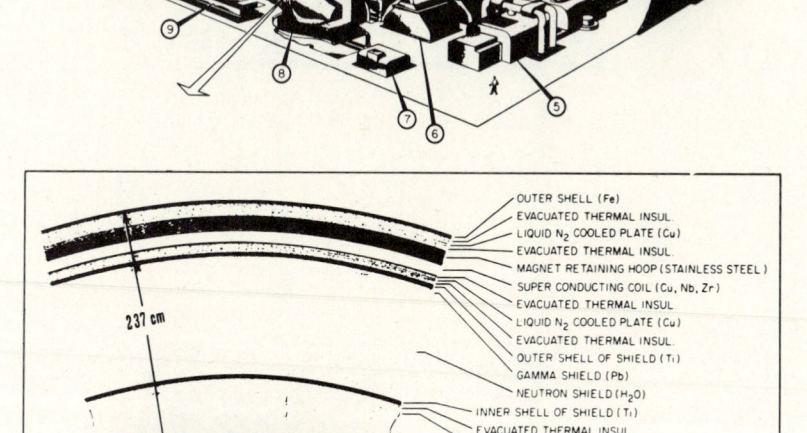

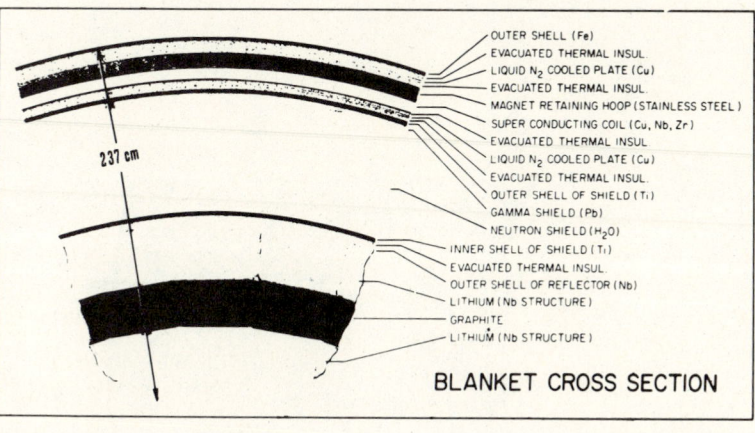

OUTER SHELL (Fe)
EVACUATED THERMAL INSUL.
LIQUID N_2 COOLED PLATE (Cu)
EVACUATED THERMAL INSUL.
MAGNET RETAINING HOOP (STAINLESS STEEL)
SUPER CONDUCTING COIL (Cu, Nb, Zr)
EVACUATED THERMAL INSUL.
LIQUID N_2 COOLED PLATE (Cu)
EVACUATED THERMAL INSUL.
OUTER SHELL OF SHIELD (Ti)
GAMMA SHIELD (Pb)
NEUTRON SHIELD (H_2O)
INNER SHELL OF SHIELD (Ti)
EVACUATED THERMAL INSUL.
OUTER SHELL OF REFLECTOR (Nb)
LITHIUM (Nb STRUCTURE)
GRAPHITE
LITHIUM (Nb STRUCTURE)

237 cm

BLANKET CROSS SECTION

1   PLASMA

2   VESSEL VACUUM SYSTEM

3   BLDG. SERVICE AND MAINTENANCE AREA

4   CRYOGENIC SYSTEMS

5   STEAM TURBINES AND GENERATORS

6   POTASSIUM TURBINES AND
    HEAT EXCHANGERS

7   TRITIUM REMOVAL SYSTEM

8   FUSION REACTOR

9   INJECTOR POWER SUPPLY

Fig. 13.53 Conceptual drawing of a fusion power plant with the
blanket cross section shown in the inset. [From H. Postma, *Nuclear
News* **14**, No. 4 (1971).]

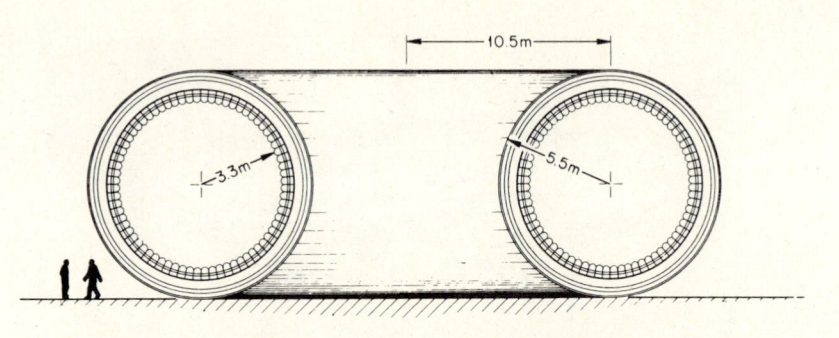

Fig. 13.54 Torroidal fusion reactor for a 5000-MW(th) plant. [From H. Postma, *Nuclear News* **14**, No. 4 (1971).]

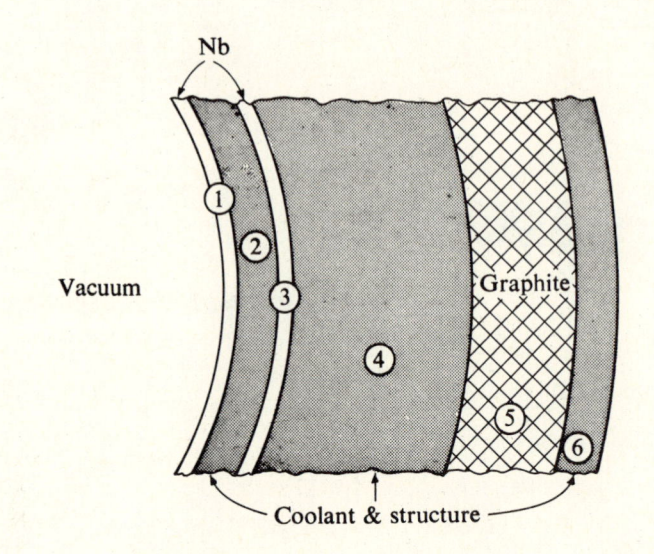

Fig. 13.55 Standard configuration of the ORNL blanket designs.

Region no.	Description of region	Thickness (cm)	Composition
1	First wall	0.5	Nb
2	Coolant & structure	3.0	94% Li or Flibe 6% Nb
3	Second wall	0.5	Nb
4	Coolant & structure	60.0	94% Li 6% Nb
5	Moderator− reflector	30.0	Graphite
6	Coolant & structure	6.0	94% Li 6% Nb

and a decay time of 35 days. This is not bad when compared to LMFBR, where the activity might be 10^{10} Ci with a decay time of 13.6 yr.

Note the modular design in Fig. 13.53, which is intended to make replacements simpler. Molybdenum and vanadium are being considered as alternative materials. The vanadium is attractive, as it will have only 1/10 the afterheat and 1/1000 the waste disposal problem of Nb.

The major environmental problem will be tritium containment. Under normal operations the in-plant radioactivity must be less than 30 percent of the natural radioactivity (36 mrem/yr) with leakage kept below 10^{-6} per day.

The probability of a catastrophic release is small, with the relative hazards considerably less than for a fission reactor.

At some future time if all power were developed by fusion [6×10^6 MW(e)] with a 10^{-6} leakage rate, there would be a release of 4×10^7 Ci/yr. If assumed, mixed with air and water on a global scale there would be an activity of 10^{-6} μCi/cm^3, which is 1/500 the dose from natural radiation.

The waste disposal problems will depend on the structure and its lifetime, which in turn are related to activation and radiation damage. Also, tritium must be kept away from water.

Inertial Confinement Fusion Reactor

Another fusion reactor concept for central station power has been proposed by workers at the Oak Ridge National Laboratory. It is shown in Fig. 13.56. In this concept frozen pellets of deuterium–tritium are injected into a vacuum in the vortex of a swirling pool of molten lithium. A pulse of laser energy ignites the pellet at the midplane, releasing fusion energy. The blast energy is attenuated in the lithium by gas bubbles introduced through the ring at the bottom of the pressure vessel. Also, the primary shock wave travels up the long jagged injection port, where the irregular profile will break the normal shock into many oblique waves to cause the attenuation.

The energy is transported by the lithium to a heat exchanger to provide a heat source for a thermal cycle. A conventional steam cycle might attain a 40 percent efficiency, but if a potassium–steam binary cycle were used with a 1000°C peak lithium temperature, the efficiency could reach 58 percent.

It was estimated in 1974 that the pressure vessel for a unit with a 40 percent efficient steam cycle and a 60 000-kW output would cost $20 per kW. The balance of the capital cost for the steam cycle ought to be equivalent to a conventional plant. The equipment for fuel recovery and pellet fabrication would be a small fraction of the total cost. It was estimated that the fuel cost for this system would be 6 percent of that for low-sulfur coal and about 30 percent that of fission reactor fuels. Thus, if this type of reactor can overcome its development problems, it appears economically viable.

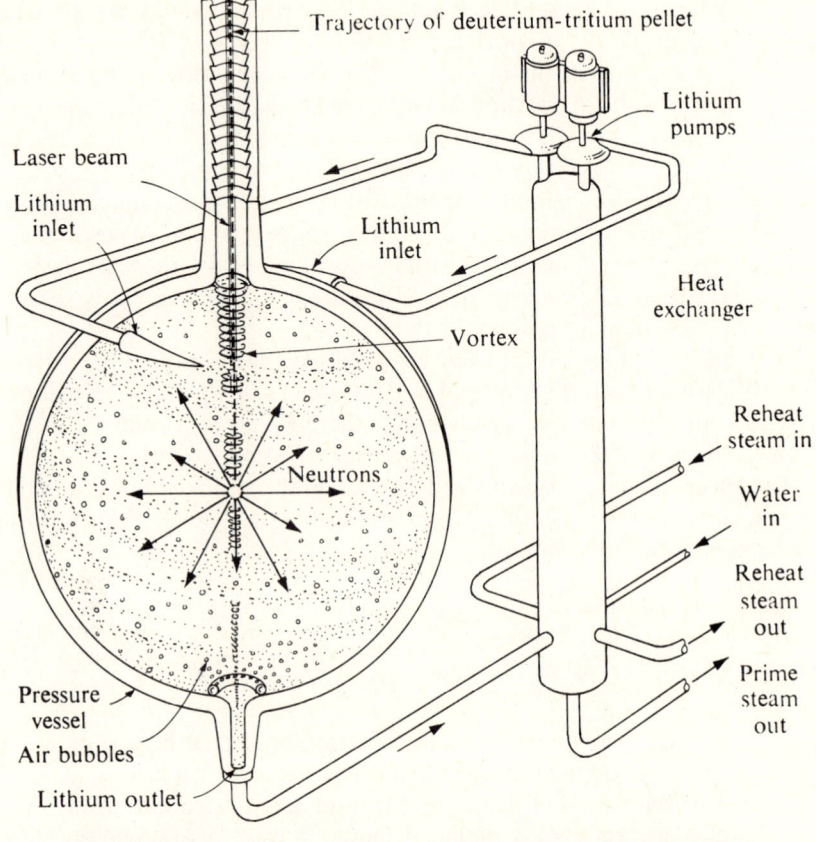

Fig. 13.56   Proposed fusion reactor where fusion takes place in an expanding laser heated plasma. [From M. J. Lubin and A. P. Fraas, "Fusion by Laser." Copyright © June 1971 by Scientific American, Inc. All rights reserved.]

As was indicated in Chapter 4, there has been a lessened interest in lasers as a means of vaporizing the inertial confinement reactor fuel and a shift toward heavy-ion driver approaches (Reference 81). The Argonne National Laboratory has proposed a synchrotron-based system, which would deliver 1-MJ pulses with xenon ions and requires eight synchrotrons, four bunchers, and 16 storage rings stacked vertically. Ions would be supplied to the storage rings, cycled in them, and diverted by strong magnetic fields through two diametrically opposed channels leading to the target volume in the fusion reactor. The simultaneous delivery would both supply the energy and provide inertial confinement for the pellet. A liniac system proposed by the Lawrence Berkeley Laboratory would require much more geographical space, needing a string of hardware 9 km in length.

Fusion–Fission Symbiosis

In a hybrid fusion–fission system, fusion neutrons would be used to breed plutonium from ^{238}U or ^{233}U from thorium. The fusion reactor blanket would contain a subcritical amount of fertile material and breed fissionable material for a thermal fission reactor. A fissile doubling time of five years might be possible in a hybrid system with the fuel being burned in a molten salt converter.

For the future there will be a multitude of challenges which must be faced by the engineer, whether it is in the development of advanced fission reactors, fusion reactors, or, as suggested above, a combination of the two.

Problems

1. HTGR fuel contains 10 moles of ThC_2 per mole of 93 percent enriched UC_2. The core contains 150 kg of uranium. The fuel burnup is to be 70 000 MWd/ton.
   (a) What is the total mass of fuel material in this core loading?
   (b) What fraction of the heavy-metal atoms is ^{235}U?
   (c) Assuming that all fissions involve ^{235}U, what mass of ^{235}U would be fissioned? Considering that some fissile atoms are consumed by nonfission captures, compute the mass of ^{235}U burned up.
   (d) What percentage of the original fissile atoms is consumed? How do you account for such a high percentage?

2. For the 1200-MW(e) direct-cycle HTGR plant of Example 4 (see Fig. 13.24), water passing through the precooler is heated from 45°C to 170°C. In the dry cooling tower, air is heated from 30°C to 115°C. The circulating pump raises the water pressure from 17 bars to 20 bars. Determine:
   (a) The required water flow rate (kg/h and m^3/s)
   (b) The airflow rate through the dry cooling tower (kg/h)
   (c) The power (kW) required by a 75 percent efficient pump

3. A steam-generating HTGR plant is designed for a net output of 1150 MW(e) when operating with a heat rate of 9000 kJ/kWh. Feedwater enters the steam generators at 250°C and 160 bars. Steam enters the turbine at 500°C and 150 bars. The helium leaves the core and enters the steam generators at 775°C and leaves them at 400°C. The cost of fuel is 60¢/10^6 kJ. The capacity factor is expected to be 75 percent. Compute the following:
   (a) The steam and helium flow rates (kg/h)
   (b) The cost of fuel for one year's operation

4. A 1000-MW(e) light-water reactor (LWR) is to be fueled with 3.75 percent enriched UO_2 and is to undergo a 3 percent burnup. The estimated initial cost of the plant is estimated at \$1.5 × 10^9 with an annual operating cost of \$8 × 10^6. The plant capacity is to be taken

at 70 percent and the annual fixed-charge rate is 20 percent. The cost of energy is $0.60/10^6$ kJ with the plant operating at a heat rate of 10 900 kJ/kWh. Determine:

(a) The annual fuel requirement (kg/yr)
(b) The annual fuel cost
(c) The cost of electricity (mills/kWh)

5. A 600-MW(e) plant of the CANDU PHWR type cost 250×10^6 and operates with a capacity factor of 82 percent. The operating and maintenance cost is the order of 1.0 mill/kWh. Consider a uranium cost of $70 per kg, an overall thermal efficiency of 31 percent, and a fuel burnup of 180 MWh/kg U. The government owned plant might have a fixed charge rate of 10 percent. Compute the following:

(a) The annual operating cost
(b) The cost of electricity (mills/kWh)
(c) The time to achieve the average burnup, if the average thermal flux is 2×10^{14} neutrons/cm²s with an average fuel temperature of 1000°C

6. A once-through steam generator (OTSG) is used with the PWR cycle shown in Example 3. Superheated steam is available at 315°C and 70 bars at the entrance to the high-pressure turbine. Assume that other conditions remain the same, except for the enthalpy after the high-pressure turbine and the fractions of bled steam. Determine new values for the thermal efficiency, the heat rate, and the net power output.

7. A dual–cycle BWR generates 650 000 kg/h of dry steam at 70 bars. 10.833×10^6 kg/h of saturated water is recycled through the secondary steam generator to produce 600 000 kg/h of secondary steam at 40 bars. Condensation takes place at 40°C. The efficiency of each section of the turbine is 85 percent and all pumps are 65 percent efficient. Steam is bled from the turbine at 10 bars to closed feedwater heaters located in both the primary and secondary feed lines (see Fig. 13P.1). Water leaving both CFWHs is at 170°C. Condensate from the heaters is trapped back to the condenser. The primary feed pump discharges water at 72 bars with a 1 bar pressure loss through the heater and another 1 bar loss between heater and steam separator. The secondary feed pump discharges at 41.5 bars with a 0.75–bar pressure drop through the heater and another 0.75–bar pressure loss through the secondary steam generator. The recirculation pump discharges at 74 bars and there is a 2–bar pressure drop in the high-pressure side of the secondary steam generator with another 2–bar pressure loss through the BWR core.

(a) Determine the enthalpy at all points in the cycle and sketch the cycle neatly on the *T–s* diagram.
(b) How many kilograms of steam are extracted per kilogram of primary steam generated?
(c) What is the thermal efficiency of the cycle?
(d) What is the net power output for the cycle?
(e) What power is required to operate the recirculation pump?
(f) What would be the annual fuel cost if the plant were to run at a 70 percent load factor? Take the energy cost at 60¢ per 10^6 kJ.

8. An HTGR is coupled to a gas turbine in a direct cycle, as shown in Fig. 13.25. The helium flow rate is 150 000 kg/h. Gas enters the turbine at 730°C and 2.5 MPa. It is discharged at 450°C and 1.0 MPa. The regenerative heat exchange cools the gas to 104°C. At the inlet to the first stage of the three-stage compressor, the pressure is 0.95

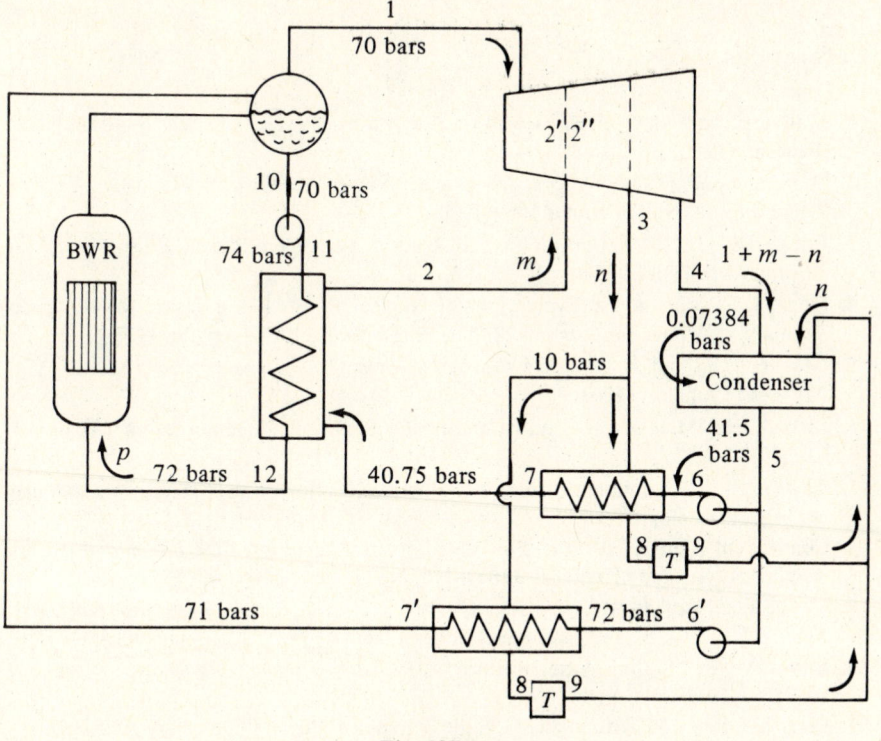

Fig. 13P.1.

MPa and the temperature is 20°C. After each stage of compression, the temperature is 67°C and intercooling returns the temperature to 20°C at the inlet of the second and third stages. The discharge pressure from the third stage is 2.55 MPa. Compression and expansion processes may be considered to be adiabatic. The generator is 97 percent efficient.

(a) List the pressure and temperature at each point in the cycle.

(b) What is the reactor power [MW(th)]?

(c) What is the net cycle output [MW(e)]?

(d) If the fuel cost is 65¢ per 10^6 kJ, how much will fuel cost to operate the plant at full load for one year? What will be the fuel cost in mills per kilowatthour?

(e) What is the overall thermal efficiency and heat rate for the plant?

(f) What would be the fuel cost in mills per kilowatthour for a plant with a single-stage compressor with the same compression efficiency as each stage in the three-stage machine and the same regenerator effectiveness?

References

1. Mason, E. A., "An Over-all View of the Nuclear Fuel Cycle," *Nuclear News* **14**, No. 2 (February 1971), pp. 35–38.
2. Steigelman, W., "The Outlook for Nuclear Power-Station Capital Costs," *Reactor Technology* **13**, No. 1 (Winter 1969–1970).
3. Edison Elec. Inst., "Report on the EEI Reactor Assessment Panel," Edison Electrical Institute, New York, 1970.
4. Hoge, J. B., "Economic Evaluations of Nuclear Power Plants Using Monte Carlo Simulation," *Power Engineering* **78**, No. 12 (December 1974), pp. 36–38.
5. Glasstone, S., and A. Sesonske, *Nuclear Reactor Engineering.* New York: D. Van Nostrand Company, 1963.
6. El-Wakil, M. M., *Nuclear Energy Conversion.* Scranton, Pa.: International Textbook Co., 1971.
7. "Advances in Thermal Performance of Nuclear Reactor Cores," *Westinghouse R and D Letter*, December 1966.
8. Connecticut Yankee Atomic Power Company, *Facilities Description and Safety Analysis*, Vols 1 and 2, Topical Report No. NYO-3250.5.
9. Connecticut Yankee Atomic Power Company, *Annual Report, July 1965–June 1966*, Topical Report No. NYO-3250-8, October 1966.
10. Elliott, V. A., "Boiling Water Reactor," *Mechanical Engineering* **89**, No. 1 (January 1967), pp. 19–26.
11. Parrish, J. R., G. M. Roy, and F. G. Bailey, "Nuclear Power Plant of T.V.A. at Brown's Ferry, Background and Description," American Power Conference, Chicago, April 25–27, 1967.
12. Bray, A. P., "Operating and Control Characteristics of a BWR," General Electric BWR Seminar, October 25–29, 1964.
13. Hoyt, H. K., "Operation and Performance of the Dresden Nuclear Power Station," American Power Conference, March, 1962.
14. Williamson, H. E., and D. C. Ditmore, "Current BWR Fuel Design Experience," *Reactor Technology* **14**, No. 1 (Spring 1971).
15. Rosenthal, M. W., et al., "*A Comparative Evaluation of Advanced Converters*," ORNL-3686, January 1965.
16. Olds, F. C., "Fusion Power Teams Begin to Tackle the Engineering Tasks," *Power Engineering* **78**, No. 10 (October 1974), pp. 34–41.
17. Cochrane, T. B., *The Liquid Metal Fast Breeder Reactor, an Environmental and Economic Critique*, Resources for the Future, Inc., Washington, D.C., 1974.
18. Technical Progress Report, 1974, Clinch River Breeder Reactor Project, Knoxville, Tenn., 1974.
19. *Clinch River Breeder Reactor Plant*, Clinch River Breeder Reactor Project, Knoxville, Tenn., October 1974.
20. Fraas, A. P., *A Potasium-Steam Binary Vapor Cycle for a Molten Salt Reactor Power Plant*, ASME Paper No. 66-GT/CLC-5, 1966.
21. Zinn, W. H., and J. R. Dietrich, "Peach Bottom Reactor," *Power Reactor Technology* **5**, No. 3 (June 1962), pp. 61–65.

22. Bechtel Corp., *Engineering and Economic Feasibility Study for a Combination Nuclear Power and Desalting Plant Summary*, Vol. 3, TID-22330, January 1966.

23. Holtom, H. T., and L. S. Galstaum, "Study of 150 MGD Desalted Water-Power Plant for Southern California," First International Symposium on Water Desalination, SWD/92, October 1965.

24. Holtom, H. T., "Integration of a Nuclear Power-Desalting Plant into the Metropolitan Water District System," 5th Annual AMU–ANL Faculty Student Conference, August 1966.

25. Homer, W. A., "Combination Nuclear Power and Desalting Plant: Engineering and Economic Feasibility Study," 5th Annual AMU–ANL Faculty Student Conference, August 1966.

26. Meneghetti, D., *Introductory Fast Reactor Physics Analysis*, ANL-6809, 1963.

27. Langley, Jr., R. A., *Evolution of the LMFBR Plant Design for Reliability and Availability*, ASME Paper No. 71-NE-3, 1971.

28. Baumann, H. F., and P. R. Kasten, "Fuel Cycle Analysis of Molten Salt Breeder Reactors," *Nuclear Applications* **2**, No. 4 (August 1966).

29. McLain, S., and J. H. Martens, *Reactor Handbook*, Vol. 4: *Engineering*. New York: Interscience Publishers, 1964.

30. Leung, P., and R. E. Moore, *Thermal Cycle Arrangements for Power Plants Employing Dry Cooling Towers*, ASME Paper No. 70-PWR-6, 1970.

31. Beall, S. E., Jr., *Uses of Waste Heat*, ASME Paper No. 70-WA/Ener, 1970.

32. Postma, H., "Engineering and Environmental Aspects of Fusion Power Reactors," *Nuclear News* **14**, No. 4 (April 1971), pp. 57–62.

33. Steiner, D., "The Nuclear Performance of Fusion Reactor Blankets," *Nuclear Applications & Technology* **9**, No. 1 (July 1970), pp. 83–92.

34. Rose, D. J., "Engineering Feasibility of Controlled Fusion— A Review," *Nuclear Fusion* **9**, No. 3 (October 1969), pp. 183–203.

35. Gough, W. C., and B. J. Eastlund, "The Prospects of Fusion Power," *Scientific American* **224**, No. 2 (February 1971), pp. 50–67.

36. Lubin, M. J., and A. P. Fraas, "Fusion by Laser," *Scientific American* **224**, No. 6 (June 1971), pp. 21–33.

37. Katterhenry, A. A., "Gas Turbine Power Plants," *Reactor Technology* **13**, No. 1 (Winter 1969–1970), pp. 7–13.

38. Böhm, E., A. Ehret, H. Geppert, W. Hauck, and K. D. Küper, "The 25 MW Schleswig-Holstein Nuclear Power Plant (Greesthacht II)," *Kerntechnik* **11**, No. 2 (February 1969), pp. 69–76.

39. Taygun, F., and H. U. Frutschi, *Conventional and Nuclear Gas Turbines for Combined Power and Heat Production*, ASME Paper No. 70-GT-22, 1970.

40. Bammert, K., et al., "Performance of High-Temperature Reactor and Helium Turbine, *Kerntechnik* **11** No. 2 (February 1969), pp. 77ff.

41. Dalle Donne, M., et al., "High Temperature Gas Cooling for Fast Breeders," *Kerntechnik* **11**, No. 2 (February 1969), pp. 99ff.

42. Gallagher, J. L., L. D. Green, and R. T. Marchese, *Nuclear Technology* **10**, No. 4 (April 1971), pp. 406–411.

43. Rosenthal, M. W., P. R. Kasten, and R. B. Briggs, "Molten-Salt Reactors—History, Status, and Potential," *Nuclear Applications & Technology* **8**, No. 2 (February 1970), pp. 107–117.

44. McCoy, E. E., et al., "New Developments in Materials for Molten-Salt Reactors," *Nuclear Applications & Technology* **8**, No. 2 (February 1970), pp. 156–169.

45. Whatley, E. E., et al., "Engineering Development of the MSBR Fuel Cycle," *Nuclear Applications & Technology* **8**, No. 2 (February 1970), pp. 170–178.

46. Scott, D., and W. P. Eatherly, "Graphite and Xenon Behavior and Their Influence on Molten-Salt Reactor Design," *Nuclear Applications & Technology* **8**, No. 2 (February 1970), pp. 179–189.

47. Bettis, E. S., and R. C. Robertson, "The Design and Performance Features of a Single-Fluid Molten-Salt Reactor," *Nuclear Applications & Technology* **8**, No. 2 (February 1970), pp. 190–207.

48. Perry, A. M., and H. F. Baumann, "Reactor Physics and Fuel Cycle Analysis," *Nuclear Applications & Technology* **8**, No. 2 (February 1970), pp. 208–219.

49. Gray, J. E., "The CANDU-PHW Programme," *Nuclear Engineering International* **19**, No. 217 (June 1974), pp. 476–477.

50. Cahill, L., L. F. Monier, G. A. Pon, and R. H. Renshaw, "Gentilly 2," *Nuclear Engineering International* **19**, No. 217 (June 1974), pp. 481–485.

51. Page, R. D., "Canadian Reactor Fuel," *Nuclear Energy Symposium*, 1974 Conference CNA/AECL, Montreal, Canada, May 1974.

52. Yaremy, E. M., "Reactor and Station Control," *Nuclear Energy Symposium*, 1974 Conference CNA/AECL, Montreal, Canada, May 1974.

53. *CANDU Nuclear Power Station*, Atomic Energy Limited of Canada, Ottawa, Ont., Canada.

54. Olds, F. C., "Demonstration LMFBR: Progress and Problems," *Power Engineering* **78**, No. 12 (December 1974), pp. 26–33.

55. Miliaras, E. S., *Power Plants with Air-Cooled Condensing Systems*. Cambridge, Mass.: The MIT Press, 1974.

56. Foster, A. R., S. A. Lamkin, and P. Kwok, *Design of Dry Cooling Towers for Use with a Direct Cycle HTGR*, ASME Paper No. 73-Pwr-7, ASME–IEEE Joint Power General Conference, New Orleans, September 16–19, 1973.

57. Bartlett, D. A., and A. R. Foster, *Natural Draft Dry Cooling Tower Performance with Varying Ambient Conditions Applied to a Direct Cycle HTGR*, ASME Paper No. 75-WA-44, ASME Winter Annual Mtg., Houston, November 30–December 4, 1975.

58. Gulf General Atomic Co., *Dry Cooling of Power Plants and the HTGR Gas Turbine System*, Gulf-GA-A12026, March 1972.

59. Koutz, S. L., J. M. Krase, and L. Meyer, *HTGR Gas Turbine Power Plant Preliminary Design*, ASME Paper No. 74-GT-104, Gas Turbine Conference, Zurich, March 30–April 4, 1974.

60. Tennessee Valley Authority, *Final Environmental Statement—Hartsville Nuclear Plants*, Vols. 1 and 2, 1975.

61. Parker, W. O., Jr., J. C. Deddens, D. W. Berger, and K. L. Hladek, "Postoperational Test and Examination—Oconee Unit I," American Power Conference, Chicago, April 21–23, 1975.

62. Fraas, A. P., "Conceptual Design of a Series of Laser-Fusion Power Plants of 100 to 3000 MW(e)," 9th Intersociety Energy Conversion Engineering Conference, San Francisco, Calif., August 26–30, 1974.

63. Klepper, O. H., and T. D. Anderson, "Siting Considerations for Future Offshore Nuclear Energy Stations," *Nuclear Technology* **22**, No. 2 (May 1974). pp. 160–169.

64. Ashworth, J. A., "Atlantic Generating Station," *Nuclear Technology* **22**, No. 2 (May 1974), pp. 170–183.

65. Tennessee Valley Authority, *Preliminary Safety Analysis Report, Bellefonte Nuclear Power Plant*, June 1973.

66. Public Service Company of Oklahoma, *Preliminary Safety Analysis Report, Black Fox Station*, December 1975.

67. Project Management Corporation, *Preliminary Safety Analysis Report, Clinch River Breeder Reactor Project*, April 1975.

68. Sege, Carol A., et al., "The Denatured Thorium Cycle—An Overview," *Nuclear Technology* **42**, No. 2 (February 1979), pp. 144–149.

69. Haffner, D. R., and R. W. Hardie, "Reactor Physics Parameters of Alternate Fueled Fast Breeder Reactor Designs," *Nuclear Technology* **42**, No. 2 (February 1979), pp. 123–132.

70. Marr, D. R., et al., "Performance of Thorium-Fueled Fast Breeders," *Nuclear Technology* **42**, No. 2 (February 1979), pp. 133–143.

71. Blake, E. M., "Fusion Power in the United States, Parts 1–4," *Nuclear News* **23**, Nos. 8–11 (1980).

72. Doncals, R. A., et al., "Clinch River Breeder Reactor Plant Core Flexibility," *CRBRP Technical Review*, Spring 1980, pp. 7–24.

73. Banal, M., "Construction of the World's First Full-Scale Fast Breeder Reactor," *Nuclear Engineering International* **23**, No. 272 (June 1978), pp. 43–60.

74. Vrable, D. L., R. N. Quade, and J. D. Stanley, *Design of an HTGR for High-Temperature Process Heat Applications*, ASME Paper No. 79-JPGC-NE-2, October 1979.

75. Baqué, P., et al., *The Creys Malville FBR Superphenix Steam Generators*, ASME Paper No. 80-C2/NE-28, August 1980.

76. *Nuclear Report* (American Nuclear Society) **3**, No. 10 (1980).

77. *Update*, DOE Office of Nuclear Reactor Programs, July–August 1980.

78. Clark, L. L., and A. D. Chockie, *Fuel Cycle Cost Projections*, Battelle Pacific Northwest Laboratory, NUREG/CR-1041, December 1980.

79. Allen, R. E., "Economics of the Nuclear Option: Can It Compete with Alternatives?" *Proceedings of the ANS Topical Meeting: A Technical Assessment of Nuclear Power and Its Alternatives*, Los Angeles, February 27–29, 1980, pp. V.1-1 to V.1-20.

80. *Status and Prospects of Thermal Breeders and Their Effect on Fuel Utilization*, Technical Reports Series No. 195, International Atomic Energy Agency, Vienna, 1979.

81. Blake, E. Michael, "Fusion Power in the United States, A Four-Part Series," *Nuclear News* **23**, Nos. 8, 9, 10, 11 (June–September 1980).

82. Agnew, Harold, M., "Gas-Cooled Nuclear Power Reactors," *Scientific American* **244**, No. 6 (June 1981), pp. 55–63.

83. Waltar, A. E., and A. B. Reynolds, *Fast Breeder Reactors*, New York: Pergamon Press, 1981.

84. Simon, Robert H., "Status of the Gas-Cooled Fast Breeder Reactor Program in the United States from the Industrial Point of View," IAEA Tech. Comm. Mtg. on Gas-Cooled Reactors, Minsk, USSR, October 12–16, 1981.

14

Selected Topics in Reactor and Fuel Cycle Technology

A few of the more important topics facing the nuclear industry are briefly described in this chapter. There are many other topics of equal or greater interest, but the ones discussed here present a typical sampling of those to be encountered in the broad field of reactor design and fuel cycle technology. There is a wealth of information available on each of the topics discussed, and the reader is encouraged to pursue them in greater detail. Some suggestions for further study are included as problems at the end of the chapter.

Thermal Discharges

Although power plants have always discharged significant amounts of heat to their surroundings, the problem has become more serious with the advent of nuclear plants for several reasons.

(1) The size of plants has jumped to be in excess of 1000 MW(e). The size of fossil units has kept pace, but the capability to produce these large capacities developed during the 1960s.

(2) The efficiency of light water nuclear plants is lower (30 to 33 percent) than that of modern fossil units (40 to 42 percent) due to the use of lower-pressure saturated steam in the nuclear units. For the same electrical output, the less efficient plant will reject more heat to its heat sink.

(3) Fossil plants discharge 10 percent or more of their heat directly up the stack to the atmosphere, where the entire heat rejection load must be borne by the condenser in a nuclear plant.

Adequate provision must be made to handle the rejected heat without adversely affecting ecological conditions in the vicinity of the plant. For a 1000-MW(e) plant, Figure 14.1 shows the massive cooling water flow rates (m³/s) required for both an LWR and an LMFBR (or HTGR), where the efficiencies are assumed to be 33.3 and 40 percent, respectively. The advantage of using an LMFBR or HTGR, where the higher peak temperatures permit greater thermal efficiency, is evident. There is 25 percent less heat rejection for the more efficient plant. If both plants were designed for a 10°C temperature rise for the condenser cooling water, the LWR would require 47.7 m³/s, whereas the LMFBR would require only 35.8 m³/s. If the LMFBR used a flow of 47.7 m³/s its cooling water would rise only 7.5°C.

As greater flows are used the impact on aquatic life is lessened, but the size and cost of cooling equipment (condensers, pumps, piping, etc.) increase rapidly. On the other hand, reduced flows will raise temperatures to the point where there may be ecological damage done to aquatic life if direct discharge to a water body is permitted. Ecological surveys must be made early in the planning at any plant site in order to know how best to handle the discharge of waste heat

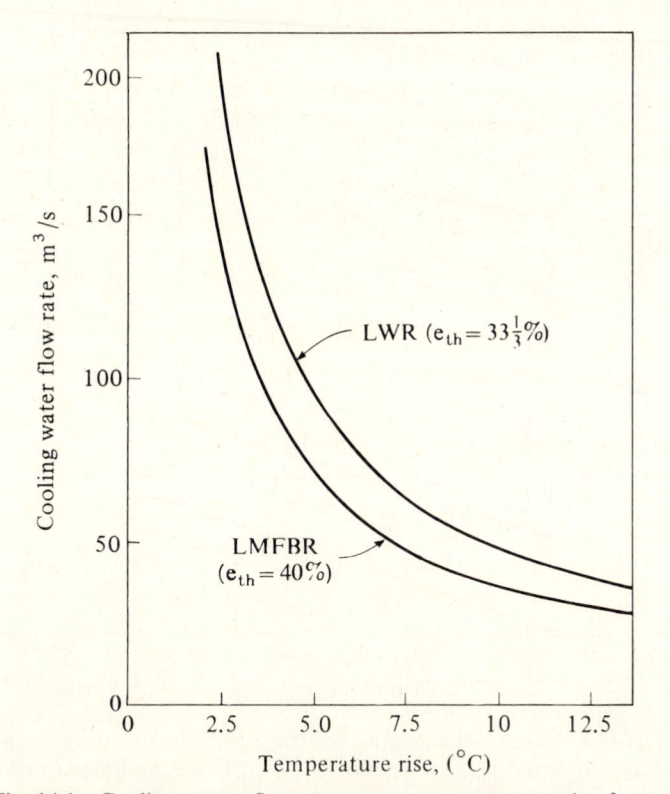

Fig. 14.1   Cooling water flow rates versus temperature rise for an LWR and an LMFBR each having an output of 100 MW(e).

to the surroundings. In the case of the Merriman study for the Connecticut Yankee Plant (Reference 19), data were accumulated for three years prior to plant operation in order to provide a comparative base for subsequent observations.

Figure 14.2 shows the ability for a hypothetical fish to function at various thermal levels. Being cold-blooded, the fish will come to thermal equilibrium with the water in which it is resident. The acclimation temperature is a temperature to which the fish has grown accustomed, while the temperature tolerance

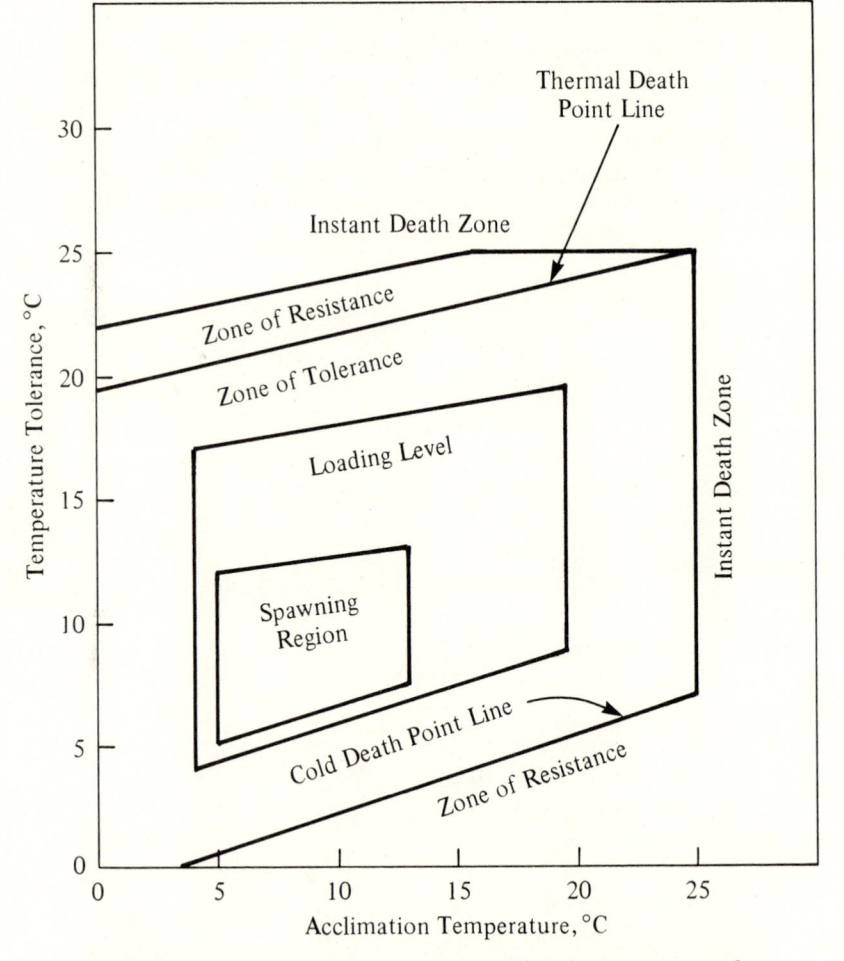

Fig. 14.2 Temperature tolerance versus acclimation temperature for a hypothetical fish. [Adapted from M. D. Triplett, "Nuclear Power Plant Cooling and Thermal Effects," Northeastern University, Boston, 1975.]

will indicate the response of the fish to a new level of temperature to which it may be instantaneously subjected. The loading level is the region within which normal growth and activity take place. Within it is a more restricted region within which spawning may occur. Outside the loading level is a zone of tolerance in which, although conditions are not optimal, life will be indefinite. The zone of tolerance is bounded at the top by the thermal death point line. When a fish is thrust into temperatures somewhat above this, the life span is limited in a region called the *zone of resistance*. Above the zone of resistance death is more or less instantaneous. At the bottom of the zone of tolerance is the cold death point line. Again, there is a zone of resistance in which the length of life is restricted and in fresh water below 0°C instantaneous death would occur, as the fish's environment would solidify.

For the species of fish shown in Figure 14.2, if the temperature to which it is acclimatized is 12°C and the temperature to which it is subjected suddenly rises to 14°C, spawning would be inhibited; but, otherwise, the fish would function and grow in a normal manner. At a new temperature of 20°C, the fish would be in the zone of tolerance and normal growth would not occur. If the temperature were suddenly raised to 22°C, death would occur after some period, and if the new temperature were to go above 24.3°C, death would be instantaneous. If the fish had been acclimatized to 17°C, a new temperature of 22°C would now be in the zone of tolerance, and the fish would survive indefinitely.

In winter a fish of this hypothetical species becomes most uncomfortable when the stream temperature drops to 2°C, although it can survive if acclimatization occurs gradually. The fish begins to migrate downstream in search of warmer water. It comes to the 12°C outfall from a power station, which is very comfortable, and the fish takes up residence in this area, soon becoming acclimated to this temperature. The power plant shuts down for refueling, and the stream temperature suddenly returns to 2°C. The fish will die shortly, as it is subjected to a temperature in the zone of resistance below the cold death point line.

It should be remembered that each species of fish has a different set of temperature tolerances. For example, the temperature that produces optimal growth for brook trout is 15.3°C (59°F), for northern pike it is 21.4°C (70°F), and for largemouth bass it is 27.6°C (81°F). A temperature producing optimal growth for northern pike will be lethal for brook trout and suboptimal for largemouth bass. To complicate the problem further, the effects on the invertebrate species and plant life must also be considered. No two sites will have identical conditions, and each must be surveyed carefully to be sure that no serious ecological damage will be caused. If such damage does occur, the thermal discharge may truly be termed thermal pollution.

Here it might be noted that there may be beneficial effects of the warm discharge from a power plant. There have been numerous reports of improved sports fishing at the outfall of many plants in the fall, winter, and spring months. The warm discharge from nuclear plants is being considered to stimulate

oyster production in Long Island Sound, to increase the growth rate of lobsters in Maine, and for shrimp aquaculture in Florida. District heating and cooling using low-level heat have been studied and experiments are under way using warm water for the irrigation of crops. In many areas streams have been impounded to create ponds or lakes for heat dissipation. These almost invariably support a sizable fish population and provide other recreational opportunities.

Careful attention must be given to the means chosen for rejection of heat from a large power station, especially to technical feasibility, economics, and its social and ecological impact. There are several schemes that may be considered for dissipation of waste heat.

Mixing followed by remote and gradual dissipation

A typical temperature rise for cooling water passing through a condenser is 10°C, but for a particular site the maximum allowable temperature rise at discharge may be 2.78°C. For the 1000-MW LWR discussed previously this would require 172.3 m^3/s instead of 47.6 m^3/s. Rather than building a bigger condenser to accommodate this flow, it is simpler to bypass the extra 125 m^3/s and allow it to mix with the condenser effluent prior to discharge. Alternatively, the discharge system may employ diffusers to promote mixing of the effluent with the passing current within a limited mixing zone to achieve the desired 2.77°C temperature rise. Because of the small temperature differential between the water and the atmosphere, a rather large surface area will be required to return the water to ambient conditions.

Rapid local dissipation using open water bodies

Here the natural buoyance of the heated water allows it to rise to the surface and exchange heat more rapidly with the atmosphere due to its higher temperature. Both sensible and latent heat transfer are more rapid. A study (Reference 18) indicates that only half as much area is required to return heated water to within 0.555°C of ambient when a 10°C temperature rise is allowed instead of a 2.78°C rise. With proper outfall design the warmed water will spread over the surface area, having little effect on the aquatic life more than a few feet below the surface. Either natural or human-made water bodies will require an appropriate mixing and dissipation zone in which the majority of the heat dissipation may take place. Beyond this zone temperature standards can be applied.

Seasonal heat storage in water bodies for dissipation in cool weather

Some bodies of water have considerable temperature stratification during the summer months. If volumes of water are sufficient, cool water may be drawn from the lower levels, discharged from the condenser at temperatures well below the surface temperature, and returned to the body of water. This returned

water will tend to seek a level underneath the warm surface layer and remain there until fall. When its temperature becomes greater than that of the surface water, it will rise and belatedly transfer its stored energy to the atmosphere. During the cooler seasons the heated water cools promptly at the surface.

Wet cooling towers and spray ponds

In some locations the size of the water body available or the ecological circumstances demand a water temperature unattainable by any of the previous methods. Evaporative cooling by wet cooling towers or spray ponds may then provide an appropriate solution. The spray pond is more or less a cross between the cooling tower and the cooling pond, where a limited surface area in contact with the atmosphere is increased by water sprays. In the cooling tower (see Figure 14.3) water is pumped to spray nozzles atop a latticework of redwood slats. The water droplets are interrupted by the latticework and the water is held up as it runs down the slats, reforms new drops, lands on another slat, and so on, until it reaches the bottom of the tower. All the while sensible heat transfer between the water and air is taking place due to temperature difference and mass transfer (evaporation) takes place due to the difference in vapor pressure between that for saturated water at the surface of the water droplets or films and that of the water vapor in the air flowing upward counter to the water. Airflow

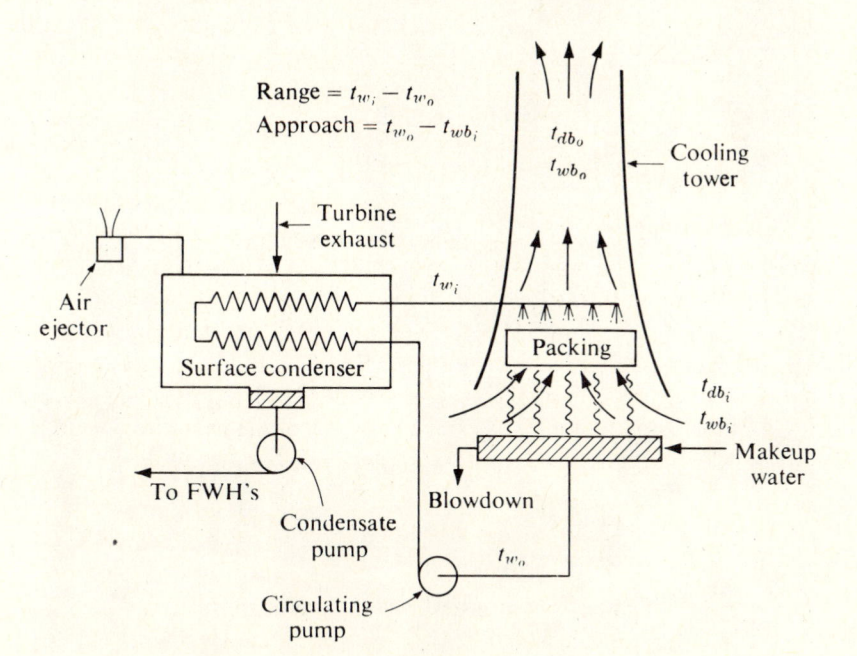

$$\text{Range} = t_{w_i} - t_{w_o}$$
$$\text{Approach} = t_{w_o} - t_{wb_i}$$

Fig. 14.3 Wet cooling tower using natural draft and a closed cooling water system.

may rely on natural convection where the heated air is less dense than the ambient air. These natural convection towers are huge hyperbolic structures, some having a base diameter in excess of 90 m and a height of 135 m or more (see Figure 14.4). Induced-draft fans are used in much smaller towers to increase the airflow, but they do add the cost of the fan and its operating power. For nuclear plants the natural draft tower costs $30 to $40 per kW and the wet mechanical draft tower costs $20 to $30 per kW. At $30 per kW for a 1000-MW plant, this amounts to $30 000 000.

The cooled water from the tower may be recirculated in a closed system or it may operate as an open once-through system with the cooled water being returned to the body of water used as the coolant source. It is well to remember that the order of 5 percent of the water flow is required for evaporation and blowdown. In the winter months the evaporative loss may produce fog and icing conditions in the vicinity of the plant. Often where insufficient river flows dictate the use of cooling towers during the warm months of the year, it may be more appropriate to use once-through cooling during the cold weather to prevent fog and icing.

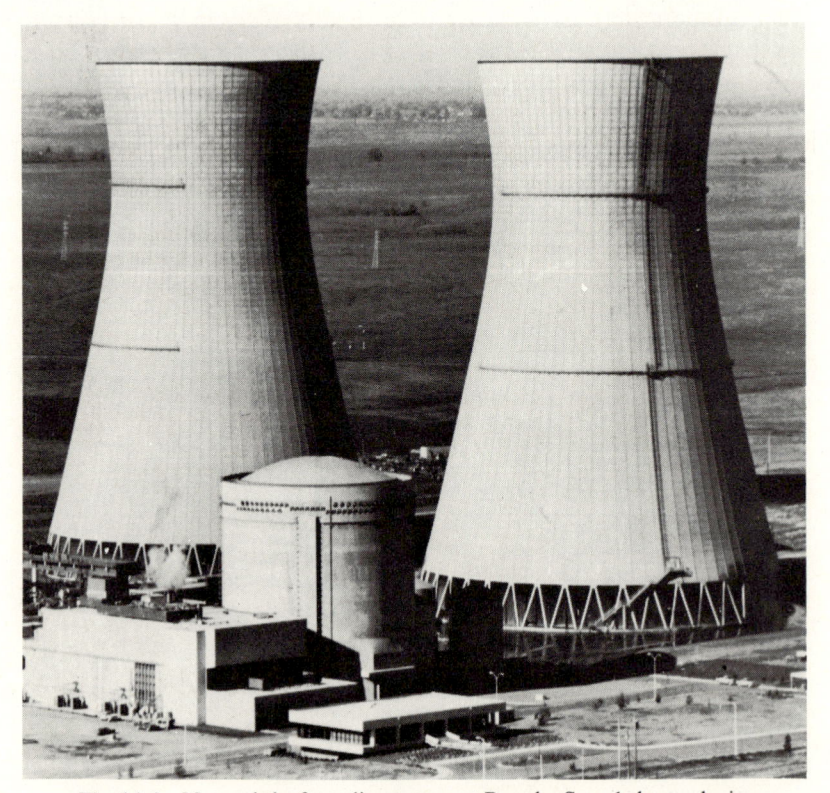

Fig. 14.4   Natural-draft cooling towers at Rancho Seco 1 shown during construction.   [Courtesy of Sacramento Municipal Utility District.]

In a wet cooling tower the drop in water temperature as it falls through the tower is called the *range*. The difference between the leaving water temperature and the inlet wet-bulb temperature is the *approach*. If a tower were to have an infinitely large water-to-air heat transfer surface area, the temperature of the leaving water, t_{w_o}, would approach the wet-bulb temperature of the air, t_{wb}, and the approach would be zero. The ability of the wet cooling tower to approach the wet-bulb temperature rather than the dry-bulb temperature gives it a distinct advantage over the air-cooled condenser or dry cooling tower, where the outlet water temperature must approach the inlet dry-bulb temperature of the cooling air.

Dry cooling tower

Where water supplies are insufficient for a wet cooling tower, the only alternative becomes a dry cooling tower.

The German GEA system is essentially an air-cooled condenser where a fan forces air to flow across the finned condenser tubes located within the tower. This system has been applied in Europe to units as large as 120 MW, but because of the size of the pipes required to convey the exhaust steam to the cooling coils of the condenser, it is doubtful that this system will be effective for large nuclear stations.

The Heller system, however, is not so limited. It is shown in Fig. 14.5. Here a jet or spray type condenser is used at the turbine exhaust. In this closed system the cooling water and condensate mix as the cool water returning from the dry

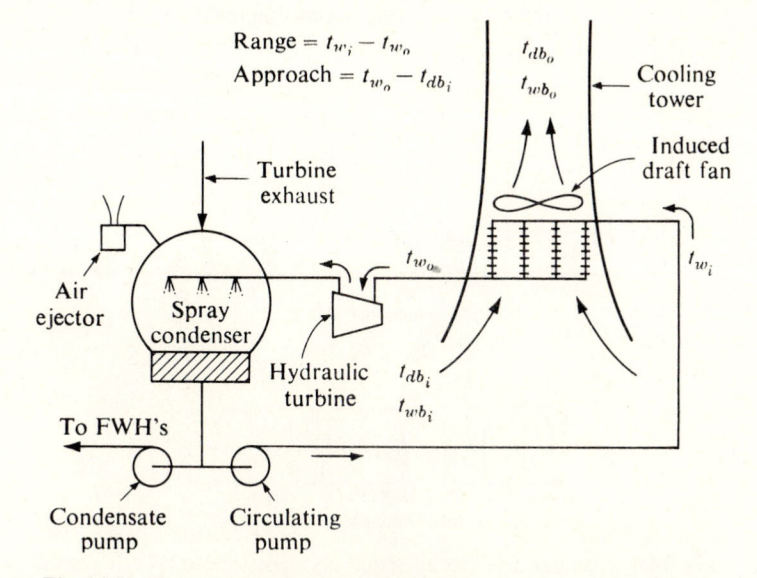

Fig. 14.5 Dry cooling tower using the Heller system with induced draft.

cooling tower sprays into the exhaust steam, causing condensation and a rise in temperature of the cooling water toward the saturation temperature at the steam generator pressure. Feedwater is pumped toward the reactor through the feed-water heaters while the circulating pump returns the cooling water to the tower. The cooling tower may use either natural draft or induced draft (as shown in Fig. 14.6). A steam-driven ejector helps maintain the vacuum by removing air and other noncondensible gases. The main circulating pump maintains the water in the tower at a positive pressure to prevent air leakage into the system. Some of the pump work may be recovered by a hydraulic turbine after the water leaves the tower on the way to the jet sprays in the condenser.

Condensing pressure and temperatures using a dry cooling tower will tend to be significantly higher and more variable than for a wet tower or a once-through system. To accommodate the higher pressures, perhaps as high as 0.33 bar compared to the usual values of 0.05 to 0.10 bar, a smaller last-stage annulus area will be required in the low-pressure turbine. During the cooler months when lower pressures are possible, this will penalize the turbine perform-ance somewhat. Because of the more variable performance from this type of

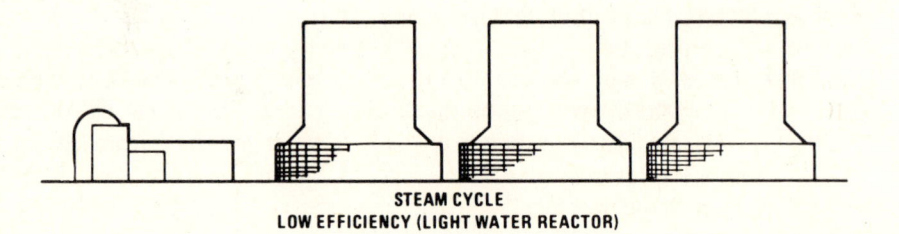

STEAM CYCLE
LOW EFFICIENCY (LIGHT WATER REACTOR)

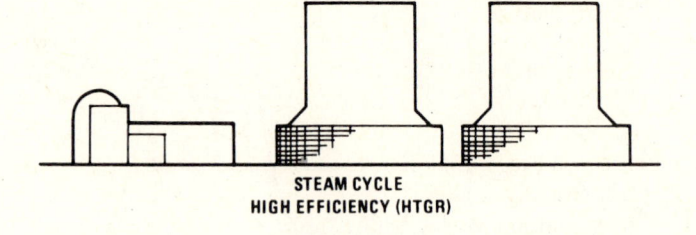

STEAM CYCLE
HIGH EFFICIENCY (HTGR)

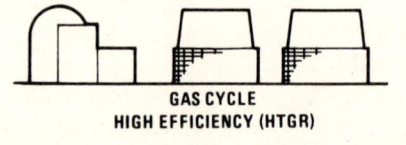

GAS CYCLE
HIGH EFFICIENCY (HTGR)

Fig. 14.6 Comparative size of steam and gas turbine HTGR plants with dry cooling towers. [Courtesy of General Atomic Company.]

system it is important that economics be studied from the point of view of the unit being integrated into the system, rather than as a baseloaded unit.

A dry cooling tower imposes a severe economic penalty on a nuclear plant with the cost estimated at $70 to $100 per kW. This cost is about the same as that for the turbine-generator or the cost of the reactor. If the cost were taken at $75 per kW, it would mean an investment of $75 000 000 for a 1000-MW plant. As more plants are built and suitable sites become scarcer, dry cooling towers may be the only answer.

Dry cooling towers for direct-cycle HTGRs

From Example 4 in Chapter 13 it may be observed that helium leaves the re-generator for the precooler at a temperature of 332°C in this particular case. Thus, it is quite possible to heat water under pressure to a temperature of 171°C, as shown in Fig. 13.24. This hot water then transmits the heat to a dry cooling tower, where air may be heated to a temperature such as the 116°C shown.

In a natural draft tower, the buoyancy of the heated air induces airflow across the heat exchanger surfaces to provide the requisite heat transfer to the airstream from the hot water. The airflow also may be enhanced by an induced-draft fan in the mechanical-draft-type tower. The use of a fan reduces the size of the tower required, but puts an additional power loss into the system. The enthalpy gain per kg of air in towers used with the direct-cycle HTGR will be 5 to 10 times that of the dry tower for a steam system. This results in a dramatic reduction in tower size, as shown in Fig. 14.6.

The dry tower adds only sensible enthalpy to the atmosphere. Thus, it will not be plagued with the problems of fogging and icing which bedevil wet cooling under certain atmospheric conditions.

The Accident at TMI-2

Shortly after 4:00 A.M. on March 28, 1979, with Unit 2 operating at 97 percent full power, a loss of feedwater in the secondary coolant system resulted in a shutdown of the steam turbine. When this happened, pressure in the primary coolant began to increase (see Fig. 14.7). Because of the high pressure, the reactor then shut down automatically, and a relief valve in the primary coolant system opened. Up to this point (which took about 8 s) normal reactor protection systems functioned properly.

Even though a reactor is shut down, a considerable amount of heat is still generated from the radioactive decay of fission products. This heat is removed by secondary cooling in the steam generators. Since the normal supply of feed-water was not available, emergency feedwater pumps automatically came on 33 s after the accident began. Unfortunately, valves had been inadvertently closed in the emergency feedwater lines and no water was fed to the steam generators. This fact went unnoticed by plant operators, and subsequently all the water in

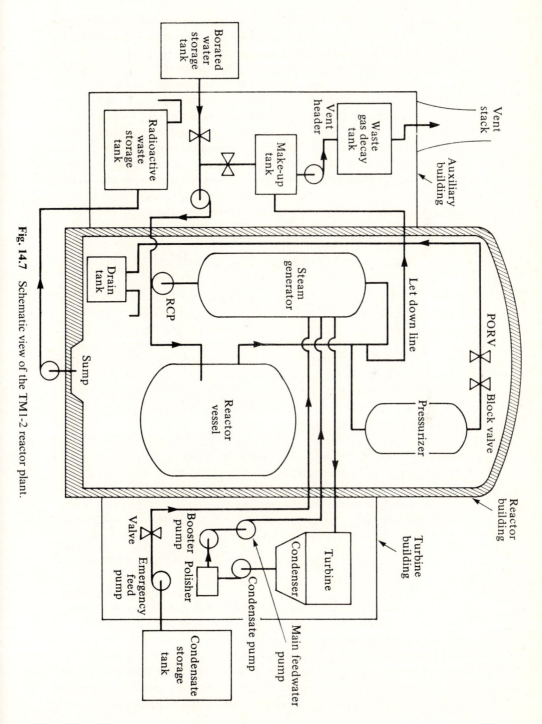

Fig. 14.7 Schematic view of the TM1-2 reactor plant.

the steam generators boiled away. Opening of the relief valve in the primary coolant system caused coolant pressure to decrease. This relief valve should have automatically closed at 22.4 MPa (3250 psi). The fact that it did not close also went unnoticed by the operators. The primary coolant pressure continued to drop and 2 min after the accident started, high-pressure safety injection valves automatically started adding water to the primary coolant system. About $2\frac{1}{2}$ min later, the operators, believing that the primary system was filling up with water, followed their operating instructions and reduced the high-pressure water injection to the system. It is believed by those investigating the accident that had the relief valve closed (or been closed) or the high-pressure injection pumps been allowed to continue, the accident could have been avoided as late as 1 h 40 min after it began with little or no damage to the core. Had either of these two actions occurred, there would have been only a minor incident.

About 2 h 20 min after the accident started, the relief valve was closed and at 3 h 40 min full high-pressure injection was restarted. In the meantime, steam formed in the reactor core, reducing heat transfer from the reactor fuel. Without adequate cooling the fuel became overheated and damaged. The full extent of fuel damage cannot be determined until the reactor is opened. However, analysis of the coolant that spilled into the containment has shown that some radioactive fission products were released from the fuel.

The water escaping from the relief valve went into the containment sump and was pumped from there to an auxiliary building tank (Fig. 14.7). The rupture disk in this tank had been ruptured previously, so reactor sump water flowed onto the auxiliary building floor. This pumping was stopped 39 min after the accident began, but other systems probably continued to pump radioactive water into the auxiliary building. At about 6:30 A.M. radioactivity levels in the auxiliary building began rising rapidly. As a result of this and rapidly rising radiation levels in the containment, a general emergency was declared. Radiation levels off-site did not start rising until after 9:00 A.M. and never did rise to significantly high levels.

The damage to the reactor core and associated parts will not be known until the reactor pressure vessel is opened and inspected. Some estimates of damage can be drawn from analysis of fission products and thermal–hydraulic analysis of the accident. It is estimated that there is some cladding failure in 90 percent or more of the fuel rods. At high temperature, zirconium reacts with water to produce zirconium dioxide, hydrogen, and heat; embrittlement of the cladding results. Estimates range that from 44 to 63 percent of the cladding has oxidized and that the upper 60 to 70 percent of the core lost its structural integrity. The loss of structural integrity permitted a portion of the core to fall into gaps between the fuel rods in the lower portion of the core.

Fuel temperatures exceeded 1925°C in 40 to 50 percent of the core and may have exceeded 2200°C in upper parts of the core. Some of the fuel may have dissolved into the zirconium–zirconium oxide cladding mixture at this temperature. Some fuel is also in very finely divided form, as evidenced by leaching of

radioactive products into the cooling water. Even though temperatures in the core rose above the 1538°C melting point of steel there is no indication that there is any damage to the reactor vessel. Some silver was detected in the containment sump water, indicating that portions of the control rods probably melted. Since control rod materials are not soluble in water, they remain in the reactor core.

At one time during the course of the accident there were reports that a hydrogen bubble inside the reactor vessel might explode. The hydrogen in the reactor resulted from radiolytic decomposition of water, and it always occurs in pressurized water reactor operation. In fact, in normal operation hydrogen overpressure is maintained in the reactor vessel to prevent accumulation of oxygen. The hydrogen concentration in the coolant system during the accident was about 200 times the level necessary to prevent accumulation of oxygen. Thus, there never was any possibility of an explosion, and reports of a possible explosion were erroneous.

The presidential commission investigating the accident found that the amount of radiation outside the plant was very low, and they concluded that there will be no detectable cases of radiation-induced health effects. The major effect as a consequence of the accident on the general population living near the plant appears to be mental stress.

The maximum dose received by any one individual in the general population was estimated to be 70 mrem (0.7 mSv). The collective dose to the general population (up to April 15, 1979) within 50 mi was estimated to be 2000 person-rem (20 person-Sv) or less than 1 percent of annual background level. A person living within 5 mi of the plant received an average dose of about 10 percent of annual background radiation.

Three TMI workers received delivered doses during the accident of about 3 to 4 rem (30 to 40 mSv). Cleanup and decontamination will require additional occupational exposure and releases to the environment.

The main source of radioactive release during the accident is believed to have been through a leak in the vent gas header to the waste gas decay tanks and/or through lifting of safety valves in the reactor coolant bleed tanks (see Fig. 14.7). During normal operation primary coolant is piped through the letdown/makeup system to the auxiliary building, where it is stored in bleed tanks. Gases released from the coolant are stored in waste gas decay tanks and bled through filters to the atmosphere. The charcoal filters are designed to absorb iodine from the gases, but they did not perform properly during the accident. The release consisted of about 13 to 17 Ci of iodine and 2.4 to 13×10^6 Ci of xenon and krypton. No detectable amounts of radioactive cesium or strontium were observed in the environment.

The radioactive iodine released from the fuel and retained in the primary system, containment, and auxiliary building ($\sim 35 \times 10^6$ Ci) decayed quickly and did not present a health hazard. During decontamination and cleanup other releases of radioactivity to the immediate environment will be done in a carefully controlled manner.

Initial cleanup of the plant began in April 1979 with decontamination of preaccident low-level-activity water stored in the auxiliary building. During the summer of 1979 construction began on a system to decontaminate the 380 000 gal of water containing intermediate levels of radioactivity. The radioactivity will be removed by resins which will be solidified and shipped to a disposal site

The schedule calls for decontamination of containment during 1981–1982, removal of the fuel by 1983, and complete decontamination in 1984. This schedule, of course, depends on the condition of the reactor containment building and fuel. It is expected during cleanup that valuable information on equipment and instrumentation survivability will be gained. Information will also be gathered on environmental conditions as they relate to fission product release and disposition and technology for decontamination and radiation dose reduction.

Initial access was gained to the containment in the summer of 1980. The first task on access to containment was an assessment of the primary system boundary condition. Techniques for nondestructive assay of fuel distribution, for assessing criticality control during cleanup, and for fuel removal, shipment, and disposition are developed as the character of the primary system boundary is known.

Also, as part of the cleanup, studies are being made on the effect of the accident on the value of property in the Harrisburg area and on the socioeconomic impact of the accident on the south-central Pennsylvania region. These studies could have effects on the future siting of many industries.

The cleanup effort will take several years and may cost 1000×10^6 for plant cleanup and repair.

Estimates place replacement power costs at about 14×10^6 per month and total refurbishment costs from $1 to 2×10^9. Other estimates for replacement with a coal-fired plant range from $2 to 2.5×10^9. Thus, refurbishment of the damaged plant appears to be an economical alternative.

It goes almost without saying that the accident showed that many changes should be made to improve reactor safety and operation. Overall, there were over 600 recommendations made to the NRC for improving safety and operation of nuclear power plants by the many groups that studied the accident. There was general agreement as to the causes of the accident and the failures (human and equipment) that took place. There has also been general agreement in broad areas where improvements needed to be made. There were, however, many differences of opinion as to relative priorities of changes, degree of changes, and ways in which changes should be made.

Broadly speaking, changes in reactor operation and design have been divided into two categories: short term (changes made immediately after the accident and before a more comprehensive improvement plan was developed), and long term (after various investigations of the accident were completed).

The short-term changes involved instrumentation and equipment additions, reanalysis of some systems, and improvements in plant operations. These

23 recommendations are described in detail in Reference 15. Examples of the short-term changes are: (a) provide position indication or flow indication for the pressurizer relief valve; (b) provide more positive isolation of the containment; (c) use a shift technical advisor who has specific training in response to non-normal operating events; and (d) reduce possible leakage of radioactive material. All the short-term changes were substantially completed by the end of 1980. The longer-term changes can be divided into seven categories: (a) plant operator qualification and training, (b) plant management, (c) plant design, (d) plant siting, (e) emergency preparedness, (f) radiation protection, and (g) NRC procedures, organization, and management. These changes and the action plan for their implementation are proposed in Reference 15.

The most significant factors causing the accident were operational aspects. Consequently, changes in plant operator qualifications and training and plant management are given the highest priority in long-term improvements. Specific changes include measures to prevent causes of accidents on the part of plant operators and improvement on the part of utility staff to recognize potential accident-causing events and take corrective actions.

Analyses are planned for all plants that should identify and correct weaknesses in current designs. These include reassessments of the reliability of some engineered safety features, such as the auxiliary feedwater, containment isolation, and decay heat removal systems. Investigation of changing reactor siting requirements to increase distance between population centers and new reactors is also being undertaken.

The Federal Emergency Management Agency has been assigned the responsibility for emergency planning and response. This action centralizes this function in one agency. It should provide improvement in emergency planning and emergency response for utilities and governments alike. An effort to increase public and media understanding of nuclear power plants and radiation is also part of the long-term improvements.

Finally, the long-term improvements include better monitoring and measurement of radioactive effluents and faster estimation of off-site exposures. This should lead to improved protection from radiation for both workers and the public.

Loss of Coolant Accidents

One of the principal concerns in any reactor safety analysis is the ability of the system to cope with a loss-of-coolant accident (LOCA). This is assumed to occur by the rupture of one of the lines leading to or from the core vessel, allowing the coolant to be discharged into the containment. In such an accident it is imperative that the fuel–cladding temperature be kept below 1200°C and that fission products in any significant quantity be prevented from leaking from the containment. Probably the worst contaminators are the iodine isotopes be-

cause of their volatility and susceptibility to bodily uptake with concentration in the thyroid.

Figure 14.8 is a schematic diagram of a typical **PWR** emergency core coolant system. The accumulator injection tanks contain water pressurized by nitrogen at 1.4 to 4.1 MPa. If the main coolant piping breaks, water from the accumulators will automatically be injected into the reactor vessel as soon as the pressure drops below the nitrogen pressure. In the United States, the borated water connection is generally in the cold leg. Some European manu-facturers prefer hot-leg injection or even injection into both legs.

In addition to the accumulator injection system, PWRs have two active emergency core cooling systems: (a) high-pressure coolant injection and (b) low-pressure coolant injection system. The high-pressure coolant injection system is employed automatically in the event of a small leak or slow depres-surization of the reactor vessel. It takes a positive suction from the water storage tank and discharges high-pressure borated water to the main coolant system.

The low-pressure coolant injection system refloods the reactor after loss of pressure and circulates the injection water through the residual heat removal system. Generally, the residual heat removal system is the same as that used for normal shutdown. The heat removal system might have to operate up to

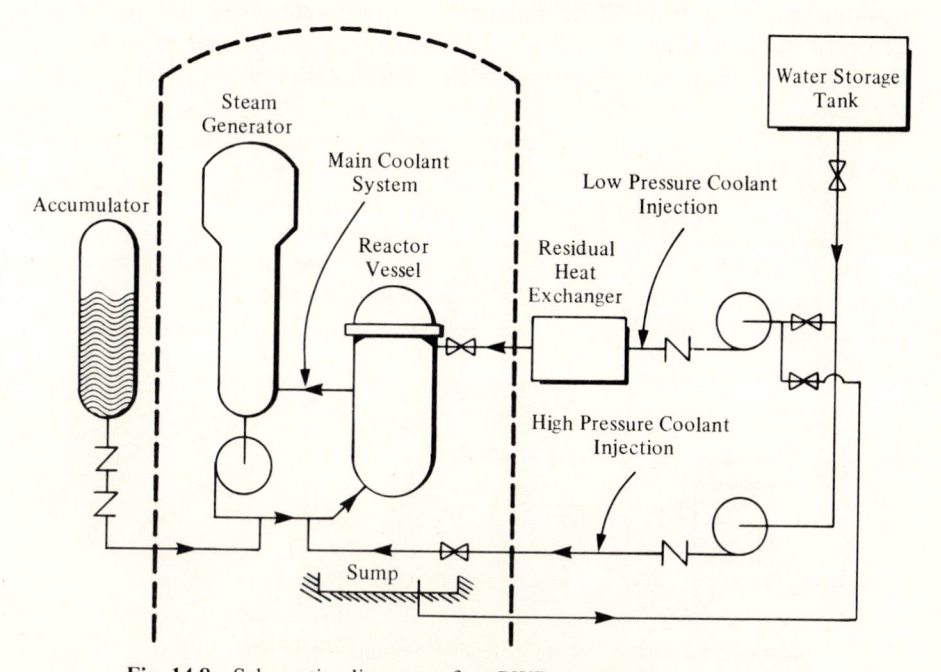

Fig. 14.8 Schematic diagram of a PWR emergency core coolant system.

90 days following a serious LOCA. Notice that both the high-pressure coolant injection pumps and the low-pressure injection pumps take their suction from the water storage tank. In addition to the ECC, each PWR contains a containment heat removal system. In the event of a LOCA or rupture of the pressure vessel, some reactor coolant and some radioactive material would escape into the containment vessel. The containment heat removal system consists of (a) a containment spray system and (b) a reactor containment fan cooler system. The containment spray system pumps water from the water storage tank to spray headers. By means of an eductor or similar system, chemical solutions are added to the spray. In addition, the spray system can also recirculate water from the sump through the residual heat exchanges to the spray header. Figure 14.9 is a schematic diagram of a containment spray system.

Sodium hydroxide is added to control the pH. It must be high enough to encourage molecular iodine absorption and low enough to discourage corrosion $(7 < \mathrm{pH} < 10)$. Sodium thiosulfate may be added to remove methyliodide. As can be seen in Fig. 14.9, once the borated water and NaOH tanks have been emptied, both the spray pumps and the decay heat pumps will recirculate the water from a sump in the reactor building with the sodium thiosulfate reaching the core at this time. In the B&W system the sump will attain an accumulated dose of 10^6 Gy in 30 days and 2×10^6 Gy in 180 days, primarily from the absorbed iodine. Activity in the containment atmosphere will be primarily due to noble gases. Zircaloy-4, Inconels, and stainless steels are not seriously corroded by these solutions. The use of aluminum and copper is restricted because of corrosion when immersed in the spray solution.

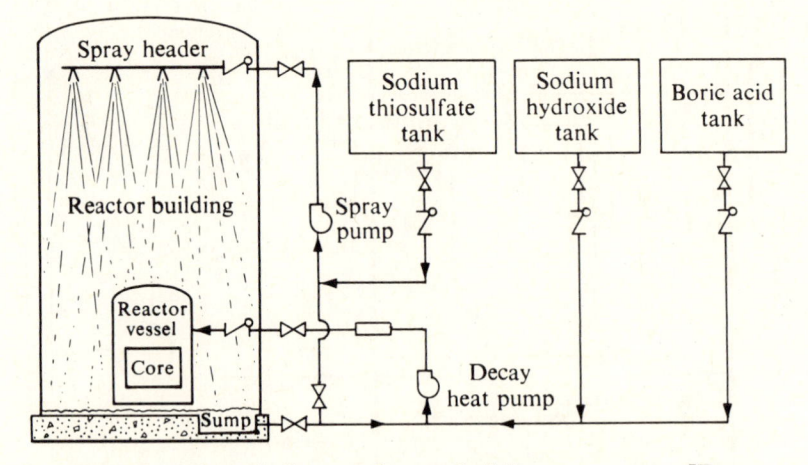

Fig. 14.9 Schematic diagram of reactor building spray system. [From W. N. Bishop and D. A. Nitti, *Reactor Technology* **10**, No. 4 (April 1971), p. 451.]

High containment pressure will initiate the containment fan cooler system. The system consists of fans located in the upper part of the containment to re-circulate containment air. Moisture separators and mist eliminators are in-cluded to remove water before it enters filters. The filters remove solid particles as well as gaseous radioactive materials. Finally, the air passes over cooling coils before it is returned to the containment atmosphere.

Figure 14.10 is a schematic diagram of a typical BWR emergency core coolant system. The systems are similar to PWR systems in that the reactor core is flooded and sprayed with water to remove heat. The high-pressure core spray system is actuated by a low water level or high drywell pressure. Makeup water from the condensate storage tank is sprayed over the fuel assemblies from a ring inside the reactor vessel. When the condensate storage tank is dry, posi-tive pump suction automatically shifts to the containment suppression pool.

If the high-pressure core spray cannot maintain an adequate water level and the reactor vessel pressure is lowered (1.7–2.1 MPa), the low-pressure core spray system is actuated. Positive suction is taken from the suppression pool, and water is sprayed over the fuel assemblies. The low-pressure coolant injection system utilizes the residual heat removal system heat exchangers. It operates

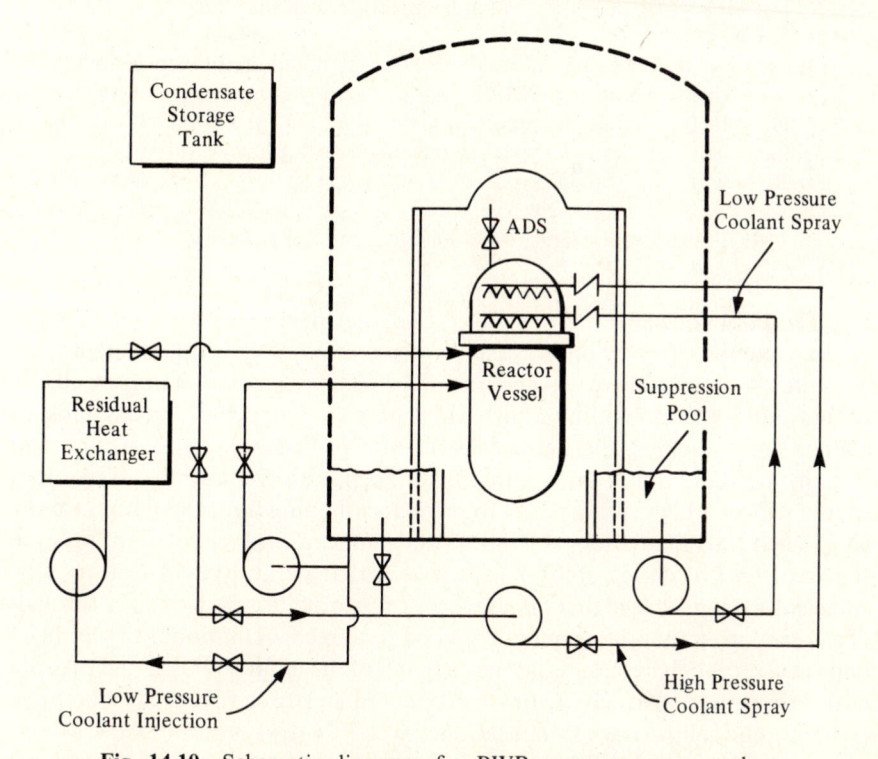

Fig. 14.10 Schematic diagram of a BWR emergency core coolant system.

in conjunction with other emergency core cooling systems to flood the reactor vessel with water and to remove residual heat from the suppression pool water.

The automatic depressurization system is an aid to the high-pressure core spray system. In the event that high pressure in the reactor vessel is not suppressed, pressure relief valves will automatically open to discharge steam into the dry well. Steam forces the water between the weir wall and primary containment to flow through horizontal openings into the suppression pool. Subsequently, steam enters the suppression pool by the same path where it is completely condensed. Low-pressure core spray and low-pressure coolant injection then provide core cooling. The automatic depressurization can be manually overridden if pressure relief is not required.

Loss of coolant accidents pose an even more serious problem in LMFBRs for several reasons:

(1) The specific power (kW/liter) is approximately 10 times that in an LWR. Thus, the temperature rise will be much more rapid during a loss of coolant accident.
(2) Cladding temperatures are higher (600–700°C) compared to LWRs (315–370°C). To compound this, the melting temperature for stainless cladding proposed for fast reactors has a lower melting temperature than the Zircaloy used in many LWRs.
(3) With loss of coolant and melting of fuel it is doubtful the LWR could achieve a critical array, but in the LMFBR loss of coolant and melting or rearrangement of the fuel may achieve a supercritical configuration. A molten critical mass of fuel might melt through the containment and keep going downward.
(4) Release of iodine would be serious, as in the LWR, but the large plutonium inventory in the core presents an even greater hazard due to its toxicity. Thus, the LMFBR may require larger exclusion areas and low-population zones.

The pool concept used in EBR-II and the Phénix reactors has its core, primary pumps, primary piping, and intermediate heat exchangers submerged in molten sodium. There are no subsurface penetrations of the tank wall. A second tank or a minimum volume vault may surround the primary tank to contain the sodium in the unlikely event of a primary tank rupture. It will contain the sodium at a level sufficient to keep the core covered. Dip tubes should stop at a high enough level to prevent siphoning liquid sodium from the core. Natural circulation is limited and some forced circulation may be required. A GE design study has indicated that natural circulation may be sufficient. It is proposed that in the event of a core meltdown with burn-through of a critical mass that the accident could be contained. Dilution by a sacrificial barrier, such as concrete, and the enlargement of the molten pool would provide both temperature reduction and tend to reduce reactivity to a subcritical configuration followed by solidification.

The loop-type LMFBR has an advantage in that greater heads are available to encourage natural circulation. However, a secondary containment must

surround all components of the primary system. A LOCA may require positive action with respect to pump operation, valve closure, and so on, to prevent serious consequences.

If sodium loss can be prevented, it appears that reliable emergency core cooling will be possible for LMFBRs.

Reactor Accident Risks

In 1972, the Atomic Energy Commission initiated a study to estimate risks to the public involving a major accident. The study is known as Wash-1400, or more commonly the Rasmussen Report. It was completed in 1974 and a draft report issued at that time with the final report following in late 1975. The objective of the study was to make a realistic estimate of the risks to the public of a major nuclear reactor accident and compare them with nonnuclear accident risks.

The design of all reactors includes engineered safety features to prevent melting of fuel and to control subsequent releases of radioactivity. To get an accidental release of radioactivity to the environment, there must be a series of failures to cause fuel to melt and to lose the systems designed to contain the radioactivity.

The study confined itself to the thousands of unlikely ways in which light-water-moderated reactors and their systems might conceivably fail and release radioactivity. There are two general categories that might lead to melting of the reactor core: loss of coolant and transients. Core melting as a result of loss of coolant accidents could conceivably occur if ECCS failed to operate. Operation of ECCS was discussed in an earlier section of this chapter. Core melting as a result of transients might occur if the decay heat systems or shutdown systems did not operate properly following a shutdown condition.

The first task in the study was to identify every way in which radioactivity could be released to the environment. A set of event trees was used to define the accident paths by which radioactivity could be released first from the core and then into the environment. Fault trees were then used to determine the likelihood of failure of the engineered safety systems in the event-tree paths. Probability values for each of the component failures were then assigned. Probabilities were developed from data on failure rates of components, human error, and testing and maintenance errors. The event trees and fault trees led to a set of accident sequences and their probability of occurence. Next, for each accident sequence the amount of radioactivity released from the fuel and how it would be transported through the containment to the atmosphere was determined. Following this, a consequence model was used to calculate the distribution of the radioactivity in the environment. Conversions from radioactive concentrations to fatalities, other health effects, and property damage were made next. This resulted in an overall risk assessment. Finally, other accident risks, such as earthquakes, dam failures, aircraft accidents, and so on, were examined

to make a comparative risk assessment. Figure 14.11 is a plot of comparative fatality risks as presented in the report.

The curves show the frequency of events per year versus fatalities. The reactor curve is for 100 power plants in operation, which is a goal for the 1985–1990 period. Fewer plants, of course, result in smaller risks. It should be observed that fatalities from nonnuclear, human-made accidents are about 10 000 times more likely than from nuclear accidents. The air crash begins to drop off at about 300 fatalities because that is about the maximum number of passengers in an aircraft.

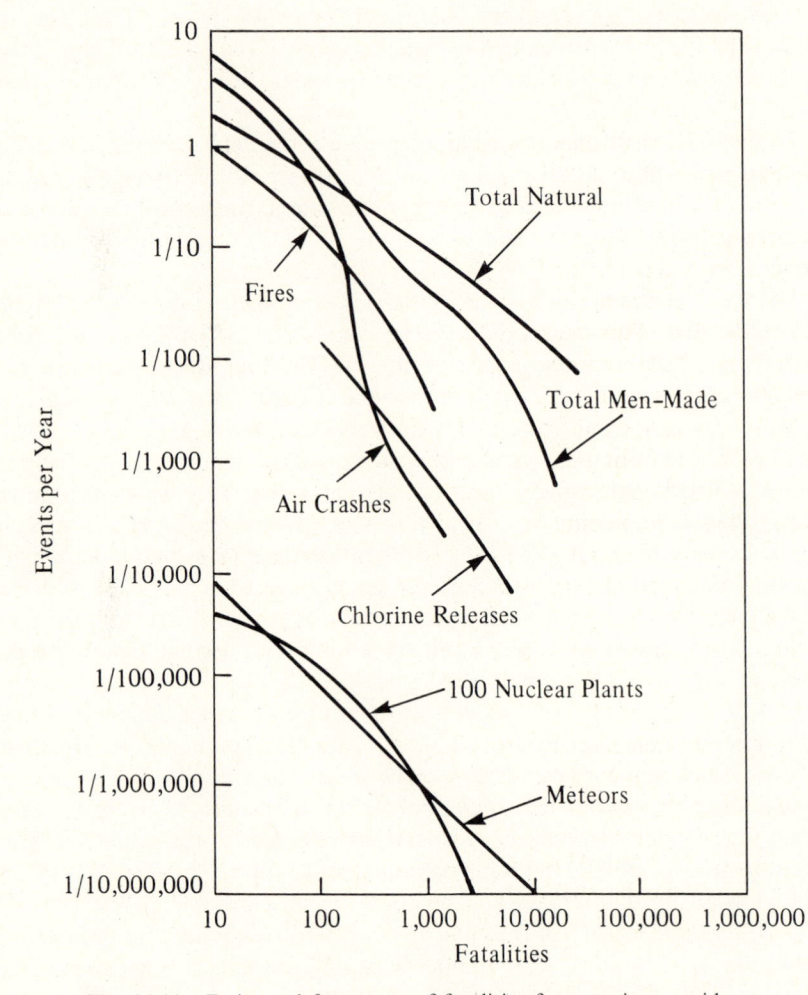

Fig. 14.11   Estimated frequency of fatalities from various accidents.
[Adapted from Wash-1400.]

Table 14.1 Predicted Delayed Health Effects from Reactor Accidents

| Effect | Chances per Year | | | | | Normal Occurrence |
	1 in 17 000	1 in 10^6	1 in 10^7	1 in 10^8	1 in 10^9	
Latent cancers	< 1	450	1300	2300	3200	64 000
Thyroid nodules	4	12 000	42 000	75 000	84 000	20 000
Genetic effects	< 1	450	1300	2300	3200	100 000

From Wash-1400.

Table 14.1 is an estimate of the delayed health effects from a nuclear accident compared with effects from normal causes. All the effects calculated were to happen within 20 years after the accident. Genetic effects would continue over many generations. Even for the worst case, the cancer estimate for a nuclear accident is only 5 percent of the normal cancer occurence.

The most likely core-melting accident is estimated as 1 chance in 17 000 per reactor per year. This means that given 100 reactors operating, an average of one accident would occur every 1.7 centuries. The most likely accident would not involve serious public injury or property damage.

In a more serious accident (but less likely), persons near the plant might have to be moved from their homes until the radioactivity decayed. Also, everything in the food chain would have to be monitored closely to prevent radioactive materials from being ingested. The most likely accident would result in property damage of about \$100 000, not counting the damage to the plant itself. An accident chance of 1 in 50 000 would result in property damage of about \$100 000 000.

The most important conclusions to be drawn from the report are the following:

(1) The risks from reactor accidents are no larger than from nonnuclear accidents; core melting is not a catastrophic accident.

(2) An accident sequence is determined by five independent factors. The probability of each of these factors is such that no single one dominates the overall probability. Thus, even if one of the factors is completely wrong, the overall probability will not be altered drastically. In other words, there is no single thing that can go wrong and cause a serious reactor accident.

Diversion-Resistant Fuel Cycles

The denatured thorium cycle

On April 17, 1977, President Carter postponed any reprocessing of reactor fuel and redirected the breeder reactor development program. This was primarily due to concern about the proliferation of plutonium based nuclear explosives. Plutonium can be separated easily from spent-enrichment-uranium fuels, and the high fissile Pu content in fast breeder designs might allow easy diversion of Pu to military purposes. This has led to marked increase of interest in the development of the Th–^{233}U fuel cycle.

Denatured fuel uses fertile ^{238}U with a low enough enrichment of ^{233}U or ^{235}U that is unsuitable for weapons use directly. The enrichment limits have been set at 12 percent for ^{233}U and 20 percent for ^{235}U. Denatured fuels could be sent to widely dispersed national reactors and additional ^{233}U be produced in highly secured Pu-fueled breeder and recycle facilities.

^{233}U is a human-made fuel, available currently only in small quantities from the development of HTGRs and the Navy's light-water-breeder program. It tends to be contaminated with highly radioactive ^{232}U ($t_{1/2} = 72$ yr). This makes it more difficult to handle, which might serve as a deterrent to its diversion. There will be, on the other hand, a penalty to using the denatured fuel (DNF) because of its greater difficulty in refabrication and handling. It appears that such systems could not have a significant impact on commercial electric generation much before the year 2000. A DOE program is assessing the use of denatured ^{233}U cycles, with the following objectives:

(1) Evaluation of proliferation resistance
(2) Development and evaluation of scenarios for the production and use of DNF in terms of utilization of resources and power costs
(3) Development of research, development, and demonstration costs and schedules for reactors and fuel cycles
(4) Development of implementation plans and assessment of commercial feasibility

Many studies have been undertaken and some results are already available. The study by Haffner and Hardie (Reference 28) considers both Pu and ^{233}U as fissile material, with either ^{238}U or ^{232}Th as fertile material in the core and both the axial and radial blankets.

For each of the two fissile fuels considered there are four possible combinations:

(1) ^{238}U in the core and both blankets
(2) ^{238}U in the core and axial blanket with ^{232}Th in the radial blanket
(3) ^{238}U in the core and ^{232}Th in both blankets
(4) ^{232}Th in all three locations

Table 14.2 Fast-Breeder-Reactor Design Parameters

Electric power	1200 MW(e)
Thermal power	3736 MW(th)
Core height	1.189 m
Core radius	1.659 m
Axial blanket thickness	0.33 m
Radial blanket thickness	0.345 m
Number of fuel assemblies	380
Number of control assemblies	20
Number of radial blanket assemblies	184
Fuel pin diameter	0.726 cm
Pins per assembly	271
Cladding thickness	0.30 mm
Core composition	
Fuel	41.43 %
Structure	14.96 %
Sodium	43.61 %
Number of fuel pins per year*	51 490
Heavy metal in core per year*	18 222 kg

* 315 full-power days.
From D. R. Haffner and R. W. Hardie, *Nuclear Technology* **42**, No. 2 (February 1979), p. 124.

Thus, there are eight possible combinations, but only three of particular interest will be compared here. The Pu–U_8/U_8/U_8 (Pu and fertile ^{238}U in the core with ^{238}U in both the axial and radial blankets) has the best breeding ratio. The Pu–Th/Th/Th cycle maximizes ^{233}U production, while the U_3–U_8/Th/Th is of special interest, as it uses denatured fuel. Table 14.2 shows the design parameters used for the study. Some results for the three combinations are shown in Table 14.3.

In this study a four-group-cross-section library was used. A reactor at its beginning of life (BOL) was burned for one cycle and half the core and axial blanket were replaced and the reactor was operated for another cycle. Since two-batch cores were used, the beginning of the second cycle was taken as the beginning of equilibrium (BOE) and the end of this cycle was considered to be the end of equilibrium (EOE). The fresh fuel was either 58.9 percent ^{239}Pu, 25.7 percent ^{240}Pu, 12.2 percent ^{241}Pu, and 3.2 percent ^{242}Pu or 100 percent ^{233}U (exclusive of ^{238}Pu). A five-batch radial blanket was planned (five cycles to reach equilibrium if one-fifth of the core is replaced at each refueling). However, since the radial blanket has little influence on reactivity, it was decided to burn the radial blanket for two cycles without changes in the blanket.

The study showed that the reactor physics parameters are relatively independent of the blanket. Note how differently k_{eff} behaves for the three different

Table 14.3 Performance of Alternative Fueled FBR Core Designs

Reactor Configuration	Pu–U$_8$/U$_8$/U$_8$	Pu–Th/Th/Th	U$_3$–U$_8$/Th/Th
k_{eff}			
BOL	1.003	1.002	1.081
BOE	1.003	1.011	1.052
EOE	1.001	1.017	1.000
Breeding ratio (MOE)	1.355	1.193	1.180
Conversion ratio	1.040	0.904	0.089
Fissile mass* (kg)			
BOL	3781	4368	3666
BOE	3802	4297	3604
Doubling time* (yr)			
Simple	18.9	56.1	35.7
Compound	13.1	38.9	24.8
Fissile mass/batch* (kg)			
Core			
Charge	1840	2185	1784
Discharge	1891	2017	1701
AB—discharge	188	178	153
RB—discharge	199	183	161
Net gain	438	193	231
Sodium void coefficient			
k_{eff} at BOL, Na in	1.0049	1.0046	1.0805
k_{eff} at BOL, Na out	1.0155	1.0103	1.0774
Δk	+0.0106	+0.0057	−0.0031

* Design assumptions: load factor = 70 percent, t_{lag} = 1 yr, t_{res} = 2.47 yr, r = 0.02, b_c = 2, b_b = 5.

From D. R. Haffner and R. W. Hardie, *Nuclear Technology* **42**, No. 2 (February 1979), pp. 123–132.

cores from BOL to EOE. The Pu–U$_8$/U$_8$/U$_8$ remains nearly constant, dropping from 1.003 to 1.001. The Pu–Th/Th/Th actually increases from 1.002 to 1.017, and the denatured U$_3$–U$_8$/Th/Th drops significantly, from 1.081 to 1.000.

Note that the greatest fissile mass is required by the Pu–Th/Th/Th and that it provides the least net increase in fissile mass per batch, 193 kg. For the Pu/U$_8$/U$_8$/U$_8$ the gain is 438 kg, and for the denatured fuel (U$_3$–U$_8$/Th/Th) it is 231 kg. The sodium void coefficients are positive for both Pu-fueled cases, but become slightly negative for the core using denatured fuel. The breeding ratio of the Pu–U$_8$/U$_8$/U$_8$ is clearly superior, being 1.355, compared to 1.193 for Pu–Th/Th/Th and 1.180 for the U$_3$–U$_8$/Th/Th. This translates to compound doubling times of 13.1, 38.9, and 24.8yr, respectively

If the Pu–U$_8$/U$_8$/U$_8$ were to change its blanket material to thorium (Pu–U$_8$/Th/Th), the compound doubling time increases to only 14.4 yr. Note that the doubling time for the Pu-fueled core with Th as the fertile material in the

Table 14.4 Material Inventories (kg of Heavy Metal) for Fast Reactor Design U_3–U_8/Th/Th

Cycle	Isotope	Core Zone 1	Core Zone 2	Axial Blanket	Radial Blanket
BOL	^{232}Th	0	0	18 418	33 160
	^{233}U	1544	2024	0	0
	^{235}U	50	48	0	0
	^{238}U	16 612	16 128	0	0
	^{239}Pu	0	0	0	0
	^{240}Pu	0	0	0	0
	^{241}Pu	0	0	0	0
	^{242}Pu	0	0	0	0
Total		18 206	18 200	18 418	33 160
BOE	^{232}Th	0	0	18 334	32 980
	^{233}U	1298	1795	79	162
	^{235}U	43	44	0	0
	^{238}U	16 278	15 910	0	0
	^{239}Pu	254	170	0	0
	^{240}Pu	9	4	0	0
	^{241}Pu	0	0	0	0
	^{242}Pu	0	0	0	0
Total		17 882	17 923	18 413	33 142
EOE	^{232}Th	0	0	18 162	32 800
	^{233}U	870	1375	235	323
	^{235}U	31	35	0	0
	^{238}U	15 596	15 462	0	0
	^{239}Pu	701	487	0	0
	^{240}Pu	42	19	0	0
	^{241}Pu	2	0	0	0
	^{242}Pu	0	0	0	0
Total		17 242	17 378	18 397	33 123
Equilibrium fuel discharge	^{232}Th	0	0	9040	6452
	^{233}U	352	600	153	161
	^{235}U	13	16	0	0
	^{238}U	7637	7625	0	0
	^{239}Pu	438	310	0	0
	^{240}Pu	33	15	0	0
	^{241}Pu	1	0	0	0
	^{242}Pu	0	0	0	0
Total		8474	8566	9193	6613

From D. R. Haffner and R. W. Hardie, *Nuclear Technology* **42**, No. 2 (February 1979), p. 130.

core is 38.9 yr. This doubling time may well exceed the lifetime of the core, making this combination not particularly attractive. The ~ 25-yr doubling time for the core with denatured fuel (U_3–U_8/Th/Th) is not remarkable, but might allow use of the concept for a fast reactor outside a highly secured site. The denatured fuel would probably be more useful in LWRs or advanced converters.

The simple and compound doubling times of the previous study were calculated using the definitions of the Advanced LMFBR Systems Design and Analysis Working Group. The definitions consider the variation in fissionable fuel charge and discharge rates from the core and blankets during the approach to equilibrium. Table 14.4 shows the heavy-metal inventory from the beginning of life to the end of equilibrium for the Pu–Th/Th/Th reactor design. It also shows the equilibrium discharge. Remember that this is only one-half of the total number of fuel and blanket assemblies in the core.

The simple system doubling time, t_2, is defined as

$$t_2 = \frac{(I + A)\,L}{C} + t_{\text{lag}} \tag{14.1}$$

and the compound system doubling time is

$$t_{2c} = t_2 \ln 2 \tag{14.2}$$

where $I = 1 + t_{\text{lag}}/t_{\text{res}}$

$$L = \frac{Ct_{\text{res}}}{Ct_{\text{res}} - rCt_{\text{res}} - 1}$$

$$A = \frac{t_{\text{res}}(1 - r)}{b_c}\left(\frac{b_c - 1}{2}\,C_c + \frac{b_b - 1}{2}\,C_b\right)$$

$$C_c = \frac{m_d^c - m}{mt_{\text{res}}} \qquad C_b = \frac{m_d^b}{mt_{\text{res}}}$$

$$C = C_c + C_b$$

t_{lag} = out of reactor time, yr

t_{res} = residence time of a core batch, yr

r = fraction of fuel lost while out of core (process losses and decay)

C = unit net fissile production rate, kg/yr per kg if fissile loaded

C_c = unit net fissile production rate in core and axial blanket (remember that axial blanket is included in the fuel assemblies, as shown in Figs. 13.41 and 13.44)

C_b = unit net fissile production in radial blanket

b_c = number of core batches

b_b = number of blanket batches

m_d^c = fissile fuel mass contained in an equilibrium discharge of the core and axial blanket

m_d^b = fissile fuel mass contained in an equilibrium discharge of the radial blanket

m = fissile fuel mass contained in one batch of fresh fuel

Example 1   The fast reactor described in Table 14.2 is fueled with denatured U_3–U_8 fuel and uses thorium in both axial and radial blankets. The residence time in the core is 2.47 yr and a 1-yr lag time is expected. There are two core batches and five blanket batches. The fissile fraction lost while out of core is 0.02. Determine the simple and compound doubling times for the fissile masses shown below.

Fissile Isotope	^{233}U	^{239}Pu	^{241}Pu
U_3–U_8 core			
Charge	1784	0	0
Discharge	952	748	1
Th axial blanket at discharge	153	0	0
Th radial blanket at discharge	161	0	0

$$I = 1 + t_{\text{lag}}/t_{\text{res}} = 1 + 1/2.47 = 1.405$$

$$C = \frac{m_d^c - m}{mt_{\text{res}}} + \frac{m_d^b}{mt_{\text{res}}} = C_c + C_b$$

$$= \frac{(952 + 748 + 1) - 1784 + 153}{1784 \times 2.47} + \frac{161}{1784 \times 2.47}$$

$$= 0.0159 + 0.0365 = 0.0524$$

$$L = \frac{Ct_{\text{res}}}{Ct_{\text{res}} - r(Ct_{\text{res}} + 1)}$$

$$= \frac{0.0525 \times 2.47}{0.0524 \times 2.47 - 0.02(0.0525 \times 2.47 + 1)} = 1.211$$

$$A = \frac{t_{\text{res}}(1 - r)}{b_c} \left[\frac{(b_c - 1)}{2} C_c + \frac{(b_b - 1)}{2} C_b \right]$$

$$= \frac{2.47(1 - 0.02)}{2} \left[\frac{(2 - 1) \times 0.0159}{2} + \frac{(5 - 1) \times 0.0365}{2} \right]$$

$$= 0.097\,97$$

$$t_2 = \frac{(I + A)L}{C} + t_{\text{lag}}$$

$$= \frac{(1.405 + 0.097\,97) \times 1.211}{0.0524} + 1 = 35.8 \text{ yr}$$

$$t_{2c} = t_2 \ln 2 = 35.8 \times \ln 2 = 24.8 \text{ yr}$$

CIVEX

An interesing proposal for a diversion-resistant fast-breeder-reactor fuel cycle was made at the Fifth Energy Technology Conference in February 1978. This fuel cycle has been named CIVEX to designate the processes designed for a civilian fuel cycle.

The interest for this process arose from the U.S. policy of indefinitely deferring reprocessing announced in April 1977. The basis for the policy and for storing spent fuel indefinitely is an assumption that it is possible to divert pure plutonium from conventional reprocessing to nuclear weapons manufacture. Although it is nearly impossible, some public opinion believes that a nuclear weapon could be made by skilled terrorists stealing a quantity of fissionable material. The U.S. policy effectively stops development of a fast breeder because the breeder depends on reprocessing for economical operation. CIVEX is one alternative method for providing Pu for the fast breeder cycle that is diversion resistant. It should be noted that CIVEX would prevent access to Pu by terrorist or other groups but would not prevent spreading of other fissionable materials.

The process is based on an objective that there be no pure Pu in the process, and that there be no feasible way to modify the process to produce pure Pu. The process used is an adaptation of the PUREX process that is used to completely separate U and Pu. In CIVEX the PUREX process is modified to coextract U and Pu and allow selected fission products to remain to keep radioactivity levels very high. The proposed CIVEX process is shown in Fig. 14.12. The dissolution and extraction steps are the same as the PUREX process. After this point the processes differ. The CIVEX process contains no separation and scrubbing process. As a result, the product from the partition step is highly radioactive. The fuel fabrication feed is U plus 20 percent Pu plus a few percent fission products. This product feed is processed in a remote fabrication operation to produce finished mixed-oxide fuel. The scrap from this process is recycled to the dissolver so as not to provide a side stream for diversion.

The excess U by-product is not to be purified by a fluoride volatility process. This process is not capable of separating Pu from fission products, but it has been demonstrated to produce uranium. Since Pu would be concentrated with the fission product residue, it would not be possible to divert the process to produce pure Pu. The U product would be used as blanket material and the highly radioactive wastes solidified and stored.

The CIVEX plant would be designed with preestablished flow rates so that equipment could be precisely matched to the various processes, thus leaving very little flexibility to change plant conditions. Existing computer codes have been used to establish the CIVEX process, and all the proposed processes have been demonstrated to be technically feasible. However, no integrated plant has been built or proposed. EBR-2, a small-scale experimental breeder reactor, did operate as a closed reactor system, but it used a very different separation process.

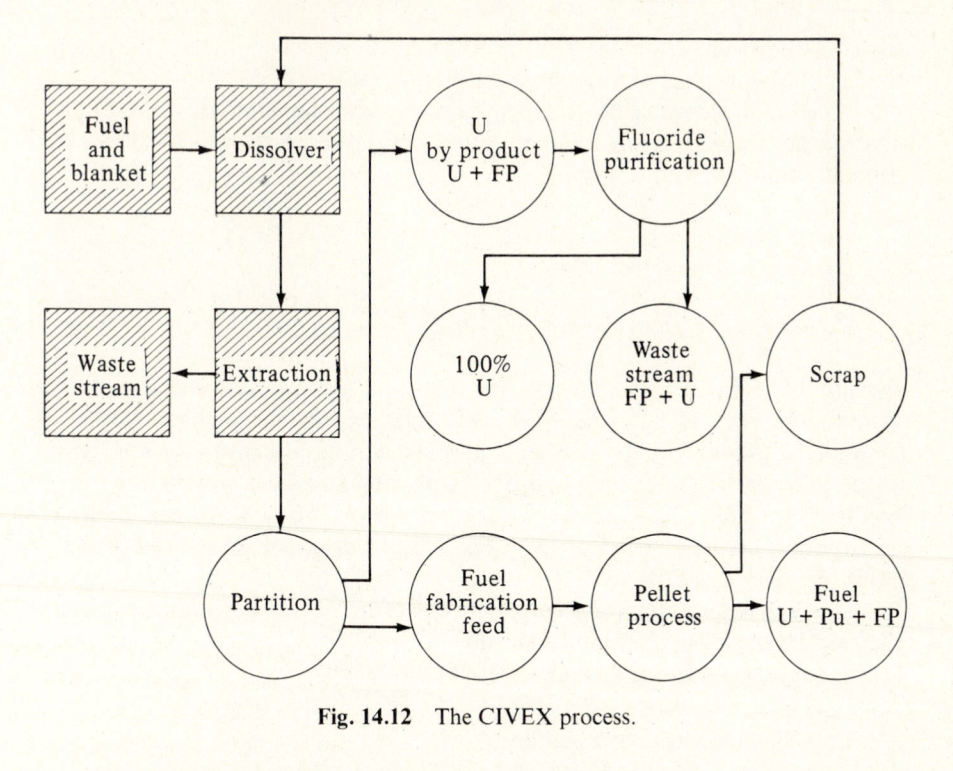

Fig. 14.12   The CIVEX process.

Proponents of the CIVEX process believe the present stored fuel from LWR reactors can be used to start the system. The output from the first plant would be fuel for the first fast reactors. In their eyes, there would then be an eventual phasing out of LWRs and a transition to a completely fast reactor system. They also believe that only a massive plant restructuring, approaching the effort required to construct a separate weapons production facility, would permit producing nuclear weapons material from a fast breeder reprocessing plant. Although this would not solve the nuclear weapons proliferation problem, it would remove many of the questions regarding diversion of nuclear materials.

Radioactive Waste Disposal Management

A most crucial consideration in the future development of nuclear power is the capability for safe storage and disposal of radioactive waste materials. Not only must the disposal methods be safe from an engineering and radiological health point of view, but the public at large must percieve them to be safe. If the latter requirement is not met, excessively restrictive legislation could prevent, or at least curtail, the further development of nuclear power. Technological development of nuclear waste management has progressed more slowly

than the overall growth of the nuclear industry and the recognition of the potential health significance of nuclear waste.

Nuclear wastes may be solids, liquids, or airborne materials produced by neutron activation, fission fragments, and transuranium isotopes. They are generally categorized as follows:

(1) High-level wastes
(2) Alpha and transuranic waste
(3) Low-level waste
(4) Naturally occurring radioactive waste

Essentially, high-level waste means the first- and second-cycle waste resulting from fuel reprocessing. Other waste falls into the other categories. Some of the more important fission products and transuranium isotopes are listed in Table 14.5, together with their half-lives, activities, and mass present in a ton of spent reactor fuel. This fuel has had a burnup of 33 000 MWd/ton, and has undergone a cooling period of 150 days to allow decay of short-lived activity. From the data it can be seen that within a century the majority of the fission fragment activity will have decayed. On the other hand, transuranic elements such as ^{239}Pu, ^{240}Pu, ^{242}Pu, ^{241}Am, and ^{243}Am will still present a serious problem and must be safeguarded in perpetuity.

Low-level waste management

Low-level waste consists of uniforms, rags, solidified low-level liquids, and so on. The liquid wastes (including decontamination fluid), produced at operating reactor plants are generally evaporated and solidified before transferral to burial sites. These solidified wastes contain long-lived fission products but very little transuranic wastes (i.e., less than 10 nCi per gram of plutonium).

The low-level waste is compacted into drums or casks and buried in shallow trenches in controlled locations. There is no maximum radiation level on wastes that go into burial, but there are state-controlled restrictions on the number of curies that can be above ground at any one time.

The philosophy of land burial has assumed that the geology and hydrology of burial sites provide a barrier to migration of radionuclides. Recent studies of some burial sites have shown this assumption to be incorrect. Although the levels of activity detected are not significant, they have indicated that conditions for burial as a long-term disposal technique must be reconsidered.

Spent fuel waste management

At present (because of former President Carter's concern over proliferation) no commercial reprocessing of spent reactor fuel is occurring. Nuclear Fuel Services, Inc., operated a reprocessing plant in West Valley, New York, from April 1966 to early 1972. Development of the Allied General Nuclear Services

Table 14.5 Representative Quantities of Fission Products and Actinides Present in Spent Reactor Fuels (33 000) MWd/Mg Burnup and a Cooling Time of 150 Days)

Class of Nuclides	Isotope	Half-Life (yr)	Activity (Ci/ton)	Mass (g/ton)
Gaseous fission products	^{3}H	12.3	800	0.083
	^{85}Kr	10.7	10 500	27
Volatile and semivolatile fission products	^{103}Ru	0.11	15	880
	^{106}Ru	1.01	180 000	5.7
	^{129}I	17×10^{6}	0.04	250
	^{131}I	0.02	2.0	0.01
	^{134}Cs	2.05	100 000	77
	^{135}Cs	3×10^{6}	1.2	1400
	^{137}Cs	30.2	106 000	1200
Solid fission products	^{89}Sr	0.14	100 000	3.5
	^{90}Sr	28.9	60 000	430
	^{91}Y	0.16	190 000	7.8
	^{93}Zr	0.95×10^{6}	2	490
	^{95}Zr	0.18	400 000	19
	^{95}Nb	0.10	800 000	21
	^{141}Ce	0.09	80 000	2.8
	^{144}Ce	0.78	800 000	250
	^{147}Pm	2.62	200 000	220
	^{155}Eu	5.01	40 000	87
Actinides in U fuel	^{237}Np	2.14×10^{6}	<1	600
	^{238}Pu	86	4000	230
	^{239}Pu	24 400	500	8100
	^{240}Pu	6580	650	2900
	^{241}Pu	13	150 000	1300
	^{242}Pu	379 000	2	510
	^{241}Am	458	750	230
	^{243}Am	7800	20	100
	^{242}Cm	0.45	35 000	10
	^{244}Cm	17.6	2000	25
Total			3 154 400	19 376

facility in Barnwell, South Carolina, was stopped in January 1975. Both of these efforts were halted because of concern over plutonium recycle and safeguards. This means that, currently, there is no system demonstration of waste disposal management for spent reactor fuel other than storage. As spent fuel is discharged from operating reactors it is being stored in water-filled stainless steel-lined concrete pools at the reactor site. These spent fuel pools will eventually be filled, but proposals for away from reactor storage are currently being studied.

Until an acceptable concentrated form for high-level and transuranic wastes is specified, and a suitable repository established, spent fuel must continue to be stored in its present form and manner.

High-level waste management

Current regulations (10 CFR Part 50, Appendix F) require that all high-level commercial and transuranic waste be converted to a stable solid within five years of processing. This solid then must be sealed in manageably sized containers and transferred to a federal repository within 10 years following separation of the fission products from the irradiated fuel. This volume of encapsulated waste is expected to be 50 to 60 ft^3/yr for each 1000-MW(e) reactor.

Several forms have been proposed for high-level radioactive wastes. They are summarized in Table 14.6. It is important that the best form and process be developed for each particular waste. Several steps must precede solidification and storage of high-level waste in the overall waste management process. They are depicted in Table 14.7. The spent fuel elements are to be stored for five years to allow short-lived nuclides to decay and heat generation to subside. The liquid waste would be stored in aboveground stainless steel-lined concrete tanks before being converted to their final solid form. The solid waste could also be stored aboveground for 10 years to minimize radionuclide decay and heat generation in the final repository. The 15-year time span from reactor

Table 14.6 High-Level Waste Forms

Type	Typical Form
In-place solidification	Concrete
Concrete	Concrete
Calcine	Direct calcine
Glass	High silica glass
Ceramic	SYNROC
Matrix	Metal with ceramic marbles

Table 14.7 Management of High-Level Wastes

Step	Time After Reactor Discharge (yr)
Store spent fuel element	0–5
Reprocess fuel	5–10
Store liquid waste	5–15
Solidify waste	10–20
Temporary storage (solid)	10–25
Permanent storage (solid)	15–up

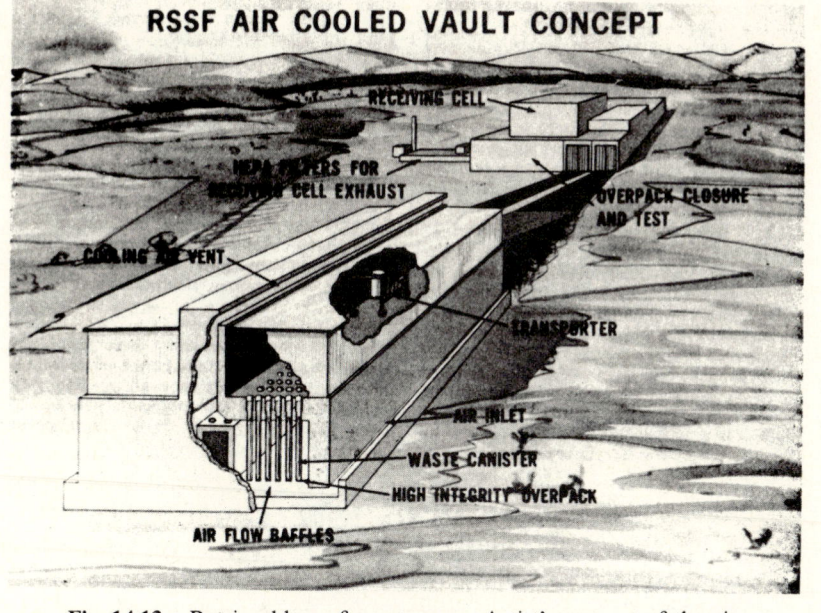

Fig. 14.13   Retrievable surface storage.   Artist's concept of the air-cooled vault system, one of three under consideration for possible use as interim storage.   [From *The Nuclear Industry—1974*, Wash-1174-74, U.S. Atomic Energy Commission.]

discharge to final storage allows the heat generation rate to decrease by a factor of about 40.   High-level liquid wastes have been successfully stored in tanks in both the United States and Europe since 1944 and have been shown to be retrievable.   One such retrievable storage concept is shown in Fig. 14.13.   A design of this type would require a land area of about 1100 acres to store all the high-level waste generated in the United States through the year 2000.   However, development of this concept was halted in September 1975.

Nearly all the solid forms proposed for high-level waste involve calcining the liquid waste from the reprocessing operation.   One interesting method has been performed on a pilot scale with spent fuel from the Point Beach reactor (Reference 2).   The process is described as spray calcination and in-can melting. The particular work cited was performed in 1979 at Pacific Northwest Laboratories.

In the process, spent fuel elements were disassembled, sheared, and dissolved in a nitric acid solution.   The portions not dissolved were sealed and stored in a retrievable storage facility.   The uranium and plutonium are separated from the high-level waste in a solvent-extraction process.   The liquid waste from the solvent-extraction is fed into the spray calciner/in-can melter shown schematically in Fig. 14.14.   Liquid high-level waste is sprayed into the top of the spray calciner through the feed nozzle.   The walls of the spray chamber are

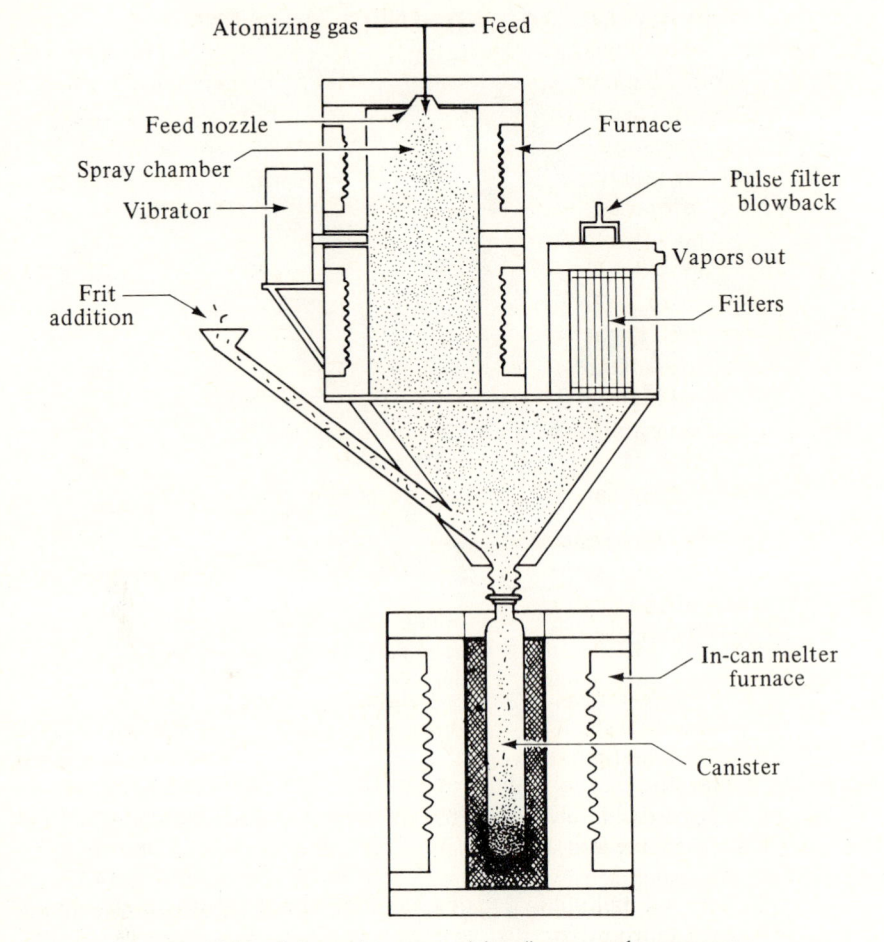

Fig. 14.14   Schematic spray calciner/in-can melter process.

heated to 300 to 800°C, depending on the waste components. Its pressure is −12 to −15 cm H_2O. As the sprayed waste falls down through the calciner it is dried and oxidized to form a very fine calcine substance. The vibrator prevents buildup of solids on the chamber walls. The filter chamber consists of very fine stainless steel filters that remove solid particles from the off-gas stream. The filters are later blown back into the spray chamber. The off-gas stream contains some radioactive particles as well as radioactive gasses. The stream is treated, decontaminated, and eventually released.

The calcine produced is next mixed with glass-forming material (called frit). The mixture falls into its canister, where it is melted at a temperature of 1050°C and then cooled. The result is a borosilicate glass product encased in

a canister which served as the melting crucible. The canisters used in the pilot facility were 244 cm high and 20 cm in diameter. The borosilicate glass has a density of about 2.8 g/cm^3 and a volume of about 50 liters. Activity and temperature, of course, depend on the burnup of the fuel.

Another interesting form for incorporating high-level waste material is the SYNROC (for "synthetic rock") concept (Reference 3). The SYNROC concept was originally proposed by Professor Ringwood of the Australian National University. The advantage cited for SYNROC is that it is more stable than other forms proposed. The SYNROC itself is composed of ZrO_2, TiO_2, Al_2O_3, $BaCO_3$, and $CaCO_3$. In the proposed process these powders are milled in alcohol, dried, precompacted, and then calcined at 1100°C.

The resulting mixture will then be crushed, mixed with about 10 percent weight of calcined high-level waste, and remilled. Hot pressing as well as cold processing and sintering have been proposed for the milling production process.

Permanent disposal

For several years research has been conducted to determine acceptable methods and locations for permanent disposal. One such location, a salt mine in Kansas, was abandoned because of potential leakage. The more promising disposal concepts appear to be:

(1) A mined geologic repository at a depth of 600 m
(2) A very deep hole about 10 000 m deep with canisters stacked in the lower 1500 m
(3) Placing high-heat-producing waste in a deep rock cavity and allowing the heat to melt the rock
(4) Placing canisters in very stable sediments in either the Atlantic or Pacific Ocean
(5) Placing the waste in solar orbit

Except for rock melting and space disposal, either spent fuel elements or encapsulated waste from reprocessing could be disposed of in any of the concepts. The concensus seems to be that a mined geologic repository is the preferred method of storage. The deep hole, seabed, and space disposal might provide additional radiological isolation, reducing the already miniscule impact associated with the mined geologic repository. This advantage is believed to be offset in all three concepts by their negative aspects.

A mined geologic repository pilot plant (the WIPP project) has been proposed for the southwestern United States near Carlsbad, New Mexico. This project has developed slowly during the last five years because of U.S. policies, but recent developments indicate that final design and development may proceed in the near future.

As currently envisioned, the WIPP project is designed to store 6 million cubic feet of contact-handled transuranic waste, 1000 canisters of remote-handled transuranic waste, and 60 canisters of experimental high-level waste. The facility will be capable of handling 1.42×10^4 m^3 per year of contact waste

on a single-shift basis. The waste handling building can handle two canisters of remote-handled waste per shift.

WIPP consists of an underground repository, four shafts connecting surface facilities to the storage level, and surface facilities. The surface facilities consist of:

(1) A waste handling building
(2) An administration building
(3) An underground personnel building
(4) A storage exhaust filter building
(5) Salt handling equipment and a salt storage area

All waste will be received at the waste handling building. There it is inspected and prepared for transport to the repository level. In addition to handling and retrieval of the remote- and contact-handled waste, the waste handling building includes a system for handling and packaging site-generated solid waste.

The underground personnel building serves as a base for all persons working in or visiting the underground facilities. It includes changing rooms for personnel working in the underground areas, rescue equipment rooms, and office space. The building is connected to the ventilation supply and service shafts by an inside corridor.

The exhaust filter building contains three 47.2-m³/s fans and three 15.6-m³/s HEPA filter banks. During normal operation two fans exhaust air from the storage area to the atmosphere. If radioactive contamination is detected, the airflow is reduced to 47.2 m³/s and is diverted through the filters before discharge to the atmosphere.

The four shafts connecting surface facilities to the storage level function as follows:

(1) Waste shaft. A 5.8-m-diameter, 693-m-deep shaft, where waste is transported from the surface to the underground repository.
(2) Ventilation and service shaft. This shaft will be used to transport personnel and materials from the surface to the storage areas. It is 4.9 m in diameter and 693 m deep.
(3) Construction exhaust and salt handling shaft. This 4.3-m-diameter, 850-m-long shaft will exhaust air during construction and will be used to transport salt to the surface during construction.
(4) Storage exhaust shaft. This shaft is 3 m in diameter and 693 m long. It will be used to exhaust ventilation air from the underground storage areas.

The repository storage area consists of eight area panels. Four are on each side of a main entry. Each of the eight panels consists of two subentries 314 m long and seven rooms perpendicular to the subentries. Each room is 91.5 m long, 10 m wide, and 4 m high.

Although final design of the project has not been authorized, much work has been done, including design of the shafts and support facilities. Site develop-

ment work will follow as the next stage in the overall project development. Regardless of the outcome of the impasse on waste management policy, the WIPP project should answer many concerns regarding safe handling and storage of nuclear wastes.

Problems

1. A river is 100 m wide and has an average depth of 3 m. The current has an average velocity of 5 km/h. A 33 percent efficient LWR takes water at 15°C and causes it to rise 10°C in passing through the condenser.
   (a) What fraction of the total river flow is diverted through the condenser?
   (b) Assuming complete adiabatic mixing when the outfall water rejoins the main stream, what will be the new steam temperature?
   (c) During a period of low water the river width decreases to 80 m and the average depth is reduced to 2 m. The average velocity is 3 km/h. The water temperature is the same and the plant operating conditions are the same. What is the percent diversion and the stream temperature after mixing under these conditions?

2. A 1000-MW(e) direct gas turbine cycle HTGR operates with a 40 percent thermal efficiency. Helium enters the precooler at 340°C and is cooled to 50°C. Pressurized water at 2.5 MPa transports the rejected heat to a dry cooling tower. Water enters the precooler at 40°C and is heated to 175°C. The airflow through the tower is at a rate of 50×10^6 kg/h. Air enters the tower at 20°C.
   (a) What is the required helium flow rate (kg/h)?
   (b) What is the water flow rate (m^3/s)?
   (c) What will be the temperature of the air leaving the tower?

3. Suppose that a 1000-MW(e) LWR uses a dry cooling tower. The plant thermal efficiency is 30 percent and the condensing temperature is 60°C. In the intermediate water loop water rises from 30°C to 55°C as it passes through the condenser. Air flowing through the dry cooling tower rises from 20°C to 50°C.
   (a) What is the water flow rate (m^3/s) in the intermediate loop?
   (b) What is the airflow rate (kg/h) through the tower?

4. Now imagine that the 1000-MW(e) LWR of Prob. 3 operates with a wet cooling tower, allowing the condensing temperature to drop to 35°C. Air entering the tower has a dry-bulb temperature of 20°C and 50 percent relative humidity. Cooling water leaves the condenser at 30°C. The approach for the tower is 4°C. Assume that saturated air at 27°C leaves the tower.
   (a) What is the range for the tower?
   (b) What is the cooling water flow rate through the condenser?
   (c) What is the air mass flow rate through the tower?
   (d) At what rate must makeup water be supplied? Assume that 150 percent of the amount required to offset evaporation is supplied. This overflow prevents an excessive salt concentration.

5. The fast reactor described in Table 14.2 is fueled with ^{233}U and uses ^{232}Th in the core and both blankets (axial and radial). The residence time in the core is 2.5 yr and the

lag time is estimated at 1 yr. There are two core batches and four blanket batches. The core contains 2100 kg ^{233}U when charged and 1850 kg at discharge. At discharge there are 160 kg ^{233}U in the axial blanket and 165 kg ^{233}U in the radial blanket. Assume a 2 percent loss of fuel while it is out of the core. Estimate the simple and compound doubling times.

6. For the solid fission products shown in Table 14.5, what will be the activity per ton of spent fuel after 10 yr? What will it be after 100 yr?

7. Discuss what might have happened at TMI had the operators not secured the high-pressure injection about $4\frac{1}{2}$ min after the accident began.

8. Describe the PUREX process.

9. Describe in detail how the CIVEX process might be integrated into a LWR fuel reprocessing cycle.

10. Contrast mined geologic disposal, seabed disposal, and very deep hole disposal from the standpoint of resource consumption for each method.

11. An analysis of core-melt accidents is not required of utilities submitting applications to build nuclear plants. Discuss the rationale for this decision.

12. Discuss the rationale for deferring development of fuel reprocessing.

References

1. "The State of Waste Disposal Technology, Mill Tailings, and Risk Analysis Models," *Proceedings of the Symposium on Waste Management at Tucson, Ariz.*, 1980.
2. Bjorklund, W. J., and M. S. Hanson, *Vitrification of Actual High-Level Waste from Fuel*, PNL-SA-78505, 1979.
3. Ringwood, A. E., *Safe Disposal of High Level Nuclear Wastes: A New Strategy*, Canberra: Australian National University Press, 1978.
4. Marshall, W., "Nuclear Power and the Proliferation Issue," Graham Young Memorial University of Glasgow Lecture, Glasgow, February 1978.
5. Starr, C., "The Separation of Nuclear Power from Nuclear Proliferation," 5th Energy Technology Conference, Washington, D.C., February 1978.
6. Levenson, M., and E. Zebrowski, "A Fast Breeder Systems Concept," 5th Energy Technology Conference, Washington, D.C., February 1978.
7. Culler, F. L., "Precedents for Diversion-Resistant Nuclear Fuel Cycles," 5th Energy Technology Conference, Washington, D.C., February 1978.
8. Flowers, R. H., K. D. B. Johnson, J. H. Miles, and R. K. Webster, "Possible Long Term Options for the Fast Reactor Plutonium Cycle," 5th Energy Technology Conference, Washington, D.C., February 1978.
9. "Regulation of Nuclear Power Reactors and Related Facilities," *Nuclear Safety* 15, No. 1 (January–February 1974), pp. 1–13.
10. Kulcinski, G. L., et al., "Energy for the Long Run: Fission or Fusion?" *American Scientist* 67 (January–February 1979), pp. 78–89.
11. Bethe, H. A., "The Necessity of Fission Power," *Scientific American* 234, No. 1 (January 1976), pp. 21–31.

12. *TMI-2 Lessons Learned: Task Force Final Report*, NUREG-0585, U.S. Nuclear Regulatory Commission, October 1979.

13. Jaffe, L., *Technical Staff Analysis Report Summary*, President's Commission on the Accident at Three Mile Island, October 1979.

14. *The Accident at Three Mile Island*, Report of the President's Commission on the Accident at Three Mile Island, October 1979.

15. *NRC Action Plan Developed as a Result of the TMI-2 Accident*, NUREG-0660, U.S. Nuclear Regulatory Commission, August 1980.

16. Nuclear Accident and Recovery at Three Mile Island, Subcommittee on Nuclear Regulation, U.S. Senate, July 1980.

17. *Three Mile Island: A Report to the Commission and to the Public*, NUREG/CR-1250, U.S. Nuclear Regulatory Commission, May 1980.

18. Lee, W. S., "Environmental Effects of Thermal Discharges-Technological Elements," Effects of Thermal Discharges (ASME Symposium), 1970.

19. Merriman, D., et al., *The Connecticut River Investigation, 1965–72*, a series of semi-annual reports submitted to the Water Resources Commission of the State of Connecticut.

20. Triplett, M. D., "Nuclear Power Plant Cooling and Thermal Effects," Northeastern University, Boston, 1975.

21. Olds, F. C., "Demonstration LMFBR: Progress and Problems," *Power Engineering* **78**, No. 12 (December 1974), pp. 26–33.

22. Miliaras, E. S., *Power Plants with Air-Cooled Condensing Systems*. Cambridge, Mass.: The MIT Press, 1974.

23. Foster, A. R., S. A. Lamkin, and P. Kwok, *Design of Dry Cooling Towers for Use with a Direct Cycle HTGR*, ASME Paper No. 73-Pwr-7, ASME–IEEE Joint Power Gen. Conf., New Orleans, September 16–19, 1973.

24. Bartlett, D. A., and A. R. Foster, *Natural Draft Dry Cooling Tower Performance with Varying Ambient Conditions Applied to a Direct Cycle HTGR*, ASME Paper No. 75-WA-44, ASME Winter Annual Mtg., Houston, November 30–December 4, 1975.

25. Gulf General Atomic Co., *Dry Cooling of Power Plants and the HTGR Gas Turbine System*," Gulf-GA-A12026, March 1972.

26. Rasmussen, N. C., *Reactor Safety Study*, Wash-1400, U.S. Atomic Energy Commission, 1975.

27. Choppin, G. R., and J. Rydberg, *Nuclear Chemistry*. Elmsford, N.Y.: Pergamon Press, Inc., 1980.

28. Haffner, D. R., and R. W. Hardie, *Nuclear Technology* **42**, No. 2 (February 1979), p. 124.

29. Pigford, T. H., The Management of Nuclear Safety: A Review of TMI After Two Years, *Nuclear News* **20**, No. 3 (March 1981), pp. 41–48.

Appendix A

Nuclear Data for Various Elements and Isotopes

Atomic Number, Z	Element or Isotope	Abundance (%)	Atomic Mass (u)	Density (g/cm³)	Half-Life	Neutron Cross Section (barns) — Absorption at 0.025 eV	Scattering Thermal	Scattering Epithermal
1	H	~100	1.007 97			0.332	38	20.4
	¹H	~100	1.007 825			0.332	38	20.4
	²H (D)	0.015 1	2.014 10			0.000 46	7	3.4
	³H (T)	—	3.016 05		12.6 yr			
2	He		4.002 6			0.007	0.8	0.83
	³He	0.000 13	3.016 03			5500		
	⁴He	~100	4.002 60			0	0.8	0.9
3	Li		6.939	0.53		70	1.4	
	⁶Li	7.52	6.015 13			945 (n, α)		
	⁷Li	92.48	7.016 01			0.033		
	⁸Li	—			0.845 s			
4	⁹Be	100	9.012 19	1.82		0.010		
	⁸Be		8.005 31		~3 × 10⁻¹⁶ s			
5	B		10.811	2.54		755	7	6.11
	¹⁰B	19.8	10.012 94			3813 (n α)	4	3.7
	¹¹B	80.2	11.009 31			<50 × 10⁻³		
	¹²B				0.019 s			
6	C		12.011 15	2.22		0.0034	4.8	4.66
	¹²C	98.89	12.000 00					
	¹³C	1.11	13.003 35			0.000 5		
	¹⁴C	—			5570 yr			

(Continued)

Nuclear Data for Various Elements and Isotopes (Continued)

Atomic Number, Z	Element or Isotope	Abundance (%)	Atomic Mass (u)	Density (g/cm³)	Half-Life	Absorption at 0.025 eV	Scattering Thermal	Scattering Epithermal
7	N		14.0067			1.88	10	9.9
	^{13}N	—			10 min	1.75		
	^{14}N	99.63	14.003 07					
	^{15}N	0.37	15.000 11					
	^{16}N	—			7.4 s			
8	O		15.9994			0.000 19	4.2	3.75
	^{16}O	99.759	15.994 91					
	^{17}O	0.037	16.999 14					
	^{18}O	0.204	17.999 16					
	^{19}O	—			29 s			
9	F		18.998 40			0.010	3.9	3.6
	^{19}F	100						
11	Na		22.989 77	0.971		0.53	4.0	3.1
	^{23}Na	100						
	^{24}Na	—			15.0 h			
12	Mg		24.312	1.74		0.063	3.6	3.4
	^{24}Mg	78.60	23.985 04			0.034		
	^{25}Mg	10.11	24.985 84			0.280		
	^{26}Mg	11.29	25.982 59			0.060		
	^{27}Mg	—			9.45 min			
13	Al		26.9815	2.70		0.230	1.4	1.4
	^{27}Al	100						
	^{28}Al	—			2.3 min			
14	Si		28.086	2.4		0.16	1.7	2.2
	^{28}Si	92.27	27.976 93			0.080		

Z	Nuclide	%	Atomic mass		Half-life			
	29Si	4.68	28.976 49			0.280		
	30Si	3.05	29.973 76			0.40		
	31Si	—			2.62 yr			
15	31P	100	30.973 76	2.34		0.20	5	3.4
17	37Cl	24.47	36.965 90			56		
18	37Ar	—	36.966 78		35.1 days			
19	K	100	39.102	0.87		2.07	2.5	2.1
22	Ti		47.90	4.5		5.8	4	4.2
23	V		50.942	6.0		4.5	5.0	3.0
24	Cr		51.996	6.92		1.82	2.3	1.9
	53Cr	9.55	52.940 7					
25	55Mn	100	54.938 1	7.2		13.2	11	11.4
	56Mn	—			2.58 h			
26	Fe		55.847	7.87		2.62	2.62	11.4
	54Fe	5.84	53.939 6			2.3	2.3	
	55Fe				2.6 yr	2.7	2.5	
	56Fe	91.68	55.934 9			2.5	1.2	
	57Fe	2.17	56.935 4					
	58Fe	0.31	57.933 3					
	59Fe	—			45 days			
27	58Co	—			71.3 days	1.9 × 10³	7	5.8
	59Co	100	58.933 2	8.71		18 and 19		
	60Co	—			8.71 / 10.5 min and 5.24 yr			
28	Ni		58.71	8.9		4.8	17.5	17.4
29	Cu		63.54	8.96		3.77	7.2	7.7
	63Cu	69.1	62.929 8			4.6		
	64Cu	—			12.9 h			
	65Cu	30.9	64.927 8			2.2		
	66Cu	—			5.10 min			

(Continued)

Nuclear Data for Various Elements and Isotopes (Continued)

Atomic Number, Z	Element or Isotope	Abundance (%)	Atomic Mass (u)	Density (g/cm³)	Half-Life	Absorption at 0.025 eV	Scattering Thermal	Scattering Epithermal
34	^{79}Se	—			3.89 min and 6.5 × 10⁴ yr	2.45 and 8.4	6	6
35	^{79}Br	50.69	78.918 33					
36	^{79}Kr	—			50 s and 34.9 h			
37	^{87}Rb	27.85			4.7 × 10¹⁰ yr	0.12		
40	Zr		91.22	6.44		0.180	8	6.2
41	^{93}Nb	100	92.906 38	8.57		1.15	5	6.5
	^{94}Nb	—			6.6 min			
42	Mo		95.911	10.22		2.7	7	6
47	Ag		107.870	10.50		63	6	6.4
	^{107}Ag	51.35	106.905 1			31		
	^{108}Ag	—			2.3 min			
	^{109}Ag	48.65	108.904 7			87		
	^{110}Ag	—			24 s			
48	Cd		112.40	8.64		2450	7	
49	In		114.82	7.28		196	2.2	
	^{113}In	4.2	112.904 3					
	^{114}In				72 s and 49 days			
	^{115}In	95.77	114.903 9		6 × 10¹⁴ yr	50 and 150		
	^{116}In	—			14 s and 54 min			
54	Xe		131.30			35	4.3	
	^{135}Xe	—			9.2 h	2.72 × 10⁶		
	^{136}Xe	8.87	135.907 2			0.15		

Z	Isotope	Abundance (%)	Atomic mass	Density	Cross section (barns)	Half-life		
62	Sm		150.35	7.7	5600			
	^{149}Sm	13.84	148.9169		40800			
64	Gd		157.25	7.94	46000			
	^{155}Gd	14.73	154.9226		61000			
	^{157}Gd	15.68	156.9239		240000			
73	Ta	—		16.6	21		11	11.3
	^{181}Ta	~100	180.9480					
	^{182}Ta	—			8200	115 days		
79	Au			19.3	98.8		9.3	
	^{197}Au	100	196.9666				5	6
	^{198}Au	—			26000	2.7 days		
82	Pb		207.19	11.35	0.170			
	^{204}Pb	1.5	203.9731		0.8	1.4×10^{17} yr		
	^{205}Pb	—				3×10^{7} yr		
	^{206}Pb	23.6	205.9745		0.025			
	^{207}Pb	22.6	206.9759		0.70			
	^{208}Pb	52.3	207.9766		0.03			
	^{209}Pb	—	208.9810			3.3 h		
83	Bi			9.8			9	9.28
	^{209}Bi	100	208.9804		0.019 and 0.015			
	^{210}Bi	—			9	2.6×10^{6} yr and 5.0 days		
	^{211}Bi	—				2.15 min		
84	^{210}Po	—	209.9829			138 days		
	^{213}Po	—	212.9928			4 μs		
88	^{226}Ra	—	226.0254		20	1620 yr		
90	^{230}Th	—	230.0331			7.6×10^{4} yr		
	^{232}Th	100	232.0382	11.5	7.56	1.45×10^{10} yr	12.5	12.5
	^{233}Th	—				22.1 min		
91	^{233}Pa	—			43	27.4 days		
92	U		238.03	19.0	7.68		8.3	8.2
	^{233}U	—	233.0395		530 (n, f) 46 (n, γ)	1.62×10^{5} yr	12.5	12.5

(Continued)

Nuclear Data for Various Elements and Isotopes (Continued)

Atomic Number, Z	Element or Isotope	Abundance (%)	Atomic Mass (u)	Density (g/cm³)	Half-Life	Absorption at 0.025 eV	Neutron Cross Section (barns) Scattering	
							Thermal	Epithermal
	^{234}U	0.0057	234.040 9		2.5×10^5 yr	105		
	^{235}U	0.714	235.043 9		7.1×10^8 yr	582 (n, f)	10	
	^{236}U	—	236.045 7		2.39×10^7 yr	112 (n, γ)		7
	^{237}U	—			6.75 days			
	^{238}U	99.28	238.050 8		4.51×10^9 yr	2.71	8.3	
	^{239}U	—			23.5 min	14 (n, f)		
93	^{239}Np	—			2.35 days			
	^{238}Pu	—	238.049 58		87.8 yr	280 (n, γ)		
94	^{239}Pu	—	239.052 2	19.6	24 360 yr	746 (n, f)	9.6	
	^{240}Pu	—	240.054 0		6760 yr	<0.1 (n, f) 295 (n, γ)		
	^{241}Pu	—			15 yr	1025 (n, f) 375 (n, γ)		
	^{242}Pu	—	242.058 7		3.79×10^5 yr	<0.2 (n, f) 30 (n, γ)		
95	^{241}Am	—	241.056 7		433 yr	710 (n, γ) 3 (n, f)		
96	^{242}Cm	—	242.058 8		163 days	25 (·, γ) <5 (n, f)		
	^{244}Cm	—	244.062 8		18.1 yr	~13 (n, γ) 1 (n, f)		
98	^{252}Cf	—	252.08		2.65 yr (α) 85.5 yr (s. fis.)	20 (n, γ)		

Appendix B

Various Convenient Constants

Avogadro constant	$6.022\,52 \times 10^{23}$ molecules/g mole (atoms/g atom)
Boltzmann constant	$1.380\,66 \times 10^{-23}$ J K^{-1}
	$8.617\,81 \times 10^{-5}$ eV K^{-1}
Gravitational constant	1 kg-m/N s^2
Planck constant	$6.626\,18 \times 10^{-34}$ J s
	$4.135\,5 \times 10^{-15}$ eV s
Rest masses	
Electron	$5.485\,80 \times 10^{-4}$ u (atomic mass unit)
Neutron	$1.008\,665$ u
Proton	$1.007\,276$ u
Speed of light	$2.997\,925 \times 10^8$ m/s

Appendix C

Useful Conversion Factors

1 angstrom (Å)	10^{-8} cm 10^{-10} m	1 joule	1 kg m^{-2} s^{-2} $0.624\,2 \times 10^{13}$ MeV 10^7 ergs 10^6 eV
1 barn	10^{-24} cm^2 10^{-28} m^2	1 Mev	$1.602\,10 \times 10^{-13}$ J $1.660\,438 \times 10^{-27}$ kg
1 becquerel (Bq)	1 disintegration/s		
1 curie (Ci)	3.7×10^{10} Bq	1 u (atomic mass unit)	931.482 MeV/c^2
1 electrostatic unit (esu)	$3.162\,3 \times 10^{-5} \; \dfrac{\text{kg}^{1/2}\,\text{m}^{3/2}}{\text{s}}$ $1 \; \dfrac{\text{g}^{1/2}\,\text{cm}^{3/2}}{\text{s}}$		
1 electron volt (eV)	$1.602\,10 \times 10^{-19}$ J 4.44×10^{-26} kWh 1.517×10^{-22} Btu	1 unit electronic charge	4.80×10^{-10} esu 1.158×10^{-14} kg$^{1/2}$ m$^{3/2}$/s
1 fission	$\sim$200 MeV (total) 8.9×10^{-18} kWh (total) 180 MeV (in fuel)	1 watt	1 J/s 3600 J/h 3.413 Btu/h
1 horsepower (hp)	0.7457 kW 2545 Btu/h 550 ft lb$_f$/s		

Appendix D

Table of Radioisotopes

Isotope	Half-Life*	Type of Decay†	Most Predominant Energy(ies) (*MeV*)
$^{1}_{0}\text{n}$	10.6 min	β^-	0.78
$^{3}_{1}\text{H}$	12.6 yr	β^-	0.18
$^{8}_{3}\text{Li}$	0.845 s	β^-	13
		2α	3.2 (total)
$^{8}_{4}\text{Be}$	$<1.4 \times 10^{-16}$ s	2α	0.047 (each)
$^{14}_{6}\text{C}$	5568 yr	β^-	0.155
$^{13}_{7}\text{N}$	10 min	β^+	1.24
$^{22}_{11}\text{Na}$	2.6 yr	β^+, γ	0.542, 1.28
$^{24}_{11}\text{Na}$	15 h	$\beta^-, \gamma_1 - \gamma_2$	1.39, 1.37–2.75
$^{28}_{13}\text{Al}$	2.27 min	β^-, γ	2.86, 1.78
$^{32}_{15}\text{P}$	14.3 days	β^-	1.707
$^{40}_{19}\text{K}$	1.2×10^9 yr	β^-	1.33 (89%)
		EC, γ	1.46 (11%)
$^{47}_{20}\text{Ca}$	4.8 d	β^-, γ	0.66, 1.3
$^{47}_{21}\text{Sc}$	3.43 days	β^-, γ	0.439, 0.160 (60%)
		β^-	0.60 (40%)
$^{48}_{23}\text{V}$	16.0 days	EC, $\beta^+, \gamma_1, \gamma_2$	0.69, 0.986, 1.314
$^{53}_{23}\text{V}$	2.0 min	β^-, γ	2.50, 1.00
$^{55}_{26}\text{Fe}$	2.6 yr	EC	
$^{59}_{26}\text{Fe}$	45 days	β^-, γ	0.460, 1.10 (54%)
			0.270, 1.29 (46%)
$^{60}_{27}\text{Co}$	5.24 yr	$\beta^-, \gamma_1, \gamma_2$	0.302, 1.33, 1.17
$^{64}_{29}\text{Cu}$	12.9 h	EC, γ	1.34 (42%)
		β^-	0.571 (39%)
		β^+	0.657 (19%)
$^{66}_{29}\text{Cu}$	5.1 min	β^-	2.63 (91%)
		β^-, γ	1.5, 1.04 (9%)
$^{65}_{30}\text{Zn}$	245 days	EC	(55%)
		EC, γ	1.12 (45%)
$^{85}_{36}\text{Kr}$	10.3 yr	β^-	0.695 (98.5%)
		β^-, γ	0.15, 0.54 (0.65%)
$^{90}_{38}\text{Sr}$	27.7 yr	β^-	0.545
$^{90}_{39}\text{Y}$	64.2 h	β^-	2.26

Isotope	Half-Life*	Type of Decay†	Most Predominant Energy(ies) (MeV)
$^{108}_{47}$Ag	2.3 min	β^-	1.77 (97%)
		β^+	0.90 (1.5%)
		EC, γ	0.45 (1.5%)
$^{110}_{47}$Ag	24 s	β^-	2.82 (40%)
		β^-, γ	2.24, 0.66 (60%)
$^{116}_{49}$In	13 s	β^-	3.29
$^{131}_{53}$I	8.08 days	β^-, γ	0.608, 0.364 (87.2%)
		β^-, γ	0.335, 0.638 (9.3%)
$^{137}_{53}$I	22 s	β^-	(94%)
		β^-, n	0.56 (6%)
$^{135}_{54}$Xe	9.23 h	β^-, γ	0.51, 0.250 (97%)
$^{137}_{54}$Xe	3.9 min	β^-	3.5
$^{132}_{55}$Cs	6.47 days	EC, γ	0.67
$^{137}_{55}$Cs	30 yr	β^-, γ	0.51, 0.66 (92%)
		β^-	1.17 (8%)
$^{133}_{56}$Ba	7.2 yr	EC, γ	0.320, 0.081
$^{140}_{56}$Ba	12.8 days	β^-, γ	1.021, 0.16–0.03 (60%)
		β^-, γ	0.48, 0.54–0.03 (30%)
$^{140}_{57}$La	40 h	β^-, γ_1–γ_2–γ_3	1.32, 0.33–0.49–1.60 (70%)
		β^-, γ_1–γ_2	1.67, 0.49–1.60 (20%)
$^{144}_{58}$Ce	285 days	β^-	0.309
$^{147}_{61}$Pm	2.6 yr	β^-	0.223
$^{192}_{77}$Ir	74.4 days	EC, β^-, γ	0.66, 11 at different energies
$^{197}_{80}$Hg	65 h	EC, γ	0.077
$^{214}_{82}$Pb	26.8 min	β^-, γ	0.67, 0.295–0.352
$^{213}_{83}$Bi	47 min	β^-	1.39 (98%)
		β^-, γ	0.959, 0.434 (2%)
		α	5.86
$^{208}_{84}$Po	2.93 yr	α	5.108
$^{210}_{84}$Po	138 days	α	5.30
$^{213}_{84}$Po	4.2×10^{-6} s	α	8.34
$^{226}_{88}$Ra	1622 yr	α	4.78 (94%)
		α, γ	4.77, 0.186 (5.7%)
$^{231}_{90}$Th	25.6 h	β^-, γ	0.308, 0.084 (44%)
		β^-, γ	0.094, 0.058, 0.026 (45%)
$^{233}_{92}$U	1.62×10^5 yr	α	4.823
$^{234}_{92}$U	2.48×10^5 yr	α	4.76 (73%)
		α, γ	4.71, 0.068 (27%)
	s.f. 2×10^{16} yr		
$^{235}_{92}$U	7.1×10^8 yr	α	4.40 (83%)
	s.f. 1.9×10^{17} yr		
$^{238}_{92}$U	4.51×10^9 yr	α	4.195 (77%)
		α, γ	4.18, 0.048 (23%)
	s.f. 8×10^{18} yr		
$^{239}_{92}$U	23.5 min	β^-, γ	1.2, 0.73

(Continued)

Table of Radioisotopes (Continued)

Isotope	Half-Life	Type of Decay†	Most Predominant Energy(ies) (*MeV*)
$^{238}_{93}$Np	2.1 days	β^-, γ	1.272, 0.044 (47%)
		β^-, γ	0.258, 1.03 (53%)
$^{238}_{94}$Pu	89.6 yr	α	5.49 (72%)
		α, γ	5.45, 0.044 (28%)
	s.f. 3.8×10^{10} yr		
$^{241}_{95}$Am	432 yr	α, γ	5.477, 0.060
$^{242}_{96}$Cm	35 yr	α	6.11 (73.7%)
		α, γ	6.07, 0.044 (26.3%)
$^{244}_{96}$Cm	17.9 yr	α	5.801 (76.7%)
		α, γ	5.759, 0.043 (23.3%)
	s.f. 1.4×10^7 yr		
$^{252}_{98}$Cf	2.65 yr (97%)	α	6.12 (82%)
			6.08 (15%)
	s.f. 85 yr (3%)		
$^{254}_{98}$Cf	s.f. 60.5 days	n	
		γ	
$^{247}_{97}$Bk	1.4×10^3 yr	α, γ	5.68 (37%)
			5.52 (58%)
$^{253}_{99}$Es	20.5 days	α, β	6.64, 0.017, 0.027
$^{257}_{100}$Fm	80 days	α	6.53 (94%)
	s.f. 100 yr		
$^{257}_{101}$Md	3 h	EC, α	7.25 (97%), 7.08 (3%)

* s.f., spontaneous fission.
† Beta energies are the maximum energies.

Appendix E

Properties of Materials for Oxide-Fueled Fast Reactors

Material	σ_{fis} (barns)	σ_a (barns)	Σ_{fis} (cm^2/cm^3)	Σ_a (cm^2/cm^3)	η	Σ_{tr} (cm^2/cm^3)
B (natural)		0.54		0.058*		
^{10}B		2.73		0.30*		
C		0				
O		0.001				
Na		0.0016		4×10^{-5}		0.08
Fe		0.010		0.000 85		0.25
SS				0.001 5		0.25
^{232}Th	0.014	0.40	0.008 15†	0.009 1†	0.076	0.18†
^{233}U	2.1	3.5	0.048 9†	0.081 5†	2.31	0.18†
^{235}U	1.3	2.71	0.048 0†	0.063 0†	1.93	0.18†
^{238}U	0.048	0.35	0.001 12†	0.008 15†	0.4	0.18†
^{239}Pu	1.8	2.36	0.045 8†	0.060†	2.4	0.18†
^{240}Pu	0.5	0.72	0.012 8†	0.018 4†	1.315	0.18†
^{241}Pu	2.5	2.51	0.051 †	0.064†	2.72	0.18†
^{242}Pu	0.35	0.71	0.008 9†	0.018 1†	1.4	0.18†

* As carbide.
† As oxide.

Appendix F

Thermal Conductivities of Various Nuclear Materials

Material	Melting Point (°C)	Density (g/cm³)	Temperature (°C)	Thermal Conductivity (W/m °C)	Remarks
UO_2 (Ref. 1)	2783	10.97	250	5.07	Unirradiated
			500	3.59	corrected to
			750	3.23	theoretical density
			1000	2.54	
			1250	2.28	
(Ref. 2)			250	2.52	Irradiated
			500	2.22	15–19 startups
			750	2.02	
UC (Ref. 3)	2316	12.97	250	22.6	Unirradiated
			500	20.8	
			750	20.3	
			1000	20.1	
Uranium (Ref. 4)	1030	19.0	50	27.5	α Phase
			250	29.7	α Phase
			500	39.0	α Phase
			705	46.0	β Phase
			781	50.5	γ Phase
Beryllium (Ref. 3)	1277	1.8	25	222	
			75	144	
Aluminum (Ref. 5)	660	2.7	0	202.5	
			100	206	
			200	215	
			300	230	
			400	249	
304 Stainless steel (Ref. 5)	1400–1455	7.82	0	13.8	18% Cr–8% Ni (austenitic)
			100	16.3	
			300	19.9	
			500	215	

Zircaloy-2 (Ref. 6)	1821	6.56	25	14.6	
			100	14.1	
			150	14.0	
			250	13.9	
			300	14.0	
Graphite, single crystal (Ref. 7)	3694	2.2	25	398	Parallel to layer planes
			25	69	Perpendicular to layer planes
Extruded (Ref. 7)	3694	1.8	25	228	Parallel to extrusion direction
			25	138.5	Perpendicular to extrusion dir.
Pyrolytic carbon (Ref. 7)	3694	2.2	25	298	Parallel to substrate
			25	29.8	Perpendicular to substrate

1. Hausner, H. H., and J. F. Schumar, *Nuclear Fuel Elements*. New York: Reinhold Publishing Corporation, 1959.

2. Daniel, R. C., and I. Cohen, "In-pile Effective Thermal Conductivity of Oxide Fuel Elements to High Fuel Depletions," WAPD-246, U.S. Atomic Energy Commission, April 1964.

3. El-Wakil, M. M., *Nuclear Power Engineering*. New York: McGraw-Hill Book Company, 1962.

4. Holden, A. N., *Physical Metallurgy of Uranium*. Reading, Mass: Addison-Wesley Publishing Co., Inc., 1958.

5. Kreith, F., *Principles of Heat Transfer*. Scranton, Pa: International Textbook Co., 1965.

6. Lustman, B., and F. Kerze, *Metallurgy of Zirconium*. New York: McGraw-Hill Book Company, 1955.

7. Walker, P. L., Jr., "Carbon—An Old But New Material," *American Scientist*, **50**, No. 2 (June 1962), pp. 259–293.

Appendix G

Properties of Coolants

Substance	T (°C)	ρ (kg/m^3)	c_p (kJ/kg K)	$\mu \times 10^3$ (kg/m s)	k (W/m °C)	Pr
Na (liquid)	100	927	1.38	0.69	85.9	1.108×10^{-5}
	400	853	1.30	0.27	71.2	4.93×10^{-6}
	700	779	1.26	0.18	60.0	3.78×10^{-6}
NaK (liquid)	100	879	1.13	0.53	25.0	2.40×10^{-5}
56 % Na–46 % K	400	811	1.05	0.23	27.4	8.81×10^{-6}
	700	754	1.05	0.15	28.9	5.45×10^{-6}
Li (liquid)	200	507	1.38	—	45.8	—
	700	461	—	—	~ 26	—
H$_2$O (sat. liq.)	0	1002	4.22	1.792	0.552	13.6
	40	995	4.18	0.655	0.628	4.34
	100	961	4.22	0.283	0.680	1.70
	200	867	4.51	0.139	0.665	0.937
	300	714	5.73	0.0964	0.540	1.019
He (gas)*	0	0.1786	5.20	0.0190	0.141	0.70
	100	0.1307	5.20	0.0233	0.171	0.71
	300	0.0851	5.20	0.0306	0.221	0.72
	600	0.0558	5.20	0.0405	0.292	0.72
CO$_2$ (gas)*	0	1.964	0.871	0.0137	0.0146	0.817
	300	0.936	1.060	0.0259	0.0405	0.678

* Density given for 1 atm.

Answers to
Selected Problems

Chapter 2

1. (a) 6.027×10^{22} atoms Al/cm^3
   (b) 3.894×10^{15} atoms ^{17}O/cm^3
   (c) 5.123×10^{22} atoms Al/cm^3
3. 1.956×10^7 m/s
5.

v/c	KE_e (MeV)	KE_D (MeV)
0.1	0.002 58	9.452
0.5	0.079 1	290.2
0.9	0.661 4	2472
0.999	10.918	40 074

8. $E = 3934$ eV
   $\lambda = 3.154$ Å
10. $\lambda_{min} = 0.124\ 1$ Å
    $E_{21} = -55.91 \times 10^3$ eV $\lambda_{21} = 0.222$ Å
    $E_{31} = -66.26 \times 10^3$ eV $\lambda_{31} = 0.187\ 2$ Å

Chapter 3

2. $v/c = 0.926$ $m_V = 52.9445\ \mu$
4. (a) $^{79}_{35}$Br
   (b) $m^*_{Se} = 78.918\ 58\ \mu$ (prior to internal conversion)
       $m_{Se} = 78.918\ 48\ \mu$ (prior to beta decay)
   (c) Vel. of internal conv. electron $= 0.5397$ c
       Vel. of beta (max) $= 0.619\ 6$
   (d) $KE_{Se} = 0.728\ 3 \times 10^{-6}$ MeV
6. (a) $^{37}_{18}$Ar
   (b) $v_{Cl} = 7.082 \times 10^3$ m/s
       $KE_{Cl} = 9.067$ eV (negligible)
   (c) $\Delta E_{31} = -3\ 497$ eV
8. $t = 12.17$ yr
   $m_{Co} = 4.393 \times 10^{-9}$ g
   $A_f = 3.7 \times 10^4$ Bq
10. $t = 4.537 \times 10^9$ yr

12. $m_{Po} = 0.652\,g$
14. $t = 5.64\,day$
16. $N_1/N_{10} = 0.435$ $N_2/N_{10} = 0.305$ $N_3/N_{10} = 0.260$
18. $N_3/N_{10} = 0.044\,62$

Chapter 4

2. 11.94 MeV
4. $E_\gamma = 9.24\,MeV$
  $KE_{55} = 832.9\,eV$
7. $E_\gamma = 2.226\,MeV$
  $v_D = 3.555 \times 10^5\,m/s$
  $KE_D = 1.319\,keV$
10. 4.79 MeV
13. 292.4 MeV
15. 2.912 MeV (threshold energy)

Chapter 5

1. Lead, 0.000 760 cm; aluminum, 0.001 41 cm
3. ^{90}Sr β in Al. Range = 0.022 cm;   half-thickness = 0.061 g/cm
5. ^{110}Ag in lead. $\mu = 1.103\,cm^{-1}$; half-thickness = 0.628 cm; $\bar{R} = 0.907$ cm
7. 35.6 percent
9. 4 V
13. 101 ± 11 counts/min

Chapter 6

2. 6.47 yr
4. 3.58 mrem/h
6. 244 mrem
8. Thyroid and lung dose = 3 mSv
10. 400 mrem
12. $0.1265\mu\,Ci$

Chapter 7

3. 127 g/s
5. 2 225 yr
7. $T_H = 453\,K$
9. Mass $^{90}Sr = 7.41 \times 10^{-8}\,\mu g$
  Mass $^{90}SrTiO_3 = 1.53 \times 10^{-7}\,\mu g$
  $P_{2y} = 190.2\,W$
11. 0.615 g

Chapter 8

1. (a) $1.66 \ n/cm^2 \ s$
   (b) $9.76 \times 10^4 \ n/cm^2 \ s$
3. 296 K
5. (a) 1.38×10^9 Bq (after 2 h)
   2.478×10^9 Bq (after 2 days)
   6.170×10^9 Bq (after 2 months)
   (b) 1.417×10^9 Bq (for 2-day foil 24 h after removal)
   (c) 1.136×10^{13} Bq (on removal)
   8.785×10^{12} Bq (after 24 h)
7. 1.144×10^6 Bq (on removal)
   2.849×10^{12} Bq (after 24 h)
9. $CR_{bare} = 13\,440$ cpm
   $CR_{Cd} = 4\,080$ cpm
11. 3.60 Ci/day
13. (a) 81.7 collisions to thermal
    (b) MSDP = 0.153 9
    (c) MR = 220.6
    (d) $\lambda_{tr_{epi}} = 1.452$
15. $\zeta = 0.8794$
    (b) MSDP = 1.757 4
    (c) 19.23 collisions to thermal
    (d) MR = 106.7

Chapter 9

1. $\eta = 1.86$ fast n's emitted/thermal n absorbed in fuel
3. $k_r = 0.776$
5. $k_r = 1.192$
7. 102.8 kW
9. (a) $r_c = 35.32$ cm
   $P = 38.59$ kW
   (b) $r_c = 34.79$ cm
   $P = 37.21$ kW
11. $r_o = 178.7$ cm
13. $x_g = \pi/B - \arctan(D_c L_r B/D_h) - 2/3\lambda_{tr}$
15. $\delta = 8.17$ cm
    $P = 48.78$ (core) $+ 6.13$ (blanket) $= 54.91$ MW(th)
17. $r_{g_{1M}} = 140.1$ cm
    $r_{g_2} = 142.2$ cm
19. $P_1 = 51.88$ MW
    $P_{1M} = 307.2$ MW
    $P_2 = 315.7$ MW
21. $\phi_{max} = 5.29 \times 10^{15} \ n/cm^2 \ s$

Chapter 10

1. $^{233}U - 0.049$ s
   $^{239}Pu - 0.033$ s
3. $0.565
5. $\rho = 0.002\ 64$
7. 5.95 inhours
9. Sm $= -0.72$ percent; Xe $= -2.8$ percent
12. $0.144
17. CR $= 0.934$
    $\Delta\rho = \$ - 49.34$
19. $F_Q = 5.93$

Chapter 11

1. $T_m = \dfrac{4AE}{(A+1)^2}$

3. 1 713 displacements/collision with 14-MeV neutron
   421 displacements/collision with 1-MeV neutron
5. 0.020 Frenkel pair/atom

Chapter 12

1. $T_{max} = 969.3°C$
3. (a) $(q/A)_{max} = 7.461 \times 10^4$ W/m^2
   (b) $P = 986.5$ W/plate
   (c) $T_0 = 192.6°C \approx T_s \approx T_{max}$
5. $S = 5.984 \times 10^8$ W/m^3
   $\Delta T_{cl} = 61.1°C$
   $T_{max} = 1193°C$
7. $T_{max} = 185.3°C$
9. (a) SEF $= 0.813 \times 10^9$ W/m^2
   (b) $q/A = 1.59 \times 10^6$ W/m^2
   (c) $F_{s_1} = 1.57$
11. $T_{c_0} = 494°C$ (steam)
    $T_{h_0} = 458°C$ (sodium)

Chapter 13

2. $\dot{m}_w = 13.71 \times 10^6$ kg/hr
   $\dot{m}_a = 83.48 \times 10^6$ kg/hr
4. Annual fuel requirement $= 36.06$ tons/yr
   AFC $= \$40.01$/yr
   $e = 55.83$ mills/kWh

6. $m = 0.0906$ $\eta_{\text{th}} = 0.313$
   $n = 0.1596$ Heat rate $= 11\,502$ kJ/kWh
   $p = 0.0684$ $P_{\text{net}} = 1179$ MW(e)
8. $P_{\text{th}} = 67\,817$ kW
   $P_{\text{elec}} = 29\,213$ kW
   AFC $= \$1.39 \times 10^6/\text{yr}$
   Fuel cost $= 5.43$ mills/kWh
   $\eta_{\text{th}} = 0.3778$ Heat rate $= 9529$ kJ/kWh
   Fuel cost (single-stage compressor) $= 6.195$ mills/kWh

Chapter 14

1. (a) 10.53 percent flow through condenser
  (b) $T_{\text{stream}} = 16.05°C$
  (c) 32.92 percent flow through condenser
     $T_{\text{stream}} = 18.29°C$
3. (a) $\dot{v}_w = 22.23$ m^3/s
  (b) $\dot{m}_{\text{air}} = 418 \times 10^6$ kg/hr
5. $t_2 = 242.4$ yr
  $t_{2C} = 168$ yr

Index